D0410792

PHYSICAL CLEANING OF COAL

ENERGY, POWER, AND ENVIRONMENT

A Series of Reference Books and Textbooks

Editor

PHILIP N. POWERS

Professor Emeritus of Nuclear Engineering
Purdue University
West Lafayette, Indiana

Consulting Editor
Energy Management and Conservation
PROFESSOR WILBUR MEIER, JR.
Head, School of Industrial Engineering
Purdue University
West Lafayette, Indiana

1. Heavy Oil Gasification, *edited by Arnold H. Pelofsky*
2. Energy Systems: An Analysis for Engineers and Policy Makers, *edited by James E. Bailey*
3. Nuclear Power, *by James J. Duderstadt*
4. Industrial Energy Conservation, *by Melvin H. Chiogioji*
5. Biomass: Applications, Technology, and Production, *by Nicholas P. Cheremisinoff, Paul N. Cheremisinoff, and Fred Ellerbusch*
6. Solar Energy Technology Handbook: Part A, Engineering Fundamentals; Part B, Applications, Systems Design, and Economics, *edited by William C. Dickenson and Paul N. Cheremisinoff*
7. Solar Voltaic Cells, *by W. D. Johnston, Jr.*
8. Energy Management: Theory and Practice, *by Harold W. Henry, Frederick W. Symonds, Robert A. Bohm, John H. Gibbons, John R. Moore, and William T. Snyder*
9. Utilization of Reject Heat, *edited by Mitchell Olszewski*
10. Waste Energy Utilization Technology, *by Yen-Hsiung Kiang*
11. Coal Handbook, *edited by Robert A. Meyers*
12. Energy Systems in the United States, *by Asad T. Amr, Jack Golden, Robert P. Ouellette, and Paul N. Cheremisinoff*

13. Atmospheric Effects of Waste Heat Discharges, *by Chandrakant M. Bhumralkar and Jill Williams*

14. Energy Conservation in Commercial and Residential Buildings, *by Melvin H. Chiogioji and Eleanor N. Oura*

15. Physical Cleaning of Coal: Present and Developing Methods, *edited by Y. A. Liu*

Additional Volumes in Preparation

PHYSICAL CLEANING OF COAL

PRESENT AND DEVELOPING METHODS

edited by *Y. A. LIU*
Professor of Chemical Engineering
Virginia Polytechnic Institute and State University
Blacksburg, Virginia

MARCEL DEKKER, INC. New York and Basel

Library of Congress Cataloging in Publication Data

Main entry under title:

Physical cleaning of coal.

(Energy, power, and environment; v. 15)
Includes bibliographies and index.
Contents: Chemical comminution for coal cleaning / Bernard M. Weiss, Robert R. Maddocks, and Philip H. Howard -- The Otisca process / D. V. Keller, Jr. -- Electrostatic benefication of coal / I. I. Inculet, M. A. Bergougnou, and James D. Brown. [etc.]
1. Coal--Cleaning. I. Liu, Y. A. (Yih An)
II. Series.
TN816.P36 1982 662.6'23 82-9773
ISBN 0-8247-1862-3 AACR2

MARCEL DEKKER, INC.
270 Madison Avenue, New York, New York 10016

Current printing (last digit):
10 9 8 7 6 5 4 3 2 1

PRINTED IN THE UNITED STATES OF AMERICA

Foreword

In recent years much attention has been given to precombustion and postcombustion cleanup as means of controlling sulfur dioxide emissions from burning coal. Most processes for controlling sulfur dioxide emissions have entailed some form of either pretreatment by chemical desulfurization or posttreatment by stack gas scrubbing (flue gas desulfurization). Coal cleaning for precombustion cleanup has been overlooked except by a few dozen experts who advocate the merit of removing ash and inorganic sulfur (mainly pyrite) as a means of improving boiler performance and, with most coals, controlling objectionable pollutants. Many coals are amenable to cleaning using advanced technology at costs ranging from 25 to 50% of those for chemical desulfurization or stack gas scrubbing. Furthermore, physical cleaning wastes containing the worrisome trace elements can be disposed of safely in stable embankments and land fills.

Beginning with the miner's crude methods of upgrading his product by removing pieces of rock and sulfur balls, coal cleaning has meant removing some of the coarse ash and pyrite but little more. This connotation has persisted through the years to the present; and the terms like *washing, preparation,* or *beneficiation* are still in use in describing the precombustion cleanup technology. As a result, coal preparation or beneficiation was not considered as adequate for controlling combustion emissions. Furthermore, the proponents of coal liquefaction and gasification were quite sure that beneficiation of the conversion feedstock was unnecessary.

Recently, we have learned that reserves of other forms of fossil fuel such as petroleum are inadequate and that a clean coal is the most economical fossil fuel. We have not traveled full circle in our use of coal but the time may not be too distant when coal will not only be the major fuel for utilities and industrial boilers but also will be used in homes again.

Thanks to the vision and the perseverance of a few researchers who never abandoned the old technology of coal preparation for the new technology of coal conversion, some significant advances have been made in cleaning coal. Instead of removing large sulfur balls, new processes are becoming available for removing micrometer-sized pyrite crystallites from powdered coal. This book describes the new methods and developments in physical coal cleaning that will permit us to replace natural gas and oil-fired utility boilers with coal-fired ones and also allow us to burn coal in home furnaces without the noxious gases and soot that many of us can remember.

Wilbert E. Warnke

Preface

This book is intended to describe in some detail the recent developments of major new physical coal-cleaning methods. These methods are based upon the differences in the physical characteristics that affect the separation of sulfur-bearing and ash-forming mineral impurities from the pulverized coal. Typical physical characteristics utilized in these methods include specific gravity, electrical conductivity, magnetic susceptibility, and surface properties. In some of the new methods being developed, chemical pretreatment is used to enhance the difference in physical characteristics to facilitate the physical separation of mineral impurities from the pulverized coal.

To ensure the best possible coverage of this book, the original inventors and leading researchers of major new methods and developments in physical coal cleaning were invited to write chapters on their areas of expertise. An attempt was made to span the entire spectrum of new methods and developments, including basic principles, process developments, engineering design, and cost estimation. The specific topics covered may be divided into four parts: (1) size reduction and impurity liberation (Chap. 1), (2) physical removal of sulfur and ash (Chaps. 2 to 7), (3) filtration and dewatering in physical coal cleaning (Chap. 8), and (4) economic assessment of selected physical coal-cleaning processes (Chaps. 9 and 10). Although emphasis is placed on new methods and developments, necessary background information, such as the advantages and disadvantages of existing methods and current industrial practice, is also summarized. Thus, this book is intended to be a self-contained and thorough survey of major new methods and development in physical coal cleaning.

The first chapter describes a new process for size reduction in physical coal cleaning which is an essential step for liberating the mineral impurities prior to their separation from the coal. The process, called chemical comminution, involves the exposure of raw coal to certain chemicals such as gaseous or liquid ammonia. This exposure leads to the fragmentation of the raw coal along the boundaries between the different microscopic constituents of coal and its mineral impurities. Note that despite the use of chemicals to achieve the size reduction of raw coal, available experimental data have shown that essentially no chemical reaction occurs between the comminuting chemical and coal substance. For this reason, chemical comminution should be primarily considered as a physical method, rather than a chemical method, and is included in this book.

The main theme is covered in Chaps. 2 to 7, in which the recent developments of major new physical coal-cleaning methods are discussed. Chapter 2 reviews the new developments in the theory and practice of gravity separation methods for cleaning coal and describes the successful experimental and commercial development of a novel anhydrous (waterless), heavy-liquid separation method called the Otisca process. Chapter 3 discusses the new electrostatic methods for cleaning coal, emphasizing the developments of new dry fulidized-bed electrostatic separation processes. Chapter 4 describes the basic principle devices of a novel magnetic separation method, called high-gradient magnetic separation (HGMS), and reviews the recent developments of HGMS processes applied to coal desulfurization. Chapter 5 outlines the theory and practice of pyritic sulfur removal from coal by froth flotation and describes a new two-stage process for floating pyrite from coal with xanthate called coal-pyrite flotation. Chapter 6 presents a novel method which can effectively deal with very fine particles during coal preparation, called oil agglomeration. The process involves the agglomeration and recovery of fine coals in an aqueous suspension by adding different water-immiscible liquids, usually oils, as collecting liquids. Chapter 7 reports the recent progress in the development and demonstration of selected fine coal beneficiation methods at one of the leading coal preparation research centers in this country, the Ames Laboratory of the U.S. Department of Energy operated by the Iowa State University. Of particular interest in this chapter is the report of experimental results obtained from various combinations of mechanical crushing, chemical comminution, gravity separation, froth flotation, and oil agglomeration methods applied to the cleaning of high-sulfur coals.

Chapter 8 describes the new developments and practical aspects of filtration and dewatering (thickening) in the wet beneficiation of fine coal and outlines the design considerations of filtration and dewatering systems in physical coal-cleaning processes.

Chapter 9 reports the results of economic evaluations of some of the new and conventional coal beneficiation methods, including (1) chemical comminution (Chap. 1), (2) wet HGMS (Chap. 4), (3) wet mechanical (conventional) beneficiation of the coarse coal fraction, (4) full-scale or thorough wet mechanical beneficiation of different size fractions, and (5) a new chemical pyrite removal method based on ferric sulfate leaching called the Meyers fine coal process. Chapter 10 presents a general methodology for the economic assessment of the use of physical coal cleaning and flue gas desulfurization (FGD) to achieve sulfur emission control.

Due to the great diversities in the background of the contributors, no attempt was made to enforce a uniform style of presentation. Rather, it is deemed desirable to preserve the style of the original presentation by the contributors, so that it may be more palatable to both pure scientists and practising engineers. The editor hopes that the reader, from whatever discipline, shares his belief that the diversity of presentation enhances the value of this book. An attempt was made, however, to achieve the best possible quality of each chapter, by inviting a large number of established researchers and practitioners to review the manuscripts. Needless to say, both the editor and publisher are most grateful to the referees for their contributions.

Because of its broad coverage and in-depth review of much of the new methods and developments in physical coal cleaning, this book provides both an introduction to the subject as well as a fairly comprehensive status report. It is hoped that this book may be of value and interest to both beginners and experienced workers in the field of coal cleaning.

Y. A. Liu

Contents

Contributors

M. A. BERGOUGNOU Professor, Department of Chemical Engineering, Faculty of Engineering Science, The University of Western Ontario, London, Ontario, Canada

JAMES D. BROWN Professor, Department of Electrical Engineering, Faculty of Engineering Science, The University of Western Ontario, London, Ontario, Canada

C. EDWARD CAPES Head, Chemical Engineering Section, Division of Chemistry, National Research Council of Canada, Ottawa, Ontario Canada

DONALD A. DAHLSTROM Vice President and Director, Research and Development, Envirotech Corporation, Salt Lake City, Utah

A. W. DEURBROUCK Manager, Coal Preparation, Coal Preparation Division, Pittsburgh Mining Technology Center, U.S. Department of Energy, Pittsburgh, Pennsylvania

RENE J. GERMAIN* Research Investigator, Research and Development Department, STELCO Inc., Hamilton, Ontario, Canada

LAWRENCE HOFFMAN President, Hoffman-Holt, Inc., Silver Spring, Maryland

ELMER C. HOLT, JR. Vice President, Hoffman-Holt, Inc., Silver Spring, Maryland

---

**Current affiliation*: Engineer, Coal Preparation Division, Coal Mining Research Centre, Edmonton, Alberta, Canada

PHILIP H. HOWARD Division Director, Life and Environmental Sciences Division, Syracuse Research Corporation, Syracuse, New York

ION I. INCULET Professor, Department of Electrical Engineering, Head, Applied Electrostatics Laboratory, Faculty of Engineering Science, The University of Western Ontario, London, Ontario, Canada

D. V. KELLER, JR. Vice President, Technology, Otisca Industries, Ltd., Syracuse, New York

RONALD P. KLEPPER Manager of Process Technology, Technology & Development, EIMCO PMD of Envirotech Corporation, Salt Lake City, Utah

Y. A. LIU* Alumni Associate Professor, Department of Chemical Engineering, Auburn University, Auburn, Alabama

ROBERT R. MADDOCKS Manager, Business Development, Catalytic, Inc., Philadelphia, Pennsylvania

KENNETH J. MILLER Project Leader, Flotation, Coal Preparation Division, Pittsburgh Mining Technology Center, U.S. Department of Energy, Pittsburgh, Pennsylvania

SUMAN P. N. SINGH Task Leader, Chemical Technology Division, Oak Ridge National Laboratory, U.S. Department of Energy, Oak Ridge, Tennessee

BERNARD M. WEISS Supervisor, Clean Energy Technology, Business Development, Catalytic, Inc., Philadelphia, Pennsylvania

THOMAS D. WHEELOCK Professor, Ames Laboratory and Department of Chemical Engineering, Iowa State University, Ames, Iowa

---

**Current affiliation*: Professor of Chemical Engineering, Virginia Polytechnic Institute and State University, Blacksburg, Virginia

Acknowledgments

It is a pleasure to thank the large number of individuals and organizations who contributed directly and indirectly to the preparation of this book.

I wish to express my deepest appreciation to the authors of manuscripts for this book. This project depended on their contributions.

I wish to thank Professor Robert P. Chambers, Department Head of Chemical Engineering at Auburn University, who encouraged the preparation of this book and provided the necessary staff support. I am very much indebted to Mr. Albert W. Deurbrouck of the U.S. Department of Energy and Professor Thomas D. Wheelock of the Iowa State University who provided much input to the contents. My thanks also go to Graham Garratt and the rest of the staff of Marcel Dekker, Inc., for their continued assistance.

I wish to express my sincere appreciation to Dr. Z. Lowell Taylor, former Department Head, and Professor Donald L. Vives, Assistant Department Head of Chemical Engineering at Auburn University, who introduced me to the field of coal desulfurization in 1975 and provided support for the initiation of my research. I would like to thank many organizations for their financial support of my studies of coal desulfurization during the past 5 years. These include the Engineering Experiment Station and Office of Vice President for Research at Auburn University, U.S. Department of Education, U.S. Department of Energy, Electric Power Research Institute, Engineering Foundation, Gulf Oil Foundation, National Science Foundation, and New England Power Service Company. I am most grateful to Mr. Wilbert E. Warnke, former coal preparation research manager of the U.S. Department of Energy, who taught me much about the practical aspects of coal desulfurization research and who prepared the Foreword.

I wish also to thank the many inspiring research associates and students in my laboratory, particularly Dr. Mike C. J. Lin and Messrs. Louis A. Cater, George E. Crow, Phillip M. Haynes, Thomas H. McCord, Donald W. Norwood, Myoung Joon Oak, Frederick J. Pehler, and Russel G. Wagner, who worked with me on coal desulfurization during the past few years.

I wish to thank the following individuals, who reviewed the first draft of this manuscript: Albert F. Baker, U.S. Department of Energy; James S. Browning, The University of Alabama; Joseph A. Cavallaro, U.S. Department of Energy; Christopher H. Cheh, Ontario Hydro; Randy M. Cole, Tennessee Valley Authority; Froster Frass, U.S. Bureau of Mines; Shiao-Hung Chiang, University of Pittsburgh; E. Bryant Fitch, Auburn University; R. J. Germain, The Steel Company of Canada; Douglas V. Keller, Jr., Otisca Industries, Ltd.; George E. Klinzig, University of Pittsburgh; J. D. Miller, The University of Utah; Kenneth J. Miller, U.S. Department of Energy; Seongwoo Min, Battelle Columbus Laboratories; Robin R. Oder, Gulf Science and Technology Company; Allen H. Pulsifer, Iowa State University; G. R. Rigby, The Broken Hill Proprietary Company, Ltd.; Hermant M. Risbud, Electric Power Research Institute; David J. Spottiswood, Michigan Technological University; David L. Springton, Envirotech Corporation; John W. Tierney, University of Pittsburgh; Shirley C. Tsai, Oxidental Research Corporation; Donald L. Vives, Auburn University; Bernard M. Weiss, Catalytic, Inc.; Thomas D. Wheelock, Iowa State University; Wilbert E. Warnke, U.S. Department of Energy; Basil J. P. Whalley,

Canada Centre for Mineral and Energy Technology; and Raymond E. Zimmerman, Paul Weir Company.

Finally, I should like to acknowledge a very special debt to my wife, Hing-Har, who so graciously undertook what must have been the seemingly endless chores associated with the preparation of this book.

Y. A. Liu

PHYSICAL CLEANING OF COAL

1

Chemical Comminution for Coal Cleaning

BERNARD M. WEISS
ROBERT R. MADDOCKS

Catalytic, Inc.
Philadelphia, Pennsylvania

PHILIP H. HOWARD

Syracuse Research Corporation
Syracuse, New York

1.1 INTRODUCTION

Chemical comminution is a size reduction process that involves the exposure of raw coal to certain low-molecular-weight chemicals that are relatively inexpensive and recoverable. The chemical breakage is unique in two ways: (1) microscopic studies [1] have demonstrated that fragmentation is strongly controlled by boundaries between macerel and mineral matter resulting in greater mineral matter (especially pyritic sulfur) liberation than mechanical crushing to a similar size, and (2) bimodal breakage occurs resulting in the generation of a small amount of fine particles. Both these features make chemical comminution attractive as a substitute or an adjunct to mechanical crushing, which is commonly incorporated into coal preparation plants. The selective liberation allows for lower sulfur value at a given recovery or a better recovery at a given sulfur value [2].

In this chapter, the currently known technical and economic information for chemical comminution is presented in order to place in perspective chemical comminution as a precombustion coal-cleaning technique. Future plans for bringing chemical comminution to commercial scale are also briefly discussed.

Chemical comminution of coal with ammonia has been studied by Syracuse Research Corporation (SRC) since 1971. This work led to the coal fragmenting techniques that have been patented by SRC [3]. These patents describe the basic process as it applies to the chemical fracturing of coal after it has been mined in the normal manner. Two other patents based on this technology apply to in situ mining of coal by chemical comminution [4].

In 1977, Catalytic, Inc. joined with Syracuse Research Corporation to undertake the continued development and commercialization of the chemical comminution process.

1.2 RESEARCH STUDIES

Extensive research on the chemical comminution process has proved that it is applicable to a wide variety of coals. Chemical fragmentation of coal is influenced by parameters such as coal rank, initial size of the coal, pretreatment, moisture content, comminution chemical, comminution pressure, and time. Studies have been carried out to determine how these parameters affect the extent of fragmentation and to examine the effects of chemical comminution on coal characteristics.

A. Experimental Apparatus and Procedure

The laboratory apparatus used to chemically comminute small coal samples is illustrated in Fig. 1.1 [5]. Unless otherwise indicated, small coal samples were floated at 1.8 specific gravity to remove

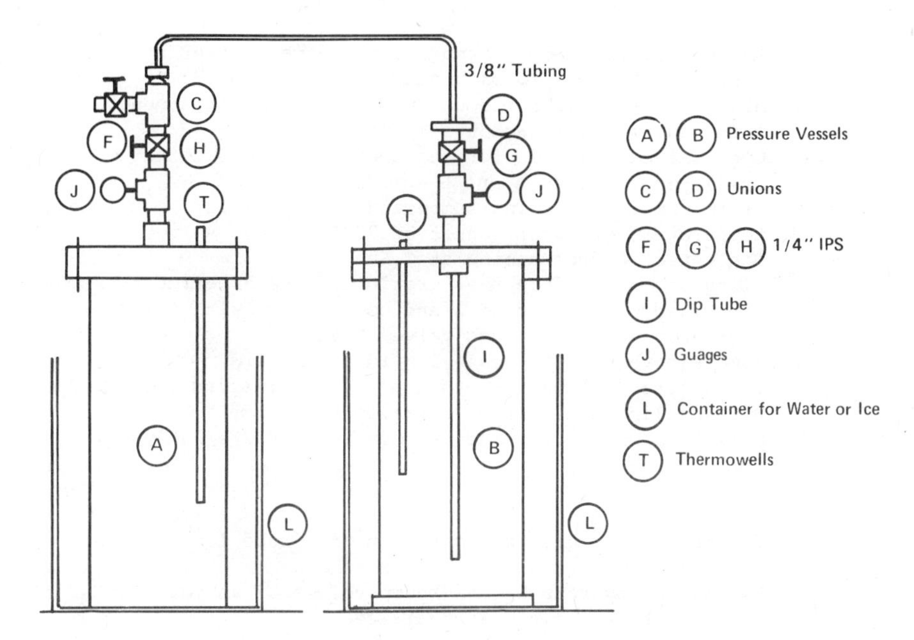

Figure 1.1 Chemical fragmentation apparatus for small coal samples. (From Ref. 5.)

large pieces of mineral matter and closely sized (e.g., 3/4 × 1/4 in.) so that the extent of fragmentation could be compared. The operation of the small bomb apparatus in Fig. 1.1 follows Procedure A [5]:

Procedure A. This procedure is followed when tests are performed in small pressure vessels (see Fig. 1.1). At start, the covers are removed from bombs A and B. The piping on covers is in place up to unions C and D. Dip tube I is in place.

Steps:

1. Weigh 300 g coal into A.
2. Close A with cover and attach piping to C.
3. Evacuate A via connection at C.
4. Close valve F.
5. Cool down B with dry ice around it.
6. Add hydrous or anhydrous NH_3 to near one-half full.
7. Close cover with attached piping, valve G open, and tighten nuts.
8. Close valve G.
9. Install connecting tubing between C and D.
10. Evacuate tubing through valve F.

11. Remove dry ice from around B and add warm water to container L so as to maintain temperature at 75°F.
12. Allow pressure to build up in B to a constant amount, maintaining water jacket temperature at 75°F.
13. Open valves H and G, forcing contents from B to A until pressures in A and B equalize.
14. Start timer (recording temperature and pressure), maintain temperature of water jacket at 75°F. Close valve H.
15. Disconnect A from tubing at C.
16. Sixty seconds before the duration of the experiment, remove water jacket from round A and cool A down with dry ice.
17. Vent A to 0 psig through valves H and F.
18. Remove cover and separate liquid from solid.
19. Remove the residual NH_3 in coal by elutriating the sample with boiling water in an elutriating column.
20. Collect sample, dry, and perform appropriate analyses.

Note:

1. In case of NH_3 gas treatment, the dip tube I has to be removed.
2. For making hydrous NH_3, a known volume of anhydrous NH_3 is added to a known weight of ice.

For large coal samples (~100 lb), the chemical fragmentation apparatus used is depicted in Fig. 1.2; and the specific testing procedures can be found elsewhere [6].

Table 1.1 summarizes the relevant information of the 15 different coals from anthracite to lignite which have been used in the chemical comminution experiments. A 50-lb sample of these coals was obtained from the Office of Coal Research, Pennsylvania State University.

The extent of chemical fragmentation under different conditions and for various coals was compared in two ways. One way was to compare the weight percent of the various coal samples passing through a 5-mesh screen. The other method consisted of plotting the results of the screen analysis of the coal samples on a Rosin-Rammler graph which corresponds to the following size distribution equation [7]:

$$R = 100 \exp \left[-\left(\frac{x}{\bar{x}}\right)^n\right]$$

In the equation, R is the residue, or percentage of oversize particles of minimum particle size x (in practice, R corresponds to the weight percentage retained on a sieve of an opening size x); $\bar{x}$ is the absolute

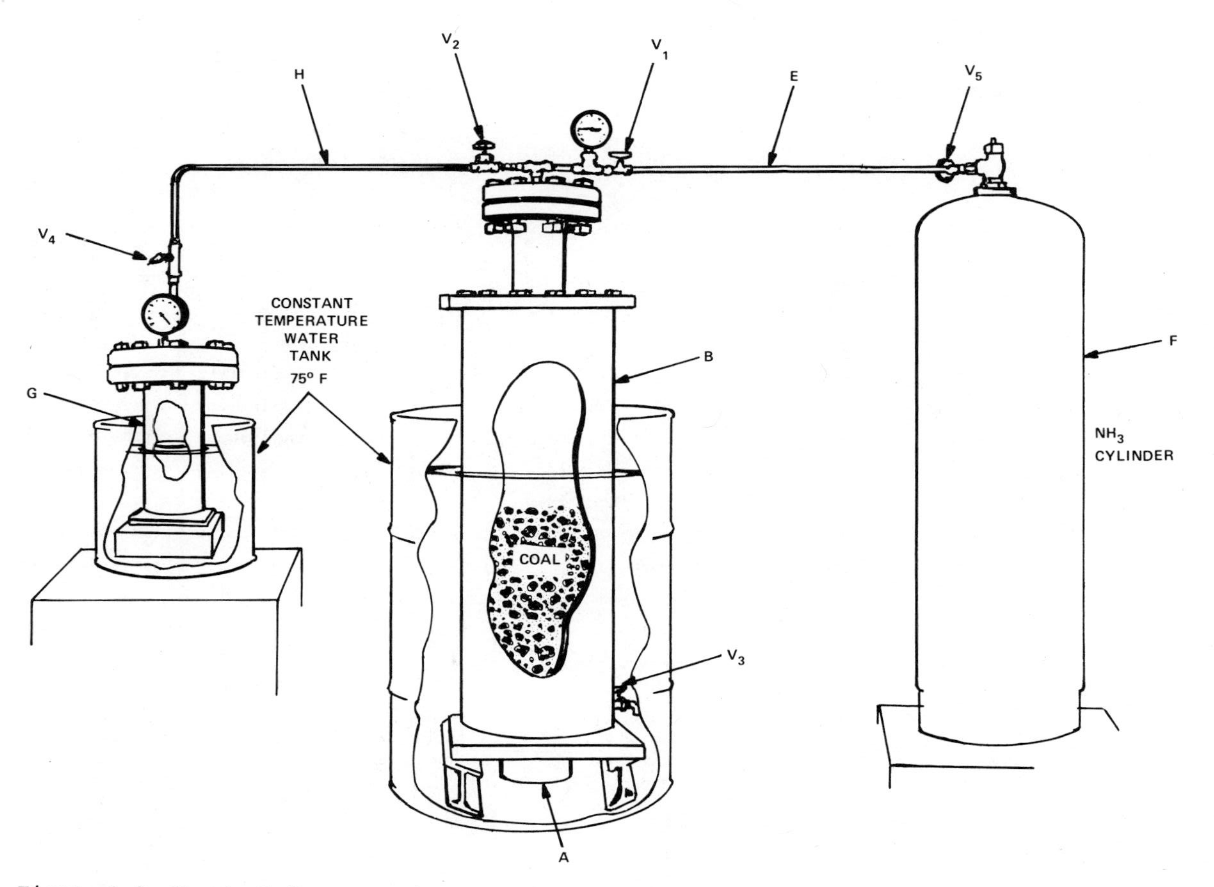

Figure 1.2 Chemical fragmentation apparatus for large coal samples. (From Ref. 6.)

Table 1.1 Samples of Coal Tested

Penn. state office of coal no.	Rank	Seam	Location	Calorific[a] value (Btu/lb)	% C[b,e]	Moisture[a,b]	Ash[b,d]	Total S[b,c]
085	Anthracite	Pennsylvania No. 8 Leader	Northumberland, Pa.	15233	92.52	1.25	8.32	1.15
379	Semi-anthracite	P & M 'B'	Sullivan, Pa.	16771	92.82	1.62	16.06	0.61
317	Low-volatile bituminous	Lower Freeport	Somerset, Pa.	15495	90.81	1.98	12.93	7.40
318	Low-volatile bituminous	Upper Freeport	Somerset, Pa.	15779	90.88	2.93	6.52	1.73
259	Medium-volatile bituminous	Upper Kittanning	Clearfield, Pa.	15822	-	0.65	10.69	1.31
258	Medium-volatile bituminous	Lower Kittanning	Clearfield, Pa.	15838	-	0.49	10.05	4.48
239	High-volatile A bituminous	Utah B	Carbon, Utah	14480	81.41	3.56	11.36	0.59
252	High-volatile A bituminous	Illinois No. 5	Gallatin, Ill.	14823	82.99	8.84	12.59	5.40
211	High-volatile B bituminous	Ohio No. 8	Gallia, Ohio	13932	77.94	2.29	15.98	5.81
216	High-volatile B bituminous	Kentucky No. 14	Hopkins, Ky.	14797	82.16	6.34	8.19	4.96
280	High-volatile C bituminous	Indiana No. 6	Sullivan, Ind.	14654	82.81	11.32	16.07	4.50
232	High-volatile C bituminous	Colorado C	Moffat, Colo.	13535	77.34	10.25	11.94	0.69
248	Subbituminous A	Adaville No. 1	Lincoln, Wyo.	13097	75.62	19.17	2.93	0.66
099	Subbituminous C	School	Converse, Wyo.	12794	72.07	24.12	24.04	0.95
245	Lignite	Upper Gascoyne	Bowman, N.D.	12329	73.56	34.19	13.51	1.53

[a]Dry mineral matter free basis.
[b]Weight %.
[c]Dry ash free basis.
[d]Dry.
[e]Ultimate analysis.
Source: Ref. 7.

size constant, which represents the size where 36.79% of the coal particles are retained on the screen (R = 36.79%); and n is the slope of the logarithmic plot of the equation.

It has been observed that the slope n of the size distributions for the chemically comminuted samples remains approximately constant as the fragmentation proceeds (Fig. 1.3). However, the value of the Rosin-Rammler size constant $\bar{x}$ varies. Therefore, a change in the value of $\bar{x}$ will give an indication of the extent of fragmentation. For the same treatment conditions, a comparison of the values of $\bar{x}$ for samples of different ranks of coal indicates their relative extents of chemical fragmentation [6]. In particular, a smaller value of the absolute size constant $\bar{x}$ corresponds to a finer size distribution.

B. Chemical Comminution Studies

Effective Comminution Chemicals. Although a number of chemicals have some comminution ability [3,4], the chemicals that appear to have the greatest effect are ammonia (gas and anhydrous and hydrous liquid) and methanol. These compounds fall in a class of chemicals containing a nonbonding pair of electrons (oxygen and nitrogen compounds) which swell [8,9] and dissolve [9] coal at ambient temperatures. Although swelling studies have not been conducted by the authors, very little coal (<0.1 wt %) was dissolved by either methanol or liquid anhydrous ammonia. The swelling effect, which has been observed with methanol-treated coal by Bangham and Maggs [8], may cause the fragmentation which occurs during chemical treatment. Keller and Smith [10] have suggested that solvent swelling is the mechanism causing the breakage. However, their only support of this contention was a theoretical discussion. The fact that coal does comminute with gaseous ammonia is not explained by this theory [11]. Another analogue between coal solvents and coal comminutants is a reduction in effect as the solvent is diluted with water. Other specific solvents mentioned as good coal solvents [9], such as n-propylamine and pyridine, have been briefly examined. These

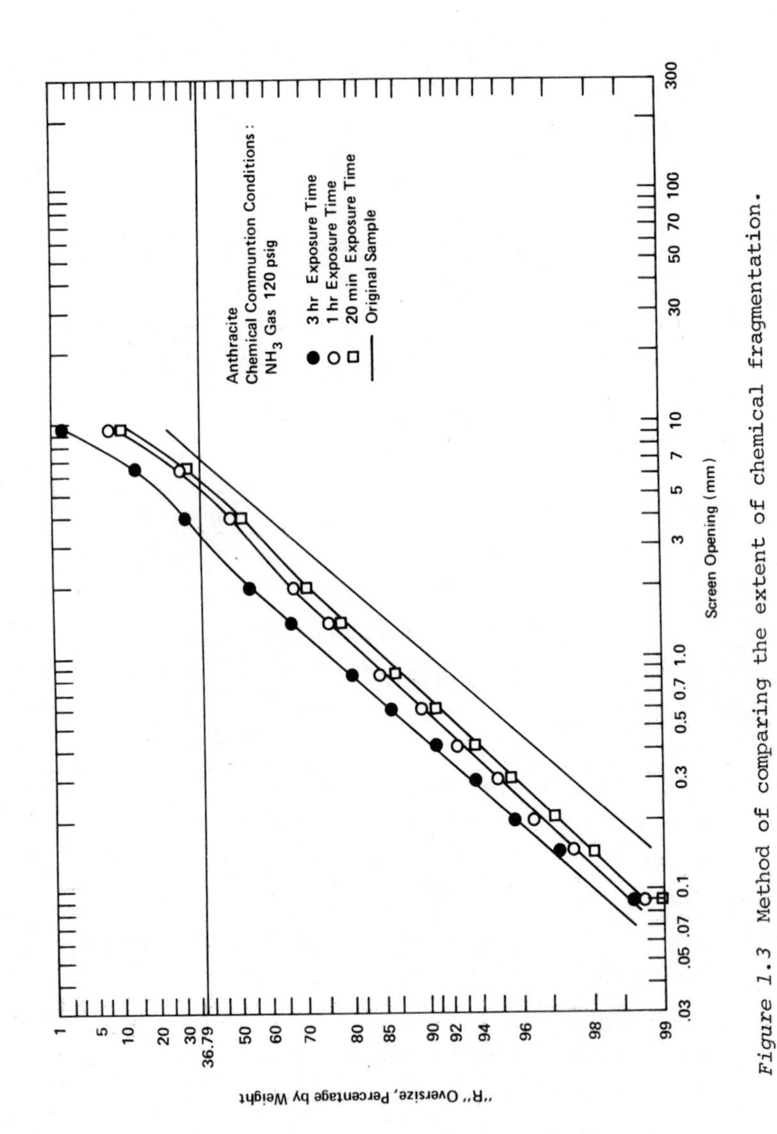

Figure 1.3 Method of comparing the extent of chemical fragmentation. (From Ref. 6.)

chemicals do cause fragmentation but are not as effective as ammonia. Since these chemicals are larger in molecular size, it is possible that molecular size is an important parameter for chemical comminution, especially if penetration of the coal is a rate-determining factor.

Effect of Coal Type. Figure 1.4 compares the fragmentation action of coals of varying rank [5] in both gaseous and liquid ammonia. For most coals, the fragmentation action of gaseous ammonia is slower than that of anhydrous liquid or hydrous ammonia solutions. However, in some instances such as anthracite, the ammonia gas is equally as effective as the ammonia liquid. This could be due to the relative

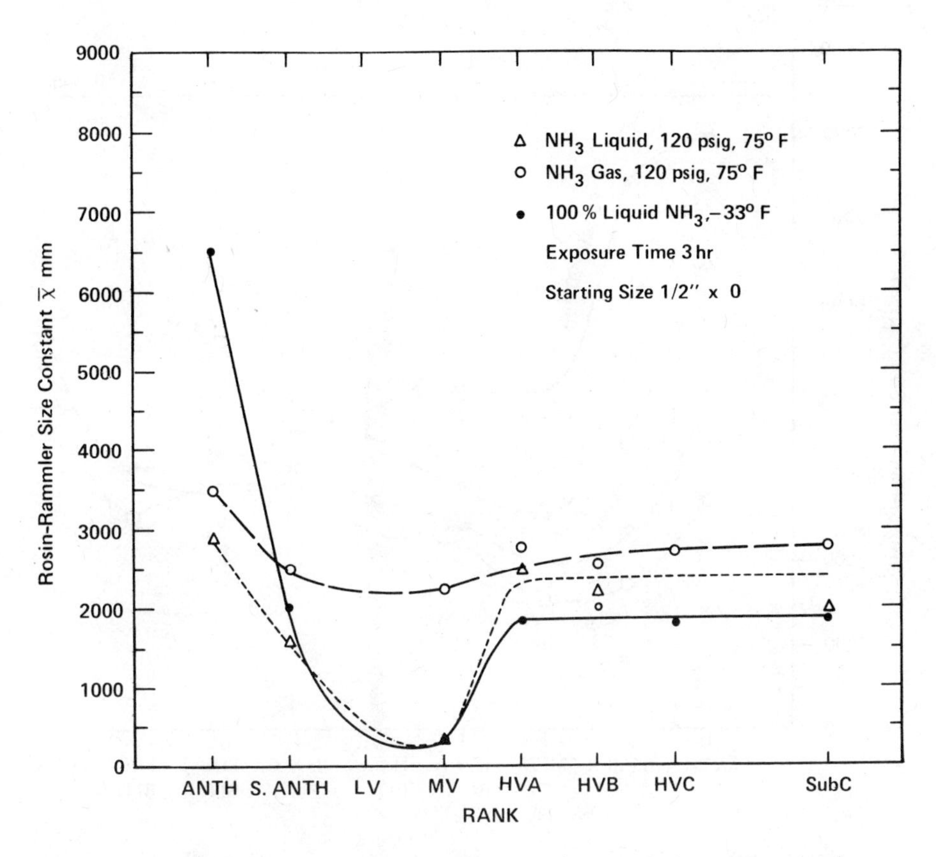

Figure 1.4 Fragmentation of coals of varying rank in gaseous and liquid ammonia. (From Ref. 5.)

ease with which gas could diffuse into the micropores of anthracite. In general, as the amount of water in the ammonia increases, the extent of fragmentation decreases. However, for high-volatile C bituminous and lower rank coals, the extent of fragmentation is greater in a 1:1 than in a 3:1 ammonia-water mixture, even though the extent of fragmentation in the former is initially smaller [6].

Effect of Comminution Pressure. Other conditions such as temperature remaining constant, an increase in the comminution pressure increases the rate at which maximum fragmentation takes place. At atmospheric pressure, little or no fragmentation occurs with gaseous ammonia. However, as shown in Fig. 1.5, increasing the pressure causes

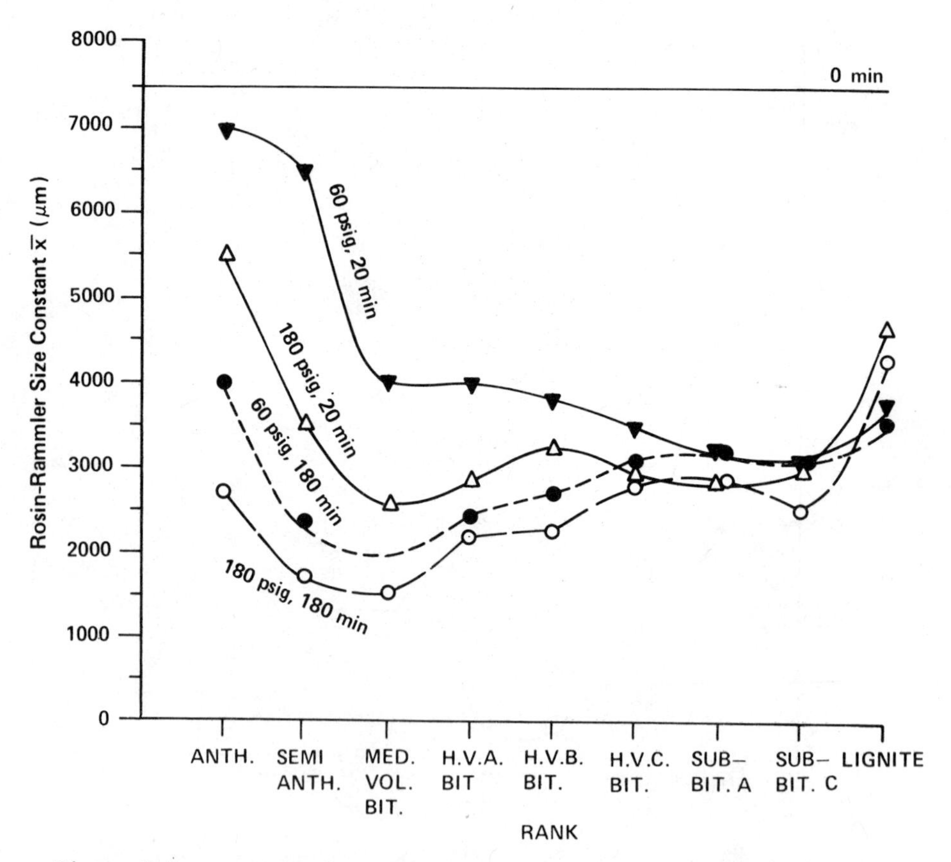

Figure 1.5 Fragmentation of coals of varying rank in gaseous ammonia at different pressures. (From Ref. 12.)

progressively greater size reduction [12]. The results in Fig. 1.5 were all obtained at room temperature (75°F).

Effect of Comminution Time. For any comminution condition, there is a limit to which the coal can be chemically fragmented. Figure 1.6 shows that increasing the comminution time increases the fineness of the samples, until a limiting comminution time is reached when further treatment does not produce any additional significant fragmentation [12].

Effect of Coal Properties. The fragmentation caused by chemical treatment is affected by such coal properties as moisture content, starting size of the coal, and preconditioning of the coal before treatment.

The moisture content of coal tends to affect fragmentation differently, depending upon whether the samples are treated in

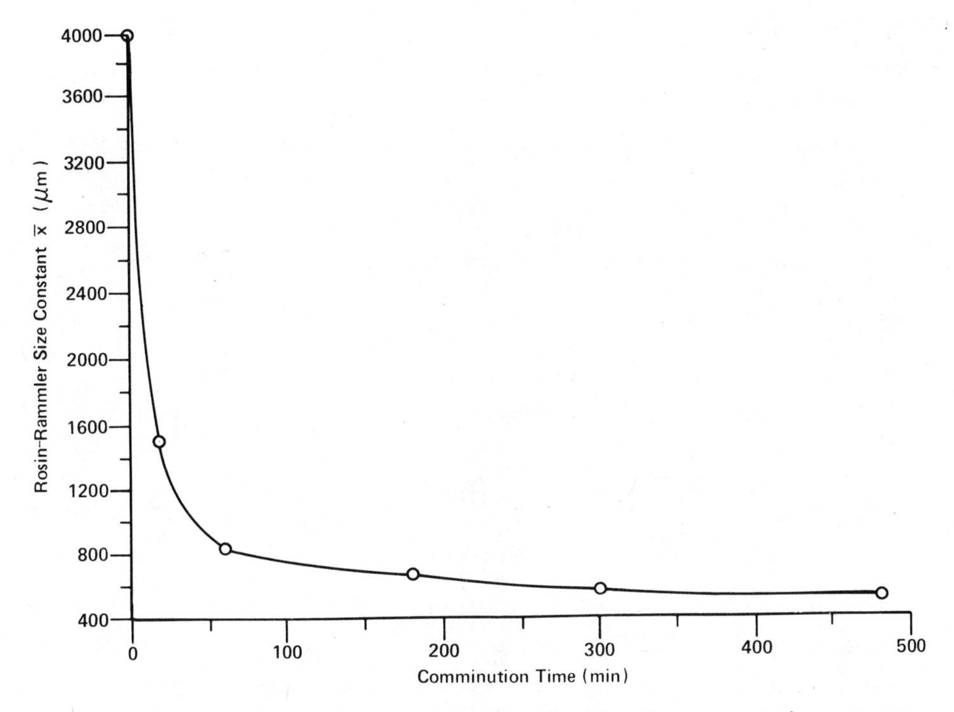

Figure 1.6 Fragmentation of low volatile bituminous coal in gaseous ammonia at 135 psig and 75°F at different communition times. (From Ref. 12.)

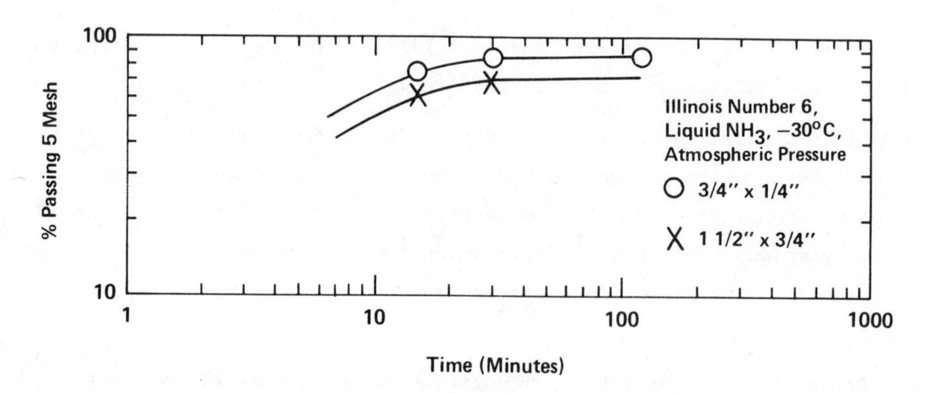

Figure 1.7 Effect of starting size on chemical comminution of an Illinois No. 6 coal. (From Ref. 5.)

gaseous or liquid ammonia. With higher rank coals, both liquid and gaseous ammonia tend to fragment slower as moisture content increases. However, with the lower rank coals, the extent of fragmentation with gaseous ammonia increases slightly with an increase in moisture content, whereas with liquid ammonia, it usually reaches a minimum at approximately 6 to 10% moisture [6,12].

As is expected, the starting size of the coal before treatment can affect the size of the treated product. This effect is illustrated in Fig. 1.7 [5].

Figure 1.8 demonstrates the importance of evacuating the reactor before chemical comminution of an Illinois No. 6 coal [5]. Without evacuation, the extent of chemical fragmentation is considerably smaller. This effect has also been noticed with other coals, although they have not been as demonstrative. For example, all the conditions used in Fig. 1.8 would have no fracturing effect on a Pennsylvania Pittsburgh coal which was examined.

C. Properties of Chemically Comminuted Coal

Nitrogen Content and Chemical Reactions. Chemical reactions between the ammonia and coal could have an adverse effect on the recovery of the ammonia and on the amount of nitrogen oxides emitted when the chemically comminuted coal is combusted. Therefore, the nitrogen content of coal before and after ammonia treatment was determined

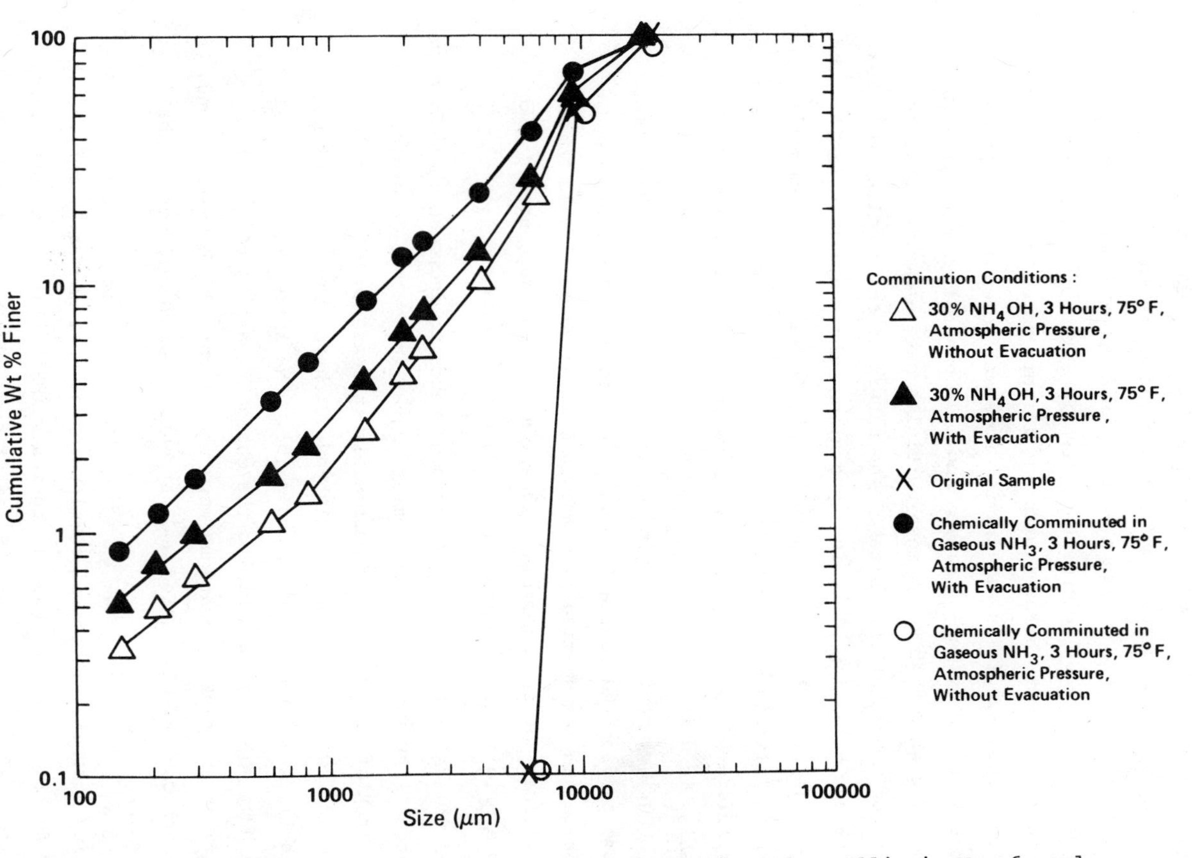

Figure 1.8 Effect of preevacuation on chemical comminution of an Illinois No. 6 coal. (From Ref. 5.)

for a variety of coal seams. The results presented in Table 1.2 [5] vary slightly for different coals, and the increase in nitrogen appears to be in correlation with a decrease in rank. No increase appears to take place with a Pennsylvania Upper Freeport coal, a slight increase (6%) with a Pennsylvania Pittsburgh coal, and approximately 20% increase with an Illinois No. 6 coal when the treated coal sample is air dried. However, some of the nitrogen can be removed by hot-water washing. In general, the results in Table 1.2 show that the 8-mesh particles have a smaller increase in nitrogen than the other sizes. The nature of the chemical reaction which may be taking place is unknown, but there are functional groups (e.g., esters) in coal which may form nitrogen compounds. With an Illinois No. 6 coal, the loss of ammonia would still be small (4.4 lb ammonia per ton of treated coal) when hot-water washing is used.

During chemical comminution, ammonia tends to leach certain elements from the coal and convert them to a water-soluble state. Lignite samples chemically comminuted in anhydrous liquid ammonia show lower copper, chromium, nickel, and manganese contents than those originally present [6].

Coal Grindability. Ammonia treatment tends to increase the Hardgrove grindability index (HGI) of coal to a varying extent depending upon the rank of the coal. For most coals increase in HGI is at least 5 to 10%, while increases as high as 50% have been noted for high volatile A and B bituminous coals [6]. The increased grindability may have some relationship to the change in ash content; but since no float-and-sink separations were used before HGI determinations, the ash content of treated and untreated coal should not be that different.

Caking Property. Chemical comminution of coal generally results in a reduction of the free-swelling index (FSI) [13]. With one highly caking coal (FSI = 9), the FSI after chemical treatment was 6 to 7 [6]. This reduction in the caking properties of chemically comminuted coal would make it an excellent feed product for aboveground

Table 1.2 Nitrogen Content of Chemically Comminuted Coal[a]

Coal	Ammonia treatment	Treatment to remove ammonia	Original ROM sample	Air dried at 60°C overnight	Air dried at 60°C and sized				Air dried at 100°C for 4 hr	Air dried at 200°C for 2 hr	Air dried at 300°C for 1 hr
					+8 mesh	8 × 20 mesh	20 × 100 mesh	−100 mesh			
Pittsburgh, Green County, Pa.	180 min, NH_3 gas, 120 psig, 75°F	Air dried, 60°C	1.43	1.51							
	30 min, 100% NH_3 liquid, 120 psig, 75°F	Rinsed with dilute HCl at 20°C, air dried		1.53							
		Elutriated with water at 100°C, air dried		1.50							
Upper Freeport, Westmoreland County, Pa.	240 min, 100% NH_3 liquid, atm. pressure, −30°F	Air dried	1.21	1.12	1.08	1.20	1.44	1.21	1.21	1.20	1.24
		Elutriated with water 100°C for ½ hr		1.21	0.97	1.15	1.37	1.22	1.24	1.37	1.38
		Elutriated with water 100°C for 2½ hr		1.25	0.98	1.21	1.39	1.16	1.24	1.26	1.11
Illinois No. 6, Franklin County, Ill.	240 min, 100% NH_3 liquid, atm. pressure, −30°F	Air dried	1.58	1.94	1.75	1.96	1.94	1.64	1.91	1.94	2.00
		Elutriated with water at 100°C for ½ hr		1.80	1.58	1.74	1.69	1.59	1.68	1.69	1.82
		Elutriated with water at 100°C for 2½ hr		1.72	1.62	1.74	1.71	1.43	1.70	1.69	1.96

[a]Numbers in percent total nitrogen.

Source: Ref. 5.

gasifiers which need noncaking or only slightly caking coals for efficient operation [6].

Mineral Liberation. Scanning electron microscope studies of the chemically comminuted coal show that fragmentation from chemical treatment is strongly controlled by maceral boundaries and other components within the material such as pyrite boundaries [1]. Microscopic comparison of mechanically crushed and chemically comminuted coals indicates that pyrite particles still joined to coal particles in the chemically comminuted coal are approximately half the size of those in the mechanically crushed sample [6]. These results appear to suggest the selective breakage along mineral matter boundaries which occurs with chemical comminution. The latter may also explain why pyrite is liberated during chemical comminution without excessive size reduction.

Washability Studies. Washability tests have been conducted on 1½-in. top size run-of-mine (ROM) samples, on two mechanically crushed samples (3/8-in. and 14-mesh top size), and on chemically comminuted samples of various coals. Percent sulfur and ash-vs.-percent recovery curves were compared along with the relative size of the coal to determine the respective efficiencies of chemical and mechanical size reductions in liberating pyrite and other mineral matters [2,6,14]. A typical comparison of the washability results of mechanically crushed and chemically comminuted samples of an Illinois No. 6 coal is discussed below.

The size distributions of the various samples are shown in Fig. 1.9. The chemically comminuted product contains only 4.5% of −100-mesh fines; whereas the 3/8-in. and 14-mesh mechanically crushed samples contain 8.7% and 21.9% of −100-mesh fines, respectively. Since the 1½-in. ROM sample contains 2.2% −100-mesh fines, chemical fragmentation has resulted in generating only a small amount of additional fines.

Figures 1.10 and 1.11 show the sulfur and ash washability curves for the sized samples of Fig. 1.9. These washability curves were generated with 100-lb samples that had been coned and quartered from the same samples according to established ASTM procedures.

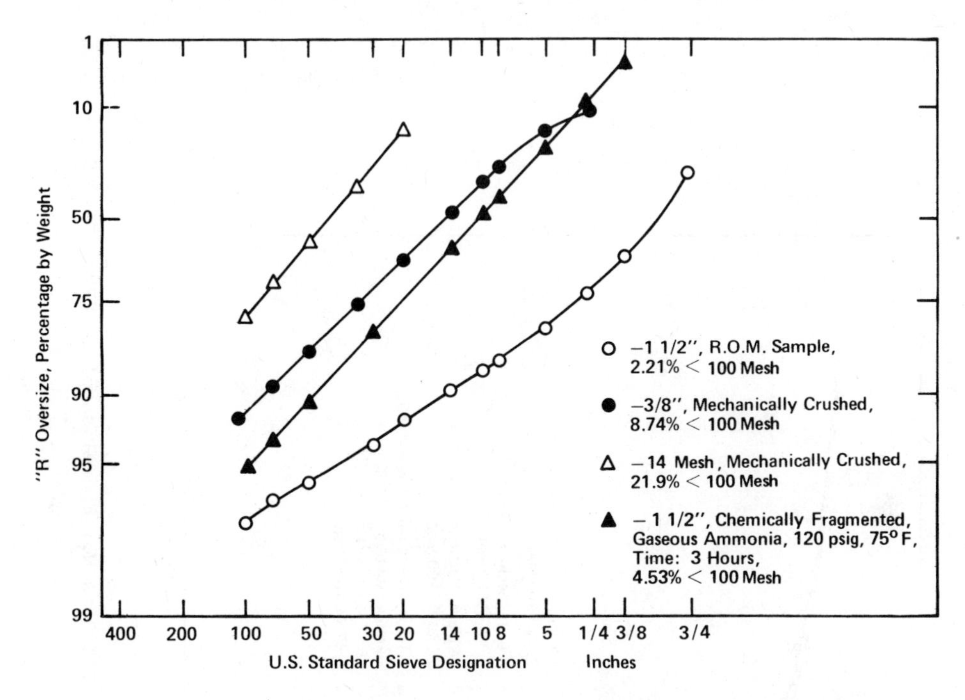

Figure 1.9 Size consist of run-of-mine, chemically comminuted, and mechanically crushed samples of an Illinois No. 6 coal. (From Ref. 5.)

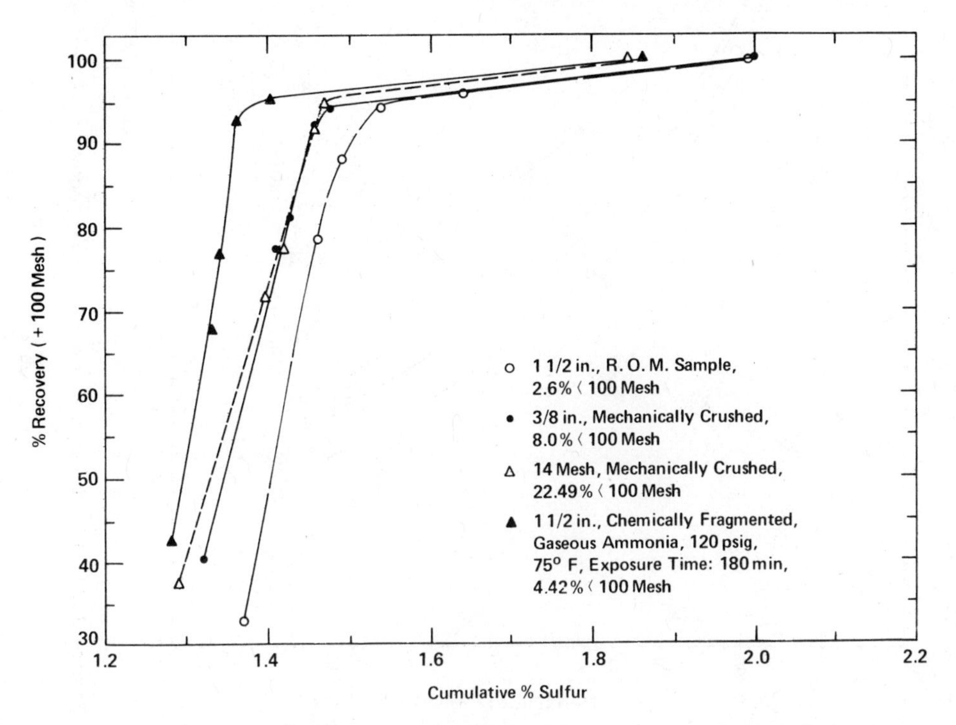

Figure 1.10 First comparison % sulfur-vs.-recovery curves for an Illinois No. 6 coal sample. (From Ref. 5.)

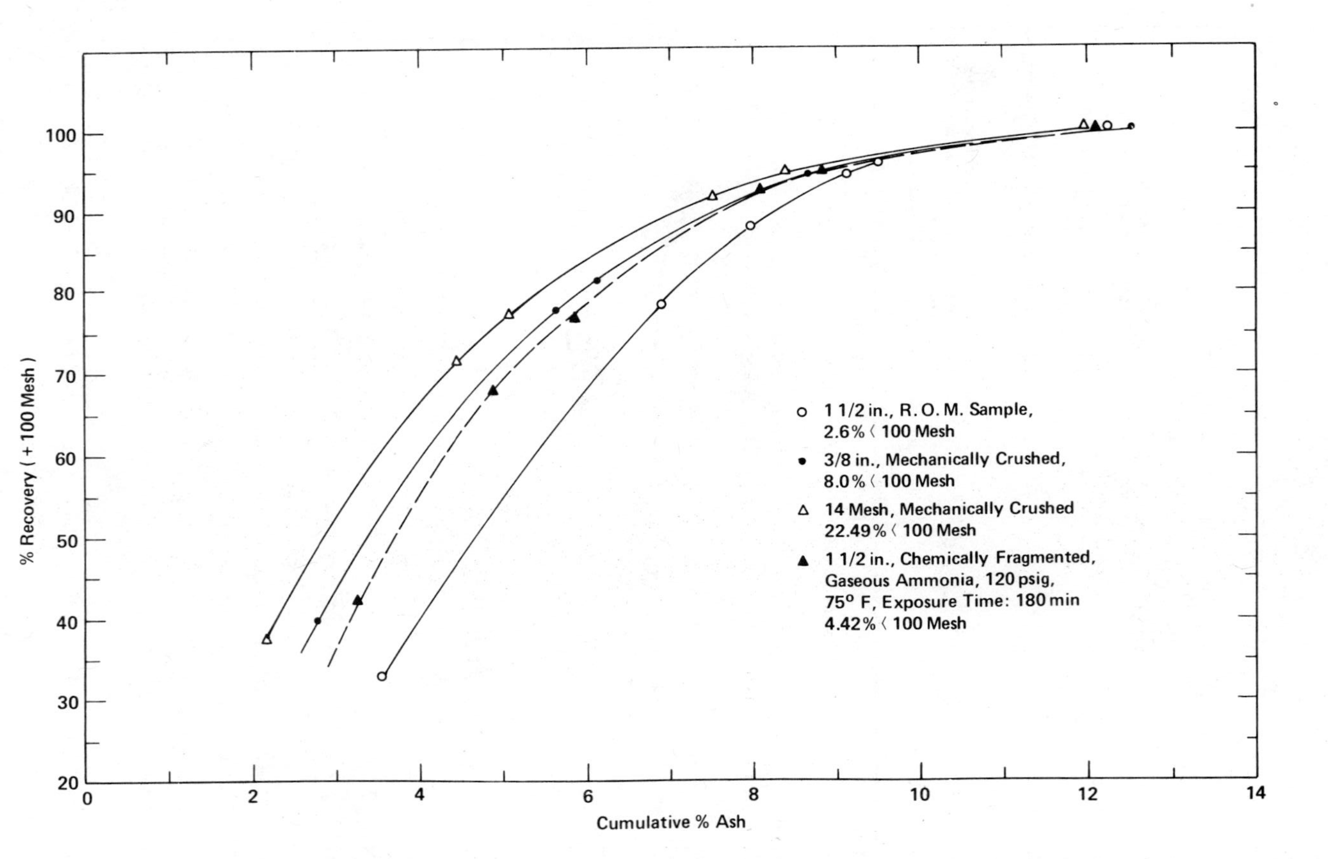

Figure 1.11 First comparison % ash-vs-recovery curves for an Illinois No. 6 coal sample. (From Ref. 5.)

After the indicated size reduction, a representative sample was analyzed by float-and-sink analysis at various specific gravities. Each point in these graphs represents the analysis and recovery of the 100-mesh product at a particular specific gravity. To consider recovery based on feed rather than product, the amount of −100 mesh must be considered.

In the comparison of cumulative sulfur recovery (Fig. 1.10), the chemically comminuted coal is superior to the other three samples. For example, at a 90% recovery of +100-mesh coal, sulfur content would be 1.35 wt % for the chemically fragmented product, 1.45 wt % for the 3/8-in. mechanically crushed coal, 1.48 wt % for the 14-mesh mechanically crushed coal, and 1.51 wt % for the 1½-in. ROM samples. Conversely, for a selected sulfur value of 1.40 wt %, weight yield recoveries would be 96% for the chemically comminuted product, 78% for the 14-mesh crushed coal, 70% for the 3/8-in. crushed coal, and 49% for the 1½-in. ROM sample. Thus, the chemically fractured coal (4.42% less than 100 mesh) liberates considerably more pyritic sulfur than mechanically ground coals, even when ground to larger than −14 mesh (22.49% less than 100 mesh) [6]. Similar results have been obtained with Pennsylvania Redstone, Pittsburgh, Upper Freeport and Lower Freeport seam coals and a Kentucky No. 9 coal [6,14,15], and with some Iowa coals [16]. Keller and Smith have reported that greater pyrite liberation does not occur with chemical comminution of a Pennsylvania Upper Freeport coal [10], but they did not consider yields in their comparison [11].

Figure 1.11 shows cumulative ash in the same four samples of Illinois No. 6 coal versus recovery of +100-mesh coal. These curves indicate that, unlike sulfur liberation, ash liberation is correlated directly with the size. The order of increasing ash liberation is given by run-of-mine < chemical comminuted < mechanically crushed to 3/8-in. top size < mechanically crushed to −14 mesh [17].

In summary, the washability studies have shown that at a fixed size and weight yield recovery, greater liberation of pyritic sulfur and comparable liberation of ash are possible with chemical comminution than with mechanical crushing.

D. Current Research and Development Programs

Two research and development programs are in their final stages or have been completed. The Syracuse Research Corporation with support from the Homer City multistream coal-cleaning plant operators (New York State Electric and Gas Corporation, General Public Utilities, and Pennsylvania Electric Co.) are conducting detailed washability studies on two Homer City coals (Helen and Helvetia Mines, Upper and Lower Freeport seams). In this study, various size fractions will be chemically comminuted separately and the product will be sized. Each separate product size will be washed separately. The developed washability curves will be compared to similarly processed mechanically crushed coal. By washing the size fractions separately, the investigation will show whether there are certain portions of the feed coal where the amount of pyrite liberation due to chemical comminution is greatest.

The majority of the experimental work on chemical comminution up to this time had been done on batch vapor-type systems. Recent experimental results obtained by SRC suggest that a liquid-mode reaction system with mild agitation considerably reduces the necessary retention time for ammonia fracturing. Using such a system for a commercial design means smaller equipment requirements and, hence, lower capital investment and operating costs. As a result of this information, Catalytic, Inc. and the Electric Power Research Institute (EPRI) jointly sponsored a developmental program at Hazen Research, Inc., Golden, Colorado. The main goal of this program was to develop a definitive data base for comparing the liquid-phase chemical comminution and the present practice of mechanical crushing as applied to ROM and middlings coal cleaning. Additionally, experimental work was undertaken on certain fundamental aspects of vapor-phase chemical comminution. Thus far, all experimental work on this program has been completed, and a final report will be published by EPRI shortly.

1.3 PROCESS DESIGN

Chemical comminution as used for coal cleaning requires the following major steps:

1. Treating raw coal at −1½ in. with a chemical agent, e.g., ammonia as a gas, liquid, or aqueous solution.
2. Removing the adsorbed ammonia from the coal, e.g., by hot-water washing.
3. Recovering ammonia from the wash water for recycling by distillation followed by compression, with or without liquefaction.
4. Delivering the comminuted coal to a physical cleaning operation as an aqueous slurry or as a partially dewatered product.

Alternate methods and conditions are applicable to the commercialization of the preceding steps and will depend on the characteristics of the coal and process conditions that will be established as the development proceeds.

A diagram of a conceptual process flowsheet of a batch-type chemical comminution process for the above operation is shown in Fig. 1.12. The chemical treatment step can consist simply of batch vessels that provide required exposure to ammonia at optimum conditions as to time, pressure, and form of ammonia. One set of conditions, based upon successful bench-scale work, includes a 2-hr exposure to ammonia gas at 120-psig pressure. This conceptual process requires a series of vessels to provide the associated operations. An example of the required associated operations and cycle time is as follows:

Operation	Cycle time, min
Charging	30
Evacuating entrained air	30
Equalizing to medium pressure with ammonia	30
Pressurizing and reacting at full pressure (100 to 125 psig) with ammonia	120
Relieving to medium pressure	30
Depressurizing to about 2 psig	30
Discharging	30
Total cycle time	300

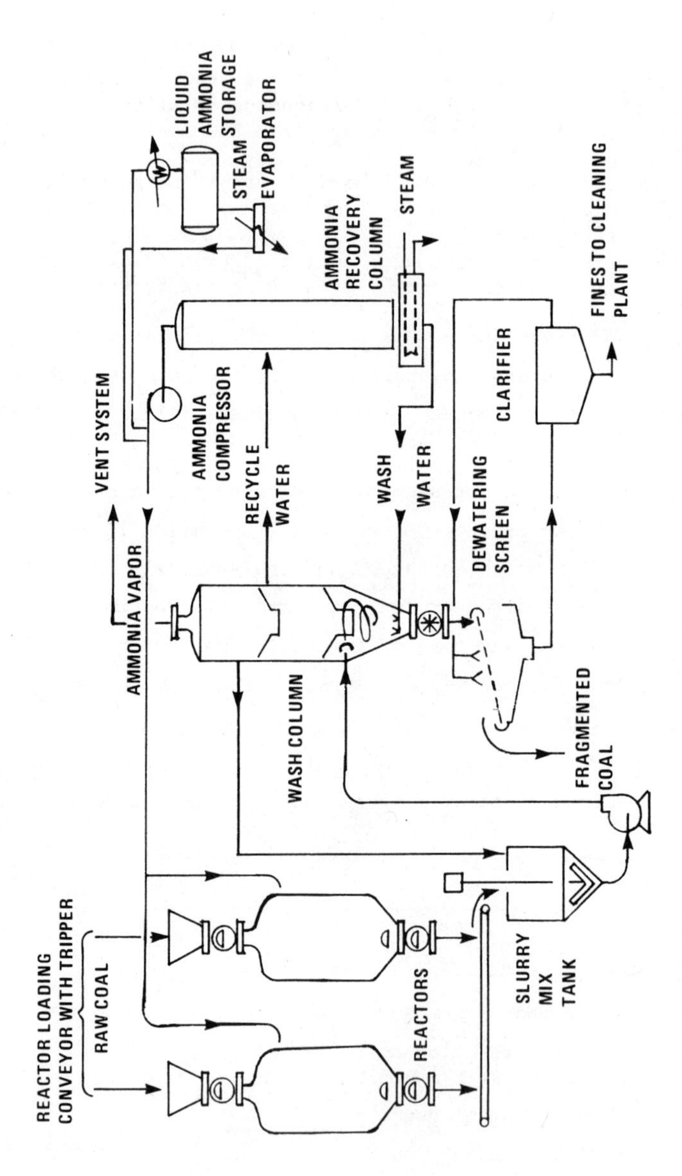

Figure 1.12 Flowsheet of a batch-type chemical comminution process using ammonia vapor.

This total cycle time is conservatively derived. It can vary from coal to coal and conceivably may be much shorter for certain coals to which the process is applicable.

To provide the foregoing cycle in a sequenced commercial operation at a coal feed capacity of 1250 tons/hr, 10 steel comminution tanks (reactors) 25 ft in diameter by 60 ft high each are required. It is visualized that with raw coal having 6% moisture, approximately 100 tons of ammonia gas will be absorbed per hour. The temperature in the reactor will rise about 50°C above the ambient temperature due to the heat of solution of the ammonia absorbed by the moisture in the coal.

Alternate methods based on continuous charging and discharging through pressure-locking devices or containers may have economic benefits (Fig. 1.13). In this flowsheet, the comminution tank (reactor) is charged using two parallel-feed hoppers. The comminuted coal is discharged to a devolatilizer where most of the ammonia vapor is separated for compression and recycled to the reactor. Coal from the devolatilizer is slurried with recycled water and pumped to a countercurrent, hot-water wash column. An aqueous ammonia stream from the wash column is sent to an ammonia recovery column, and the coal product stream is screened and sent to the washing plant for separation of impurities.

Use of aqueous ammonia for simplifying continuous charging and as a means of providing mechanical agitation to minimize comminution time may be economically advantageous (Fig. 1.14). Coal is fed continuously to the comminution tank (reactor) as a slurry and then discharged to a flash vessel from which a large portion of the ammonia is separated, compressed and recycled. The coal from the flash vessel is slurried in a mixing tank and pumped to a wash column for treatment similar to that described in Fig. 1.13.

The system for removing the residual ammonia from the coal by hot-water washing consists of one or more steel wash columns or extraction towers. The discharged coal from the reactors at near ambient temperatures would be slurried into a continuously recycled

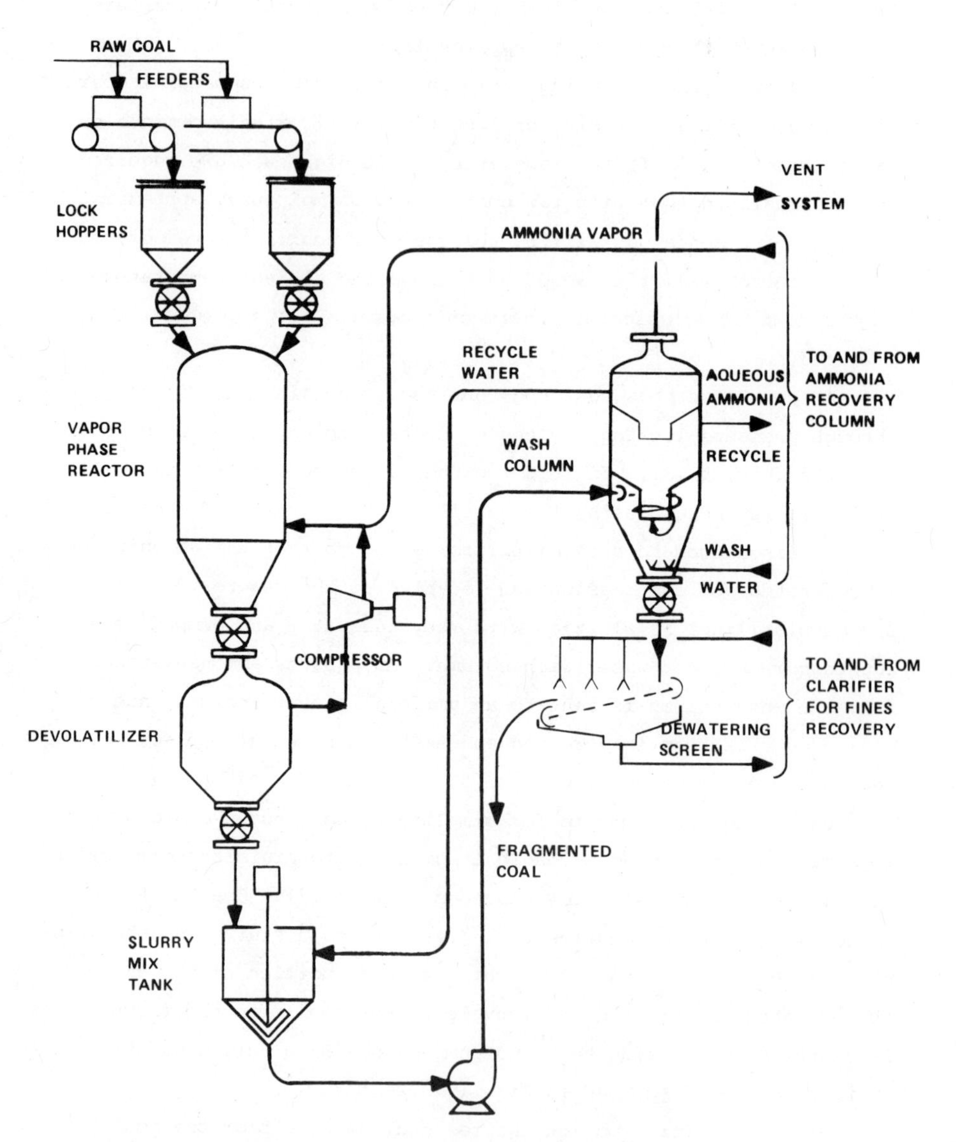

Figure 1.13 Flowsheet of a continuous chemical comminution process using ammonia vapor (lock hopper feed design).

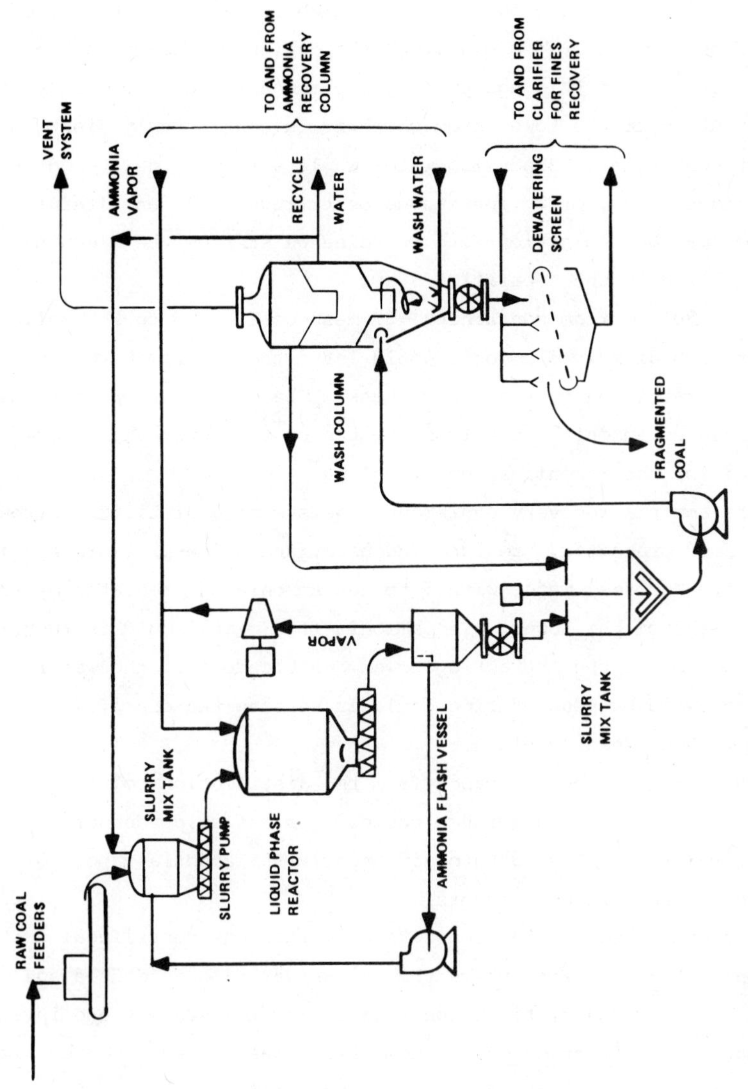

Figure 1.14 Flowsheet of a continuous chemical comminution process using ammonia liquid (slurry feed design).

stream of water which would feed the towers as a 35% solids slurry at near mid-height. The tower is sized to permit the feed coal to drop through the lower section countercurrent to a stream of hot water that extracts the ammonia. The high relative volatility of this ammonia-water system minimizes the design requirements and operating costs of this equipment. The ammonia-free wet coal settles on the bottom of the tower from which it is continuously discharged. The wet coal with accompanying water would either be dewatered on vibrating screens, or by continuous centrifuges. As an alternate, the coal may be reslurried with recycled water from the cleaning plant as feed to the separation step.

A recycle stream containing —30-mesh coal of 15 to 20% solids and 5 to 10% dissolved ammonia would leave the towers above the feed point for returning to the slurry tank. The excess water containing the recovered ammonia flows from the top of the tower and becomes the feed for the ammonia recovery column.

The ammonia recovery equipment consists of a distillation tower with a feed preheater, reflux condenser, and a low-pressure steam reboiler. The feed water with 5 to 10% ammonia is preheated by the exit vapor from the tower. The ammonia-free exit liquid is recycled to the bottom of the extraction tower. A bleedoff of the water, taken to avoid buildup of fine coal, is sent to the clarification system in the cleaning plant.

The partially cooled vapor from the distillation column at near atmospheric pressure is further cooled in a reflux condenser with cooling water to about 95°F to give nearly 95% ammonia vapor for feed to the ammonia compressors.

The compressor system consists of multistage centrifugal units with interstage coolers, where the excess water is condensed and recirculated to the distillation tower. For temperature and inventory control, a fraction of the compressed gas is liquefied by either refrigeration or further compression and cooling.

1.4 ECONOMIC EVALUATION

The economic benefits of chemical comminution in coal preparation in combination with normal washing procedures result from the more effective removal of the pyritic sulfur. This pyritic sulfur removal is accomplished for coals that are selectively responsive to the treatment, at significantly coarser size consistency than that from using traditional crushing methods. Chemical comminution not only produces larger particle sizes, but also tends to release greater amounts of impurities than comparable mechanically created particle size distributions. These conditions are reflected in the improved product recovery for a given sulfur content in the cleaned coal product and in the reduction in the capital and operating costs of the washing plant for treating coal containing less fines.

The economic advantages of chemical treatment are shown by comparing conceptual capital and operating costs for a washing plant for chemically comminuted feed coal with one treating mechanically ground feed coal. In both cases the washing plant is considered as a relatively simple, two-train, heavy-medium bath and cyclone separation system. Only one steam coal product is being produced. Cleaning is provided for down to 100-mesh coal particle size material. No desliming or thermal drying facilities are included.

The basis of operation of the washing plant for both cases is two 8-hr shifts per day and 330 days/year for a total yearly operation of 5280 hr. In the mechanical crushing case, the instantaneous feed rate is 750 tons/hr. This material is crushed from 1½ to 3/8-in. top size before entering the washing plant. No storage facilities are involved.

For the chemical comminution case, coal is fragmented chemically from 1½- to 3/8-in. top size in a 24 hr/day operation. No mechanical crushing is required, but intermediate storage and load-out facilities costing $1.2 million are included. The instantaneous flow rate through the washing plant is 562.5 tons/hr. Estimated capital investment and annual operating cost data for both washing plants were obtained from the Roberts and Schaefer Company, a major coal preparation plant contractor [18].

Table 1.3 Economic Comparison of Coal Cleaning Mechanical Crushing vs. Chemical Comminution (for 8000 ton/day Clean Coal Production)

Basis: Raw coal (−1½ in.)
Totals = 2.1 wt % S; ash = 12.5 wt %
Btu value = 12,500/lb
Value = $/MM Btu
Illinois No. 6 coal

Washing plant:
Feeds mechanically crushed to 3/8 in.
Product, sulfur = 1.4 wt %
Operation = 330 days/year

		Mechanical crushing		Chemical comminuted	
Clean coal yield[a]	wt %	73		95	
Ash content	wt %	5.2		8.8	
Btu value	per lb	13,500		13,000	
Feed coal	tons/day	12,000		9,000	
Fines, 100 x 0 mesh	tons/day	1,000		400	
Rejects	tons/day	3,000		600	
Product	tons/day	8,000		8,000	
Capital investment					
Plant facility	$\$10^6$				
Chemical treatment		-		11.0	
Washing plant		10.5		8.6	
Contingency		2.1		4.0	
Plant facility investment		12.6		23.6	
Total capital investment[b]		22.9		33.6	
Annual operating cost					
Coal	$\$10^6$/year	99.0		72.8	
Production costs[c]		5.3		6.0	
Refuse disposal (resale)[d]		1.3	(12.4)	0.3	(1.6)
Fixed costs		1.2		2.3	
Total operating cost (resale)[d]		106.8	(93.1)	81.4	(79.5)
Average revenue required					
Utility economics[e]					
Per year ($\$10^6$)		109.1	(95.4)	84.3	(82.4)
Per MM Btu		1.52	(1.33)	1.22	(1.19)
Commercial economics[f]					
Per year ($\$10^6$)		111.6	(97.9)	86.4	(84.6)
Per MM Btu		1.55	(1.36)	1.25	(1.22)

[a]Yield includes 100-mesh coal only.

[b]Includes land, interest during construction, start-up, and working capital requirements.

[c]Production costs include chemicals, labor, maintenance, and other physical separation costs not included elsewhere. Based upon $1.35/ton coal feed in the mechanical crushing case [18].

[d]Resale case gives a credit to refuse rather than charging a disposal cost. Value of the mechanically cleaned refuse is 47¢/MM Btu with 30 wt % ash constant and 28¢/MM Btu for the chemical comminution case with 47% ash.

[e]Includes 12% return on utility investment, 10% interest with 65% debt financing. Also includes income tax at 50% rate.

[f]Based upon 15% DCF with 65% debt financing.

Test results for the chemically fragmented and mechanically crushed Illinois No. 6 coal, previously plotted in Figs. 1.10 and 1.11 [2], are used for the comparison. Additional assumptions made in the economic evaluation together with the estimated process costs are summarized in Table 1.3.

The economic differences shown by the comparison are as follows:

1. The capital investment for the plant producing 8000 tons/day clean coal using chemical comminution is $11 million higher than with mechanical ground coal at 3/8 in.
2. The additional investment cost of the comminution plant is offset by the reduced amount of feed coal required for the washing plant due to higher product recovery and simpler washing equipment required for the comminuted product.
3. The higher operating costs of chemical comminution are more than offset by the lower raw coal consumption. This is true even with a credit allowance for the heating value remaining in the rejected coal from a mechanical crushing plant.
4. The unit price per MM Btu of coal product necessary to realize an acceptable return on investment favors the chemical case: $1.25/MM Btu chemical comminution versus $1.55/MM Btu mechanical crushing on a commercial economic basis, or $1.22/MM Btu chemical comminution versus $1.52/MM Btu mechanical crushing applying utility economic factors.

Comparison on the basis of maintaining the same recovery of coal for the mechanical crushing operation as for the chemical treatment with the resulting reduced sulfur removal shows that the penalties from the higher sulfur in the product also results in a less economical overall condition. In this case the poorer steam generator conditions and the heavier load on flue gas scrubbing systems lead to added capital and operating charges with resulting overall economy for the chemically comminuted coal.

Table 1.3 gives the cost per million Btu for disposing of refuse or considering it of value as feed coal to further recovery. It appears unlikely that 3000 tons/day of refuse obtained from mechanically crushed material would be disposed of without recovery of further Btu values. Recovery of this refuse probably would not be considered in the chemical comminution case. Nevertheless, chemical comminution appears more attractive with or without consideration of refuse recovery.

The 100 × 0 mesh material, when subjected to further cleaning, results in low recoveries (below 50 wt %) and requires three to four times the capital and operating costs of coarse coal. The additional capital and return on investment would have to be justified in the mechanical crushing case.

Other coals such as those from the Pennsylvania Redstone and Kentucky No. 9 seams would show substantially the same advantage for chemical treatment.

An economic analysis simulating the conditions of an actual multistream coal-cleaning plant located at Homer City, Pennsylvania, operated by the Pennsylvania Electric Co., also shows an advantage for chemical treatment [19]. A plant capacity at 30,000 tons/day of a blend of Pennsylvania Upper and Lower Freeport coals with a top size of 1½ in. was used as a basis.

Two plant concepts were compared, i.e., plant A based on the traditional method of mechanical crushing to a top size of 3/8 in. with a capital cost of $35 million and plant B based on chemical fragmentation with ammonia vapor, having a total cost of $53 million. Size reduction in each plant is followed by identical state-of-the-art heavy-medium coal separation techniques. All other facilities are the same in the two conceptual plants.

The comparison indicated that the above capital cost differential of $18 million would yield an additional return on investment of $3,700,000/year over and above increased operating costs. In other words the *added* investment for chemical comminution would yield a 20% return on that portion of the total investment. This is due to the various benefits that the coarser, high-yield comminuted coal would develop [19].

1.5 CONCLUSIONS

Chemical comminution, as a promising cost-effective approach for meeting increasingly stringent pollution standards while using cheaper lower-grade coals, warrants continued development.

In combination with conventional coal cleaning processes, chemical comminution offers a means of increasing the removal of pyritic sulfur in some coals. The special economic value of the process lies in its ability to improve the yield of desulfurized coal product without, at the same time, reducing particle size and creating troublesome excessive fines.

Bench-scale data indicate that the process will largely eliminate the expensive equipment needed for processing fine coal by permitting the use of relatively simply hydrocyclone-type equipment which effectively cleans coarser coal. These savings largely offset the initial construction cost of the chemical comminution processing unit.

Realistic cost analyses of coal-cleaning plants utilizing chemical comminution for size reduction have shown that the increased product yield or decreased sulfur content, or both, would make the final preparation system more profitable than an equivalent plant processing mechanically crushed coal. A further advantage is that chemically comminuted coal can be readily stored and easily shipped, thus eliminating problems associated with mechanically ground fine coal.

For certain eastern coals, chemical comminution may be the key to producing a fuel sufficiently clean for direct combustion. For any particular coal, environmental acceptability will depend on its calorific value and proportions of pyritic and organic sulfur. For "cleaned coals" not meeting environmental standards, the process promises to reduce the amount of postcombustion desulfurization required to meet future air pollution standards. Furthermore, "cleaned coals" should result in increased boiler reliability and availability by reducing power plant maintenance costs.

Other applications of chemical comminution include coal preparation to provide a feed product for gasification, liquefaction, coking, coal-oil mixtures and blending schemes.

ACKNOWLEDGMENTS

Some of the research studies reviewed in Sec. 1.2 were sponsored by the U.S. Energy Research and Development Administration (now the Department of Energy) under contracts 14-32-0001-1777 (1976) and EX-76-C-01-2520 (1978).

We wish to thank Mr. Arnold Hanchett, a consultant to SRC, for his valuable contributions to the work discussed here. Many of the initial plant designs were his and the final designs and economic estimates have been considerably influenced by Mr. Hanchett's participation.

REFERENCES

1. R. T. Greer, Coal Microstructure and the Significance of Pyrite Inclusions, *Scanning Electr. Microscopy*, Vol. I, Proceedings of the Workshop on Materials and Component Characterization/Quality Control with the SEM/STEM, IIT Research Institute, Chicago, Ill., Mar. 1977.

2. R. S. Datta, P. H. Howard, and A. Hanchett, *Feasibility Study of Precombustion Coal Cleaning Using Chemical Comminution*, FE-1777-4, Final Report ERDA Contract 14-32-001-1777, Syracuse Research Corporation, Syracuse, N.Y., Nov. 1976.

3. R. G. Aldrich, D. V. Keller, Jr., and R. G. Sawyer, Chemical Comminution and Mining of Coal, U.S. Patents 3,815,826 (1974) and 3,870,237 (1975).

4. R. G. Aldrich, D. V. Keller, Jr., and R. G. Sawyer, Chemical Comminution and Mining of Coal, U.S. Patents 3,850,477 (1974) and 3,918,761 (1975).

5. P. H. Howard and R. S. Datta, Chemical Comminution: A Process for Liberating the Mineral Matter from Coal, in *Coal Desulfurization: Chemical and Physical Methods, ACS Symposium Series, 64* (T. D. Wheelock, ed.), American Chemical Society, Washington, D.C., 1977, pp. 58-69.

6. R. S. Datta and P. H. Howard, *Characterization of the Chemical Comminution of Coal*, Final Report ERDA, Contract No. EX-76-C-01-2620, Syracuse Research Corporation, Syracuse, N.Y., Mar. 1978.

7. J. W. Leonard (ed.), *Coal Preparation*, 4th ed., AIME, New York, 1979, p. 7-5.

8. W. Francis, Physical Considerations, in *Coal: Its Formation and Composition*, 2d ed., Edward Arnold, London, England, 1961, Chap. XI.

9. I. G. C. Dryden, Solvent Power for Coals at Room Temperature, *Chem. Ind.*, 502 (June 2, 1952).

10. D. V. Keller, Jr. and C. D. Smith, Spontaneous Fracture of Coal, *Fuel, 55*, 273-280 (1976).

11. P. H. Howard, R. S. Datta and A. Hanchett, Chemical Fracture of Coal for Sulfur Liberation, *Fuel, 55*, 346 (1977).

12. R. S. Datta, Chemical Fragmentation of Coal, NCA/BCR Coal Conference and Expo IV, Louisville, Ky., Oct. 18-20, 1977.

13. J. W. Leonard, W. E. Lawrence, and W. A. McCurdy, Coal Characteristics and Utilization, in *Coal Preparation* (J. W. Leonard, ed.), 4th ed., AIME, New York, 1979, pp. 3-9 to 3-11.

14. P. H. Howard, A. Hanchett, and R. G. Aldrich, Chemical Comminution for Cleaning Bituminous Coal, Clean Fuels from Coal Symposium II, Institute of Gas Technology, Chicago, Ill., June 23-27, 1975.

15. R. S. Datta, P. H. Howard, and A. Hanchett, Pre-Combustion Coal Cleaning Using Chemical Comminution, NCA/BCR Coal Conference and Expo III, Louisville, Ky., Oct. 19-21, 1976.

16. S. Min and T. D. Wheelock, Cleaning High Sulfur Coal, NCA/BCR Coal Conference and Expo III, Louisville, Ky., Oct. 19-21, 1976.

17. P. H. Howard and R. S. Datta, Desulfurization of Coal by Use of Chemical Comminution, *Science, 197*(4304), 668-669 (1977).

18. J. M. Clifford, personal communication, Roberts and Schaefer Company, Chicago, Ill., Jan. 26, 1979.

19. Catalytic, Inc., Philadelphia, Pa., technical report, *Chemical Comminution: An Improved Route to Clean Coal*, 1977.

2

The Otisca Process: An Anhydrous Heavy-Liquid Separation Process for Cleaning Coal

D. V. KELLER, JR.

Otisca Industries, Ltd.
Syracuse, New York

2.1 INTRODUCTION

Raw coal in the natural state is a physical mixture of carbonaceous material with a density range of 1.20 to 1.70 g/cm^3 and mineral matter with a density range of greater than 2.0 g/cm^3 and extending up to that of iron pyrite with a density of 5.0 g/cm^3 [1]. There is reasonable evidence that a fraction, if not all, of the mineral matter in coal is distributed throughout all of the size ranges down to the submicrometer range. The mining and/or crushing of raw coal to some maximum diameter usually results in a size distribution of particles which is reasonably well approximated by a Rosin-Rammler distribution function [2]. Crushed raw coal, therefore, consists of discrete particles of coal, mineral matter, and mixtures of the two bound together as they originally existed in their natural state. Physical separation of these components based on the density differential between the carbonaceous material and mineral matter has employed separation devices utilizing air fluidization, vibratory motion, or liquids of an intermediate density such as organic liquids (usually referred to as *heavy liquids*), salt solutions, and fine heavy minerals dispersed in flowing water (called *heavy media*). Other separation devices, such as tables, jigs, spirals, and hydrocyclones, utilizing a combination of frictional and/or gravity or centrifugal forces, are used to affect an apparent density differential between the coal and mineral matter. Since air fluidization, vibratory motion, heavy media and other combined processes have been discussed fully in coal preparation articles and textbooks [1,3-7], this chapter will be primarily concerned with physical separations involving heavy liquids. Particular emphasis of the discussion will be placed on an anhydrous, heavy-liquid separation process for cleaning coal, called the *Otisca process* [8].

The use of a heavy liquid, such as carbon tetrachloride, of an intermediate specific gravity to separate two or more phases of specific gravities, one above and the others below that of the parting fluid, has been recognized for about 100 years [9-14] as the near perfect means of achieving an absolute separation of the solids. This fact becomes apparent if one considers the separation of a cubic centimeter of a hypothetical block of coal with one small rock parting contained therein, after the coal-rock system has been reduced to 1-μm-diameter spheres. If one assumes the two rock-coal parting interfaces generate spheres of half coal and half rock, then the middling or misplaced material of this system only constitutes about 0.02 wt % of the entire mass which is recovered either as coal or rock. The exciting prospects of using a heavy-liquid separation system to beneficiate coal were studied intensely in the 1930s by the DuPont Company and put into a pilot plant for cleaning anthracite at the Western Breaker, Shenandoah, Pennsylvania, shortly thereafter [10]. With the exception of this pilot plant and a small number of continuous float-and-sink separators used for quality control, heavy liquids have not been used commercially in coal preparation [3].

The most recent application of the heavy-liquid separation process for cleaning coal is called the *Otisca process* [8], which has been under continued laboratory investigation and pilot-plant development since 1972. The Otisca process is a waterless, heavy-liquid separation process employing CCl_3F as a parting liquid with certain physical and chemical characteristics which make it particularly advantageous for expanding the familiar laboratory float-and-sink test (washability analysis) into the realm of commercial coal cleaning. In this process, a raw coal of any maximum particle diameter and its complete size distribution and with surface moisture up to 10 wt % is subjected to separation in a static bath of the parting liquid. The products of separation, namely, coal and its mineral matter, are transported directly to their respective evaporators where essentially all of the parting liquid is recovered for reuse. The two primary advantages of this process are that the raw coal feed does not have to be screened or sized in any manner; and even so, separation produces a product coal with its chemistry and weight, and Btu yield, which closely approximate the results of the classical washability test for the same raw coal and size distribution throughout the entire distribution of sizes.

The parting liquid used in this has two key characteristics: (1) it does not react to any known extent with the coal product or refuse material, and (2) it permits complete dispersion of the coal product particles throughout the separation bath, even though surface moisture is present. The first characteristic, namely, the lack of liquid-coal reaction [15-19], may be cited as one of the major reasons for the failure of previous attempts at commercializing the heavy-liquid process for cleaning coal. For example, if a reaction does occur between the heavy liquid and coal or its mineral matter, complete heavy-liquid recovery becomes a very expensive, complex process. The cost of the heavy liquid alone prohibits its loss from the process. The complete dispersion of the coal particles from the refuse material will ensure near theoretical recovery of the coal product with a minimum amount of mineral matter, which is the objective of the process.

The separation behavior in a heavy liquid is generally defined by Stokes' law where the velocity of separation in a static bath is proportional to the square of the particle radius times the density differential between the particle and parting liquid, and is related inversely to the viscosity of the liquid phase. In the case of coal separation, with a suitable parting liquid and given sufficient time in a static bath, all but a very small fraction of the particles will either float or sink irrespective of size above the limit for Brownian motion. The need, therefore, does not exist for the screening of a raw coal into special size ranges to perform a heavy-liquid separation. The use of certain additives in very small concentrations in the parting liquid provides a dramatic control over the middling concentration and the concentration of water and included slimes on the coal product. Separation mechanisms and velocities can also be controlled by the presence of certain other additives.

In this chapter, the basic principles and recent developments of the density differential separations utilizing anhydrous, heavy liquids as applied to coal cleaning are first described. The design, construction and testing of a pilot plant, which demonstrates the commercial viability and advantages of coal cleaning using the Otisca process, are then discussed. The commercial viability of the process depends upon producing more coal product of a better quality at a lower cleaning cost than can be achieved presently by commercial hydrobeneficiation techniques. The advantages of the process include very low noise levels and dust-free environment in the cleaning plant, a nearly water-free coal product, a refuse material of less than 12 wt % moisture which is nearly free of coal and can be readily compacted, the elimination of all water treatment streams such as "black water," settling ponds, water refinement procedures, and in general, a low plant maintenance effort. The projected capital investment for fine coal cleaning utilizing the process is in the range of $21,000 per raw coal ton per hour, and the estimated operating costs total about $1.47 per raw coal ton. Details of the process cost estimation along with the large-scale commercial facilities under construction using the process are also described.

2.2 DENSITY DIFFERENTIAL SEPARATIONS USING ANHYDROUS HEAVY LIQUIDS

A. Background

The introduction section cited an ideal mixture of coal and mineral matter to demonstrate density differential separation of coal in a heavy liquid. Such a coal, however, is rarely found in nature; and what is observed is a very complex mixture of carbonaceous macerals of varying density and of various minerals also with varying density [1]. Coal is a porous gel structure where the solid carbonaceous material surrounding the pores has a density considerably higher than the apparent density of the bulk. The pores have an average diameter of about 4 nm which results in a total surface area in the range of 200 m^2/g depending on the rank of the coal [20]. The small diameter of the pores permits the condensation of atmospheric water into the capillary system establishing what has come to be known as the *inherent* moisture content of coal. The apparent density of this composite system, i.e., that mass of pure coal one might observe for 1 cm^3 of ash-free coal, varies from a little under 1.30 for lower-rank coals to more than 1.7 g/cm^3 for anthracites [1] where the porosity is much smaller than that cited above. Of particular interest to this discussion is the separation of bituminous coals where the density of the ash-free coal is in the range of 1.30 to 1.40 g/cm^3, and it will be this material which establishes the buoyant force that permits the coal product recovery using a density differential separation process.

The mineral matter found in the raw coal is distributed in all size ranges from massive partings with the dimensions of meters down to highly dispersed particles with dimensions below the micrometer range [21]. The mineral chemistry and size distribution vary with the geologic history of the particular coal seam; but for the most part, the system includes sedimentary rocks such as illites, and various sulfur-bearing minerals such as iron pyrites [22-24].

The practical case of mining coal from a massive coal body produces a raw, mined product with a particle size distribution which usually conforms to a Rosin-Rammler size distribution function, where the maximum particle diameter can be in excess of 30 cm or as small as 0.6 cm depending on the mining techniques and may include particles smaller than 1 μm. There is considerable evidence indicating that iron pyrite is distributed throughout a large number of bituminous coals in a manner which also follows the Rosin-Rammler size distribution [25-27]. However, the average pyrite particle diameter for a number of seams lies in the range 20 to 200 μm [25]. The latter the average size, lies in the range 20 to 200 μm [25]. The latter suggests that a considerable concentration of pyrite can occur imbedded as a physical mixture in the larger particles of the raw coal such that they may never be made available to physical separation. In other words, the buoyant force of the surrounding coal is sufficient to carry along the imbedded pyrite microparticles with the coal product. Quantitatively, one may consider a two-phase system of coal with a unique specific gravity of 1.30 and iron pyrite with a specific gravity of 5.0 where the iron pyrite is imbedded in the coal. Under these conditions, the apparent specific gravity of the two-phase particle increases with the volume fraction of pyrite as would the concentration of the pyritic sulfur in that particle. This relationship is demonstrated in Table 2.1 [28]. It is important because in an ideal density differential separation of coal and refuse, free pyrite with a specific gravity of 5.0 will obviously separate as refuse (mineral matter) instantaneously. However, it is now evident that the product coal with a substantial amount of pyrite could also be recovered at parting liquid densities between 1.30 to 1.60 g/cm^3. One should observe that there is no regard for particle diameter in this relationship since only the volume fraction and density differential are involved.

Table 2.1 Variation of Sulfur Concentration in an Ideal Coal/Iron-Pyrite System as a Function of Volume Fraction of Iron Pyrite

Volume fraction iron pyrite	Combined specific gravity	Pyritic sulfur, wt %
0.1	1.67	15.99
0.07	1.559	11.49
0.05	1.485	8.99
0.04	1.448	7.38
0.03	1.411	5.68
0.02	1.374	3.89
0.01	1.337	2.00

Source: Data courtesy of Otisca Industries, Ltd., Syracuse, New York.

The specific case of iron pyrite has been chosen here as an illustration because there is substantial evidence that the pyrite particle sizes are distributed in an orderly fashion. The same evidence is not available for the remainder of the mineral matter; however, there is some circumstantial evidence that this may be the case. For example, the variation in ash content of a number of eastern bituminous coals is a linear function of the specific gravity of separation for a particular size range; and the slope of this linear relationship decreases as the size range decreases all with a nearly constant intercept. The fact that a very large fraction of the mineral matter in bituminous coals responds as an independent phase is further suggested by the observation that the heating value of coal is usually a linear function of the ash content often with a correlation coefficient squared in excess of 0.99. Furthermore, this is supported by the observations from recent laboratory investigations that the ash in most bituminous coals can be reduced by physical separation to less than 1.5 wt % ash and a number of selected coals to less than 0.5 wt % ash provided the particle size at separation is small enough. The density of the coal particles as a result will vary as a function of the volume concentration of the mineral matter in the coal matrix, as will the ash and sulfur concentrations of the coal product recovered from a separation at a unique density of the parting liquid. Furthermore, the coal content combined as a fixed portion of the mineral matter also functions to decrease the density of those mineral particulates and affect even more misplaced material, which, incidentally, causes a Btu loss to the refuse. Keller [28] pursued this analysis to completion and then demonstrated its applicability for establishing a theoretical washability curve, which agreed reasonably well with the data produced from the careful separation of a 70 μm × 0 raw coal sample by standard washability methods. The details of this procedure are reported in Ref. 28.

The conclusions of this study indicated that for the separation of two different samples of the same coal with the same size distribution, the chemistry and quantity of the product coal separated at some fixed density of parting liquid should be identical irrespective of how the size distributions were achieved. Variations in the product coal chemistry can only be achieved through variations in the size distribution when identical coals are compared at a fixed parting liquid density. The variation in particle size distribution must usually be decreased to achieve lower ash and pyritic sulfur. This follows because size reduction tends to separate mixtures of coal and mineral matter into particles with a higher concentration of pure components as illustrated by the hypothetical case described above.

The change in the pure coal density with varying concentrations of mineral matter (and the mixture of coal and mineral matter) which was explored above can be further extended to include the presence of moisture of 1 g/cm^3. The latter is almost always present in significant quantities varied from 2 to 40 wt % in raw coal if the separation is to be conducted in an organic, heavy-liquid bath. Organic liquids used for the separation of coal are universally immiscible in water. Consequently, a surface water envelope surrounding either the coal particle or the mineral matter particle, as well as inherent moisture in the coal, must be included in the apparent density which the mixture of water, coal and mineral matter presents to the parting liquid system. Ten to fifteen wt % surface moisture in some fine raw coals can completely prevent separation in an organic, heavy-liquid system due to the formation of immiscible liquid bridging.

B. Separation Behavior

The separation of two solids of different density by any true liquid of an intermediate density under ideal conditions is governed by the Stokes' law [29] as given by Eq. (2.1):

$$v = \frac{2gr^2(\rho_1 - \rho_2)}{0.09\eta} \tag{2.1}$$

where the velocity of the separating particle (v) in centimeters per second is directly proportional to the radius (r) in centimeters of the particle squared times the density differential between that of the particle (ρ_2) and that of the liquid (ρ_1), and inversely proportional to the viscosity (η) of the liquid system centipoise (cP). The proportionality constant g has the value of 980.6 cm/sec^2. In an organic liquid ($\rho_1 = 1.5\ g/cm^3$) with a viscosity of 0.7 cP, a 0.5-cm-diameter coal particle ($\rho_2 = 1.30\ g/cm^3$) will separate at a velocity of 389 cm/sec; while a particle of 0.01-cm diameter separates at a velocity of 0.16 cm/sec. Again, the density of the particles must be regarded as an apparent density including water, coal, and mineral matter.

Clearly, the separation medium acts to a certain extent as a classifier when the separation involves a number of particles with a distribution of diameters. Furthermore, when separating coal from mineral matter, the solid flows are in opposite directions: coal tending to rise (v = +) and mineral matter tending to fall (v = —). Therefore, although the heavy-liquid separation bath is usually regarded as being a static system, there is a complex flow pattern established during the separation of particles. The dynamics of the process result in an increase in the viscosity of the liquid system, which is counterproductive to the separation velocities according to Eq. (2.1).

The increase in viscosity of a parting bath can be demonstrated by increasing the volume percent (vol %) of the solids in a bath and measuring the resulting plastic viscosity (η_ρ), defined as the slope of the line established when the shear stress (σ) of the system is measured at various shear rates ($\dot{\gamma}$). For example, the results obtained for a coal with a diameter distribution of 74×10^{-4} cm $\times$ 0 dispersed in an organic liquid with a viscosity of 0.4 cP follow Eq. (2.2):

$$\eta_\rho = 0.32 \text{ vol } \% - 2.39 \tag{2.2}$$

for solid concentrations in excess of 8 vol % [30]. The correlation coefficient squared was 0.9 which may be regarded as rather good, considering the difficulty of measuring viscosities in a system where the particles tend to separate from the system during the viscosity measurements. The term *plastic viscosity* was interjected into the discussion because various studies, such as that by Whitmore [31], have demonstrated that slurry systems often behave as a Bingham plastic fluid [32]. The latter has the characteristic that although the relationship between shear stress and shear rate is linear, the intercept of the plastic viscosity curve is not always zero at a zero shear rate as would be predicted for a true Newtonian fluid. As the solids content of the slurry increases from 0 vol %, the intercept of the plastic viscosity curve at zero shear rate increases from zero, defining a unique yield stress (σ_y) for the system. This leads to the relationship

$$\eta_\rho \dot{\gamma} = \sigma - \sigma_y \tag{2.3}$$

Whitmore [31] further suggested that the existence of a sufficient value of yield stress may, in fact, establish a barrier or resistance to particle separation. For example, a value of the yield stress, as measured in an operating heavy-medium bath (magnetite-water), was in the range of 92 dyn/cm^2 which might inhibit the separation of particles smaller than 1 cm. A corresponding value for 33 vol % solids in an organic medium was reported as 0.1 dyn/cm^2 [30]. The extreme difference between the two yield stress values could be due to the ability of the polar liquid such as water, to transmit interparticle forces with more efficiency than the nonpolar organic liquid. The resulting effect is that finer particles in a higher concentration can be separated in the nonpolar liquid than would be expected in polar liquid systems.

In conclusion, it is evident that the physical chemistry of the liquid used as a parting medium (i.e., the density, viscosity, and polarizability in relationship to the solids under separation) controls the velocity of separations.

C. Anhydrous Heavy Liquids

The ideal anhydrous heavy-liquid parting fluid should have the following characteristics [9-14]: an appropriate density which is readily varied between 1.25 and 1.70 g/cm^3, unreactive with the materials to be separated, low viscosity, low boiling point, low heat of vaporization, nonflammable, nontoxic, noncorrosive, and inexpensive. The liquids which have been proposed for mineral separation by Tveter and his co-workers [10-12] are presented in Table 2.2 along with some of their more important properties [33,34]. Most generally, parting liquids have been obtained from the class of C_1-C_3 halohydrocarbons where the most common and characteristic of this group is carbon tetrachloride. The establishment of an appropriate density between 1.2 and 2.0 g/cm^3 is usually accomplished by mixing various combinations of these liquids or hydrocarbons, as long as the second liquid meets the cited criteria and the two are compatible with each other.

From the list shown in Table 2.2 and a chemical handbook, it becomes evident that the requirements of low viscosity, low boiling point, low heat of evaporation, and low flammability and low corrosion properties are met by most of the liquids in this class. Furthermore, they all should be regarded as more or less toxic; yet through careful engineering, this problem can be abated, as was pointed out by Foulke [10] in the review of the history of DuPont's heavy-liquid process for cleaning anthracite. Of the liquids shown,

Table 2.2 Common Parting Liquids and Pertinent Properties

Compound	Formula	Specific gravity	Normal boiling point, °C	Freezing point, °C	Viscosity, cP at 25°C
Methylene iodide	CH_2I_2	3.31	182[a]	6.1	2.6
Acetylene tetrabromide	$CHBr_2$-$CHBr_2$	2.95	243.5[a]	0.1	9.6
Bromoform	$CHBr_3$	2.89	149	8	1.8
Tribromofluoromethane	CBr_3F	2.75	108	−73.9	1.5
Methylene bromide	CH_2Br_2	2.48	97	−52.6	.97
Methylene chlorobromide	CH_2BrCl	1.92	68.1	−88	.63
Pentachloroethane	CCl_3-$CHCl_2$	1.67	161.0	−22	2.33
Perchloroethylene	CCl_2=CCl_2	1.61	121.0	−22.4	.86
Carbon tetrachloride	CCl_4	1.59	76.5	−23.0	.90
Trichloroethylene	CCl_2=$CHCl$	1.46	87.1	−86.8	.55
Methyl chloroform	CCl_3-CH_3	1.33	74.1	−30.4	.80
Ethylene dichloride	CH_2Cl=CH_2Cl	1.25	83.5	−35.7	.79

[a]Decomposing temperature

Source: Data courtesy Otisca Industries, Ltd.

therefore, all appear to be appropriate parting fluids, if one could disregard the cost of the liquids. The cost of the liquid is involved to the degree that if complete recovery was affected, i.e., total recycle, then the parting fluid would fall under the capital cost of the plant and be readily recovered over a period of time. If, on the other hand, due to some chemical reactions and/or poor engineering, the liquid becomes lost, then that loss becomes an operating cost and the cost of the liquid may become a major factor in the overall production costs. The first DuPont test [10] in the early 1930s clearly demonstrated this point, and the process was abandoned due to the extremely high liquid loss and the resulting toxic environment. Subsequent tests by this group shortly thereafter in a 100 ton/hr anthracite preparation plant demonstrated that a parting liquid loss in the range of 0.50 to 0.75 lb/ton in a continuous test of about 3000 tons, could be achieved through more refined engineering and applied surface chemistry. More recently, Tippin and Tveter [14] have also suggested that heavy liquid losses in mineral separators can be maintained in the range of 0.5 to 1 lb/ton, which is sufficiently low to enable justification for the process on an economic basis.

The fact that some halohydrocarbons are dissolved into coal in a solvent-solute relationship was pointed out in the detailed studies of Dryden [15,16], Van Krevelin [17], and others [18]. For example, chloroform not only extracts various fractions of the coal but may be imbibed into the coal structure with a significant heat of adsorption such that its removal becomes most difficult, if not impossible. In the case of the halohydrocarbons, the presence of large excesses of the halogen atoms in the finished coal product constitutes a loss in an economic sense due to the parting-liquid loss, and will also create a serious corrosion problem in the generation stream. The DuPont process carefully avoided this problem by establishing a water envelope for each coal particle in the system. The parting liquids and additives were specifically chosen so as to perpetuate this layer.

The use of methanol must be avoided as a density modifier for organic, heavy liquids due to the fact that methanol is imbibed into the coal structure [19], which causes the coal structure to swell, thus depressing the natural density of the coal by increasing the coal volume at constant mass. The result is a twofold effect: the alcohol is extracted from the parting-liquid mixture into the coal causing the density of the parting liquid mixture to increase, while the density of the coal is changed due to the methanol adsorption. Clearly, this could not be a reasonable basis for a continuous process.

D. Interaction of Variables

The preceding section indicated that the applicability of the heavy-liquid separation to the removal of mineral matter including pyritic sulfur was related directly to the size distributions of the mineral matter, pyritic sulfur, and raw coal and to the specific gravity of the parting bath. The driving force for the separation, i.e., the density differential as modified by the square of the particle size, is retarded by the plastic viscosity parameter, as the particle size is decreased and/or the density differential $(\rho_2 - \rho_1)$ is decreased. Separations within a reasonable time frame become more difficult.

The complex interaction of variables in the heavy-liquid separation may be best illustrated from a study of the change in product coal chemistry with the Btu-yield recovery under different separation conditions [30]. Note that the variation in product coal as recovered pounds of sulfur per million Btu (lb S/MMBtu) reflects not only the loss in pyritic sulfur from the coal during separation but also the reduction in other mineral matters. The latter follows because the heating value of the coal product, Btu/lb, increases as the concentration of mineral matter decreases. The objective of perfect physical separation is to separate all of the available carbon and hydrogen which is reflected in the percent Btu yield (% Btu yield), as given by the weight percent yield of product coal multiplied by the Btu/lb in the product coal divided by the Btu/lb of the raw coal (all on a dry basis). The variation of both variables, lb S per MMBtu and

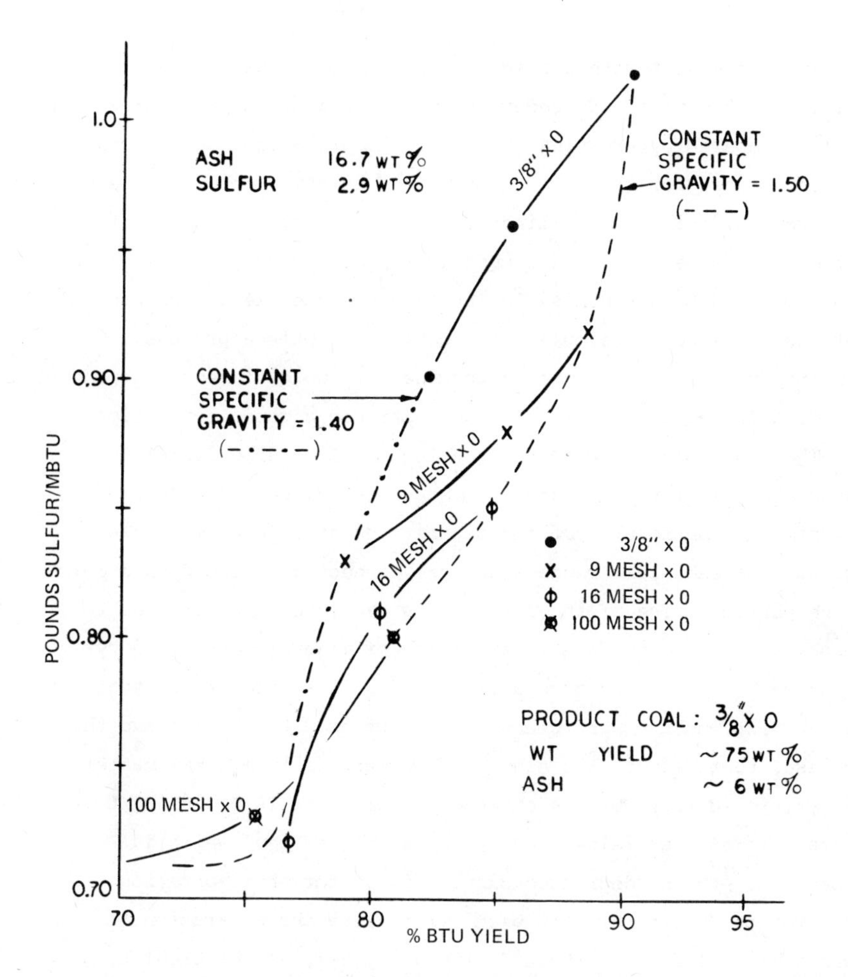

Figure 2.1 Bench-scale laboratory simulation of the Otisca process for the beneficiation of a Pennsylvania Lower Freeport coal with various top sizes at different specific gravities. (Data courtesy of Otisca Industries, Ltd.)

% Btu yield, for certain particle size distributions with a given maximum particle diameter at particular separation densities is shown in Fig. 2.1. All of the curves illustrated are for the same raw coal which is from a Lower Freeport seam in Pennsylvania with 16.7 wt % of ash and 2.9 wt % of total sulfur. The coal product ash for the 3/8 in. (9.55 mm) × 0 separation at 1.50 g/cm^3 was in the range of 6 wt % with a weight yield of about 75%. The entire series of tests were made under identical separation procedures with samples prepared by crushing the raw coal to 3/8 in. × 0, separating sufficient coal for the three separations by ASTM techniques, and then stage-grinding the remainder to 9 mesh (2 mm) × 0 and continuing in the same manner to the 100 mesh (0.15 mm) × 0 samples. These were then separated by a bench-scale laboratory procedure (described below) that was developed to simulate the commercial Otisca process. The chemistry of the products was determined by two different independent outside agencies and the results demonstrated a good correlation.

The four solid curves representing the four size distributions illustrate all of the effects that have been discussed. Consider, for example, the results of the separations at 1.50 g/cm^3 shown in Fig. 2.1 for each size range which are connected by the dashed curve. As the particle size of the raw coal is decreased from 3/8 in. to 9 mesh, there is considerable rejection of mineral matter and pyritic sulfur with little loss of coal, resulting in an almost constant Btu yield. Next, as the particle size is decreased to 16 mesh and then 100 mesh, there is still a significant decrease in mineral matter and pyritic sulfur; but the extremely small size is beginning to generate lower Btu yields, indicating a loss of coal which is probably due to the decreased driving force for the separation. A separation density of 1.50 g/cm^3 will allow the separation of essentially all free pyrite and mineral matter. In particular, the separation of the 3/8 in. × 0 size fraction at a density of 1.50 g/cm^3 produces a refuse of 5300 Btu/lb with ultrafine coal dispersed throughout, even though the specific gravity of the material is well in excess of 2.0 g/cm^3 and the sulfur content in excess of 7 wt %.

The mineral matter in that fraction of the coal which is just recovered at separation density of 1.50 g/cm^3 is lost when an identical sample is separated at 1.45 g/cm^3 as shown by the middle point in each of the four solid curves representing their respective size distributions. In each case, the reduction in pounds of sulfur per million Btu also involves a substantial fraction of coal as shown by the corresponding loss in Btu yield. If one accepts the specific gravity of this coal to be about 1.30 g/cm^3, then the driving force for the separation at a constant particle diameter and viscosity is reduced by a factor of 2 by decreasing the density of the parting liquid from 1.50 to 1.40 g/cm^3. This magnification becomes most evident in the particles with diameters below 100 mesh and results in exceedingly low, if not poor, recoveries as is illustrated by the large loss in the Btu yields. The corresponding weight percent of the 100 mesh × 0 particles varies in each size range as follows: 5.4 wt % in 3/8 in. × 0, 10.9 wt % in 9 mesh × 0 and 22.1 wt % in 16 mesh × 0.

In summarizing this example, and for the moment, disregarding moisture effects, one may assume raw coal of a constant size range, e.g., 9.6 to 4 mm, as consisting of particles of pure coal, pure mineral matter, and mixtures of the two. Separation in a heavy liquid of a density of 1.50 g/cm^3 will permit the recovery of a product coal which consists of the pure coal and that fraction of the mixtures of each whose volume percent of mineral content is low enough to produce a particle density that is less than 1.50 g/cm^3. The size reduction of this same coal and separation at a density of 1.50 g/cm^3 tends to produce more ash-free coal and coal-free refuse and a smaller concentration of the mixture of the two, that is, a reduction in lb S per MMBtu at constant Btu yield.

Counterproductive to the size reduction process is the fact that the driving force for the separation is severely decreased. If one considers the driving force proportional to the particle radius squared times the density differential while neglecting the viscosity effects, the driving force for a 3-mm particle with a density of 1.49 g/cm^3 and a 0.15-mm particle with the same density is about 0.1 to 2×10^{-4}, respectively. The 100-mesh particle has three orders of magnitude less driving force for the separation which results in recoveries more subject to outside retarding forces.

2.3 THE OTISCA PROCESS

A. Pilot Plant Site [8]

The first Otisca process pilot plant was designed in 1975 to be placed as an integral part of the existing Island Creek Coal Company mining facility, the North Branch Mine, in Bayard, West Virginia. The existing facility mines Upper Freeport coal, which was scalped, stored, and then dry sieved over 5/16 × 3 in. slot screens into two streams. The oversize stream proceeded to an existing Daniels washer. A fraction of the undersize material was used as an input to the pilot plant by diverting this stream into a 40-ton raw coal bin. The coal product and reject material from the pilot plant were separately transported by belt conveyors to the respective belts for reject material and coal product existing in the commercial plant.

The pilot plant was installed in a new steel building on a 15-cm (6-in.) concrete slab (12.2 × 15.2 × 3.6 m) which housed the entire plant including a 3.6 × 6.1 m operating station and chemical laboratory. The coal storage bin and a portable steam boiler were located outside the building. Space is available inside the building to install a permanent boiler station.

B. Process Pilot Plant

A schematic flowsheet of the Otisca process pilot plant is illustrated in Fig. 2.2. In what follows, the process details are described according to solids, vapor, and liquid flows. Relevant process parameters and pilot-plant start-up problems are also discussed.

Solids Flow. The raw coal stored in a 40-ton bin was metered into the input rotary valve by a variable-speed screw feeder which conveyed between 5 and 20 tons/hr into the processing system. The raw coal then entered the conditioner where the raw coal was exposed to the parting liquid, CCl_3F, and certain additives. The slurry was then carefully mixed and the temperature of this mixture also established. The additives caused the surface moisture, carried in with the raw coal, to be transferred to the mineral matter and, as such, reject with moisture intact. The advantage of the water shift is to carry

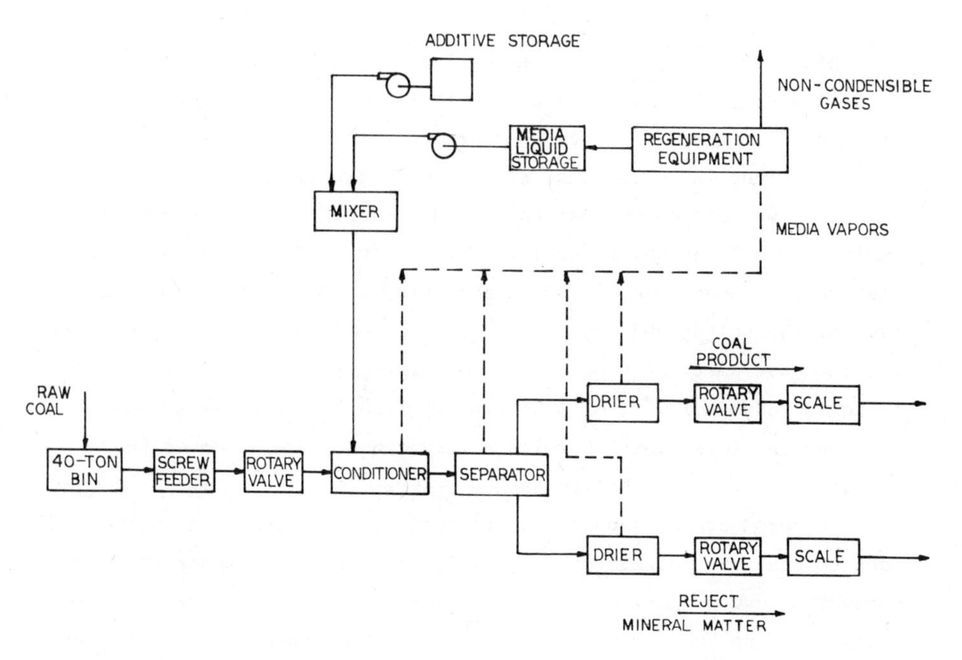

Figure 2.2 Otisca process schematic drawing. (Data courtesy of Otisca Industries, Ltd.)

along a quantity of fine mineral matter, e.g., silts and slimes, such that the product coal contained 1 to 5 wt % less ash than in the case without raw coal conditioning.

The parting liquid, CCl_3F, has a boiling point of 74.9°F, a heat of vaporization of 77.51 Btu/lb, a specific gravity of 1.50 at 60.8°F and a low viscosity of 0.4 cP. It is nontoxic, noninflammable, and noncorrosive. It is also an extremely stable chemical. These physical properties of the parting liquid for the most part establish the operating parameters of the process. In particular, input raw coal with a temperature above 75°F will cause some of the parting liquid to be evaporated, when mixed with the warm coal in the conditioner. This will reduce the temperature of the coal to a desirable separation temperature, e.g., 60 to 68°F. During winter operations, if the input raw coal should be above freezing, the desirable operating temperature can be achieved with the available sensible heat of the parting liquid or auxiliary heaters in the conditioner.

After conditioning, the raw coal slurry enters the separation bath (1.2 × 1.2 × 1.8 m) where the parting of the product coal (float) and mineral matter (sink) takes place at about 1.50 g/cm^3. Since the entire process is continuous, i.e., raw feed is continually entering the bath, while product coal and refuse mineral matter are continuously being withdrawn, to call the bath static is, in a sense, a misnomer. However, the dynamics are only incidental to the separation of the solids and not to any other intended pumping mechanism. For the most part, each coal particle, irregardless of size, "sees" a liquid surrounding it of a density of 1.50 g/cm^3; and responds to this buoyant force which permits a very close correlation of the results with the washability data.

The products of separation are removed mechanically from the bath to their respective parting-liquid evaporators. Indirect-fired conductive evaporators, such as hot-water-heated hollow screws, operating at about 200°F, serve to raise the temperature of the coal and vaporize all of the included parting liquid with a boiling temperature of 74.9°F. The products are ejected through their respective rotary seal valves back to ambient conditions, where they are ready for utilization. The temperature of the solids rarely exceeds 110°F.

The simplicity of the solids stream treating all sizes from any maximum diameter to zero, i.e., no screening, no fines removal, and no multiprocessing steps for the respective size ranges, is the principal reason that this process is inexpensive relative to the competitive processes. For example, in a complete heavy-media system, one needs to treat at least three size ranges in essentially three separate plants. The product coal from the Otisca process can be almost devoid of surface moisture. This follows because the original surface moisture on the raw coal has been transferred to the refuse mineral matter, which is readily handleable by standard landfill disposal. Thickeners and their underflow eliminations, a common problem in hydrobeneficiation, are unnecessary in this process.

Vapor Flow. The entire Otisca process system was designed and constructed to be completely airtight so as to contain all of the parting-liquid vapors evolved during conditioning, separation, and evaporation. This vapor is conducted through ducts, a particle filter and then to a gas recovery apparatus, which converts all of that vapor back into the liquid state for reuse in the process. This is accomplished through a series of compression, condensation and sorption steps such that the remaining residues, principally noncondensible gases (air), which are eliminated from the system, are free of parting-liquid vapor. The regeneration system is controlled to operate under conditions to insure that each unit process in the system is at a vacuum of about 0.2 in. of water. As a result, any system leakage results in an infiltration of air in the system rather than CCl_3F vapor loss from the system. The parting-liquid condensate is essentially pure and is placed in a storage tank until it is required for conditioning raw coal.

Since the parting liquid is expensive, costing about $0.46 per lb (1978) in the bulk, one is fully justified in taking extreme care in system design and construction to reduce liquid and vapor losses to an absolute minimum. Present design criteria established by pilot plant testing cite less than 0.25 lb per raw coal ton as a typical parting-liquid (vapor) loss. Process emission loss from the mechanical equipment should account for less than 0.05 lb per raw coal ton, while losses due to adsorption of the molecules into the coal matrix are less than 0.2 lb per raw coal ton, or less than 100 parts per million (ppm). This chloride residue resulting from using the parting liquid, CCl_3F, is less than the existing chloride concentration in coal of between 700 and 900 ppm [21,22]. Pilot plant testing has also indicated that the additive consumption is in the range of 0.4 lb per raw coal ton.

Liquid Flow. A subtlety of the process is a liquid flow through the separation bath so as to completely regenerate the parting liquid over a fixed period of time, depending on the rates of feed to the bath and products from the bath. Under such conditions, solutes and/or entrapped ultrafine coal particles (sometimes called *near gravity material*) are continuously removed from the bath. The raw coal slurry feed is made up with freshly condensed parting liquid from the vapor regeneration system; while the coal products are also removed from the bath as a slurry, the liquid of which came from the bath. At equilibrium, a continuous exchange of liquid results.

Start-up Problems. The major problems encountered in the first pilot-plant start-up were related to the fact that three individual flows, i.e., solids, liquid, and gas, had to be interbalanced not only through control devices but in the operator's mind before comfortable continuous operation could be achieved. Continuous solids flow through the plant was achieved in a matter of hours from start-up. Continuous flow of raw coal from the raw coal bin was never achieved with a high degree of independent reliability; although with considerable personal attention, flow could be maintained. This was due to the fact that the 5 mesh × 0 coal had a rather high moisture content. The gas recovery station and liquid handling equipment required several weeks to place in balance simply due to the fact that no operational experience existed prior to this start-up. During this period, numerous minor design features were modified and several added in order to improve performance.

Operational procedures were established for the balancing of all three flows together as a continuous process only when the behavior of the solids in each unit operation was well understood. This fell primarily into the category of operational experience rather than mechanical plant start-up problems. The time required to experience, identify, design, and correct a problem was typically 4 to 6 weeks.

Upon completion of the initial testing of the plant and a long-duration test, the pilot plant was moved to Florence, Pennsylvania where it has been operated for three years. Current start-up procedures involve little more than filling the raw coal bin, parting liquid and additive tanks, and the bath and turning the apparatus on. One operator runs the plant while the second manipulates the raw coal and products to their respective ends and accumulates analytical samples.

C. Process Data

Bench-Scale Process Simulation. The detailed laboratory studies of coal separation by the Otisca process began during 1972 and have continued without interruption to date. The first Otisca semicontinuous pilot plant was operational in 1973 and the 20 tons/hr continuous pilot plant became operational in 1976 and has been operational ever since. A 125 tons/hr plant was completed for American Electric Power late in 1979 and is in the start-up phase. Throughout this entire period, a well-staffed and well-equipped laboratory has been directed toward the development of a separation simulation procedure of the commercial Otisca process for laboratory coal samples that anticipates each aspect of the larger facility. These investigations have not only explored in detail all of the aspects of the separation phenomena under various conditions, but have also extended into other areas within the plant operation, i.e., coal conditioning before separation, liquid content before evaporation, evaporation, parting-liquid content of the products, parting-liquid chemistry, etc. Each of these variables has a significant effect on the process engineering of the plant, as well as its economics. Since each raw coal has a unique chemistry and set of separation characteristics, it is worthwhile to examine the results of a number of batch-scale simulations of this process such that some of these data can be compared with the actual pilot-plant tests.

The bench-scale simulation procedure is somewhat similar to the one used in laboratory washability studies in that the coal is separated in the laboratory under near ideal conditions of temperature and volume percent of solids. The key differences between this procedure and that of washability lie in the preparation of the sample before separation and the extent of separation. Most washability procedures begin with a removal of the –100-mesh, or 325 mesh × 0, material by a careful water washing (desliming) of the raw coal. The coal is then sized into narrow size fractions and air-dried to eliminate the surface moisture before separation in organic heavy fluids to an ideal sink-float endpoint irrespective of time. Occasionally, centrifugation is used to hasten this separation procedure. The bench-scale simulation of the Otisca process utilizes raw coal of the whole size distribution at a surface moisture content anticipated during the operation of a commercial cleaning plant. Separation is conducted over some finite time period (minutes) also consistent with anticipated plant operations. The cumulative results from the washability test are almost identical to those from the bench-scale Otisca process simulation for separations of the same coal and size distribution at the same separation density, the difference being the higher yields observed in the simulation due to the −100-mesh material that is recovered in the simulation. A comparison of the bench-scale simulation tests directly with pilot-plant results are found in a later section under "Recent Pilot-Plant Results."

Typical results of the bench-scale simulation of the Otisca process shown in Tables 2.3 to 2.9 illustrate some of the flexibility of this process in the separation of various size distributions from greater than 1 in. × 0 to as fine as 200 mesh × 0; and in some cases, directly comparisons are made. The various tests also explore a number of different seams and separations at different liquid densities. The Otisca process can be conducted at any separation liquid density from 1.27 to 1.6 g/cm^3. The propensity of data with liquid separation densities at 1.50 g/cm^3 reflects the most direct design case which is now being applied to commercial practice.

In examining the data in Tables 2.3 to 2.9, it should be recognized that each of the samples was taken by personnel responsible to the particular mine owner of that particular coal. In no cases, have "head-on" comparisons been conducted to examine the relationship between the whole hydrobeneficiation process with the Otisca process. Truly significant comparisons, as have been shown, can only be achieved through tests where one raw coal sample is riffled into two portions each receiving a different separation procedure, independently if possible, and also all chemical analysis performed independently, preferably by two separate laboratories. Such a bench-scale test was conducted [28] on the separation of a commercially ground, 200 mesh (74 μm) × 0 Pennsylvania Upper Freeport coal, subjected to

Table 2.3 Bench-Scale Simulation of the Otisca Process for Various Upper Freeport Coals from Pennsylvania

County	Fayette			Clearfield		Indiana		Indiana	
Analysis	Feed	Product	Product	Feed	Product	Feed	Product	Feed	Product
Size distribution		1 in. × 0	5 mesh × 0		3/8 in. × 0		200 mesh × 0		3/8 in. × 0
Specific gravity of separation		1.50	1.50		1.50		1.50		1.40
Volatile matter, wt %	27.66	28.59	29.01	26.46	30.18	24.30	27.80		
Fixed carbon, wt %	60.36	64.54	65.09	59.33	62.45	52.60	64.00		
Ash, wt %	11.98	6.87	5.90	14.21	7.37	23.10	8.20	29.20	6.20
lb/MMBtu	8.85	4.75	4.04	11.24	5.30	19.70	5.73	28.10	4.28
% Reduction/MMBtu		46.30	54.40		52.80		70.90		84.80
Total sulfur, wt %	3.44	1.16	0.78	2.51	1.61	2.72	1.00	2.13	1.16
lb/MMBtu	2.54	0.80	0.53	1.98	1.16	2.32	0.70	2.05	0.80
% Reduction/MMBtu		68.50	79.10		4.15		69.80		61.00
Pyritic sulfur, wt %	2.98	0.62	0.13		1.12	2.01	0.31	1.42	0.40
lb/MMBtu	2.20	0.43	0.09		0.81	1.71	0.22	1.37	0.28
% Reduction/MMBtu		80.50	95.90				87.o0		79.60
Btu/lb	13,534	14,458	14,598	12,644	13,896	11,724	14,314	10,401	14,498
Weight yield, wt %		89.30	86.30		79.60		60.10		51.10
Btu yield, %		95.40	93.10		87.50		73.40		71.20

Source: Data courtesy of Otisca Industries, Ltd.

Table 2.4 Bench-Scale Simulation of the Otisca Process for Various Seams of Pennsylvania Coal

Seam	Lower Freeport			Brookville		Waynesburg		Pittsburgh	
County	Indiana			Butler		Washington		Westmoreland	
Analysis	Feed	Product	Product	Feed	Product	Feed	Product	Feed	Product
Size distribution		3/8 in. × 0	100 mesh × 0		30 mesh × 0		30 mesh × 0		28 mesh × 0
Specific gravity of separation		1.50	1.50		1.50		1.50		1.50
Volatile matter, wt %				35.88	42.14	33.88	38.56	28.30	33.22
Fixed carbon, wt %				47.38	54.17	50.52	51.07	51.67	57.80
Ash, wt %	17.20	7.00	8.30	16.74	3.69	15.60	10.37	20.03	8.98
lb/MMBtu	13.40	4.79	5.77	13.77	2.52	12.40	7.68	16.96	6.59
% Reduction/MMBtu		64.30	56.90		81.70		83.10		61.10
Total sulfur, wt %	2.84	1.69	1.12	9.23	2.05	1.77	1.03	1.21	0.81
lb/MMBtu	2.21	1.16	0.78	7.59	1.40	1.41	0.76	1.02	0.59
% Reduction/MMBtu		47.50	64.70		81.60		45.90		42.20
Pyritic sulfur, wt %	1.81	0.83	0.38	8.41	0.76	1.23	0.41		
lb/MMBtu	1.40	0.57	0.26	6.92	0.52	0.97	0.30		
% Reduction/MMBtu		59.30	81.40		92.50		68.70		
Btu/lb	12,844	14,606	14,382	12,160	14,618	12,584	13,504	11,807	13,621
Weight yield, wt %		79.40	72.40		68.70		72.60		70.20
Btu yield, %		90.30	81.10		82.60		77.90		81.00

Source: Data courtesy of Otisca Industries, Ltd.

Table 2.5 Bench-Scale Simulation of the Otisca Process for Various Coals from Pennsylvania and Ohio

Seam	Lower Kittanning			Pittsburgh		Pittsburgh No. 8		Waynesburg-II	
State	Pennsylvania			Pennsylvania		Ohio		Ohio	
County	Indiana			Washington		Belmont		Belmont	
Analysis	Feed	Product	Product	Feed	Product	Feed	Product	Feed	Product
Size distribution		3/8 in. × 0	9 mesh × 0		3/8 in. × 0		60 mesh × 0		60 mesh × 0
Specific gravity of separation		1.40	1.40		1.40		1.50		1.50
Volatile matter, wt %				38.61	39.61			36.39	41.17
Fixed carbon, wt %				51.24	55.25			43.92	48.81
Ash, wt %	22.10	5.00	4.40	10.15	4.99	30.25	4.83	19.69	10.02
lb/MMBtu	18.60	3.30	2.91	7.64	3.54	30.19	3.42	16.80	7.55
% Reduction/MMBtu		82.30	84.40		53.70		88.70		55.10
Total sulfur, wt %	4.86	1.82	1.64	2.06	1.54	4.51	3.41	1.95	1.31
lb/MMBtu	4.10	1.21	1.08	1.55	1.13	4.50	2.41	1.67	0.99
% Reduction/MMBtu		70.50	73.70		27.10		46.40		40.90
Pyritic sulfur, wt %	3.11	0.66	0.59	1.30	0.81	2.79	1.68	1.81	0.37
lb/MMBtu	2.62	0.44	0.39	1.05	0.57	2.78	1.19	1.55	0.28
% Reduction/MMBtu		83.20	85.10		45.70		57.20		81.90
Btu/lb	11,863	15,022	15,133	13,278	14,141	10,131	14,131	11,687	13,263
Weight yield, wt %		67.90	58.70		86.70				
Btu yield, %		86.00	74.90		92.40		93.80		73.90

Source: Data courtesy of Otisca Industries, Ltd.

Table 2.6 Bench-Scale Simulation of the Otisca Process for Various West Virginia Coals

Seam	Upper Kittanning		Powellton		Stockton-Lewiston		Upper Freeport		Peerless	
County			Boone		Boone		Preston		Fayette	
Analysis	Feed	Product	Feed	Product	Feed	Product	Feed	Product	Feed	Product
Size distribution	3/8 in. × 0		3/8 in. × 0		3/8 in. × 0		3/8 in. × 0		60 mesh × 0	
Specific gravity of separation		1.50		1.50		1.50		1.40		1.50
Volatile matter, wt %	21.74	23.71	33.30	35.73	30.79	36.07	27.64	31.00	33.35	40.60
Fixed carbon, wt %	59.49	68.75	52.36	60.91	47.78	57.40	54.44	63.96	55.75	55.01
Ash, wt %	18.77	7.54	14.34	3.36	21.43	6.53	17.72	5.04	10.90	4.39
lb/MMBtu	15.55	5.38	11.05	2.26	18.40	4.67	14.25	3.38	7.99	2.98
% Reduction/MMBtu		65.40		79.50		74.60		76.20		62.70
Total sulfur, wt %	0.58	0.66	1.28	0.86	0.89	0.78	2.66	1.02	3.59	1.54
lb/MMBtu	0.48	0.47	0.99	0.58	0.76	0.56	2.14	0.69	2.63	1.05
% Reduction/MMBtu				41.50		26.50		68.00		60.30
Pyritic sulfur, wt %	0.07	0.11	0.93				1.98	0.47	2.53	0.41
lb/MMBtu	0.06	0.07	0.72				1.59	0.32	1.86	0.28
% Reduction/MMBtu								80.10		85.00
Btu/lb	12,072	14,018	12,976	14,858	11,647	13,971	12,433	14,890	13,636	14,736
Weight yield, wt %		76.60		87.20		52.40		65.80		77.70
Btu yield, %		88.90		99.80		62.90		79.90		84.00

Source: Data courtesy of Otisca Industries, Ltd.

Table 2.7 Bench-Scale Simulation of the Otisca Process for Various Seams of Coal from West Virginia and Kentucky

Seam	Lower Kittanning		Middle Kittanning		Splash Dam		No. 9	
State	West Virginia		West Virginia		Kentucky		Kentucky	
County	Taylor				Pike			
Analysis	Feed	Product	Feed	Product	Feed	Product	Feed	Product
Size distribution		3/8 in. × 0		3/8 in. × 0		3/8 in. × 0		1 in. × 0
Specific gravity of separation		1.50		1.50		1.50		1.50
Volatile matter, wt %			23.72	23.68	28.19	34.79	36.98	39.44
Fixed carbon, wt %			62.61	68.86	51.83	57.70	47.19	52.50
Ash, wt %	29.98	8.62	13.67	7.46	19.98	7.51	15.83	8.06
lb/MMBtu	21.43	6.21	10.21	5.19	16.56	5.30	13.40	6.14
% Reduction/MMBtu		71.00		49.20		68.00		54.20
Total sulfur, wt %	1.29	0.69	3.69	1.91	0.95	0.83	4.59	2.40
lb/MMBtu	1.15	0.50	2.75	1.33	0.79	0.59	3.89	1.82
% Reduction/MMBtu		56.80		51.70		24.80		53.00
Pyritic sulfur, wt %	0.84	0.13	2.90	1.11	0.44	0.13	3.60	1.02
lb/MMBtu	0.75	0.09	2.16	0.77	0.36	0.09	3.05	0.77
% Reduction/MMBtu		87.50		64.30		74.50		78.40
Btu/lb	11,192	13,876	13,395	14,375	12,068	14,178	11,787	13,126
Weight yield, wt %		71.00		79.00		71.50		78.00
Btu yield, %		88.00		84.80		84.00		87.00

Source: Data courtesy of Otisca Industries, Ltd.

Table 2.8 Bench-Scale Simulation of the Otisca Process for Various Seams of Midwestern Coal

Seam	No. 6		No. 5		Louilia	
State	Illinois		Illinois		Iowa	
Test size	Bench		Bench		Bench	
Analysis	Feed	Product	Feed	Product	Feed	Product
Size distribution	30 mesh × 0		60 mesh × 0		30 mesh × 0	
Specific gravity of separation		1.50		1.50		1.50
Volatile matter, wt %	38.22	49.80			39.43	42.91
Fixed carbon, wt %	47.62	44.00			37.23	51.40
Ash, wt %	14.16	6.20	9.22	3.83	23.34	5.69
lb/MMBtu	11.53	4.50	7.03	2.76	21.91	4.25
% Reduction/MMBtu		61.00		61.10		80.60
Total sulfur, wt %	3.72	1.64	1.89	1.08	4.00	1.83
lb/MMBtu	3.03	1.19	1.37	0.77	2.85	1.37
% Reduction/MMBtu		60.70		43.80		52.00
Pyritic sulfur, wt %	2.43	0.60	1.22	0.43	2.88	1.11
lb/MMBtu	1.98	0.44	0.88	0.31	2.05	0.83
% Reduction/MMBtu		78.00		65.10		59.50
Btu/lb	12,277	13,774	13,116	14,015	10,652	13,374
Weight yield, wt %		87.10		91.60		68.50
Btu yield, %		97.80		97.90		86.00

Source: Data courtesy of Otisca Industries, Ltd.

Table 2.9 Bench-Scale Simulation of the Otisca Process for Various Seams of Western Coal

Seam	Lignite				Stray	
State	North Dakota		New Mexico		Montana	
County			McKinley		Rosebud	
Test size	Bench		Bench		Bench	
Analysis	Feed	Product	Feed	Product	Feed	Product
Size distribution	3/8 in. × 0		3/8 in. × 0		3/4 in. × 0	
Specific gravity of separation		1.50		1.50		1.50
Volatile matter, wt %			39.88	42.81	52.11	44.26
Fixed carbon, wt %			44.51	50.33	34.26	46.76
Ash, wt %	11.81	7.38	15.61	6.86	11.93	8.97
lb/MMBtu	10.82	6.57	13.36	5.30	10.43	7.46
% Reduction/MMBtu		39.30		60.30		28.50
Total sulfur, wt %	1.91	0.42	0.86	0.69	1.81	0.96
lb/MMBtu	1.75	0.37	0.74	0.53	1.58	0.80
% Reduction/MMBtu		78.90		28.40		49.40
Pyritic sulfur, wt %			0.15	0.00	1.01	0.15
lb/MMBtu			0.13		0.88	0.12
% Reduction/MMBtu				100.00		86.40
Btu/lb	10,917	11,240	11,686	12,952	11,433	12,025
Weight yield, wt %		96.30		82.60		91.70
Btu yield, %		99.10		91.50		96.40

Source: Data courtesy of Otisca Industries, Ltd.

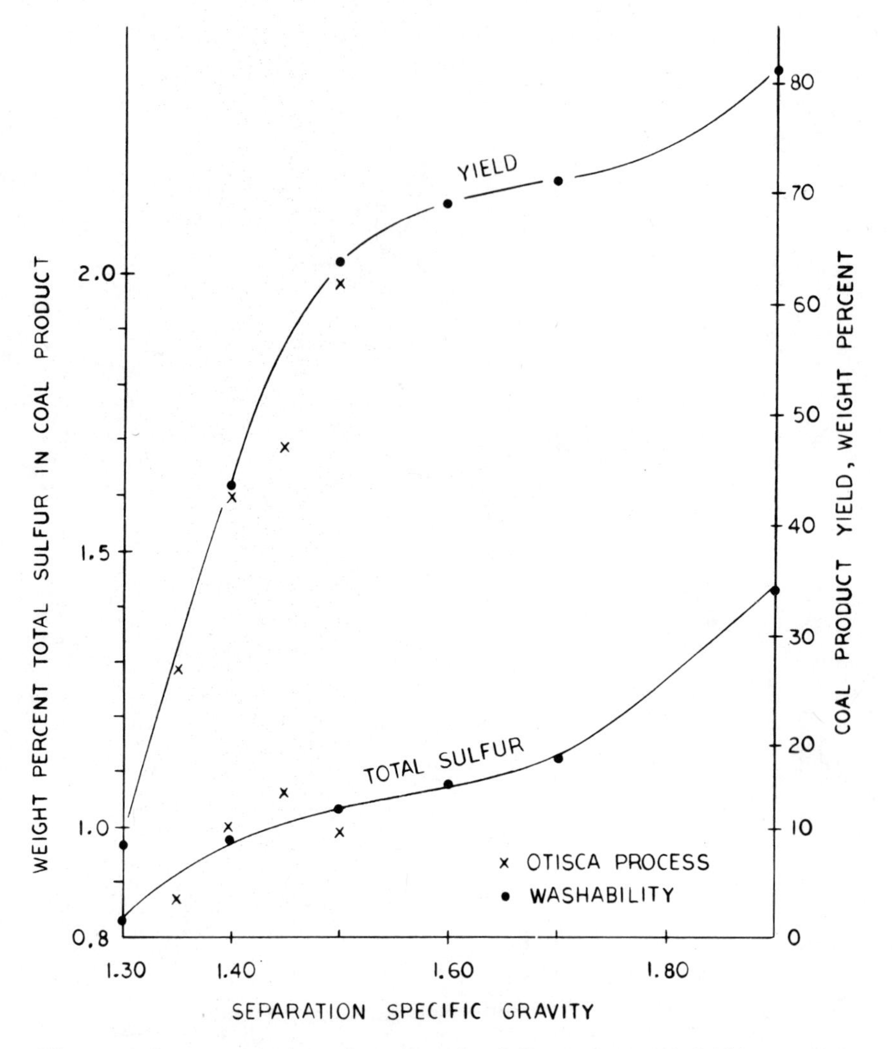

Figure 2.3 A comparison between the laboratory washability curves and bench-scale simulation of the Otisca process for an Upper Freeport coal from Pennsylvania pulverized to 200 mesh × 0. (From Ref. 28.)

both the standard washability test and the bench-scale simulation of the Otisca process conducted. The coal in this particular test had a size distribution as follows: 25.6 wt % greater than 74 μm; 32.8

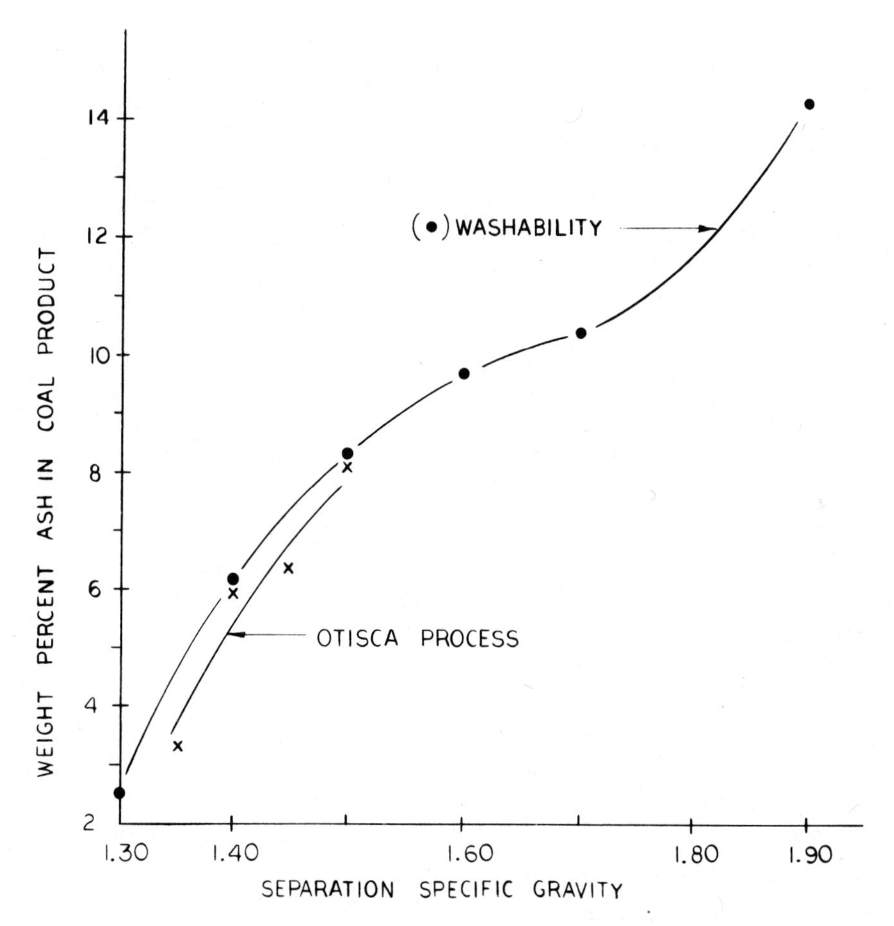

Figure 2.4 A comparison between the laboratory washability curves and bench-scale simulation of the Otisca process for an Upper Freeport coal from Pennsylvania pulverized to 200 mesh × 0. (From Ref. 28.)

wt %, 74 × 37 μm, and 41.6 wt % less than 37 μm. In all cases, a third group independently performed the chemical analysis of the products. The results are compared in Figs. 2.3 and 2.4 where, as expected, there is little difference in the results of the different procedures.

Initial Pilot-Plant Operations. The size consist and proximate

Table 2.10 Size Distribution of Otisca Pilot Plant (1976) Feedstock

Size	Wt %	Cumulative wt %
3/8 in. × 5 M[a]	6.5	6.5
5 × 10 M	14.7	21.2
10 × 30 M	39.7	60.9
30 × 60 M	19.2	80.1
60 × 100 M	6.0	86.1
100 × 200 M	4.3	90.4
200 × 400 M	4.1	94.5
400 M × 0	5.5	100.0

[a]M = International Standard (lOS Sieve Number).

Source: Ref. 8.

analysis of the West Virginia Upper Freeport slack coal fed to the pilot plant during September of 1976 are shown in Table 2.10. A detailed washability analysis was not conducted on the raw coal received throughout the 6-month testing period for the pilot plant. A washability analysis was conducted on the North Branch coal in 1974 and these data are given in Table 2.11. The washability did have a size distribution and sulfur and ash concentrations in the raw coal which was somewhat similar to the coal used in the pilot-plant test. It is evident from a comparison between Tables 2.10 and 2.11 that the washability represents a higher concentration of 1/4 in. × 28 mesh, i.e., 70.28 wt %, than was observed in the more recent sample, i.e., about 60.9 wt %. The opposite is the case for the 28 mesh × 0 where the washability sample only contained 29.7 wt % and the test sample contained 39.1 wt %.

From a detailed study of the separation characteristics of this coal, a curve of cumulative weight ash and sulfur of product coal versus separation specific gravity was prepared as shown in Fig. 2.5. Results of a proximate analysis of the raw slack coal and product coal, as separated at 1.50 specific gravity in a bench-scale simulation test, are shown in Table 2.12.

Table 2.11 Washability Characteristics of Otisca Pilot-Plant (1976) Feedstock[a]

Screen analysis				
Passing	Retained on	Wt %	% ash	% sulfur
	2 in.	19.02		
2 in.	¼ × 3 in. slot	30.48		
¼ × 3 in slot	28 m	35.49	24.50	2.40
28 m	100 m	9.98	17.81	2.29
100 m	0	5.03	21.00	2.04

Sink	Float	Wt %	% ash	% sulfur	Cumulative wt	Cumulative ash	Cumulative sulfur
¼ in. × 3 in. slot 28 mesh fraction — 70.28% of ¼ × 0 sample							
	1.40	57.16	7.20	1.10	57.16	7.20	1.10
1.40	1.45	10.67	15.00	1.82	67.83	8.43	1.21
1.45	1.50	3.81	19.60	2.54	71.64	9.02	1.28
1.50	1.55	1.94	23.70	3.10	73.58	9.41	1.32
1.55		26.42	66.53	4.98	100.00	24.90	2.30
28 × 100 mesh fraction — 19.76% of ¼ × 0 sample							
	1.40	65.64	6.34	1.14	65.64	6.34	1.14
1.40	1.45	8.33	14.84	1.72	73.97	7.30	1.21
1.45	1.50	3.97	19.15	2.10	77.94	7.90	1.25
1.50	1.55	2.58	22.30	2.36	80.52	8.36	1.29
1.55		19.48	56.84	6.46	100.00	17.81	2.29
100 mesh × 0 fraction (raw) — 9.96% of ¼ × 0 sample							
			21.00	2.04			

[a]Special run-of-mine sample taken from the raw coal feed belt at the North Branch processing plant for washability to determine quality at the different sizes. Report dated June 17, 1974.

Source: Data courtesy of Otisca Industries, Ltd.

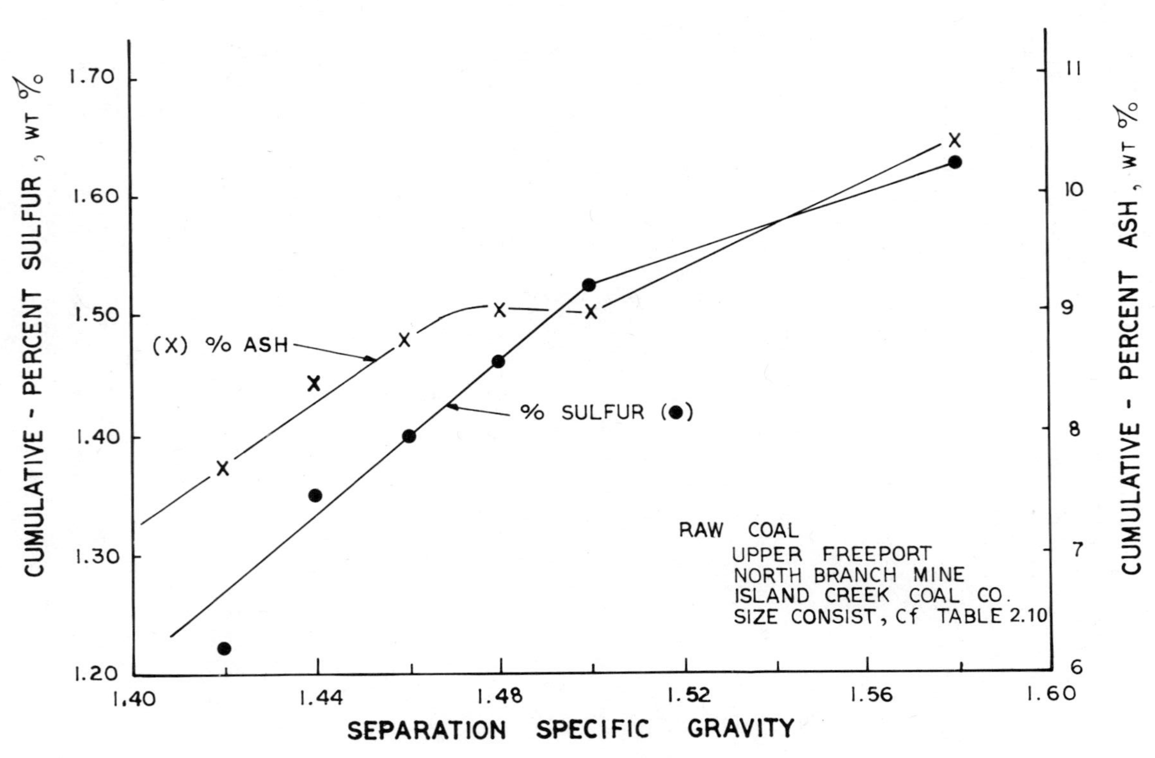

Figure 2.5 Coal quality vs. specific gravity for Otisca process separations using laboratory bench-scale simulation [8]. (Data courtesy of Otisca Industries, Ltd.)

Table 2.12 Proximate Analysis of Product Coal from Bench-Scale Simulation of Pilot Plant (1976) Feedstock

Analysis	Raw coal	Product coal of bench-scale simulation
Volatile matter, wt %	17.70	19.12
Fixed carbon, wt %	60.65	71.22
Ash, wt %	21.65	9.66
lb/MMBtu	18.33	6.93
Total sulfur, wt %	2.63	1.53
lb/MMBtu	2.23	1.10
Pyritic sulfur, wt %	1.54	0.62
lb/MMBtu	1.30	0.44
Organic sulfur, wt %	1.05	0.87
lb/MMBtu	0.89	0.62
Btu/lb	11,812	13,933
But/lb (MAF)	15,076	15,423
Weight yield, wt %	-	74.90
Btu yield, %	-	88.30
Specific gravity of separation	-	1.50

Source: Data courtesy of Otisca Industries, Ltd.

The analyses for raw coal, product coal, and reject material from the initial pilot-plant operations from May of 1976 until the completion of a 30-hr run in December of 1976 are shown in Table 2.13. The data shown in Table 2.13 for December 17, 1976, represent an average of 12 samples. During these sampling periods, measurements on the percent of misplaced material, as well as parting-liquid concentrations in the products and gas streams, were also conducted.

Several points should be emphasized in the data shown in Table 2.13 which demonstrate some of the interesting aspects of the Otisca process. Clearly, the first few hours of operation (May 25, 1976) could not be characterized as a well-organized and -operated continuous process. However, according to the analytical data, the separation process was performing well within design specifications. This

Table 2.13 Raw Coal and Product Analyses from the Initial Otisca Process Pilot-Plant Operations

	Moisture	Ash wt % (dry)	Sulfur wt % (dry)	Yield
Date: 5/25/76				
Raw coal	3.96	20.33	1.90	Not
Product	2.76	9.23	1.30	taken
Reject	2.29	61.20	3.20	
Date: 9/29/76 (I)				
Raw coal	6.57	23.34	2.30	74%
Product	4.36	9.99	1.28	
Reject	8.10	64.28	4.92	
Date: 9/29/76 (II)				
Raw coal	6.13	20.01	2.36	74%
Product	4.85	9.71	1.38	
Reject	8.54	57.50	4.84	
Date: 11/11/76				
Raw coal	5.75	22.8	2.60	74%
Product	4.69	9.48	1.42	
Reject	7.52	56.88	5.80	
Date: 11/15/76				
Raw coal	6.19	20.76	2.62	74%
Product	4.53	9.68	1.46	
Reject	6.34	61.36	6.12	
Date: 12/8/76				
Raw coal	5.41	17.93	2.40	73%
Product	5.28	9.74	1.30	
Reject	6.54	56.70	5.80	
Date: 12/13/76 (I)				
Raw coal	5.51	17.10	2.30	74%
Product	4.55	9.54	1.44	
Reject	5.94	56.38	6.72	
Date: 12/13/76 (II)				
Raw coal	5.27	20.78	2.62	74%
Product	5.05	9.31	1.38	
Reject	5.54	56.38	5.80	
Date: 12/17/76 (ave.)				
Raw coal	5.3 ± 0.4	20.60 ± 1.2	2.70 ± 0.1	73 ± 1%
Product	3.60 ± 0.7	9.77 ± 0.6	1.51 ± 0.1	
Reject	6.29 ± 1.2	57.50 ± 3.0	6.14 ± 0.7	

Source: Data courtesy of Otisca Industries Ltd.

is particularly enforced if one compares these data with those shown in Fig. 2.5, which illustrate the variation of weight percent ash and sulfur with specific gravity.

The increase in sulfur concentration of the product coal from the summer period of 1.35 ± 0.1 wt % to a value of 1.50 ± 0.1 wt % in December was, in all probability, due to the fact that the separation bath temperature during the summer was in the range of 68 to 72°F; while the corresponding temperature during December was in the range of 55 to 65°F, representing an increase of about 0.2 specific gravity units. The initial design of the existing pilot plant did not contain conditioning heaters to preheat the raw coal so as to stabilize the separation bath temperature. Current commercial facilities using the Otisca process, however, have been equipped with conditioning preheaters.

The wide variations in the ash and sulfur concentrations of the reject material shown in Table 2.13 reflect the wide variations in the raw coal feed chemistry. Through these variations, the coal product chemistry over the 6-month period, on the other hand, held to within rather narrow limits; for example, sulfur, 1.4 ± 0.15 wt % and ash, 9.70 ± 0.4 wt %. The narrow variation in the chemical analysis of the coal product over such a long period represents one of the advantages of the Otisca process.

One of the key purposes of the initial pilot-plant operations was to ascertain the extent of parting-liquid losses during continuous operation. This was approached from two directions. Firstly, during the initial 6-month operations, the measurement of losses due to improper welds, joints, flanges, shaft seals, etc., was conducted using a General Electric gas detector having a sensitivity in the range of a few parts per million. Losses at seals were of a very minor consideration. Losses in the overboard noncondensible stream and the losses with product coal and reject material were conducted using gas chromatographic techniques.

The final assessment of the parting-liquid losses was conducted over a 30-hr separation test where a material balance was conducted of the parting liquid in the system. The results of that balance indicated that the losses were 1.2 ± 0.7 lb per raw coal ton, which was in reasonable agreement with the losses determined through examination of the points where, in fact, the liquid was lost.

It should be recognized that the losses of 1.2 lb/ton were directly due to two sources. Firstly, the parting-liquid gas recovery facility was not of an optimum design due to economic considerations. Gas chromatographic studies indicated that about 1 lb/ton could be collected from this point using the best available technology. Coal and refuse adsorb some fraction of the parting liquid depending on the particular coal. Extensive studies indicate that a loss due to sorption is in the range of 100 parts per million (ppm) or 0.2 lb per raw coal ton. The most recent designs project parting-liquid losses in the range of 0.25 lb per raw coal ton.

Recent Pilot-Plant Results. At the conclusion of the first series of pilot plant tests discussed in the previous section, the pilot plant was moved from Bayard, West Virginia, to Florence, Pennsylvania, in a fully independent facility dedicated to providing semi-works scale data for plant design. During reassembly of the plant in Florence, numerous new developments were incorporated based on the experiences of the first tests. Anyone interested in the results of an Otisca plant operation on a particular coal at a scale of 100 to 150 tons per shift, can rent the facility for any period. The system provides coal owners with a first-hand observation of the advantages of the process, while also providing large samples for end product testing. The Florence facility also serves as a valuable means of testing new developments in unit processes in the plant while operation is conducted continuously.

In an examination of the more recent results from the Otisca process pilot-plant operation, one should recognize that a discussion of results of beneficiation of any coal is beset with many problems from sampling to a comparative set of standards. The results of this operation are in similar question. However, if one accepts the fact that ASTM procedures [35] were carefully followed throughout data production, and the processing plant product streams were sampled every 15 min to accumulate an average 4-hr sample, an improved confidence level is attained. Furthermore, the data from each accumulated sample of 4 hr of plant operation were compared to a second 4-hr period of samples, as well as a bench-scale simulation of separation of the raw coal from that same time period. Variations occurred, however, always within anticipated limits.

Table 2.14 illustrates a comparison of the coal product analyses from a bench-scale simulation with that of a 4-hr pilot-plant run period of the same raw coal. In most cases, the bench-scale simulation is a fair representation of how the plant product will appear. The key variable in this assessment is the moisture content of the raw coal fed to the pilot plant and its associated slimes.

In 1979 Electric Power Research Institute [36] awarded Otisca Industries a contract to operate the pilot plant for eight hours using a 3/8 in. × 0 raw Upper Freeport coal, such that production samples could be taken of the products and the distribution curves and probability error (E_{pm}) of the separation process constructed from the results. The probable error (E_{pm}) results were as follows:

Composite feed	Organic efficiency, %	Probable error (E_{pm})	Specific gravity
3/8 × 1/4 in.	100	0.008	1.48
1/4 in. × 28 mesh	99	0.015	1.49
28 × 100 mesh	98	0.175	1.57
100 × 325 mesh	96	0.260	1.80
3/8 in. × 325 mesh	98	0.023	1.49

Figure 2.6 illustrates the cumulative results from the 3/8 in. × 325 mesh material.

D. Projected Process Economics

Based on pilot-plant operational experience and recent engineering studies (1979), the following projected economies apply to a nominal 250 tons/hr plant designed to beneficiate a typical 3/4 in. × 0 raw coal size consist.

Table 2.14 Analyses of the Product Coal from Bench-Scale Simulation and the Otisca Process Pilot-Plant Operations

Seam	Upper Freeport				Bakerstown			
State	West Virginia				West Virginia			
County	Grant				Grant			
Test size	Bench scale		Pilot plant		Bench scale		Pilot plant	
Analysis	Feed	Product	Feed	Product	Feed	Product	Feed	Product
Size distribution		2 in. × 0		2 in. × 0		$1\frac{1}{2}$ in. × 0		$1\frac{1}{2}$ in. × 0
Specific gravity of separation		1.50		1.50		1.50		1.50
Volatile matter, wt %	17.72	20.84	16.82	19.23	17.85	20.15	19.18	22.17
Fixed carbon, wt %	52.22	69.55	51.42	71.56	63.32	70.45	63.84	69.74
Ash, wt %	30.06	9.61	31.76	9.21	18.83	9.40	16.98	8.09
lb/MMBtu	28.20	6.79	31.10	6.50	15.21	6.73	13.52	5.81
% Reduction/MMBtu		75.90		79.10		55.80		57.00
Total sulfur, wt %	1.91	1.29	2.37	1.34	0.69	0.67	0.53	0.57
lb/MMBtu	1.79	0.91	2.32	0.94	0.55	0.48	0.42	0.41
% Reduction/MMBtu		49.10		59.20		12.70		2.40
Pyritic sulfur, wt %	1.71	0.85	2.28	0.75	0.43	0.24	0.00	0.00
lb/MMBtu	1.61	0.60	2.23	0.53	0.35	0.17	0.00	0.00
% Reduction/MMBtu		62.70		76.20		50.90		-
Btu/lb	10,642	14,162	10,202	14,160	12,381	13,975	12,562	13,928
Weight yield, wt %		57.60		55.70		73.90		70.00
Btu yield		76.70		77.30		83.40		78.00

Seam	Upper Freeport				Harlem-Elklick			
State	West Virginia				West Virginia			
County	Tucker				Grant			
Test size	Bench scale		Pilot plant		Bench scale		Pilot scale	
Analysis	Feed	Product	Feed	Product	Feed	Product	Feed	Product
Size distribution		1½ in. × 0		1½ in. × 0		1 in. × 0		1½ in. × 0
Specific gravity of separation		1.50		1.50		1.50		1.50
Volatile matter, wt %	17.92	22.54	17.24	20.97	21.58	22.78	19.69	23.57
Fixed carbon, wt %	53.72	70.40	47.73	70.91	62.21	66.55	54.38	67.62
Ash, wt %	28.36	7.06	35.08	8.12	16.21	10.67	25.98	8.81
lb/MMBtu	26.46	4.88	36.43	5.58	12.59	7.66	22.88	6.16
% Reduction/MMBtu		81.60		84.70		93.20		73.10
Total sulfur, wt %	1.93	1.45	1.66	0.87	2.68	1.56	1.77	1.21
lb/MMBtu	1.80	1.00	1.73	0.60	2.08	1.12	1.56	0.85
% Reduction/MMBtu		44.40		65.30		46.20		45.50
Pyritic sulfur, wt %	1.05	1.04	1.31	0.41	2.55	1.25	1.26	0.66
lb/MMBtu	0.98	0.72	1.36	0.28	1.98	0.90	1.11	0.46
% Reduction/MMBtu		26.50		79.40		54.50		58.60
Btu/lb	10,717	14,457	9,615	14,563	12,878	13,918	11,354	14,306
Weight yield, wt %		66.20		64.00		78.30		68.00
Btu yield		89.30		97.00		84.60		86.00

Source: Data courtesy of Otisca Industries, Ltd.

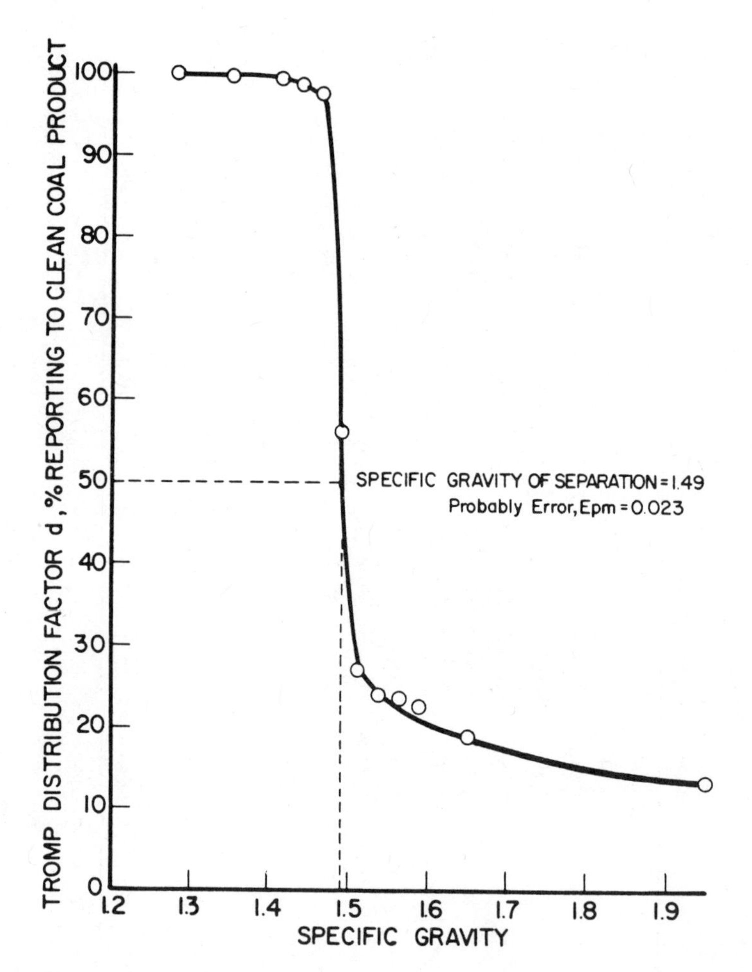

Figure 2.6 Distribution curve for Otisca pilot-plant 3/8 in. × 325 mesh composite feed. (Data courtesy of Otisca Industries, Ltd.)

First cost for the Otisca process equipment is in the range of \$21,000 per raw coal ton per hour. Thereafter, Otisca process equipment for a 250 raw coal tons/hr plant would cost \$5,250,000. This figure includes the cost of a complete building to house the process equipment, laboratory data for plant design, production laboratory, start-up costs, and operator training. Costs for auxiliary facilities, such as storage, crushing station, electrical substation, maintenance shop, load-out facilities, and site work, must be added to the above figures.

The approximate capital cost of a 250 raw coal tons/hr hydrobeneficiation facility treating all sizes 3/4 in. × 0, including dewatering to 6 wt % moisture, is considerably more than that of an Otisca plant.

Electrical energy consumed by the motors associated with an Otisca plant is approximately 9.6 kWhr/ton of raw coal. Therefore, if electrical power costs 3.0¢/kWhr, the process electricity would cost 29¢/raw coal ton.

If no energy conservation techniques are applied to the plant, the process heat required to recover the parting liquid is approximately 42,400 Btu/raw coal ton. Assuming the use of No. 2 fuel oil, with additional capital investment, energy conservation could be applied which would result in a significant reduction of the cost of this energy from 21¢/raw coal ton to as low as 10¢/raw coal ton.

Assuming a very conservative stance, the process materials cost approximately 51¢/raw coal ton with an additive consumption of 0.4 lb/raw ton at $0.70/lb and parting-liquid losses of 0.5 lb/raw coal ton at $0.46/lb. It should be recognized that for certain coals, additives may not be needed at all in using the Otisca process.

Maintenance is expected to be a small fraction of that typically associated with a conventional coal beneficiation plant, as an Otisca plant is more akin to a chemical plant than a mineral beneficiation facility. The noncorrosive characteristics of the heavy liquid, coupled with the fact that most relative velocities (and hence, abrasion velocities) in the plant are extremely low, yield an expected maintenance expense which is considerably less than that of a conventional mineral beneficiation facility. The validity of this comparison is reinforced by the fact that there are no vibrating devices such as screens or relatively high-speed devices such as centrifuges included in an Otisca plant. Maintenance costs were estimated to average 0.21¢/raw coal ton, or an annual expense of 5% of capital cost.

It is expected that four men will be required to operate a 250 raw coal ton/hr Otisca plant. Assuming labor at $15.00/man-hr, labor cost would be less than 25¢/raw coal ton.

In summary, direct operating costs for labor, maintenance, energy, and process materials totals $1.47/raw coal ton. Capital cost for Otisca process equipment is in the range of $21,000 per raw coal ton per hour.

E. Commercial-Scale Installations

In 1977, demonstration work at the Florence Development Center led to the signing of a contract with the American Electric Power for the Otisca Industries to construct and operate a 125 tons/hr Otisca Demonstration Plant at the American Electric Power's Muskingum Mine near Beverly, Ohio. The facility, which is presently in the start-up phase, is essentially a complete coal preparation facility in that it has its own site services, raw coal storage, raw coal recovery, crushing station, crushed coal storage, Otisca separation facility, refuse storage and handling bin, and a product coal stacking conveyor. The total budget for the construction, start-up, and operation of the plant is $6.7 million.

Plant start-up is projected for the fourth quarter of 1979, with Otisca responsible for the design, construction, and operation of the plant. It is important to note that the $6.7 million budget covers the installation and operating cost of a significant amount of instrumentation and/or equipment that would not be found in a commercial coal beneficiation plant, whose objectives usually do not include the monitoring and generating of a significant amount of detailed operating data. The projected capital costs referred to in the last section (Sec. 2.3D) were based, for the most part, on the anticipated costs of this program in January 1979.

A commercial plant of 75 tons/hr has been contracted and will be located in Maryland.

2.4 CONCLUSIONS

The complexities of preparation of a coal with a size distribution of essentially 4 mm × 0 with 15 wt % of that size distribution smaller than 0.2 mm and 5 wt % smaller than 37 μm is a well-established fact in the coal industry. Materials handling, dusting, excessive bath viscosities, and very inefficient separations are typical of these problems. The solution to these problems always results in an increase of coal preparation costs, often dramatically. On the other hand, it is also well accepted that as the size distribution of most coals is decreased, the potential for pyritic sulfur and ash reduction increases.

The Otisca process appears to offer a solution to the posed dilemma in that the process is able to recover more fine coal with less misplaced material at a lower processing cost than alternate processes which are currently available. An important feature of the Otisca process is that both bench-scale process simulation and pilot-plant test results closely duplicate the theoretical washability data. In contrast, when transferring the theoretical washability data to the conventional wet beneficiation systems, such as the heavy-media separation by a water-magnetite suspension, a significant loss in separation efficiency (often up to 30%) can be experienced.

Further, in this line of comparison and, while it may have already become obvious to the reader, it should be emphasized that an Otisca coal-cleaning plant is significantly different from the typical hydrobeneficiation plant. The Otisca process plant is completely enclosed; and consequently, noise and dust levels are considerably less than those encountered in a conventional plant. The chemicals used in the Otisca plant are nonflammable, nontoxic, virtually odor free, and noncorrosive, which allows for inexpensive construction of material handling and electrical equipment. Thermal or mechanical driers necessary to remove process water from the coal are eliminated as are requirements for large quantities of process water. Hence, the following unit operations are not required: clarifiers, settling ponds, water treatment, bag houses, vacuum filters, filter presses, centrifuges, cyclones, screens, magnetic separators, fluidized-bed driers, venturi scrubbers, multistory structures, and dewatering screens. Finally, by comparison to a conventional hydrobeneficiation plant, the Otisca plant flowsheet is extremely simple in that the raw coal does not have to be classified in order to provide a suitable feed to a multiplicity of separating unit operations.

Plant atmospheric emissions of the nontoxic parting liquid can be reduced to extremely low values through proper seal design, product liquid removal, and noncondensible gas incineration. The parting-liquid losses under these conditions could be reduced to much less than 0.05 lb/raw coal ton. The nontoxic additives constitute a fixed loss which may vary between 0 and 1 lb/raw coal ton, depending upon the particular coal being separated.

The raw coal feed to an Otisca plant may vary from any reasonable top size by zero, as the ash and pyritic sulfur releases may dictate. For example, batch investigations have indicated that separations of settling-pond coal fines, essentially 100 wt % –200 mesh (below 74 μm) are practical.

The lack of classification equipment and size-sensitive separation operations in an Otisca plant leads to the obvious fact that the Otisca process can routinely tolerate wide fluctuations in raw coal feed rates, size distributions and chemical analyses.

There appears to be little question that the Otisca Process can at present be moved into commercial cleaning of coal with a sound advantage in product coal yield and economics. Further pilot-plant testing and commercial-scale demonstration are expected to increase these advantages.

ACKNOWLEDGMENT

The author would like to acknowledge the support and confidence provided by numerous members of the coal and power industry who uniquely have recognized the need for better methods of fine coal cleaning and have indulged generously toward the support of a solution to this problem. American Electric Power and Island Creek Coal Co. were among the first supporters, followed closely by Penelec and many others.

The author would also like to acknowledge C. D. Smith, author of the innovative engineering and plant design for the Otisca process, and the Otisca laboratory personnel responsible for the data developed in this chapter.

REFERENCES

1. H. Tschamler and E. de Ruiter, Physical Properties of Coals, in *Chemistry of Coal Utilization* (H. H. Lowry, ed.), Wiley, New York, 1963, Chap. 2.
2. R. T. Greer, Coal Microstructure and Pyrite Distribution, in *Coal Desulfurization: Chemical and Physical Methods* (T. D. Wheelock, ed.), *ACS Symp. Series*, No. 64, Amer. Chem. Soc., Washington, D.C., 1979, Chap. 1.
3. J. W. Leonard (ed.), *Coal Preparation*, 4th ed., AIME, New York, 1979.
4. J. A. Cavallaro and A. W. Deurbrouck, An Overview of Coal Preparation, in *Coal Desulfurization: Chemical and Physical Methods* (T. D. Wheelock, ed.), *ACS Symp. Series*, No. 64, Amer. Chem. Soc., Washington, D.C., 1977, Chap. 4.
5. A. W. Deurbrouck and J. Hudy, Jr., *Performance Characteristics of Coal Washing Equipment: Dense Medium Cyclones*, U.S. Bureau of Mines, Report of Investigation, No. RI 7673, 1972.
6. J. Hudy, Jr., *Performance Characteristics of Coal Washing Equipment: Dense Medium, Coarse Coal Washing Equipment*, U.S. Bureau of Mines, Report of Investigation, No. RI 7154, 1968.
7. D. C. Wilson, Dry Table-Pyrite Removal from Coal, in *Coal Desulfurization: Chemical and Physical Methods* (T. D. Wheelock, ed.), *ACS Symp. Series*, No. 64, Amer. Chem. Soc., Washington, D.C., 1977, Chap. 8.
8. D. V. Keller, Jr., C. D. Smith, and E. F. Burch, Demonstration Plant Test Results of the Otisca Process Heavy Liquid Beneficiation of Coal, paper presented at SME-AIME Annual Meeting, Atlanta, Ga., Mar. 1977.
9. J. D. Sullivan, Heavy Liquids for Mineral Analysis, U.S. Bureau of Mines, Technical Paper, No. 381, 1927.
10. W. B. Foulke, Sink and Float Separation Commands New Attention, *Eng. Min. J.*, *139*, 33 (1938).

11. J. S. Browning, Heavy Liquids and Procedures for Laboratory Separation of Minerals, U.S. Bureau of Mines, Information Circular, No. IC 8007, 1961.

12. E. C. Tveter and W. L. O'Connell, Heavy Liquids for Mineral Beneficiation, unpublished report, Dow Chemical Company, Mar. 1963.

13. L. A. Roe and E. C. Tveter, Application of Heavy Liquid Processes to Minerals Beneficiation, *Trans. Soc. Min. Eng., 226,* 141 (1963).

14. R. B. Tippin and E. C. Tveter, Heavy Liquid Recovery Systems in Mineral Beneficiation, *Trans. Soc. Min. Eng., AIME, 241,* 15 (1968).

15. I. G. C. Dryden, Solvent Power for Coals at Room Temperature, *Chem. Ind.,* 502 (1952).

16. I. G. C. Dryden, Action of Solvents on Coals at Low Temperature, *Fuel, 30,* 39 (1951).

17. D. W. Van Krevelin, Chemical Properties and Structure of Coal, XXVIII, Coal Constitution and Solvent Extraction, *Fuel, 44,* 229 (1965).

18. N. Y. Kirov, J. M. O'Shea, and G. D. Sergeant, The Determination of the Solubility Parameters of Coal, *Fuel, 64,* 415 (1967).

19. D. V. Keller, Jr., and C. D. Smith, Spontaneous Fracture of Coal, *Fuel, 55,* 273 (1976).

20. J. Thomas, Jr., and H. H. Damberger, *Internal Surface Area, Moisture Content and Porosity of Illinois Coals: Variations with Rank,* Illinois State Geological Survey Publication, Circular No. 493, 1976.

21. H. F. Yancey and M. R. Geer, Properties of Coal and Impurities in Relation to Preparation, in *Coal Preparation* (J. W. Leonard and D. R. Mitchell, eds.), 3d ed., AIME, New York, 1968, Chap. 1.

22. J. H. Gluskoter, R. R. Ruch, W. G. Miller, R. A. Cahill, G. B. Dreher, and J. K. Kuhn, *Trace Elements in Coal: Occurrence and Distribution,* Illinois State Geological Survey, Publ. No. 499, 1977.

23. J. H. Gluskoter, Mineral Matter and Trace Elements in Coal, in *Trace Elements in Fuel* (S. B. Babu, ed.), *Adv. Chem. Series,* No. 141, Amer. Chem. Soc., Washington, D.C., 1975, Chap. 1.

24. M. Th. Mackowsky, Mineral Matter in Coal, *Coal and Coal Bearing Strata* (D. G. Murchism and T. S. Westoll, ed.), Oliver and Boyd, London, 1968, pp. 309-321.

25. J. T. McCartney, H. J. O'Donnell, and S. Ergun, *Pyrite Size Distribution and Coal Pyrite Particle Association in Steam Coals,* U.S. Bureau of Mines, Report of Investigation, No. 7231, 1969.

26. D. R. Bomberger and M. Deul, Study of Fine Coal Cleaning Processes by Automatic Microscopy, *Trans. Soc. Min. Eng., AIME, 229,* 65 (1969).

27. W. C. Grady, Microscopic Varieties of Pyrite in West Virginia Coals, *Trans. Soc. Min. Eng., AIME, 262,* 268 (1977).

28. D. V. Keller, Jr., A Theoretical Approach to Washability Curves Is Compared to the Otisca Process Separation of Fine Coal, *Proceedings of Engineering Foundation Conference on Clean Combustion of Coal,* Rindge, N.H., July/Aug., 1977; EPA Publ. No. EPA-60017-78-073, Apr. 1978, p. 131.

29. J. R. Kleeman, Equations for the Terminal Settling Velocities of Spheres, *Chem. Eng., 82,* 102 (1975).

30. D. V. Keller, Jr., The Otisca Process: The Physical Separation of Coal Using a Dense Liquid, paper presented at Ohio University Coal Preparation Workshop, Columbus, Ohio, Oct. 1978.

31. R. L. Whitmore, Coal Preparation: The Separation Efficiency of Dense-Medium Baths, *J. Inst. Fuel, 31,* 422 (1958).

32. M. J. Wohl, Designing for Non-Newtonian Fluids, *Chem. Eng.,* 5 (1968).

33. C. W. Weast (ed.), *Handbook of Chemistry and Physics*, Chemical Rubber Co., Cleveland, Ohio, 1977

34. J. A. Riddick and W. B. Bunger, *Organic Solvents*, Wiley Interscience, New York, 1970.

35. American Society of Testing Materials, *Annual Book of ASTM Standards,* Part 19, ASTM, Philadelphia, Pa., 1977.

36. D. V. Keller, Jr., and A. Rainis, The Otisca Process: A Pilot Plant Study of Dense Liquid Separation, Electric Power Research Institute, CS-1705, Project 1030-15, Feb. 1981.

3

Electrostatic Beneficiation of Coal

ION I. INCULET
M. A. BERGOUGNOU
JAMES D. BROWN

The University of Western Ontario
London, Ontario, Canada

3.1 INTRODUCTION

Small particles can be charged by triboelectrification, conductive induction, or corona charging. The charges developed depend on the particle characteristics such as composition, crystal structure, and surface state, as well as environmental factors such as temperature and humidity. After charging, particles of different types can be separated according to charge to mass ratio using forces in an electric field. Changes in the method of charging, the surface preparation, or the environmental factors can be used to alter the degree of separation of particles of different types. Experimental techniques and apparatus have been developed for application of electrostatic beneficiation principles to separation of ash and pyrite from coal as well as separation of the coal macerals themselves.

The electrostatic separation methods are dry processes and therefore offer many advantages, both in terms of energy efficiency and environmental pollution, over conventional wet separation processes. Only the moisture present in the mined coal need be driven off and the energy loss associated with the evaporation of large quantities of water absorbed during a flotation separation is avoided. The Hat Creek coals in Western Canada contain a significant quantity of clay minerals which form a gelatinous mass during any coal washing process, presenting a formidable water pollution problem. This kind of problem can be completely avoided by electrostatic separation techniques. The environmental problem becomes one of containing dust particles using conventional electrostatic precipitators.

3.2 PRINCIPLES OF ELECTROSTATIC SEPARATION

The basis of any electrostatic separation or beneficiation process for finely divided matter is the interaction between an external electric field and the electric charges acquired by the various particles.

In any such processes, while the external electric field can be easily produced and controlled, the selective charging of the particles to be separated usually requires most of the developmental effort prior to the building of an industrial installation.

A. Generation of Charges

A review of what is known of the electrical properties of coal points to three types of selective electrification processes which will very likely find application in a beneficiation facility:

1. Triboelectrification
2. Conductive induction charging
3. Ionic and/or electronic bombardment from a corona generator in combination with conductive discharges

*Triboelectrification.** The principle is shown schematically in Fig. 3.1. Let us assume that the particles placed on the surface D are made to repeatedly contact one another as well as the surface D. This is typical of a Syntron vibrator or a traveling electric field [1] conveying system.

The *work function* is defined as the difference between the energy of an electron at the Fermi level inside the surface of a solid and an electron at rest in vacuum outside the solid. Assume that W_A, the work function of the surface of particle A, is smaller than W_D, the work function of the surface D; and W_D is smaller than W_B, the work function of the surface of particle B. Then, upon contact and separation, particle A will become positively charged,

---

*In a multibody system, assuming that no more than one of the surfaces in contact is conductive.

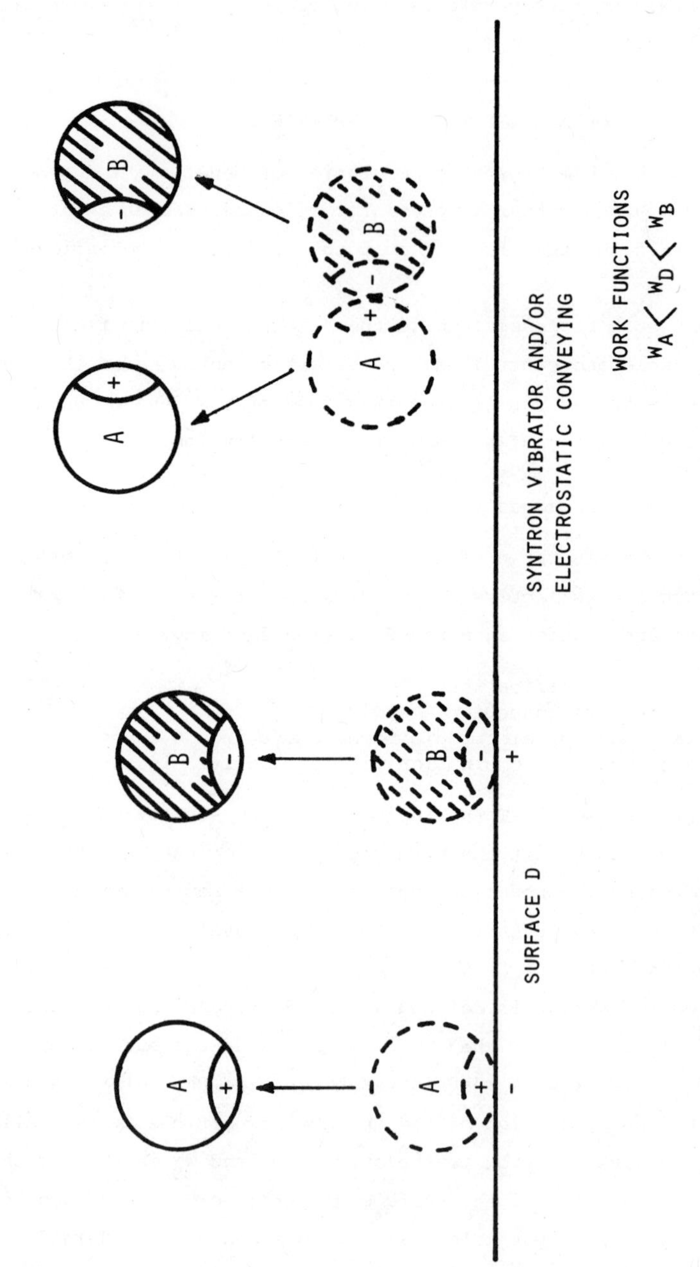

Figure 3.1 Triboelectrification between particles A and B and surface D.

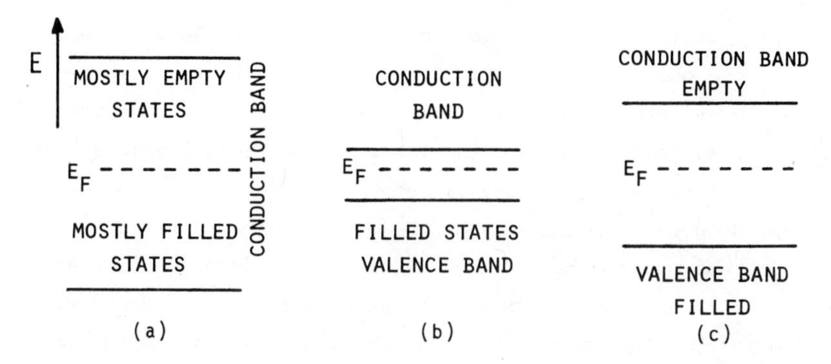

Figure 3.2 Energy-level diagrams for (a) metals, (b) semiconductors, and (c) insulators showing the Fermi level E_F.

particle B negatively charged; and the surface D will acquire electrons from particle A and give electrons to particle B.

The triboelectrification of various materials in contact in vacuum has been extensively studied and proven to be a very effective way of charging [2-4]. The principle of electrification by contact is that the Fermi levels of particles in contact must equilibrate. The *Fermi level* is defined as that level at which the probability of finding an electron is 0.5. The position of the Fermi level for metals, semiconductors and insulators relative to the conduction and valence bands is shown in Fig. 3.2. As shown in Fig. 3.3, if two

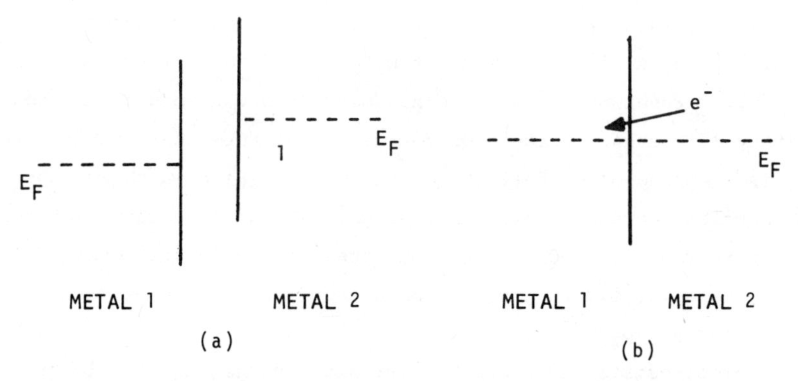

Figure 3.3 Contact electrification of two metals: (a) before contact; (b) after contact (electrons flow from 2 to 1 until equilibration of Fermi levels is achieved).

metals with different Fermi levels come in contact, electrons will flow from the one with the higher Fermi level (metal 2) to the one with the lower level (metal 1). This simple picture becomes more complex for semiconductors and insulators. The Fermi levels at the point of contact must equilibrate, but the total charge transferred will depend on the conductivity of the particles and the time of contact. Thus many contacts and a significant relaxation time are necessary for complete charging of insulators and semiconductors.

Two further complications of the charging process are the leaking back of charge through the small distances as the contact is broken off and the presence of surface states which may alter significantly the Fermi level at the surface of a particle. The first effect of the leaking back of charge is probably responsible for the generally poorer electrostatic separation observed in the presence of high humidity. The presence of surface states can be used to alter separation characteristics. This is a possibility which has not yet been thoroughly explored; however, a patent has been granted to Kali and Salz AG [5] based on treatment of coals with fatty acid glycerides to enhance separation of pyrite from coal. The state of electrostatic separation might be compared to early flotation separation before the exploration of the effects of surface active agents. The possibilities for changing separation efficiencies by surface agents appear to be unlimited.

Coal macerals are, in general, insulators when dry but show a much higher conductivity than might be expected when saturated with water [6]. Crushing of coal also has been reported to decrease conductivity [7] perhaps because of decreased moisture content but perhaps also because of oxidation. Mettus et al. [8] have shown that the electrical conductivity had predominantly ionic character in the temperature range of 300 to 800°C and that the activation energy of the electrons was 1.7 to 2.0 eV, which should be an approximate value for the energy gap.

In electrostatic separation, vitrinite charges positively and fusinite, pyrite and ash negatively. This implies that the Fermi level of vitrinite is higher than that of fusinite, pyrite, or ash components [9,10].

Triboelectrification is extremely sensitive to surface state, and any transient surface changes at the point and time of contact will have profound effects on the charge transfer. (A monomolecular layer of adsorbed gas can substantially alter the work function of the material.) Generally, two processes are involved: (1) a charge transfer during contact, and (2) a charge backflow just before separation. In practice, two good conductors will triboelectrify poorly because of charge backflow. As an example, consider Fig. 3.4a and b [11]. These figures show the results of charge transferred (vertical axis) when contacting borosilicate glass or quartz with gold in a uniform electric field between parallel plates to which the potential (horizontal axis) was applied. The potential is considered (+) when the electric field at the point of contact is directed from the gold to the glass (or quartz). The experiments were performed at a 10^{-9} Torr vacuum and over a temperature range from −45 to 130°C. The solid lines enclose the range of observed values for charge transfer in a number of replicate experiments.

Except for the borosilicate-gold experiment at high temperature (Fig. 3.4a A), the polarity of charge transfer is independent of the direction of the applied field. This indicates that triboelectrification is the dominant charge transfer mechanism. The magnitude of the "pure" triboelectric effect can be seen from the values obtained in the absence of an electric field (i.e., along the ordinate axis in the figure). While similar data are not available for coal macerals and ash minerals, the above effects would very likely take place.

Electric Charging by Conductive Induction. The principle is shown in Fig. 3.5 which is a schematic representation of a classical electrostatic ore drum separator used in industry now. Typically, the raw ore is fed over the drum and the beneficiated material is collected in the bins below. The positive high-voltage electrode attracts or "induces" greater negative charges on the highly conductive particles than on insulating particles or those of lower conductivity. The

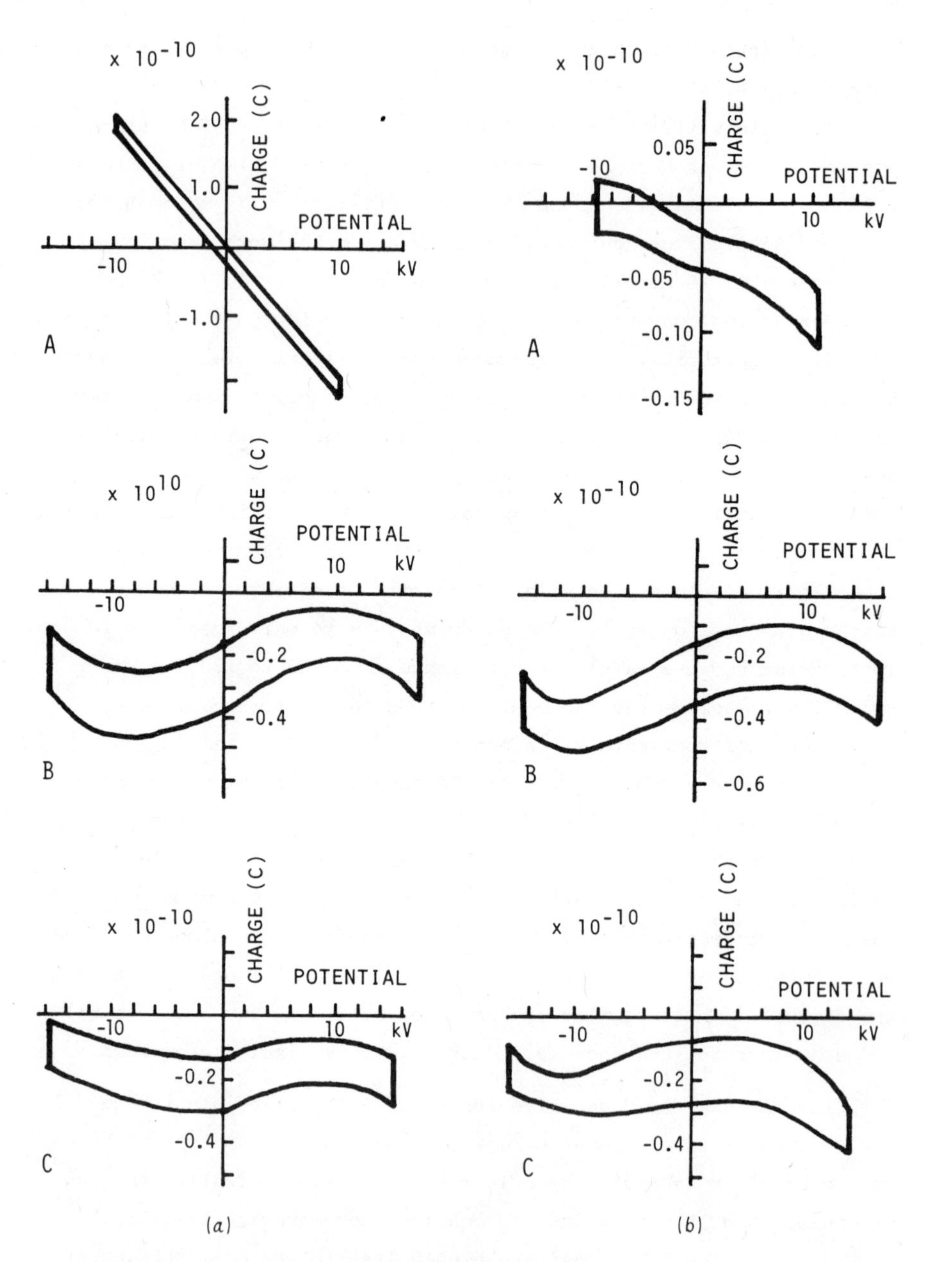

Figure 3.4 (a) Electric charges developed on gold contacting borosilicate glass plotted against voltage: A, 125 to 130°C; B, 25°C; C, −55 to −19°C. (b) Electric charges developed on gold contacting quartz plotted against voltage: A, 125 to 130°C; B, 25°C; C, −45 to −17°C. [From Ref. 11. © (1971) *Proceedings of the Third Conference on Static Electrification*, Institute of Physics. Reprinted with permission.]

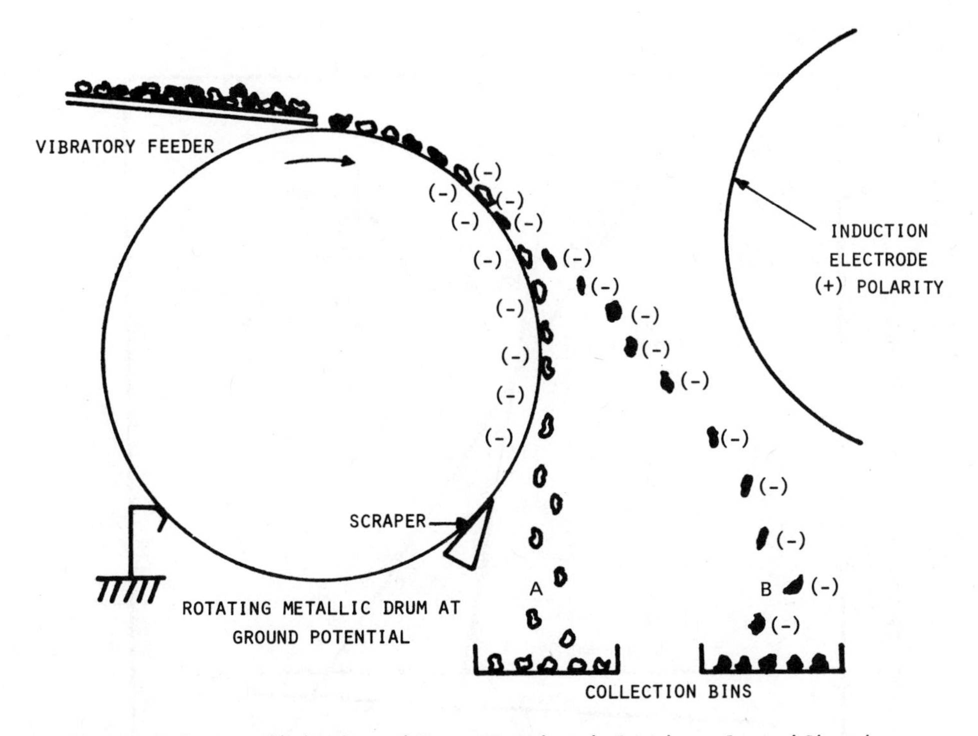

Figure 3.5 Beneficiation with conductive induction electrification. It is assumed that the electrical conductivity of the B particles is substantially higher than that of the A particles.

resultant electrical forces pull the negatively charged particles away from the drum.

Upon separation from the drum, while still under the influence of the electric field, the conductive particles will retain their negative charge and be deflected toward the collecting bin.

Referring to Fig. 3.4, conductive induction effects dominate only in the case of the gold-borosilicate glass experiment at high temperature. This leads to the linear dependence of charge transfer and the change in the polarity of the charge with the change in direction of the electric field observed in Fig. 3.4a A. (Note that the electrical conductivity of glass increases by some three orders of magnitude between 0 and 100°C.) A smaller effect is perhaps discernible in the gold-quartz at the high temperature.

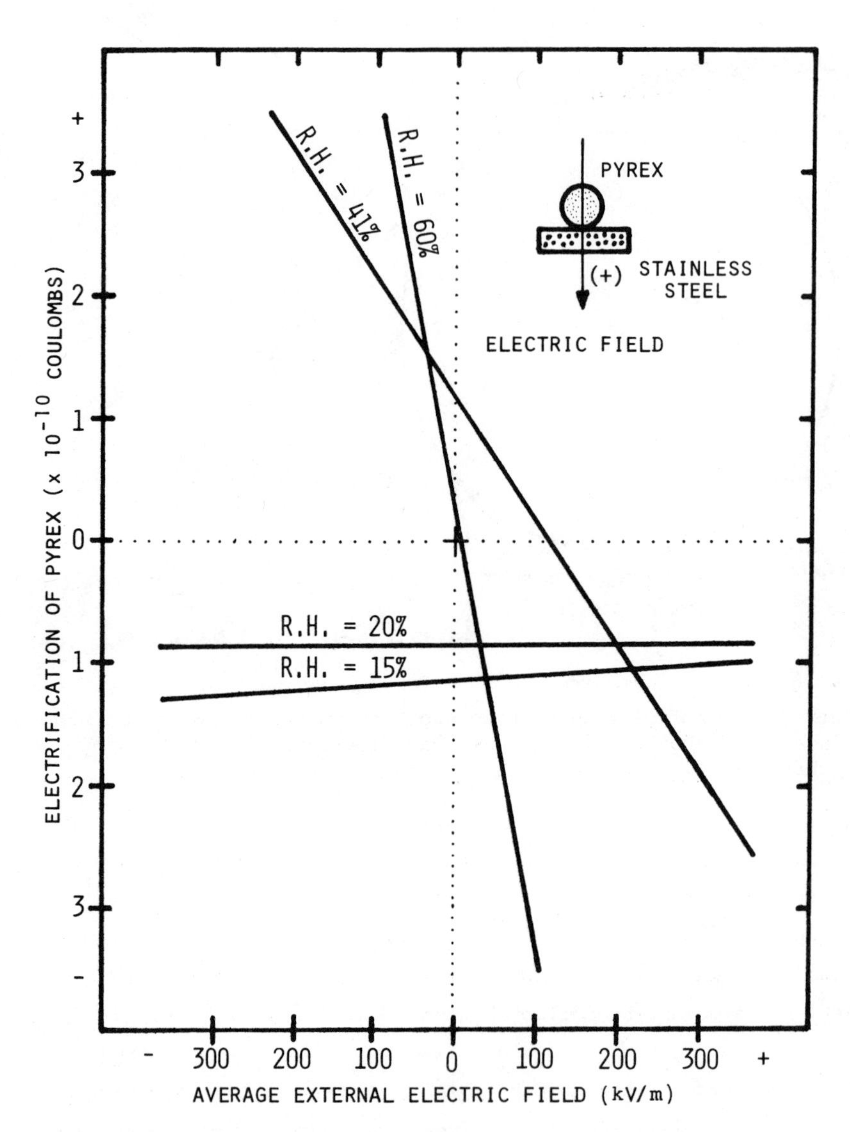

Figure 3.6 Relative humidity (RH) and external electric field influence on the electrification of Pyrex glass in contact with stainless steel. [From Ref. 15. (© 1973) *Tenth International Mineral Processing Congress*, Institution of Mining and Metallurgy. Reprinted with permission.]

These results suggest that by proper choice and control of operating temperatures, or mere fast surface temperature conditioning, one could effectively charge some particles by conductive induction while not charging others.

Of direct practical interest in the understanding of electrostatic beneficiation of ores are experiments carried out in air at various relative humidities [12]. Figure 3.6 shows the electrification of pyrex glass in contact with stainless steel in air at four relative humidities (RH). At low relative humidities, the triboelectric charging is effective and dominant over the electrification by conductive induction as shown by the independence of charge transferred on electric field. At relative humidities in excess of 60%, conductive induction dominates and charge transfer depends strongly on applied field.

Ionic and/or Electronic Bombardment from a Corona Generator in Combination with Conductive Discharges. Ionic or electronic bombardment from corona wires may be used effectively for electric charging of ore particles. To achieve a selective discharge, the ore falls over a rotating drum maintained at ground potential. The conductive particles lose their charge and fall into the first collection bin (see Fig. 3.7). The charge on the insulating particles remains and the particles stay attached to the drum held by the image charge until they are scraped off in the second collection bin.

B. The External Electric Field for Separation of Particles

Free Fall in an Electric Field. Once the particles are charged selectively, and assuming that they are allowed to fall freely under gravity forces in an applied horizontal electric field, they will acquire a horizontal acceleration component. The intensity and direction of the horizontal acceleration are determined by the polarity and magnitude of the charge, by the direction and magnitude of the electric field, and by the mass of the particle. Positively charged particles will move in the direction of the field, negatively

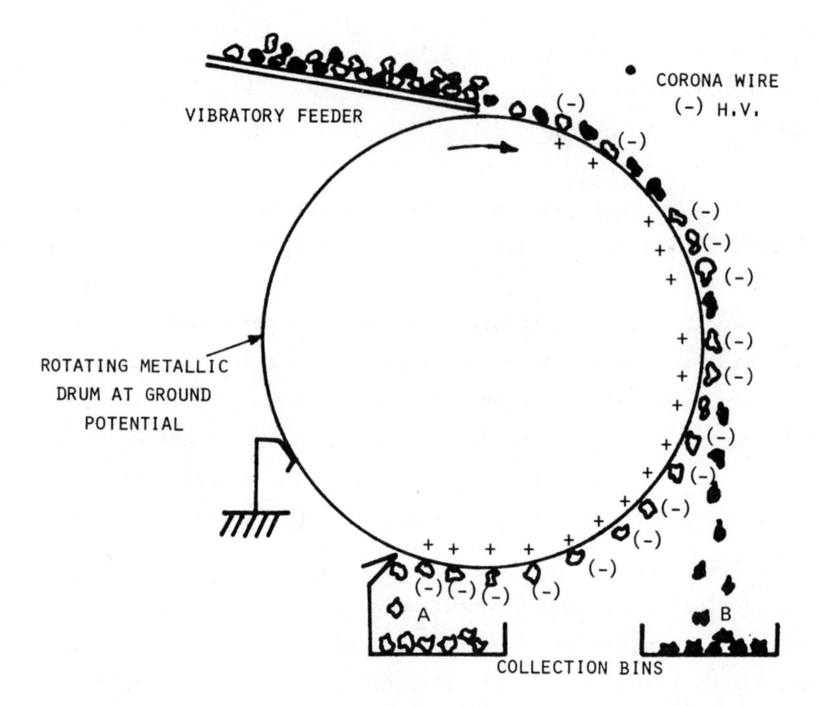

Figure 3.7 Beneficiation with ionic or electron bombardment in combination with conductive discharge. It is assumed that the B particles lose their (—) charge faster than the A particles.

charged particles will move in a direction opposite to that of the field. The initial horizontal acceleration of each particle will be proportional to its charge-to-mass ratio and to the intensity of the electric field.

From the above, it is obvious that when one uses the triboelectrification process, all positively charged particles will move in a direction opposite to that of the negatively charged particles regardless of their charge-to-mass ratio. Given sufficient time and space to spread, there will be a complete separation of the two types of particles. Hence, particles with different surface work functions may be very effectively separated.

If one chooses conductive induction to charge the particles, the separation is based primarily on the difference in the electrical conductivities of the materials at the time of processing.

With proper design and strategically located collection bins, both triboelectrically and conductive induction-charged particles may be also partly separated according to their size on account of the specific charge to mass ratios.

It is of interest to note that when charged particles are placed in a uniform electric field, the distortion of the field produced by the concentration of the field lines on the particles results in an enhancement of the resultant forces (Appendix 3.1 shows the calculation of such forces under ideal conditions).

In summary, once particles are charged by one of the three methods outlined above, the external electric field can selectively separate:

1. Positively from negatively charged particles
2. Charged from uncharged particles
3. Particles of various (charge/mass) values.

Fluidization in an Electric Field. Fluidization is a physical process which makes a powder behave like a fluid. The basic ideas can be best illustrated by an example.

Figure 3.8 shows a laboratory glass column sectioned into two parts by a horizontal partition called a *grid support plate*. The latter is, in general, in a laboratory situation, a porous glass or metallic plate. At the start of the experiment, a batch of powder (typically but not exclusively between 50 and 150 μm in particle size) is deposited on top of the porous plate. If gas is admitted very sparingly to the bottom chamber and through the porous plate, it diffuses upward through the powder finding its way in between the particles and out of the column. Because the velocity is low, the drag of the gas on the particles is not large enough to move them. This powder state is known as a *fixed bed* of powder. If the gas flow rate is gradually increased, a critical value is reached where the drag of the gas on the particles is equal to their weight. The superficial velocity at which this takes place is called the *minimum fluidization velocity*, V_{mf}. At V_{mf}, particles become suspended in the gas so that they do not support each other any more as was the case in a fixed bed. They are separated by a cushion of gas and do

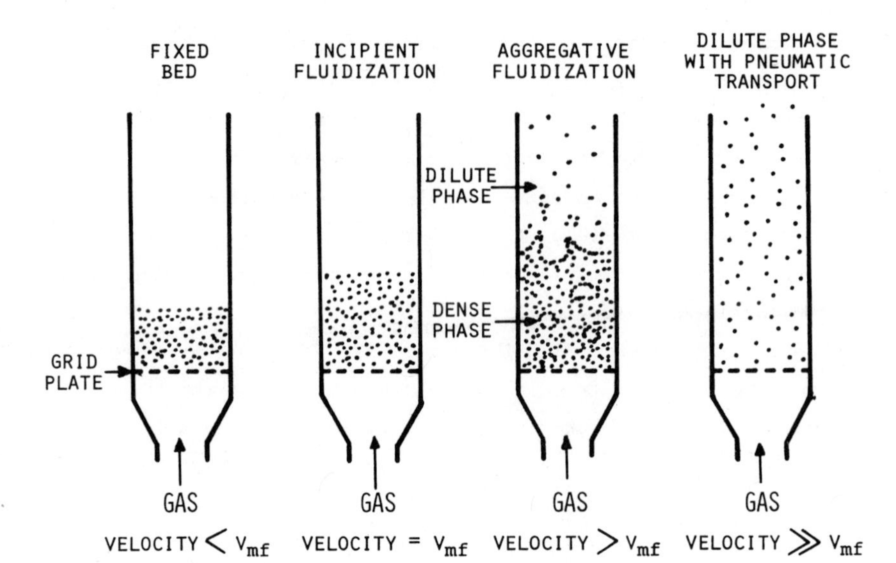

Figure 3.8 Various kinds of gas-solid fluidization.

not fly away from each other. Like minuscule "flying saucers" they fly close to each other, "wing to wing," colliding with each other, staying together for a fraction of a second, then flying apart again. During these collisions, if the temperature and humidity conditions are right the particles will become charged with opposite charges by triboelectrification. At V_{mf}, the pressure drop of the gas through the bed is equal to the weight of the bed per unit area of bed horizontal cross section. This state of the bed is known as *incipient fluidization*. The bed behaves like a liquid, although it is only an intimate mixture of gas and solids. This pseudoliquid has a horizontal free surface, can be poured from vessel to vessel and can be transported in vertical pipes. Mechanically speaking, the fluidized mass looks like and behaves like a quicksand, except that the mobilizing fluid is a gas instead of a liquid.

At gas velocities several times greater than the minimum fluidization velocity, the excess gas segregates into gas pockets or bubbles which ascend rapidly through the fluidized mass in the way vapor bubbles do in a boiling liquid. These bubbles churn the bed in a violent manner and are responsible for the excellent mixing and heat transfer properties of fluidized beds. This type of fluidization is called *aggregative fluidization*. At still much higher multiples of the minimum fluidization velocity, the bubbles disappear and the particles become widely dispersed in the flowing gas. Eventually all particles would be entrained out of the column. This state is called *dilute-phase fluidization*. Further information on fluidization can be found in textbooks [13,14].

Fluidization, although barely 50 years old, has found a multitude of applications in the petroleum and chemical industries. It is now invading the mining, metallurgical and power industries. In the work reported here, its unique properties have been married with the ones of electrostatics to carry out particle separations which could not be obtained as easily by other means. Fluidization is, indeed, an ideal means of "individualizing" and charging electrically fine particles while subjecting them to the influence of an electric field.

Under certain conditions the repeated collisions generate contact or triboelectrification. Positive and negative electric charges appear in equal total amounts, but of various magnitudes, on the particles, according to the specific surface properties of each particle and to the number of collisions it experiences.

An electric field traversing the bed will exert forces in the direction of the field for positively charged particles, and in a direction opposite to the field for those that are negatively charged. Normally, the electric field is vertical and is produced by a high voltage applied to the porous metallic plate, serving as the fluidizing grid, or to any conductive grid superimposed on the regular fluidizing plate. The collection electrode, which is generally connected to the ground potential, is located above the fluidization area. Individual particles are subjected to three forces: (1) the drag force of the fluidizing gas directed upward, (2) the gravity force, and (3) the electric force, which may be directed upward or downward, according to the polarity of the charge and the direction of the electric field.

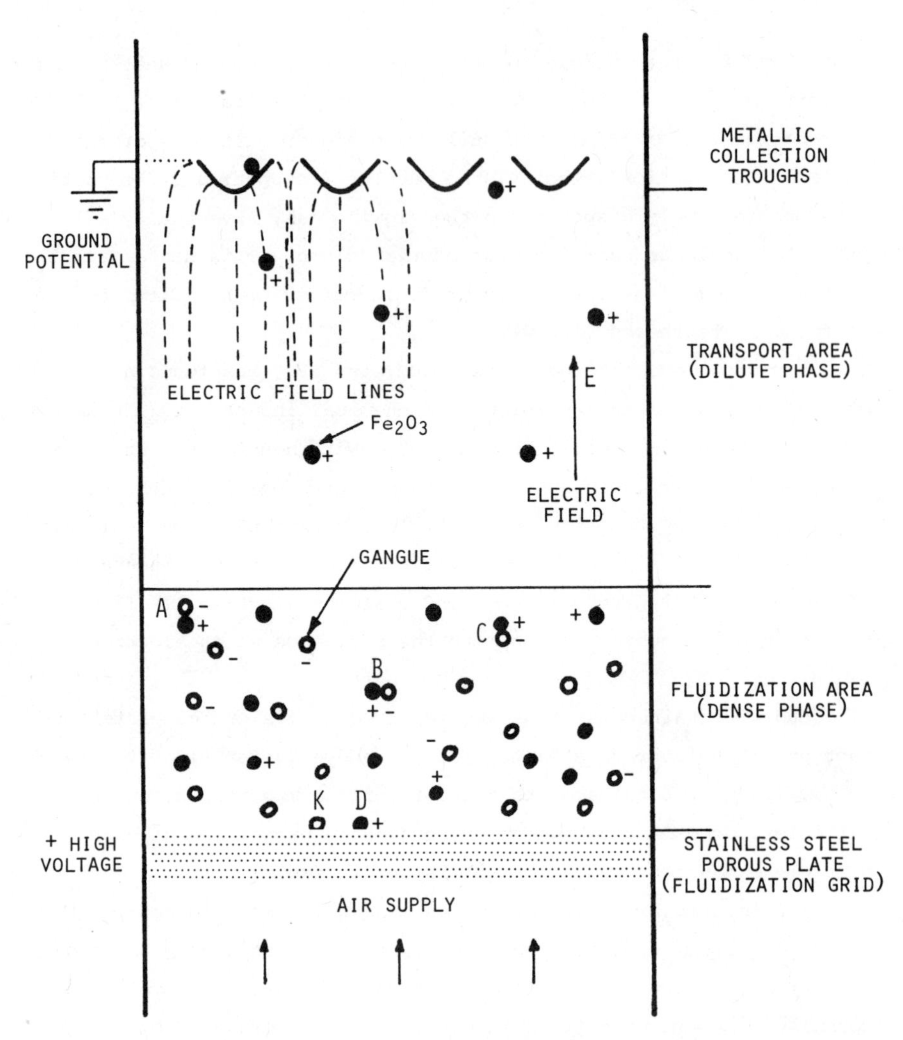

Figure 3.9 Electric charges developed in a fluidized-bed apparatus with troughs for the beneficiation of iron ore; positively charged magnetite particles are collected by the troughs. [From Ref. 15. (© 1973) *Tenth International Mineral Processing Congress*, Institution of Mining and Metallurgy. Reprinted with permission.]

The resultant of these three forces must be directed upward for the particles to be collected and downward for those intended to remain in the bed. Efforts to optimize the resultant forces consist primarily in the control of the electric forces. The latter are equal to the product of the local electric field and the charge accumulated on the particles.

As a specific example with an ore, Fig. 3.9 [15] shows the separation of the magnetite from gangue. Assuming that the materials and polarity are as shown, there are several different conditions, identified by points A, B, C, D, and K, under which such electric charges will develop. At A, the contact area between a particle of Fe_2O_3 (a semiconductor) and a particle of gangue (an insulator) is traversed by an electric field directed from the Fe_2O_3 to the gangue. At B, the electric field is tangential to the surface of contact. At C, the direction of the electric field relative to the Fe_2O_3 and gangue is reversed in comparison with A. At D, the Fe_2O_3 particle contacts the high-voltage porous plate, and at K, the gangue particle contacts the high-voltage porous plate. In all cases, an electric charging of the various particles occurs and the phenomenon is governed by tribo- and conductive induction electrification laws.

With the knowledge presently available of the predictability of triboelectrification, it is impossible to be specific as to the surface preparation or the ambient temperature and humidity which will give the optimum separation for a particular mineral ore, e.g., coal.

3.3 EQUIPMENT

The electrostatic beneficiation of coal is still in a developmental stage. Various types of equipment have been tried in laboratories around the world. In what follows, two types of equipment, considered by the authors as the most promising for application to coal, are described. These are the free-fall separation tower in combination with fluidized-bed triboelectrification for particles around 50 to 500 μm in diameter and the dilute-phase loop for particles smaller than 50 μm. A new experimental piece of equipment, a vibrated fluidized-bed electrostatic separator currently being developed, is described in Appendix 3.2. Although no experiment has been performed on coal, this type of equipment has been successfully used to triboelectrify and separate other ores, e.g., iron ore and brucitic ore.

Drum-type separators are not considered, at this time, as appropriate for separation of very fine particles. Operating experience with other fine ores has shown that particles in the micrometer range are difficult to confine on the drum surface on account of the hydrodynamic drag of the neighbouring air on the rotating drum and of electrostatic dispersion forces in case of corona charging.

The experimental equipment described below has been developed at The University of Western Ontario under the sponsorship of the Department of Energy, Mines, and Resources of Canada.

A. Separation Tower

The tower (Fig. 3.10) is of parallelepipedic shape (76 × 92 × 370 cm high) and is built from 0.63-cm-thick acrylic sheeting. The top section contains a fluidized-bed charging system, where coal is triboelectrified. The fluidized bed is made out of acrylic tubing and has an inside diameter of 14 cm. The inside walls of the bed are lined with 0.16-cm-thick copper sheeting because experiments show that copper is an ideal material for charging. The bed is fluidized by a porous plate grid made of 0.63-cm-thick P.S.S. grade F sintered stainless steel plate. The porous plate is covered with 400-mesh copper screening. In the center of the grid plate, a 0.32-cm hole feeds the charged coal particles through a copper tube and down to the separation area in between the electrodes. A valve is used to plug or unplug the hole when needed. The bed is always grounded.

The separation system consists of two parallel vertical electrodes (183-cm high and 44-cm wide), spaced 61-cm apart. The electrodes have collection baffles on the inside surfaces, the role of which is to collect particles sliding down the electrodes. The coal particles, during their free fall from the fluidized bed above, migrate towards the electrodes under the influence of the electric field. They are collected either on the baffles located on the electrodes or in 12 troughs or trays placed at the bottom of the tower.

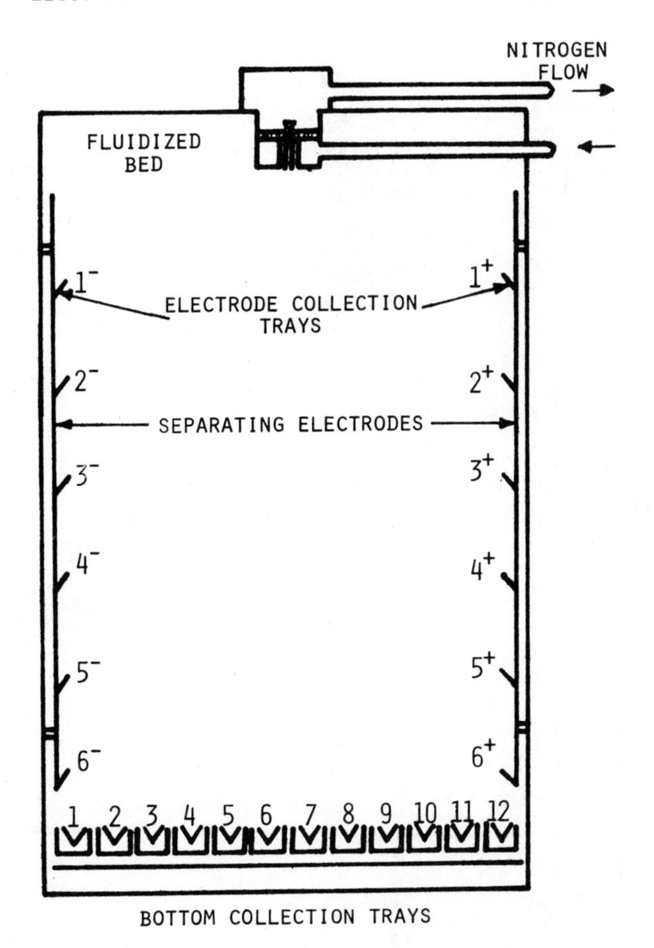

Figure 3.10 Electrostatic separation tower,

The troughs are 58 cm long and 5 cm wide (Fig. 3.11). They are made of aluminum, are electrically insulated, and can be connected to an electrometer to measure the charge of the material collected in the tray.

The tower is equipped with gas lines which are used to purge it and fluidize the charging bed with nitrogen.

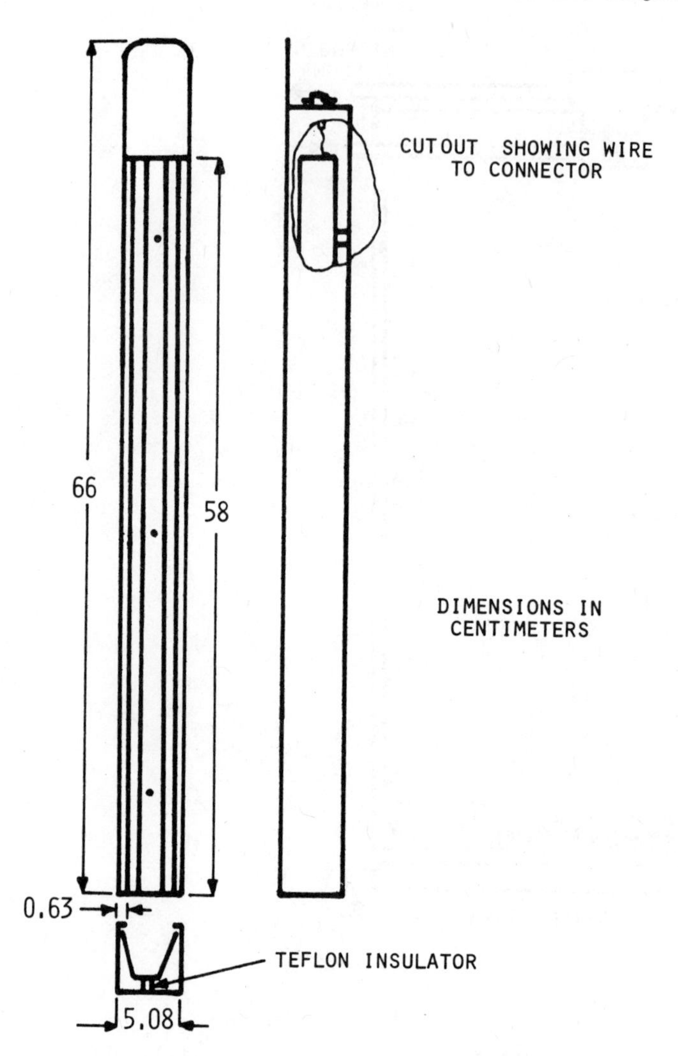

Figure 3.11 Collection troughs.

B. Dilute-Phase Electrostatic Loop

The function of the loop is to charge particles by triboelectrification while in suspension in a gas and to separate them (Fig. 3.12). As coal particles are transported around the loop, they collide with the walls and other particles and become charged.

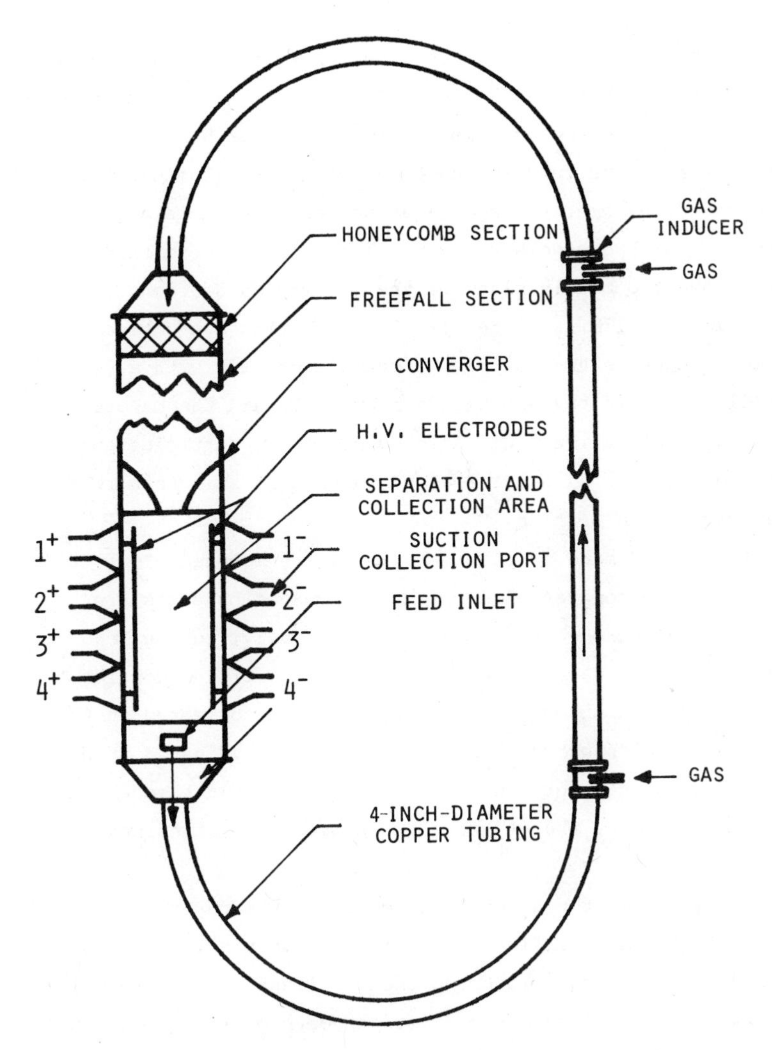

Figure 3.12 Dilute-phase electrostatic loop. [From Ref. 11. (© 1979) *IEEE*. Reprinted with permission.]

The loop is 8 m high. Propulsion of the coal is obtained from two gas inducers powered by bottled nitrogen, which blow gas along the walls of the loop pipe to impart upward momentum to the solid-gas suspension. The added nitrogen is removed continuously from the loop through the suction particle recovery ports. The pipe particle raceway on the right side of the loop is made of 10-cm-diameter copper tubing. On the left side of the loop, the charged coal falls down into a diffuser and then enters a parallelepipedic separation section (25 × 7.5 cm). At the top of this section, a copper honeycomb helps to charge the particles further and to create a quasilaminar flow of gas. From the honeycomb, the solid-gas suspension enters a copper flow converger which channels the particles into a ribbon thin falling curtain down the middle plane of the separation section. The converging element eliminates the problem of having a negative ash particle going towards the positive electrode collide with a positive coal particle going in the opposite direction toward the negative electrode. This situation would also occur if the particles entered the separation section as a large cloud instead of as a thin ribbon along the axis of the section.

The electrodes, made of aluminum plate, are mounted along the narrow sides of the separation section. Four suction ports are drilled in each electrode and are connected by vinyl tubing to filter thimbles and a vacuum source. Particles drifting towards the electrodes are sucked through these ports and are collected in the thimbles. The suction pressure at the thimbles is adjusted such that the gas introduced into the inducers is evacuated through the thimbles. In this way, the pressure in the loop is maintained constant.

The wide sides of the separation section are made of thin plastic sheeting. Thinness here is important to minimize the leakage of the electric field through the sides and the resultant nonuniformity of the field which would disturb the drift pathways of the particles.

3.4 PROCEDURE

A representative sample of about 500 g was obtained by quartering the large sample received from the mine. It was then dried in a vacuum oven at 50°C for 12 hr to remove moisture and prevent oxidation. The coal was simultaneously crushed and sieved under a nitrogen atmosphere to a grain size of less than 180 μm in a specially designed grinder [16]. The grinder produced coal of relatively uniform particle size, free of fines and with clean, sharp, oxidation-free fracture surfaces. After grinding and sieving, the sample was further subdivided into 100-g lots which were stored under vacuum until needed.

A. Tower Operation

After installing the troughs at the bottom of the tower, the coal sample was placed in the charging bed in the upper part of the tower. Dry, oxygen-free nitrogen was then admitted to the tower to purge it and to the charging bed to fluidize it. After 20 min of fluidization, the nitrogen to the tower was turned off, the electrodes were energized, and the bed valve opened. A small amount of fluidization gas was necessary at times to facilitate the discharging of the coal from the bed through the dump hole. A vibrator was used on the bed to further ease the movement of solids out of the bed. The mass and electric charge of the separated coal fractions were determined.

B. Loop Operation

The loop was first purged with high-purity nitrogen for ½ hr. Nitrogen flow into the loop was then increased to establish the desired operating conditions while, at the same time, adjusting the vacuum suction through the filter thimble particle recovery system to maintain constant pressure in the loop. The electrodes were energized and the vibrator started to shake dust loose from the loop walls. The prepared coal sample was fed in slowly (10 g at a time), while the separated particles were collected in the thimbles. After recovery of the separated particles from the thimbles, the samples were weighed and stored for analysis.

3.5 SELECTED LABORATORY RESULTS OF ELECTROSTATIC BENEFICIATION OF COALS

Electrostatic beneficiation experiments have been carried out on the following coals:

1. Lingan coal mined in Eastern Canada
2. Conemaugh anthracite coal from Pennsylvania
3. Luscar and Hat Creek coals mined in Western Canada

Table 3.1 Compositions of Feed Coals

Coal type	Constituents	Weight %
Lingan	Moisture	4.2
	Ash	4.3
	Volatile matter	35.0
	Fixed carbon	56.5
	Sulfur	1.9
Conemaugh	Sulfur	2.5
	Ash	8.0
Luscar	Ash	7.5
Hat Creek	Moisture	23.0-25.0
	Ash	36.0-39.0
	Sulfur	0.4-0.8
	HHV	3800-4500 Btu/lb (wet basis)

Analyses of the important coal constituents from the standpoint of the separation objectives are given in Table 3.1. For the Lingan coal, separation of sulfur, ash, and coal macerals was studied. For the Pennsylvania coal, only pyritic (inorganic) sulfur separation was considered; while for the Western Canadian coals, the main consideration was the separation of ash.

A. Lingan Coal Separation

Pyrite. The results of separation of Lingan coal in the electrostatic tower, with respect to sulfur concentration in the fractions collected in the bottom and electrode trays described in Fig. 3.10, are given in Fig. 3.13. Particles rich in pyritic sulfur are attracted to the positive electrode while those with low sulfur are attracted to the negative electrode. The mass of coal deposited in each tray is shown in Fig. 3.14. Note that the sulfur-rich and sulfur-poor fractions collected at the positive and negative electrodes account for only 7.5 and 21%, respectively, by weight of the

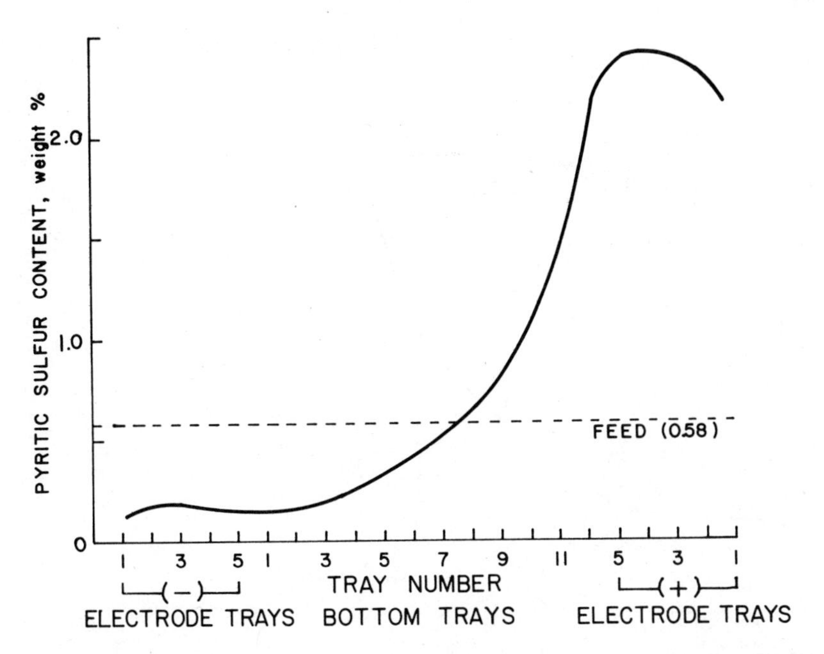

Figure 3.13 Pyritic sulfur distribution in the 22 trays of the electrostatic tower.

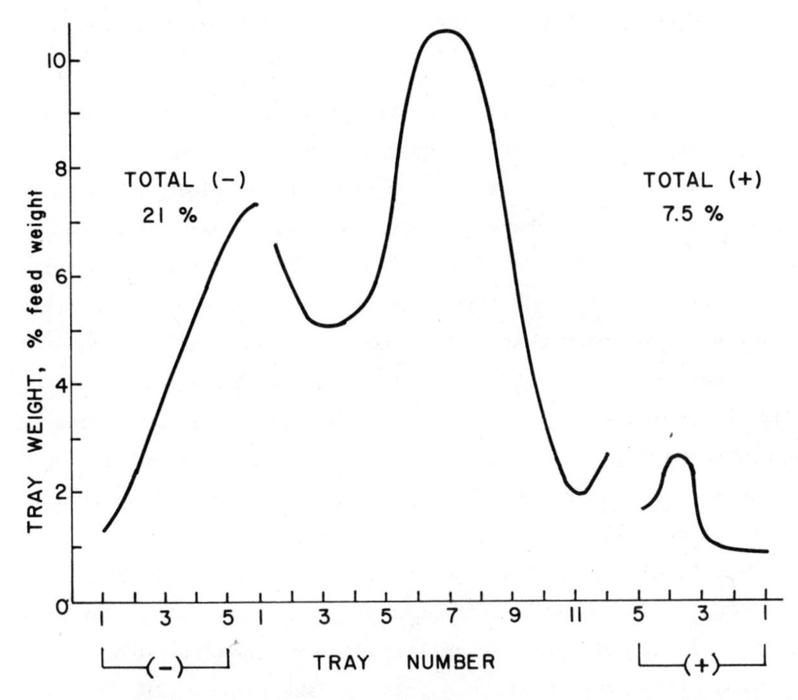

Figure 3.14 Mass distribution of the extract in the 22 trays of the electrostatic tower.

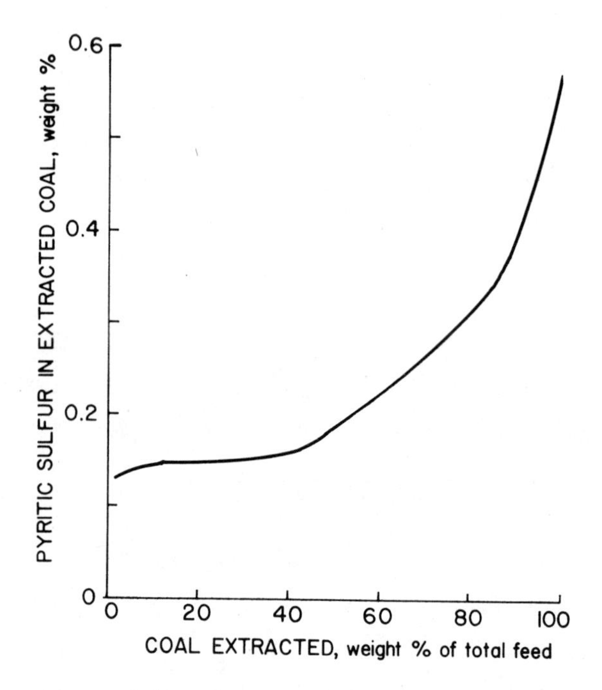

Figure 3.15 Lingan coal beneficiation in the electrostatic tower.

original feed. A large proportion of the original coal falls in the middle trays, however, separation according to sulfur content is evident here also. The results suggest that higher field strengths, or perhaps more effective charging, could improve separation. On the basis of the data of Figs. 3.13 and 3.14, the beneficiation curve of Fig. 3.15 was constructed. Even for the separation achieved, the beneficiation curve shows that at 80% coal recovery (and presumably ~ 80% Btu recovery), the sulfur content is reduced by a factor of approximately two on a once-through process. Presumably by better charging, stronger fields and a recycling of selected fractions through the separation process, better recoveries and/or lower sulfur contents could be achieved.

The results of separation experiments in the electrostatic tower and loop are presented in the bar graph in Fig. 3.16. Seventy-nine percent of the material was collected at the negative electrode, the remaining 21% of the positive electrode (see Fig. 3.12). Taking

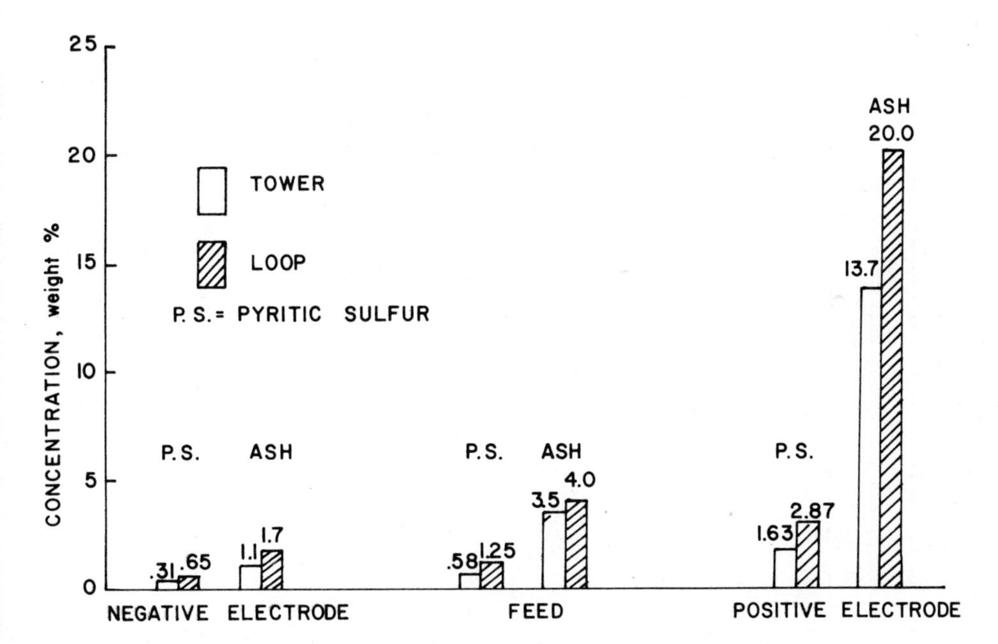

Figure 3.16 Comparison of results from the electrostatic tower and dilute phase electrostatic loop by x-ray analyses of the collected coal fractions.

into account the differences in feed sulfur concentration due to variations in the coal from batch to batch, the sulfur separations for the tower and loop are quite comparable. For the comparison, the tower sulfur concentrations are derived from the beneficiation curve (Fig. 3.15) using the 79% point as the split between the positive and negative fractions.

Ash and Macerals. Figure 3.16 also shows the separation of the ash in the tower and loop. The ash separation is somewhat better than the sulfur with a concentration factor in the ash-rich portion of about 5 rather than 2 to 3 in the case of sulfur. The other important observation is that the ash is concentrated in the same fractions as the sulfur is concentrated, suggesting that removal of both sulfur and ash simultaneously is feasible.

A 1000-point petrographic analysis on three fractions of the coal separated in the tower gave the results shown in Fig. 3.17.

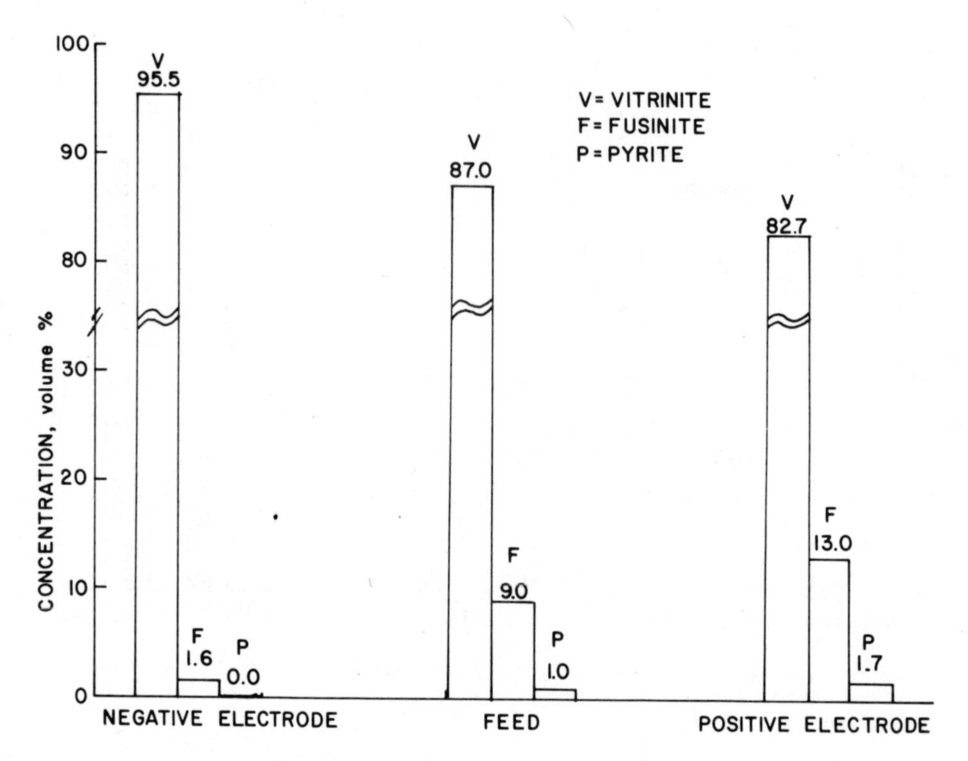

Figure 3.17 Petrographic analysis (1000 points): coal fractions separated in the electrostatic tower.

Separation of the coal macerals is observed with the vitrinite tending to be positively charged and the fusinite and semifusinite tending to be negatively charged. Some separation of the minor constituents was also observed. It may be possible by the recycling of separated fractions through the process to produce small quantities of almost pure coal macerals (especially vitrinite), or to use the separation to produce blends with specific maceral contents for coking purposes.

B. Conemaugh Anthracite Coal

This coal was studied primarily to evaluate electrostatic processes for sulfur removal. The coal was separated in the electrostatic loop in nitrogen at room temperature using the various potentials

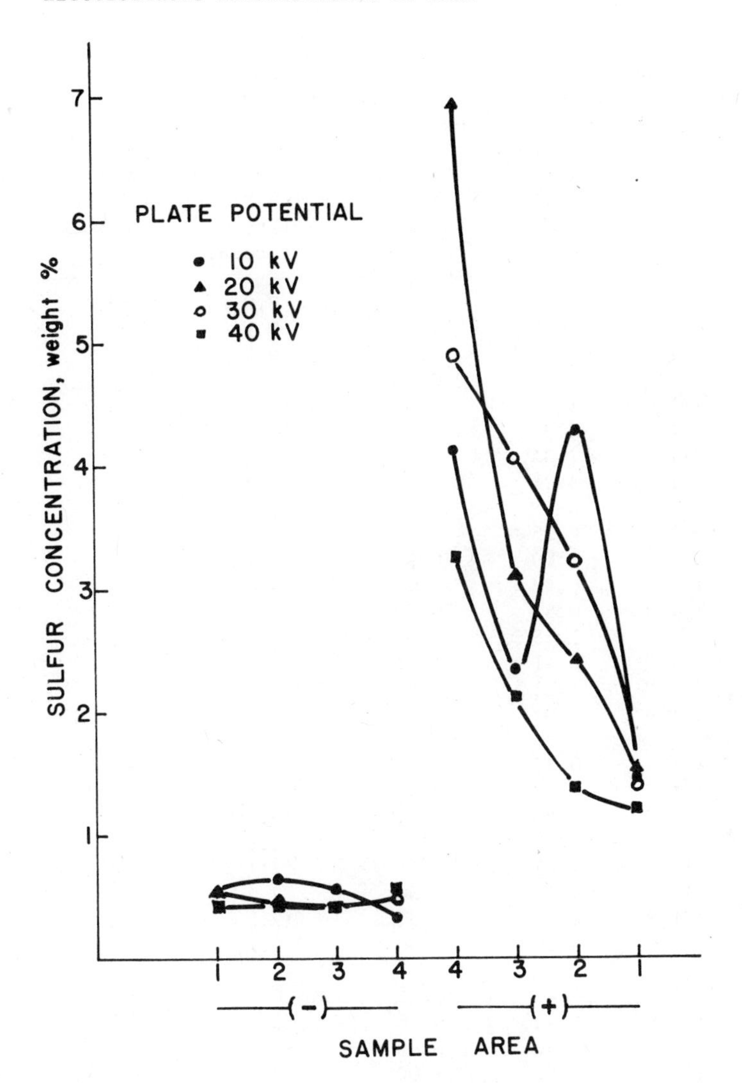

Figure 3.18 Pyritic sulfur distribution on plates of the electrostatic loop.

on the high-voltage electrodes (Fig. 3.12) for extraction of the particles from the loop. Results are shown in Fig. 3.18. The sample areas referred to in this figure correspond to the suction collection ports of Fig. 3.12. Table 3.2 lists the pyritic sulfur, total sulfur, and weight collected at the individual sample ports

Table 3.2 Sulfur and Weight Distribution of Product Separated at Various Locations on the Plates of the Electrostatic Loop

Plate potential, kV	Sample location	Concentration pyritic sulfur, wt %	Concentration total sulfur, wt %	Weight per plate, % of total	Weight per electrode, %
10	(−) 1	0.55	1.00	8.8	57.3
	(−) 2	0.65	1.10	4.1	
	(−) 3	0.55	1.05	9.7	
	(−) 4	0.35	0.80	34.7	
	(+) 1	1.30	1.65	10.0	42.7
	(+) 2	4.32	4.62	12.0	
	(+) 3	2.34	2.64	10.6	
	(+) 4	4.13	4.43	10.1	
20	(−) 1	0.55	1.00	8.5	73.4
	(−) 2	0.50	0.98	6.1	
	(−) 3	0.42	0.95	21.2	
	(−) 4	0.53	0.99	37.6	
	(+) 1	1.55	1.91	6.6	26.7
	(+) 2	2.41	2.71	5.9	
	(+) 3	3.11	3.41	6.6	
	(+) 4	6.94	7.24	7.6	
30	(−) 1	0.42	0.95	7.7	66.0
	(−) 2	0.47	0.96	6.3	
	(−) 3	0.40	0.87	19.2	
	(−) 4	0.53	0.99	32.8	
	(+) 1	1.41	1.80	7.1	34.0
	(+) 2	3.24	3.54	7.6	
	(+) 3	4.07	4.37	12.2	
	(+) 4	4.90	5.20	7.1	
40	(−) 1	0.41	0.89	8.1	59.8
	(−) 2	0.42	0.95	5.3	
	(−) 3	0.40	0.86	23.7	
	(−) 4	0.55	1.00	22.7	
	(+) 1	1.22	1.60	10.3	40.2
	(+) 2	1.40	1.76	8.4	
	(+) 3	2.15	2.45	14.1	
	(+) 4	3.30	3.60	7.4	

as a percentage of the feed. At the positive electrode, a marked increase can be seen in the sulfur concentration from the upper to lower collection ports, indicating a smaller charge to mass ratio for the negatively charged high sulfur particles probably due to a

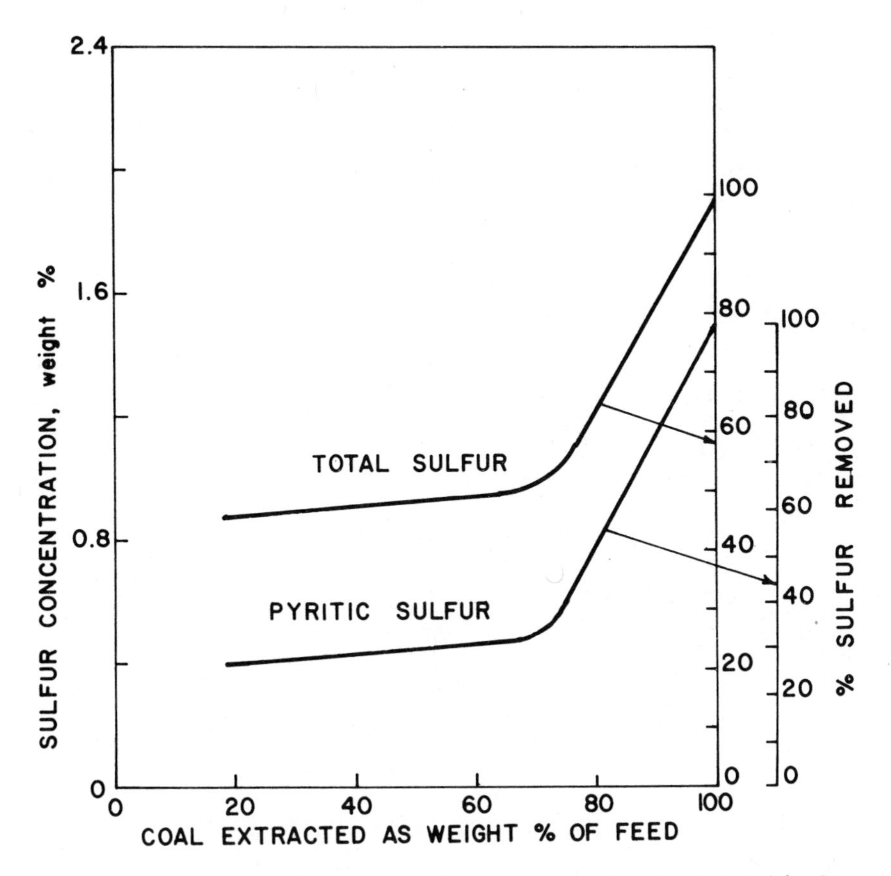

Figure 3.19 Conemaugh coal beneficiation in the electrostatic loop.

high density. The other trend is that the highest sulfur concentration in the extracted particles appears to occur at 20 to 30 kV. On the negative side, the majority of the mass is deposited at the lowest port in all cases and perhaps there is a tendency toward a decrease in sulfur concentration with increasing voltage. The 30-kV data have been converted to a beneficiation curve for both pyritic and total sulfur (Fig. 3.19). The lower linear portion of the curve is for the coal extracted at the negative electrode with the steeper linear portion the result of adding in the material from the positive electrode. The feed material contained 1.3 wt % pyritic sulfur and 1.7 wt % total sulfur. The small differences between the feed values and the 100% values on the beneficiation curve represents a loss of coal of about 10 to 15% by weight due to incomplete extraction during the experiment. This material left in the loop has a low sulfur content. For this single-pass separation, a reduction of better than 70% in pyritic sulfur content can be achieved with a recovery of 70% of the processed coal.

C. Luscar and Hat Creek Coals

The main problem with the Western Canadian coals from the Luscar and Hat Creek deposits is a high ash content which must be reduced to permit more effective utilization of these coals. The results of a separation experiment on Luscar coal in the electrostatic tower are shown in the form of a beneficiation curve in Fig. 3.20. The same characteristic of a linear portion of low slope related to coal extracted on the negative electrode with low ash content and the high-slope region related to the coal extracted at the positive electrode with high ash content. Roughly 60% of coal is extracted with a significantly reduced ash content on the basis of this single-pass experiment.

To illustrate the benefits of a recycle process and to show that beneficiation can be improved by recycling fractions of the separated material, the separation summarized by the beneficiation curve for Hat Creek coal is shown in Fig. 3.21. Hat Creek is an enormous deposit of coal in British Columbia, Canada, which unfortunately has an extremely high ash content of approximately 40 to 50 wt %. Much of this is in the form of fine clay particles which present major water pollution problems if the coals are cleaned by conventional washing techniques. A further undesirable attribute of the clays are large quantities of water of crystallization which soak up energy during combustion reducing the Btu value for the coal. Removal of ash therefore improves the energy value of the coal not only because of the increased hydrocarbon concentration but also because of the reduced energy requirement for breakdown of the clay minerals.

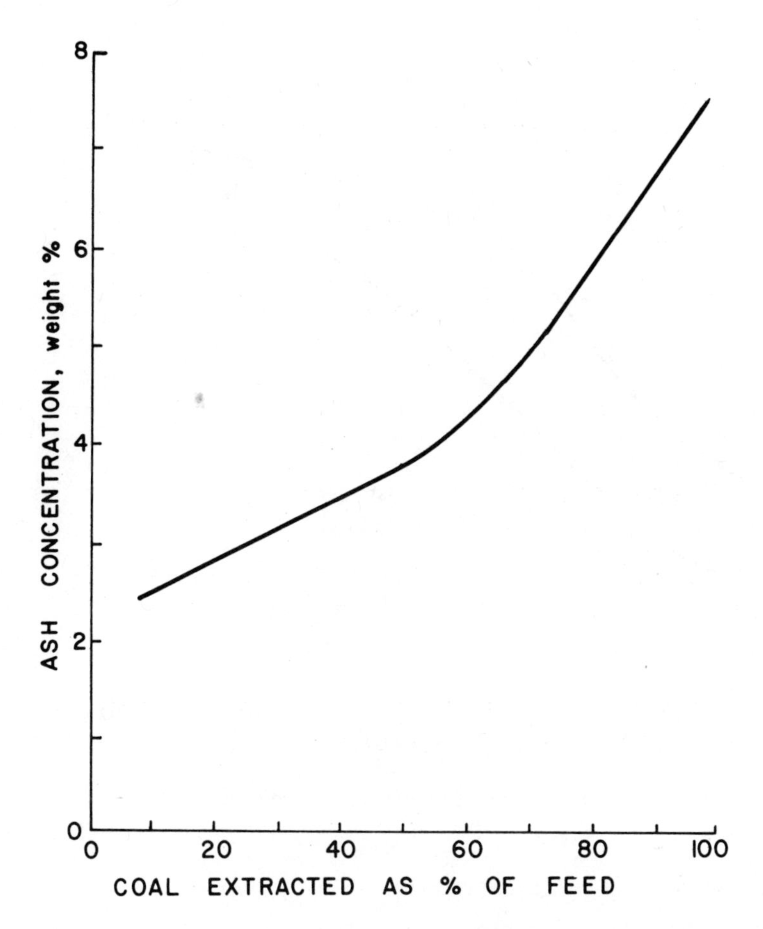

Figure 3.20 Ash content of Luscar coal versus percentage coal extraction in the electrostatic separation tower.

The curve labeled first cycle (Fig. 3.21) is a beneficiation curve calculated from a separation of the original coal in the electrostatic tower. The portions at the positive electrode, the negative electrode, and the middle trays from this separation were combined into low-ash, medium-ash, and high-ash fractions. The low-ash and high-ash fractions were separated a second time and the results combined with the medium-ash portion from the first cycle

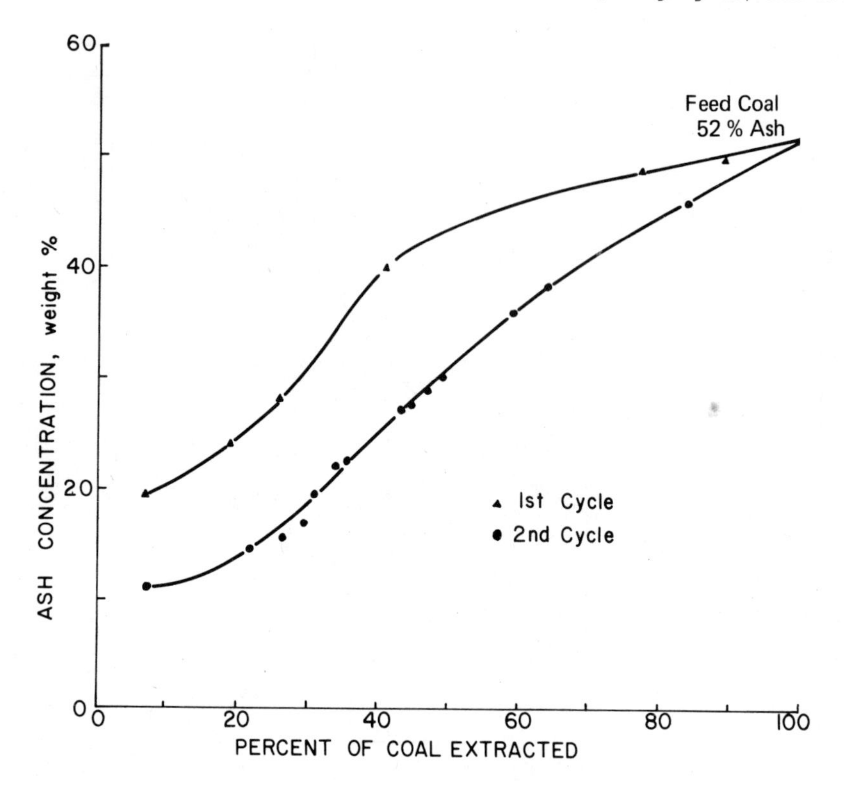

Figure 3.21 Beneficiation curve for Hat Creek coal.

to obtain the second cycle beneficiation curve of Fig. 3.21. A significantly improved separation is obtained.

3.6 CONCLUSIONS

Electrostatic methods applied to coal in a fluidized state are the newest in the field of separation. As such, they require considerable further development before they can be used on a commercial basis. At present, work is under way at The University of Western Ontario to demonstrate these techniques on a pilot-plant scale involving multiple recycling of fractions. The field of applications of these methods is extensive.

In power generation, there has been increased emphasis on using coals even of lower grade. Environmental considerations make it mandatory to reduce the SO_2 emissions.

For combustion, coals are generally crushed to –200 mesh, the size at which 80% of the pyrite is generally liberated. Classical wet beneficiation techniques, i.e., flotation, are expensive for fine coals of low quality, require drying and lead to almost intractable water pollution problems when clays are present. Moreover, water is simply not available in many promising mining areas such as deserts or the Arctic. As a consequence, a dry beneficiation process is highly favored over a wet one.

As envisaged now, coal will be shipped from the mine in a coarse state to avoid the problem of handling the coal fines during transportation. At the power station the coal will be crushed, dried with flue gas, and beneficiated by dry electrostatic techniques. The work presented here shows the potential of these techniques for pyrite and ash removal before combustion. They could also be useful to beneficiate coal before gasification, pyrolysis, and liquefaction.

Another area of application for the electrostatic beneficiation process is the eventual separation of the individual macerals. The manufacturers of metallurgical coke blend different coals to obtain the desired coke strength. Increasing the vitrinite content of coal and reducing the fusinite would improve the coking properties. Being able to produce metallurgical coke from one mine eliminates transportation costs and reduces the dependence of the metallurgical industry on sources of increasingly scarce and costly coking coals.

APPENDIX 3.1: FORCES ON CHARGED PARTICLES IN AN ELECTRIC FIELD*

A. Forces on Conductive Spherical Particles of Radius r Placed into a Uniform Electric Field E_o in Air at Standard Pressure and Temperature

Referring to Fig. 3.22:

1. The electric field is distorted as shown. On the surface of the particles, the field is perpendicular to it and varies sinusoidally, reaching a maximum of three times the value of E_o along the z axis.

---

*Extracted from Ref. 17 by permission of the *Journal of Electrostatics*. Notations used are explained in Sec. E of this appendix.

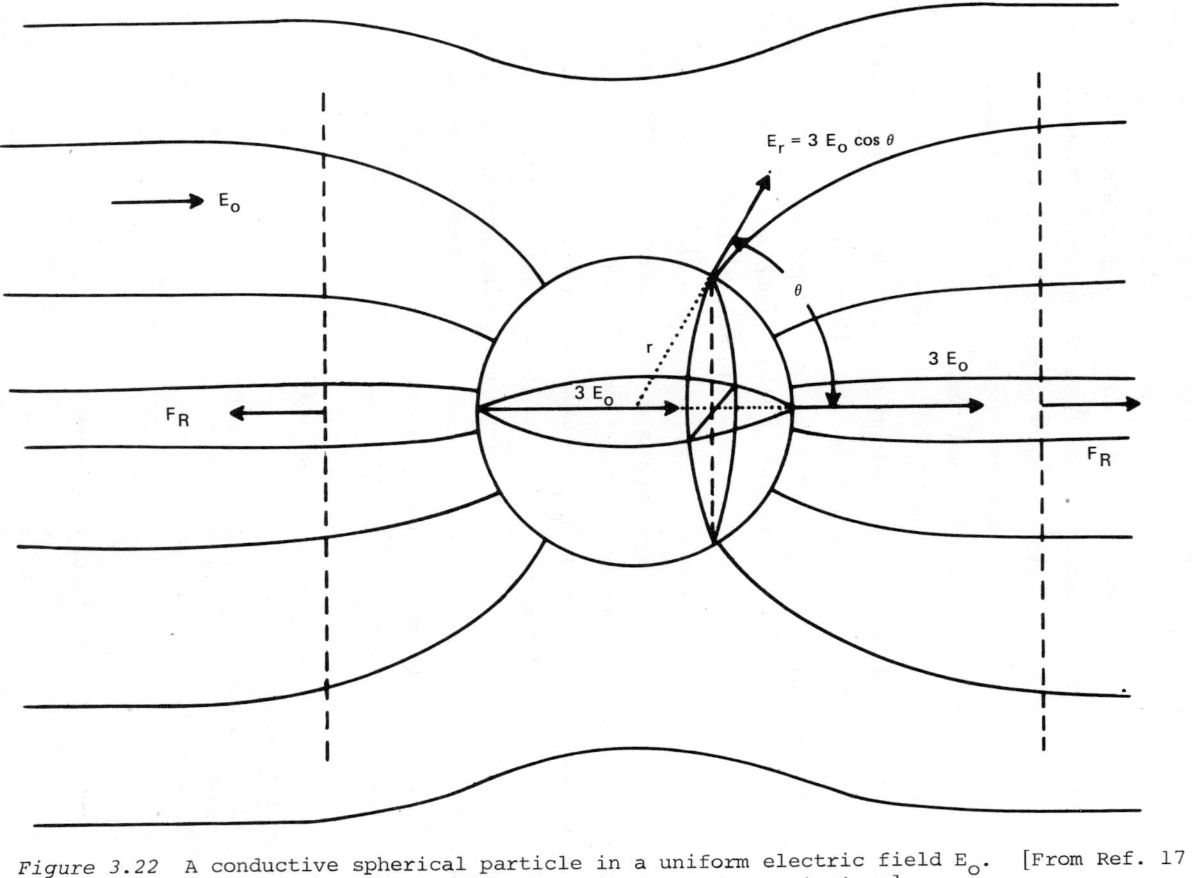

Figure 3.22 A conductive spherical particle in a uniform electric field E_o. [From Ref. 17 © (1977/1978) *Journal of Electrostatics*. Reprinted with permission.]

2. Two equal and opposite resultant electrostatic forces F_R appear which tend to pull the particle apart.
3. If the particle is a liquid, it will change its shape becoming elongated, and if the surface tension is exceeded by the electrostatic forces, the particle will break into two or more particles--some positively, and an equal number negatively, charged.

B. Maximum Force on a Conductive Spherical Particle of Radius r Charged by Ion Bombardment to Saturation and Placed in a Uniform Electric Field E_o

Referring to Fig. 3.23, where it is assumed that the saturation charge is represented by the situation in which no field lines of E_o land on the charged particle and that the resultant electric field intensity at the upstream point $E_{P_u} = 0$:

1. Under these conditions, the resultant electric field intensity on the surface of the particle can be described by the sinusoidal expression $E_r = 3E_o$ $(1 + \cos \theta)$ and has a maximum value of $6E_o$.
2. The total force on the particle = $24\pi\varepsilon_o r^2 E_o^2$

C. Size of Conductive* Spheres of Radius r Which Electrostatic Forces Can Lift under Ideal Conditions Using a Value of 3 MV/m† as the Dielectric Strength of the Air

Equating the electrical force with the gravitational force, one finds the maximum radius for a spherical particle of iron to be $r_{max} \approx 1$ mm, and for a spherical particle of water, $r_{max} \approx 8$ mm.

---

*For a dielectric material, a correction factor is introduced in the calculation of the saturation charge. Neglecting the tangential components of the electric field at the surface of a dielectric sphere placed in a uniform electric field, the radial component may be approximated by $E_r = [2(\varepsilon_r - 1)/(\varepsilon_r + 2) + 1](E_o \cos \theta + 1)$. Assuming $\varepsilon_r = 4$, $E_r = 2E_o(\cos \theta + 1)$. Hence, at the maximum force conditions, the value of the uniform electric field into which the charged sphere may be placed is greater than for the conductive sphere case.

†One must realize that depending on the radius of the particle, this value may be ten or more times greater.

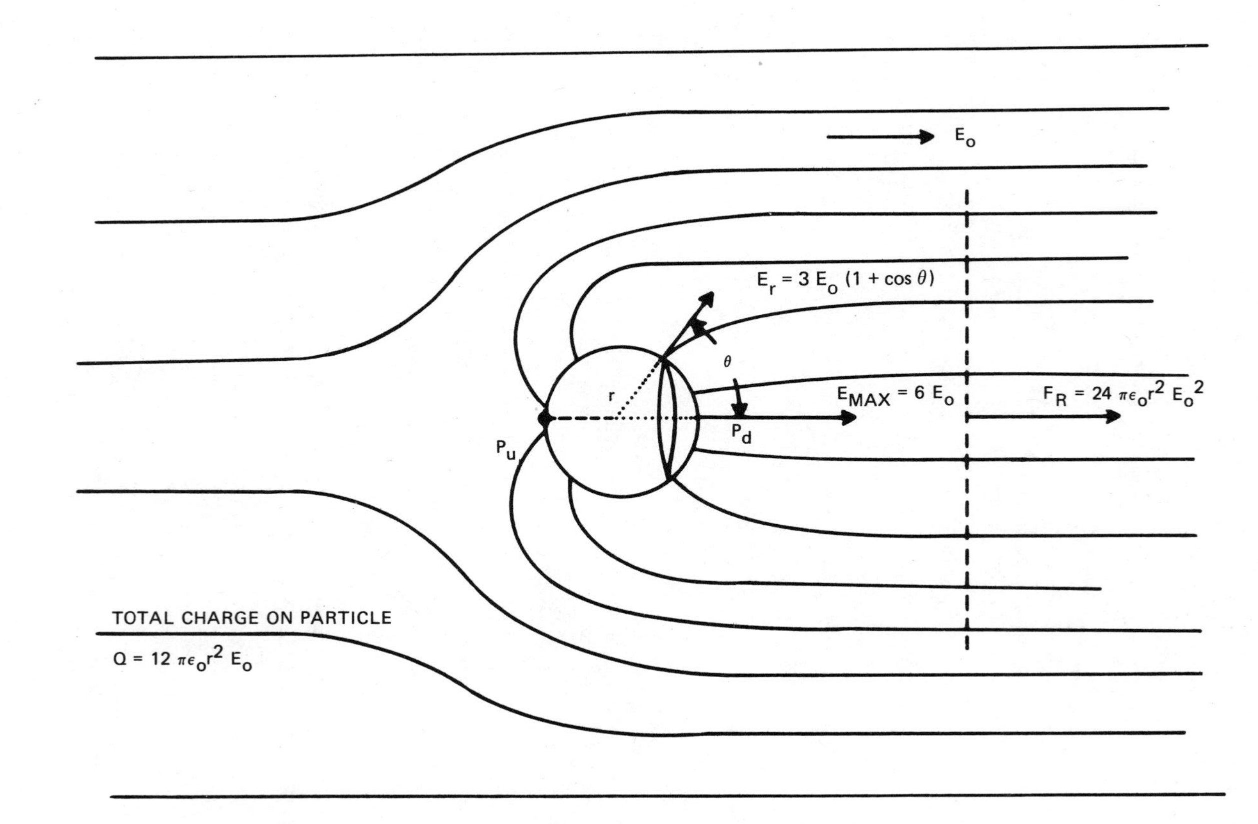

Figure 3.23 A "fully" charged conductive spherical particle in a uniform electric field. [From Ref. 17. © (1977/1978) *Journal of Electrostatics*. Reprinted with permission.]

At the other end of the scale, very small particles such as a 10-μm radius conductive droplet of water may be lifted by electrostatic forces which are 400 times greater than that of gravity.

Correspondingly, a 10-μm radius spherical particle made out of steel may be lifted by means of an electrostatic force which is 50 times greater than that of the force of gravity.

D. Dielectrophoretic Forces (Electrostatic Forces on Particles in Nonuniform Electric Fields)

These arise from the polarization of the matter and the particles move toward the region of increasing field strength if the permittivity is greater than that of the medium. Contrary to the forces on charged matter, their direction is independent of the direction of the electric field.

In an idealized case of a very small neutral body which is linearly, homogeneously and isotropically polarizable, in a conservative static field at equilibrium, the force $F = (p.\nabla)\ E_o$, where p is the dipole vector.

E. Notation

- E_o Electric field in a distortion-free region
- E_r Electric field at the surface of a particle of radius r (depends on θ)
- E_{P_u} Electric field at a point on the circumference of a particle pointing in the opposite direction to the electric field (upstream point)
- F_R Electrostatic forces on a particle in an electric field which tend to pull them apart
- r Radius of the particle
- ε_o Permittivity of free space
- ε_r Dielectric constant of the material of the particle

APPENDIX 3.2: THE VIBRATED FLUIDIZED-BED ELECTROSTATIC SEPARATOR

Although no work was done on coal with the system described below, it is felt that it should be mentioned briefly because it is potentially useful for coal separation. Many very fine powders (30 μm

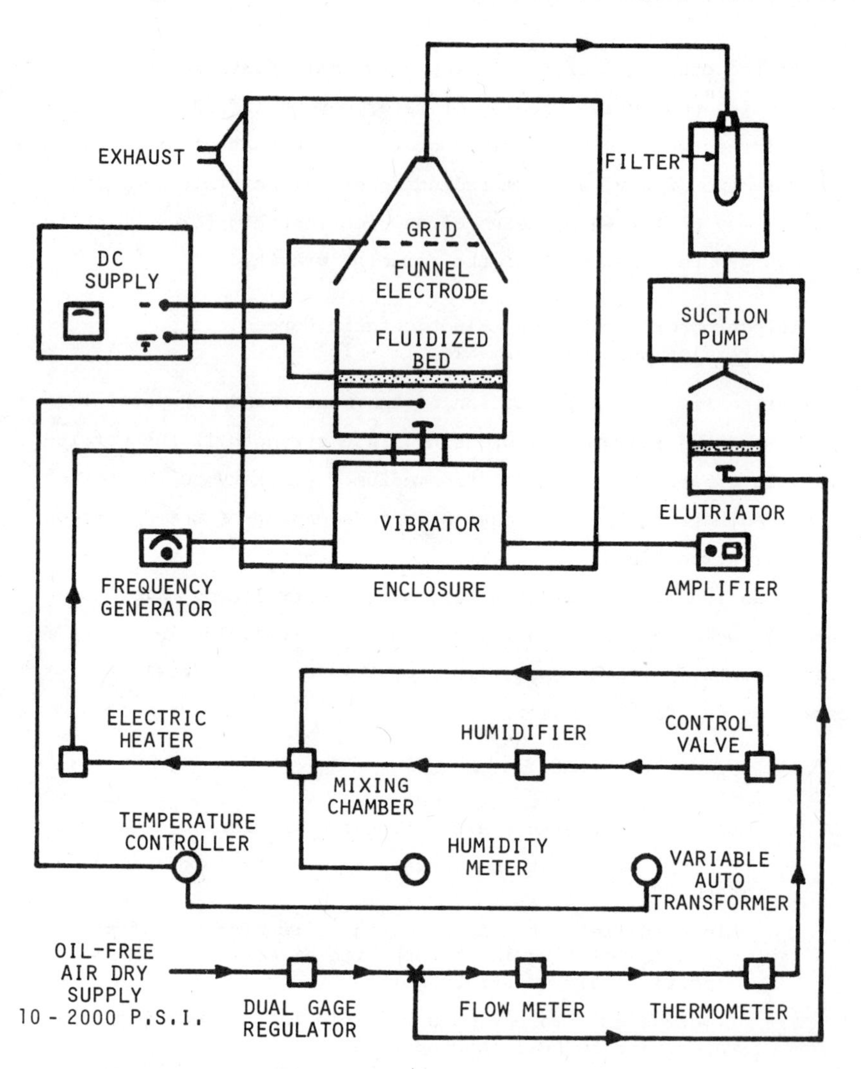

Figure 3.24 A schematic diagram of a vibratory fluidized-bed electrostatic separator. [From Ref. 18. © (1978) *IEEE*. Reprinted with permission.]

average size and under) cannot be successfully fluidized because of Van der Waals and other interparticle forces. Instead of aerating themselves, they ball up and give very bad bed-channeling and extremely poor fluidization. Such particles can be effectively fluidized through a combination of gas at low flow rates and mechanical vibration of selected frequency and amplitude. Alternately, stirring could be used to break up interparticle forces.

To test the potential of vibrated fluidized beds for electrostatic separation, experiments were carried out on a synthetic mixture of iron and glass powder (50:50 by weight) of particle size ranging from 15 to 45 μm.

Figure 3.24 illustrates the experimental setup. A cylindrical fluidized bed is mounted on an electromechanical vibrator. A funnel-shaped suction collection electrode is suspended above the fluidized bed and is equipped with a high-voltage grid inside the funnel. The fluidized bed, 10 cm in diameter, is made of plastic material and has a stainless steel porous plate gas distributor. The funnel-shaped electrode is connected to a vacuum system and the grid inside the funnel to a dc high-voltage power supply. The porous plate distributor is linked to the other electrode of the power supply. The particles extracted by the electric field were entrained through the funnel electrode into the vacuum system where they were trapped in a cellulose filter.

Table 3.3 Testing Parameters

Parameter	Setting
Voltage	200 ± 2, 230 ± 2, 260 ± 2 kV/m
Temperature	20 ± 1, 50 ± 1, 80 ± 1, 110 ± 1 °C
Absolute humidity	0 to 0.1, 15 ± 1, 30 ± 1 $\frac{\text{gr moisture}}{\text{lb dry air}}$[a]
Vibration frequency	≃ 300 Hz
Air velocity	2.10 ± 0.05 cm/sec (≃ 10 l/min)
Conditioning time	6 min
Separation time	4 min
Air gap	2 ± 0.1 cm
Electrode separation	10 ± 0.1 cm (plate to grid)
Polarity of field	Plate grounded, grid negative

[a]One grain moisture/lb dry air = 1/7000 kg moisture/kg dry air.

Source: Ref. 18. © (1978) *IEEE*. Reprinted with permission.

Table 3.4 Factorial Experiment Results

		Humidity, gr of moisture/lb dry air					
		0		15		30	
Temperature, °C	Voltage, kV/m	Extract iron content, wt %	Iron recovery, %	Extract iron content, wt %	Iron recovery, %	Extract iron content, wt %	Iron recovery, %
20	200	86.3	44.6	58.3	41.7	45.3	41.2
	230	84.8	61.7	56.0	57.0	43.3	53.8
	260	81.5	73.8	51.4	70.0	42.9	68.1
50	200	84.8	55.4	85.3	56.2	82.6	63.1
	230	79.2	68.0	82.8	72.1	74.3	77.0
	260	74.2	75.6	77.6	79.6	66.2	88.4
80	200	82.1	45.7	83.7	56.2	83.1	63.4
	230	79.3	60.8	80.6	72.8	80.0	79.8
	260	76.1	69.8	73.9	77.9	74.9	82.0
110	200	77.7	60.5	80.4	58.8	81.0	62.4
	230	70.3	69.2	74.9	71.0	72.1	70.8
	260	64.1	73.2	66.7	76.1	68.7	80.3

Source: Ref. 18. © (1978) *IEEE*. Reprinted with permission.

Table 3.3 shows the parameters which were controlled during the experiments. The averaged results are presented in Table 3.4 and are defined in terms of the extract iron content and iron recovery percentages. The extract iron content is the concentration of iron in the extract in weight percent. The iron recovery is the percentage of the total iron in the charge recovered in the extract.

Most of the iron particles were charged positively and were attracted by the negative polarity of the funnel electrode.

For the experiments reported above, conductive induction on the distributor and triboelectrification contributed both to charge the iron positively (reversing polarities resulted in much lower extract iron contents). Some glass particles were extracted with the iron either because they had been charged positively or because they were still agglomerated with the iron particles by interparticle forces or because they were physically entrained by the fluidization gas. Altogether, it is possible to conclude that vibration extends the benefits of fluidized electrostatic separation techniques to very fine particles.

APPENDIX 3.3: RELATED PATENTS

1. H. M. Sutton, W. L. Steele, and E. G. Steele, Process of Electrical Separation, U.S. Patent 1,116,951, Nov. 10, 1914.
2. J. E. Lawver, Beneficiation of Nonmetallic Minerals, U.S. Patent 2,805,760, Sept. 10, 1957.
3. O. M. Steutzer, Electrostatic Separation Means, U.S. Patent 3,249,225, May 3, 1966.
4. R. W. Madrid, Electrostatic Separation of Round and Nonround Particles, U.S. Patent 3,477,568, Nov. 11, 1969.
5. N. E. Oglesby, Process and Apparatus for Grading and for Coating with Comminuted Material, U.S. Patent 2,328,577, Sept. 7, 1943.
6. T. Bantz, Method and Apparatus for Electrostatic Separation, U.S. Patent 2,106,855, Feb. 1, 1938.
7. L. I. Blake and L. N. Morscher, Process of Electrical Separation of Conductors from Non-Conductors, U.S. Patent 668,791, Feb. 26, 1901.

REFERENCES

1. S. Masuda, S. Mori, and T. Hoh, Applications of Electric Curtain in the Field of Electrostatic Powder Coating, Conference of Electrostatics Society of America, University of Michigan, Ann Arbor, Mich., 1975.

2. D. K. Davies, The Generation and Dissipation of Static Charge on Dielectrics, *Static Electrification--Proceedings of the 1967 Conference, Institute of Physics and the Physical Society,* Conference Series No. 4, 1967, pp. 29-36.

3. I. I. Inculet and E. P. Wituschek, Electrification by Friction in a 3×10^{-7} Torr Vacuum, *Static Electrification--Proceedings of the 1967 Conference, Institute of Physics and the Physical Society,* Conference Series No. 4, 1967, pp. 37-43.

4. I. I. Inculet, Static Electrification of Dielectrics and at Materials' Interfaces, in *Electrostatics and Its Applications* (A. D. Moore, ed.), Wiley, New York, 1973, Chap. 5.

5. Kali and Salz AG, Electrostatic Extraction of Pyrites from Crude Coals, Belgian Patent 818,039 (Chemical Abstract 84-108376), 1974.

6. A. G. Duba, Electrical Conductivity of Coal and Coal Char, *Fuel, 56*(4), 441-443 (1977).

7. T. M. Khrenkova and V. Lebedev, Nature of Change in Properties of Coal during Crushing, *Khim. Tverd, Topl., 11,* 1 (1975).

8. A. A. Mettus, N. I. Ol'shanskaya, and I. K. Sorokina, Electrical Properties of Ekibastuz Coals, *Electrofiz. Metody Obrab. Redkomet. Syr'ya,* 29-36 (Chemical Abstract 83-13136), 1972.

9. I. I. Inculet, M. A. Bergougnou, and J. D. Brown, Electrostatic Separation of Particles below 40 μm in a Dilute Phase Continuous Loop, *Trans. IEEE IAS, IA-13*(4), 370-373 (1977).

10. M. A. Bergougnou, I. I. Inculet, J. Anderson, and L. Parobek, Electrostatic Beneficiation of Coal in a Fluidized State, *J. Powder and Bulk Solids Tech., 1*(3), 22-26 (1977).

11. I. I. Inculet and W. D. Greason, Effect of Electric Fields and Temperature on Electrification of Metals in Contact with Glass and Quartz, *Static Electrification--Proceedings of the Institute of Physics,* Conference Series No. 11, 23-32, 1971.

12. W. D. Greason, Effect of Electric Fields and Temperature on the Electrification of Metals in Contact with Insulators and Semiconductors, Ph.D. thesis, Faculty of Graduate Studies, The University of Western Ontario, London, Canada, 1972.

13. F. A. Zenz and D. A. Othmer, *Fluidization and Fluid-Particle Systems,* Reinhold, New York, 1960.

14. D. Kunii and O. Levenspiel, *Fluidization Engineering*, Wiley, New York, 1960.

15. I. I. Inculet and M. A. Bergougnou, Electrostatic Beneficiation of Fine Mineral Particles in a Fluidized Bed, *Proceedings of the Tenth International Mineral Processing Congress*, London, Paper 11, 1973, pp. 1-14.

16. J. D. Brown, I. I. Inculet, and M. A. Bergougnou, An Apparatus for Simultaneous Grinding and Sieving of Coal, *CIM Bulletin*, *68*(764), 82-83 (1975).

17. I. I. Inculet, Electrostatics in Industry, *J. Electrostatics*, *4*(2), 175-192 (1978).

18. C. Kiewiet, M. A. Bergougnou, J. D. Brown, and I. I. Inculet, Electrostatic Separation of Fine Particles in Vibrated Fluidized Beds, *Trans. IEEE IAS*, *IA-14*(6), 526-530 (1978).

4

High-Gradient Magnetic Separation for Coal Desulfurization*

Y. A. LIU†

Auburn University
Auburn, Alabama

---

*The terms *coal cleaning*, *coal beneficiation*, and *coal desulfurization* are used interchangeably in this chapter.

†Present affiliation: Virginia Polytechnic Institute and State University, Blacksburg, Virginia.

4.1 BACKGROUND AND OVERVIEW

A. High-Gradient Magnetic Separation (HGMS): Basic Principles and Devices

High-gradient magnetic separation (HGMS) is a relatively new technology which promises to be a practical means for separating micrometer-sized, weakly magnetic materials on a large scale and at much faster flow rates than are possible in ordinary filtration. The technology is also applicable to separating nonmagnetic materials which can be made to associate with magnetic seeding materials such as magnetite. It was developed in 1969 for the wet separation of weakly magnetic contaminants (iron pyrite, titanium oxide, etc.) from kaolin clay to upgrade its quality and brightness [27,29,71,74]. A typical HGMS unit used in this application is illustrated

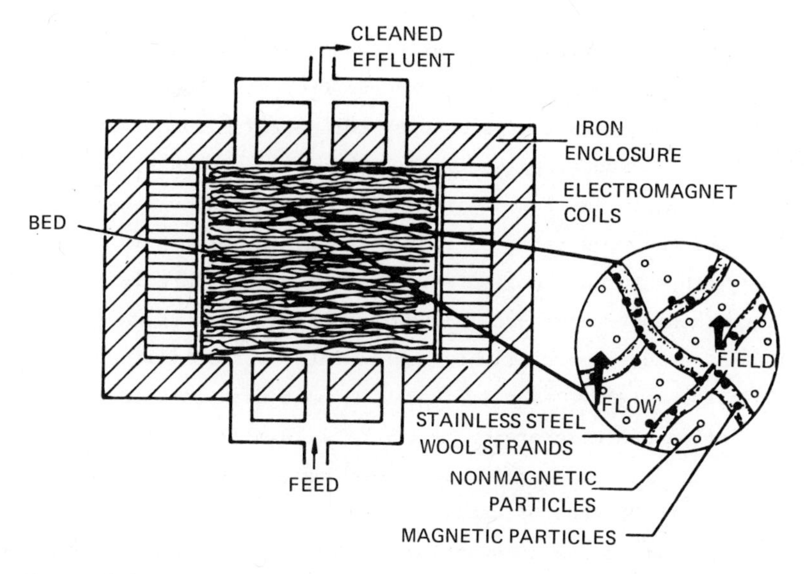

Figure 4.1 A cyclic electromagnetic HGMS unit. (From Ref. 74. © 1976 IEEE.)

schematically in Fig. 4.1 [74]. The electromagnet structure consists of energizing coils and a surrounding iron enclosure. The coils in turn enclose a cylindrical working volume packed with fine strands of strongly ferromagnetic materials such as ferritic stainless steel wools. With this design, a strong field intensity of typically 20 kilo-Oersted (kOe) or 1.591 million ampere per meter (MA/m) can be generated and distributed uniformly throughout the working volume. Additionally, by placing in the uniform field the ferromagnetic packing materials (which increase and distort the field in their vicinity), large field gradients of the order of 1 kOe/μm can be produced.

In the wet beneficiation of kaolin clay, the HGMS unit is used in a batch or cyclically operated process like a filter. The kaolin feed containing the weakly magnetic contaminants in low concentrations is pumped through the magnetized stainless steel wool packing (matrix or canister of the separator) from the bottom of the magnet while the magnet is on. The magnetic materials (mags) are captured and retained inside the separator matrix; and the nonmagnetic materials (nonmags, tailings, or tails) pass through the matrix and are collected as the beneficiated products from the top of the magnet. After some time of operation, the separator matrix becomes saturated with the attracted magnetic materials. The feed is then stopped and the separator matrix is rinsed with water. Finally, the magnet is turned off and the mags retained inside the separator matrix are backwashed with water and collected. The whole procedure is repeated in a cyclic fashion. Large-scale HGMS units like that shown in Fig. 4.1 are being used commercially to beneficiate over 4 million tons of kaolin clay per year. The typical HGMS unit used in this commercial application has a separator matrix of 2.13 m (7 ft) in diameter and 0.508 m (20 in.) in length. At a residence time of 30 sec, it can process up to 30 tons of kaolin clay of 30 wt % slurry concentration per hour. Its power consumption is about 400 to 500 kW. Depending upon the kaolin clay brightness requirements, the total investment and operating cost was estimated in 1976 to vary from 68¢ to $3.00/ton processed [27].

In general, if the cyclic HGMS units illustrated above are used in other applications where the magnetic materials occupy a large fraction of the feed stream, the downtime for backwashing would be considerable, possibly necessitating the use of one or more backup separators. To overcome this problem which is inherent in batch operations, semicontinuous or continuous HGMS units using rotating or reciprocating multiple matrix systems have been developed. Figure 4.2 shows a cutaway view of one such continuous unit, called the *carousel separator*, currently being used commercially for mineral processing applications [67,73]. The basic working element of such a separator is a revolving magnetizable ring, called a *carousel*, as shown in Fig. 4.3 [41]. Figure 4.2 shows that multiple separator matrices are mounted in the ring which passes continuously through a magnet station (also called a magnetic head). This station is an elongated and steel-enclosed solenoid. The coils of this solenoid have been turned up at either end to allow the matrix ring to continuously move into and out of the solenoid. Separation takes place in the matrix ring while it is passing through the magnet station. Magnetic particles are captured in the matrix and carried out of the

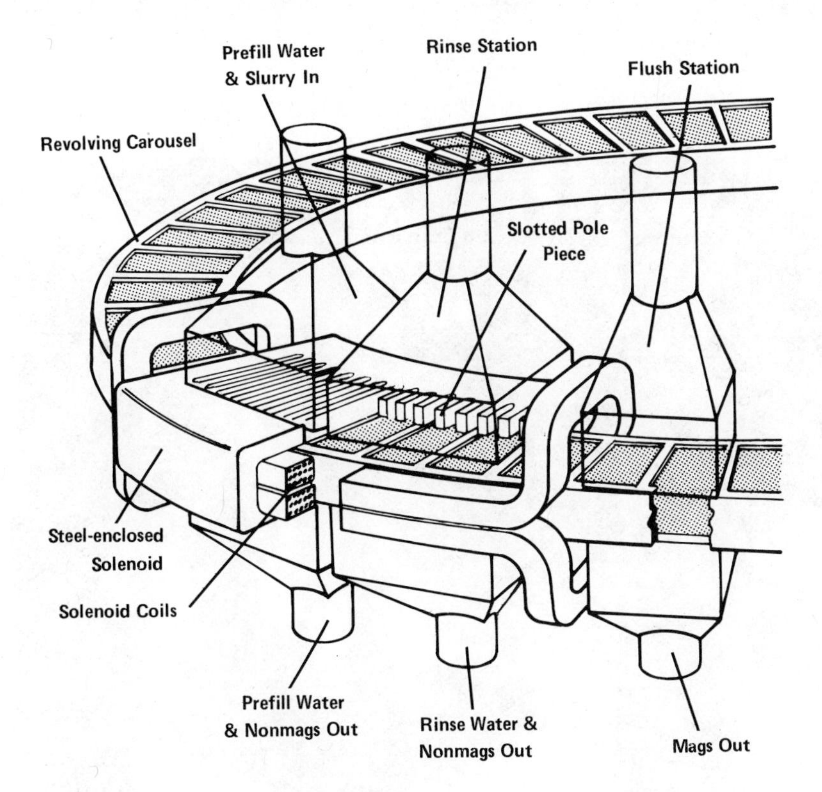

Figure 4.2 A cutaway view of a large-scale carousel separator. (From Ref. 73. © 1979 IEEE.)

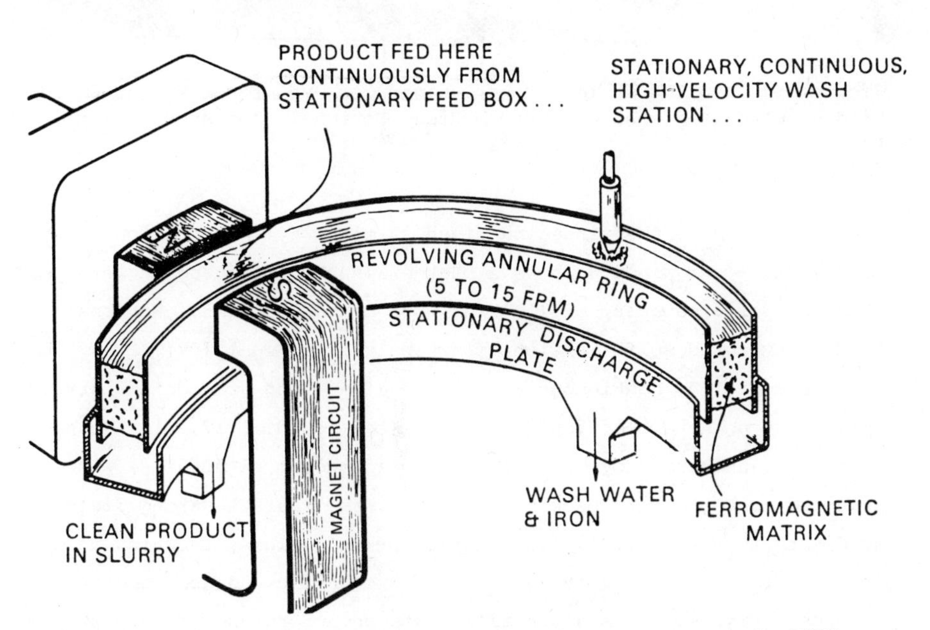

Figure 4.3 The basic elements of a carousel electromagnetic HGMS unit. (From Ref. 41.)

Figure 4.4 A pilot-scale carousel separator used for desulfurization of dry pulverized coal by the Oak Ridge National Laboratory. (From Ref. 97.)

magnet station to the rinse and flush stations by the rotation of the matrix ring. Figure 4.4 shows a photograph of a pilot-scale carousel HGMS unit of a capacity of 1 ton/hr used by the Oak Ridge National Laboratory for desulfurization of dry pulverized coal [97].

In addition, continuous HGMS units of an open-gradient or stream-splitter type have been designed and tested [3,10,11,17,23]. These continuous units employ specially designed magnet structures, without using separator matrix packing materials, to generate strong field gradients in a relatively large, open working volume, in which a flowing slurry is effectively split into magnetic and nonmagnetic streams. Figures 4.5 and 4.6 illustrate one prototype of such stream

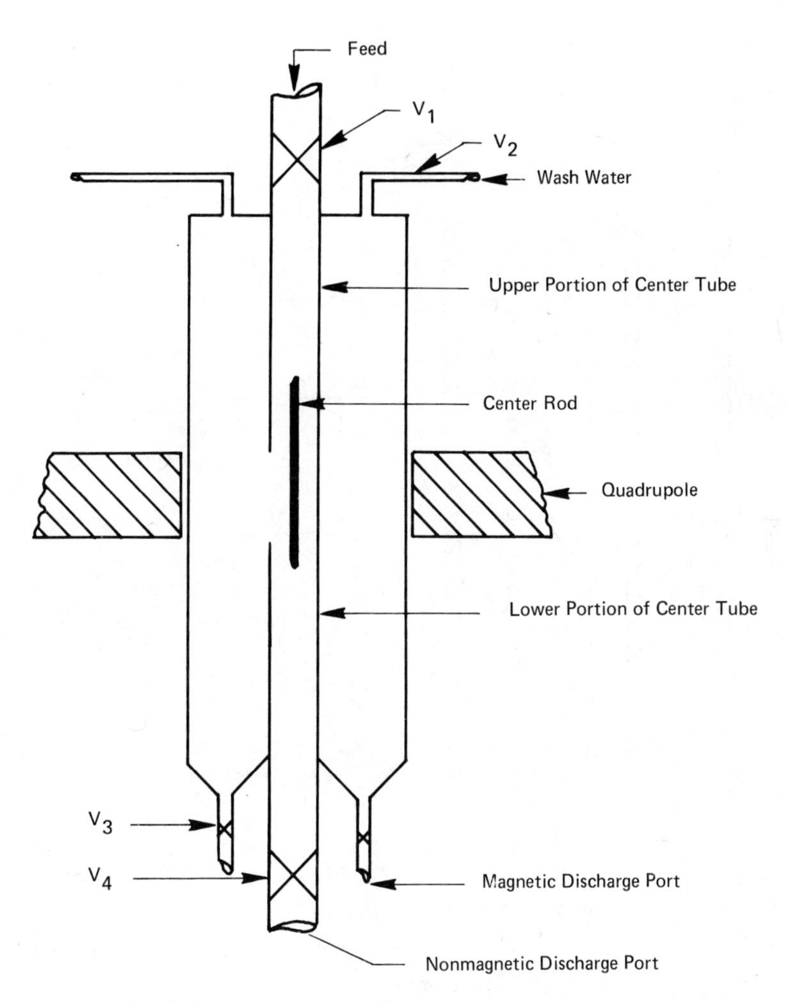

Figure 4.5 A schematic diagram of a prototype quadrupole separator built by the Bethlehem Steel Corporation. (From Ref. 3.)

splitters, called the *Bethlehem quadrupole separator* [9]. As shown, the feed stream is introduced through the upper portion of the center tube and its magnetic particles are radially deflected through the gap and into the lower annulus formed between the two tubes. Coherent flow of the nonmagnetic particles is hydraulically maintained by properly adjusting the inlet and outlet valves so that there is a net influx of wash water through the gap. This wash water is introduced

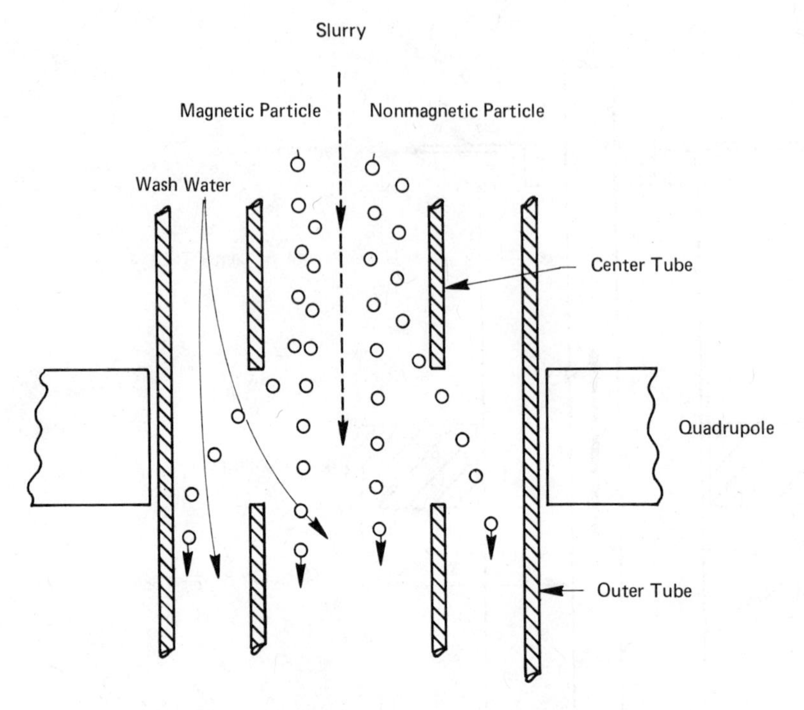

Figure 4.6 The separation zone of the prototype quadrupole separator built by the Bethlehem Steel Corporation. (From Ref. 3.)

into the system through the upper portion of the annulus. Nonmagnetic particles are collected and discharged through the lower section of the center tube, and magnetic particles discharge through the lower annulus. The functions of the center rod are to blank off the area of the field which is too low in intensity to sufficiently attract the magnetic particles, and also, through its shape, to help maintain a coherent flow across the gap.

The ability of superconducting coils to produce strong magnetic fields over large working volumes without an excessive expenditure of power makes them ideal candidates for use in magnetic separation [19]. Figure 4.7 shows a schematic diagram of a laboratory-scale, cyclic superconducting HGMS unit of a maximum field intensity of 50 kOe built by the Magnetic Corporation of America (MCA) [88]. The HGMS unit is built in a stainless steel dewar of 50.8-cm inside

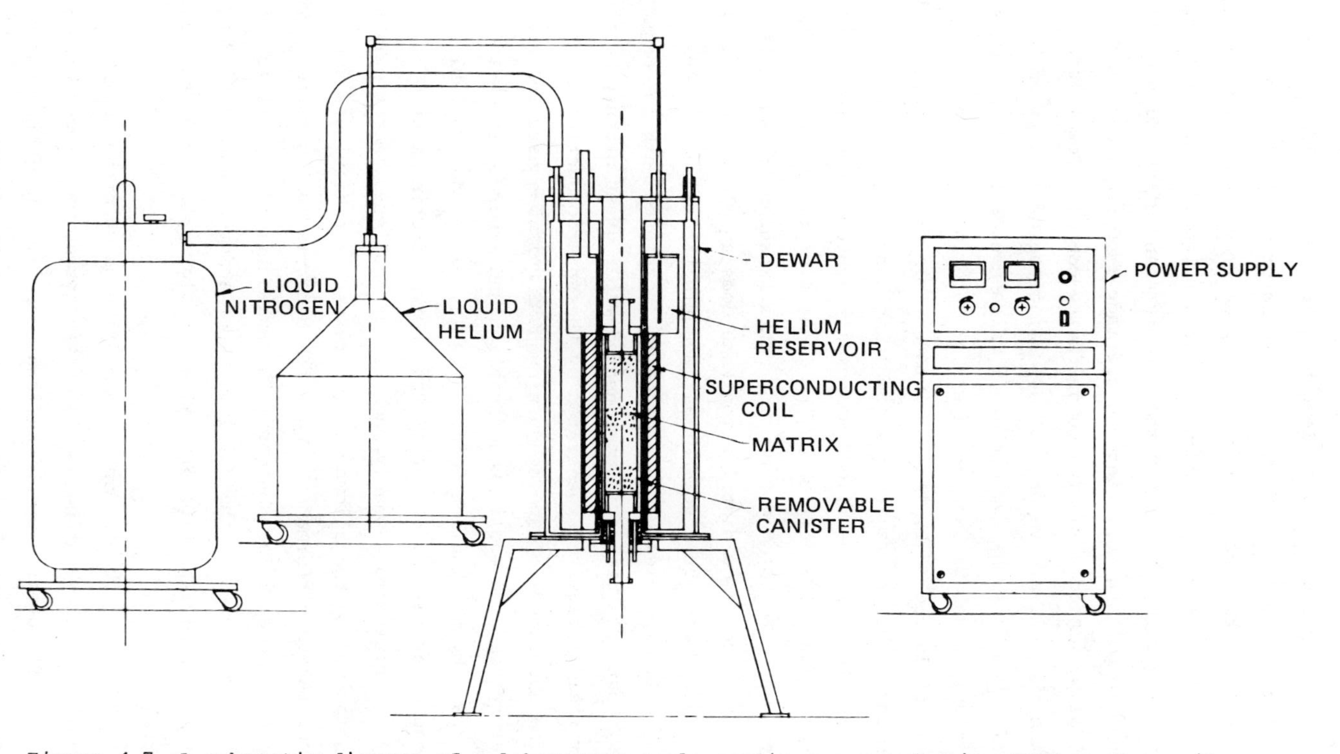

Figure 4.7 A schematic diagram of a laboratory-scale cyclic superconducting HGMS system. (From Ref. 90.)

diameter and 12.2 m (4 ft) in length, with a bore of 12.7-cm inside diameter which houses the removable separator matrix (canister). The superconducting coil windings are made from a nobelium-titanium (NbTi) alloy imbedded in a high-conductivity copper. Power leads for continuous system operation without replacement of the liquid helium. The superconducting coil windings are made from niobium-titanium (NbTi) alloy imbedded in high-conductivity copper. Power leads for charging the coil are vapor cooled, making use of the system helium boiloff. The inner helium container which houses the superconducting coil is surrounded by several layers of superinsulation and by a thermal shield cooled by liquid nitrogen. In order to minimize the helium loss, the entire inner assembly is suspended from the top of the dewar with a low heat-leak suspension system. Note that in the superconducting HGMS unit shown, no iron pole pieces are included at the top and bottom of the separator matrix (canister), and this provides for better flow distribution than does an electromagnetic HGMS unit with an iron-bound solenoidal magnetic structure as illustrated in Fig. 4.1. The disadvantage of having no closed path for magnet flux without pole pieces is compensated, to a great extent, by the higher field intensities generated by the superconducting system.

The above cyclic superconducting HGMS unit was used in extensive laboratory tests on kaolin beneficiation and mineral processing carried out by the English China Clays [9,102]. The available results have shown several advantages of the superconducting system in effectively separating more weakly magnetic or smaller-diameter particles, and in increasing the allowable flow rates for achieving a satisfactory separation performance with a fixed matrix configuration. Two cyclic units of similar designs with maximum field intensities of 50 to 70 kOe were used in the development of superconducting HGMS processes for desulfurization of coal-water slurry [89] and dry pulverized coal [99].

In Ref. 107, the concept of a continuous superconducting HGMS unit using a reciprocating double-matrix system was proposed as a means to fully utilize the high-field capabilities of the superconductor and its theoretically zero resistance to electric current.

In this continuous unit, the reciprocating double-matrix system is designed such that when one separator matrix has become saturated with the attracted magnetic materials in the feed stream, a second matrix is automatically inserted into the magnetic field while the first matrix is moved outside of the magnetic field and its attracted materials are being flushed out. The use of such a reciprocating matrix system eliminates the idle time (lost processing time) involved in charging and discharging the coils that is commonly associated with operating cyclic superconducting HGMS units. It also eliminates the downtime due to flushing (washing) the separator matrix when the flushing period is not longer than the feeding (separation) period. A prototype unit of this reciprocating separator, shown schematically in Fig. 4.8, was built and installed at the English China Clays for use in clay and mineral processing applications [9,88]. This reciprocating moving-matrix concept was later adapted at Auburn University for the development of a novel continuous superconducting fluidized-bed HGMS process for dry coal desulfurization, and a number of equipment innovations to the reciprocating moving-matrix design were developed to make it suitable for separating gas-particulate mixtures [99].

The use of superconducting coils greatly enhances the applicability of continuous quadrupole separators, previously illustrated in Figs. 4.5 and 4.6, to effectively separate magnetic particles of smaller sizes and lower susceptibilities [10,11,17,23]. An attractive superconducting quadrupole separator design which has been tested for the beneficiation of coal-water slurry was proposed by Cohen and Good [10,11,23]. This separator consists of a superconducting quadrupole magnet installed in a vertical cylindrical cyrostat with a hollow central portion, and a specially designed curved channel for mineral slurry and wash water stream placed around the outer circumference of the cryostat. Figure 4.9 shows a photograph of the curves channel being mounted outside of the cylindrical cryostat, and Fig. 4.10 gives a cutaway view of the quadrupole magnet and curved channel [10]. The curves channel is designed to generate a secondary circulation which imposes on the bulk flow of solid

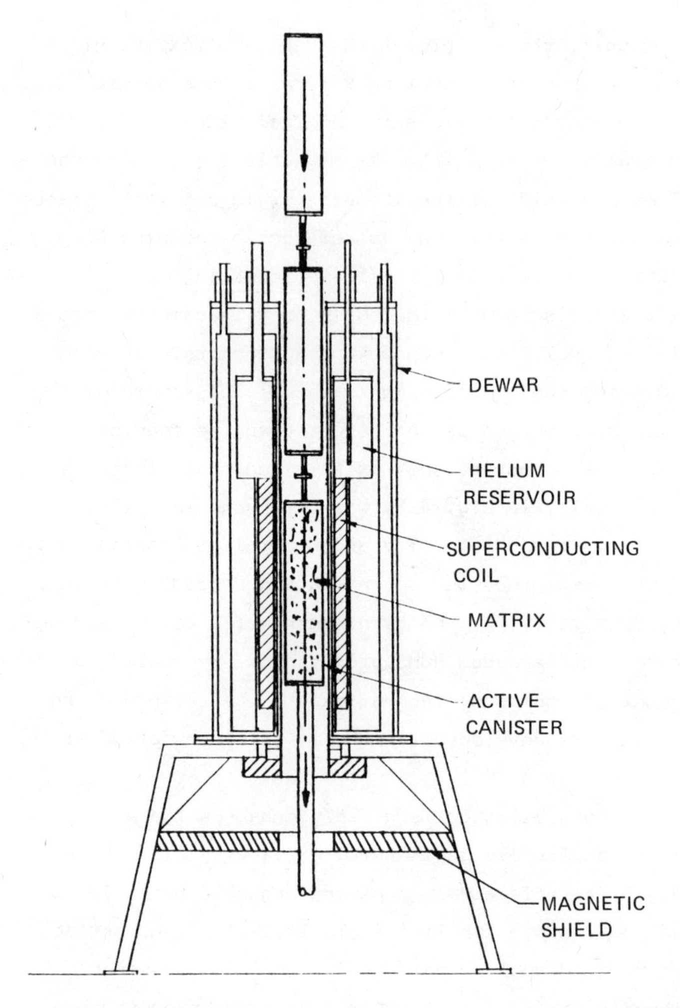

Figure 4.8 A schematic diagram of a prototype reciprocating superconducting HGMS unit. (From Ref. 88.)

particles in the slurry. In particular, in the curved channel, the top of the flowing stream contains few particles and moves faster. It experiences a greater centrifugal force and thus moves to the outer wall of the channel. Similarly, the lower part of the stream flows forward more slowly and moves to the inner wall of the channel. The secondary circulation results in a spiral flow of the fluid which sweeps all particles in suspension close to the magnet. The

Figure 4.9 A portion of a superconducting quadrupole separator with a curved channel for mineral slurry and wash water placed around the outer circumference of the cryostat. (From Ref. 10. © 1976 IEEE.)

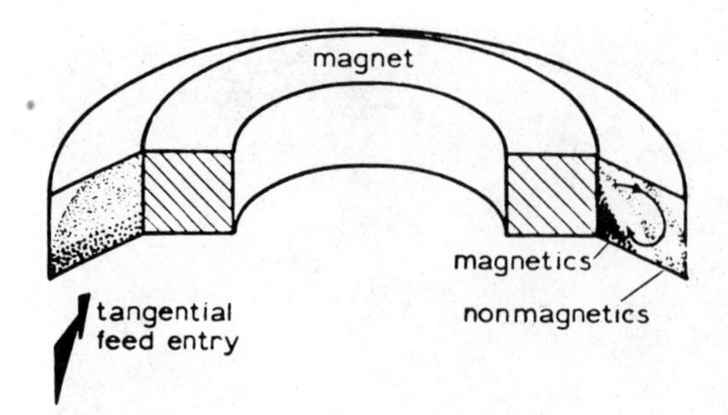

Figure 4.10 A cutaway view of the quadrupole magnet and curved channel shown in Fig. 4.9. (From Ref. 10. © 1976 IEEE.)

nonmagnetic particles that are not in suspension move along the floor of the channel. Because the channel is sloping outwards, those nonmagnetic particles are not swept into the base of the inner wall but travel near the outer part of the channel. The magnetic particles move to the inner wall of the channel. At the exit from the channel, the magnetics leave as a concentrated stream. The bulk of the water and nonmagnetics leave the channel at higher velocity and are collected separately. A prototype unit of the proposed continuous quadrupole separator was used in feasibility studies of magnetic beneficiation of water slurries of pulverized coals [10].

Further discussion of the basic principles and devices of HGMS can be found from several recent state-of-the-art reviews available in the literature [45,46,48,69-74,103-105].

B. HGMS Applied to Coal Desulfurization: An Overview

Because of its very low costs and outstanding technical performance demonstrated in the kaolin application, HGMS was recently adapted to solving many separation problems related to minerals and chemical processing industries [70-74]. A promising application of HGMS is the magnetic removal or inorganic sulfur and ash-forming minerals (ash) from coal. Previous experimental investigations have indicated clearly that most of the minerals impurities in coal, which contribute to its pyritic sulfur, sulfate sulfur, and ash content, are weakly magnetic (called *paramagnetic*). Those sulfur-bearing and ash-forming minerals, if sufficiently liberated as discrete particles, can be separated normally from the pulverized diamagnetic, or practically nonmagnetic, coal by magnetic means [16,33,34]. Indeed, HGMS was successfully adapted in 1973 in a bench-scale study to remove sulfur and ash from a finely pulverized Brazilian coal suspended in water [93-95]. Since then, the technical feasibility of utilizing cyclic and continuous HGMS units for the magnetic separation of sulfur and ash from water slurries of pulverized United States coals has been demonstrated in a number of experimental studies [20,43,57,66,89,106], with substantial amounts of sulfur and ash removal being achieved. In addition, studies on the quantitative modeling and correlation of desulfurization performance [50,57,89,93] as well as the conceptual design and cost estimation of HGMS processes for desulfurization of coal-water slurries [32,44,89,106] have been reported.

In 1977, the first successful application of HGMS to dry coal desulfurization was reported by Auburn and the Oak Ridge National Laboratory (ORNL) [51]. By using either an air-entrained flow separation or a recirculating air fluidization, HGMS effectively removed up to 94% of the inorganic sulfur and 70% of the ash from classified size fractions (typically −100 +200 mesh) of several dry pulverized Pennsylvania coals, with an average mass recovery of cleaned coal being over 85 wt %. In subsequent studies reported by ORNL [24,97], the technical feasibility of using both laboratory cyclic and pilot-scale carousel HGMS units for desulfurization of classified size fractions of several dry pulverized coals by gravity feeding assisted by a slow-velocity air stream was demonstrated. Recently, the recirculating air fluidization was further developed and demonstrated as an effective approach to HGMS for desulfurization of utility boiler feed coals (typically 70 to 80% −200 mesh) without any external size classification, ultrafines removal and coal drying prior to magnetic separation [54,65,99]. Continued developmental work on dry coal desulfurization processes is presently being pursued using carousel electromagnetic and reciprocating superconducting HGMS units.

The technical feasibility of adapting HGMS as an effective alternative to conventional precoat filtration in the solvent-refined coal (SRC) process was first demonstrated in a bench-scale study done by Hydrocarbon Research, Inc. (HRI) in 1975 [26]. HGMS effectively removed up to 90% of the inorganic sulfur and 35% of the ash from the liquefied SRC filter feed slurry of an Illinois No. 6 coal. Since then, similar success of HGMS for desulfurization of filter feed slurries of other liquefied coals was reported in a number of bench- and pilot-scale studies [2,13,55,79]. Several practical considerations seem to suggest a significant potential of utilizing HGMS to remove mineral residue from liquefied coal. The most important one is that hydrogenation reaction in coal liquefaction processes will generally reduce a major portion of the pyritic sulfur to the highly magnetic pyrrhotites, particularly the ferrimagnetic (strongly magnetic) monoclinic pyrrhotite.

Preliminary studies have also been reported on the applications of HGMS for the removal of sulfur and ash from pulverized coal dispersed in fuel oil [12] or suspended in methanol [57]. In the first application, HGMS would also improve the settling stability of the resulting coal-oil mixture (COM) by removing the heavier compounds of iron and sulfur and trace elements; it may be developed as an effective fuel preparation method for COM combustion being considered for replacing coal burning in utility power plants. In the second application, the use of methanol as a carrier liquid for pulverized coal would eliminate the necessity of drying the product coal after magnetic beneficiation.

A number of physical methods such as gravity separation (heavy-medium separation) and froth flotation have been tested as pretreatment steps to remove certain amounts of sulfur and ash from pulverized coal prior to its further beneficiation by HGMS [20,25,106]. The latter would serve to deep clean the rougher product from a heavy-medium separator or the froth concentrate from a flotation cell. Of particular interest is a pilot-scale study which demonstrated the technical feasibility of combining froth flotation and HGMS for coal beneficiation during a 120-hr sustained operation [20].

Physical methods such as thermal treatment and microwave radiation (dielectric heating) have been used to convert pyrite to more magnetic compounds of iron, thereby increasing the efficiency of the subsequent magnetic beneficiation [1,8,16,96,111]. In addition, the selective adsorption of some specially prepared, dispersions of colloidal magnetic particles (called *magnetic fluids*) on the surfaces of either organic coal constituents or inorganic mineral impurities has been tested as a means to alter the magnetic characteristics of pulverized coal and its mineral impurities and prepare them for efficient magnetic separation [86]. Also, several chemical methods have been investigated for converting pyrite and ash-forming minerals (ash) in coal to strongly magnetic forms [106]. One new method is the Magnex process [36-39,80-82] which exposes a dry pulverized coal to gaseous iron carbonyl [$Fe(CO)_5$] under conditions whereby the pyrite and ash become strongly magnetic and can be efficiently removed from coal by magnetic separation.

C. Major Operating Variables and Conditions and Main Performance Indexes of HGMS Applied to Coal Desulfurization

Based on the preceding discussion of the basic principles and devices of HGMS along with the introduction of applications of HGMS to coal desulfurization, it is possible to summarize the major operating variables and conditions of HGMS applied to coal desulfurization as follows:

1. Characteristics of feed coal
   a. Type of feed coal
   b. Chemical analyses of feed coal: total sulfur, inorganic sulfur and ash contents, trace element analysis, etc.
   c. Physical properties of feed coal: particle size distributions and magnetic susceptibilities of the pulverized feed coal and its mineral impurities, moisture content, calorimetric (Btu) value, fraction of magnetically susceptible materials in the feed coal, etc.
   d. Liberation characteristics of feed coal and its mineral impurities: proper grinding level for liberating the mineral impurities from coal
   e. Physical and/or chemical pretreatment of coal, e.g., thermal and/or chemical pretreatment to enhance the magnetic susceptibilities of pyrite and ash-forming minerals in coal

2. Method of magnetic separation
   a. Method of feed preparation: dry or wet feed, slurry or solids concentration, size classification of feed, ultrafines removal from feed, use of dispersant, etc.
   b. Method of separation: cyclic (batchwise) or continuous
   c. Temperature of separation: low-temperature or high-temperature separation
   d. Type of recycle: partial or complete recycle of separated products (magnetic refuse, cleaned products)
   e. Method of flushing (washing), e.g., pressurized water flushing
3. Characteristics of magnetic field
   a. Field intensity: electromagnetic or superconducting
   b. Field gradient: matrix gradient or open gradient
4. Characteristics of flow field
   a. Type of carrier fluid: water, methanol, fuel oil, air, flue gas, etc.
   b. Method of fluid transport: gravity flow (gravity feeding), upward pumped flow, air-entrained flow, recirculating air fluidization, etc.
   c. Fluid velocity or residence time
   d. Fluid viscosity
5. Characteristics of separator matrix (for matrix-gradient separations)
   a. External matrix configuration: inside diameter and apparent matrix length
   b. Internal matrix configuration: type, size, surface area and magnetic characteristic (e.g., saturation magnetization) of matrix packing material, packing density or void volume percentage
   c. Matrix loading factor (separator matrix throughput), e.g., amount of feed coal processed per unit amount of matrix packing for a given separator matrix configuration

In the literature (e.g., Refs. 47, 51, and 61), the technical performance of HGMS applied to coal desulfurization has been examined with reference to the following indexes.

1. Grade: percent reduction of total sulfur, inorganic sulfur and ash contents of the feed coal, as well as percent reduction of sulfur emission level (lb S/MM Btu) or sulfur dioxide emission level (lb SO_2/MM Btu) of the feed coal.
2. Recovery: percent recoveries of the amounts of coal, Btu, total sulfur, inorganic sulfur, and ash of the feed coal in the magnetically cleaned product and magnetic refuse stream.
3. Capacity of the separator: typically expressed by the matrix loading factor defined as the amount of feed coal processed per unit amount of matrix packing for a given separator matrix configuration for matrix-gradient separations, and

expressed by the amount of feed coal processed per unit cross-sectional area of the separator per unit time for open-gradient separations.

D. Scope of the Chapter

The objective of this chapter is to present a comprehensive discussion of the recent development and current status of HGMS processes for the removal of sulfur and ash from coal. Section 4.2 summarizes the currently available technical and economical information on HGMS processes for desulfurization of coal-water slurry. In Sec. 4.3 the recent developments of HGMS processes for desulfurization of dry pulverized coal are discussed. Section 4.4 describes the reported experimental studies on HGMS for the removal of mineral residue from liquefied coal, and discusses the potential of utilizing HGMS in the solvent-refined coal (SRC) and related coal liquefaction processes. Section 4.5 summarizes the available results obtained from laboratory studies of HGMS for desulfurization of a coal-oil mixture (COM) and a coal-methanol slurry. In Sec. 4.6, the use of several physical methods as pretreatment steps for magnetic desulfurization is discussed. Included in the discussion are gravity separation, microwave treatment (dielectric heating), thermal treatment, froth flotation, and magnetic fluids. Section 4.7 describes the use of chemical methods to enhance the performance of magnetic desulfurization with particular emphasis on the Magnex process.

4.2 HIGH-GRADIENT MAGNETIC SEPARATION FOR DESULFURIZATION OF COAL-WATER SLURRY

In this section, the currently available technical and economical information on HGMS for desulfurization of coal-water slurry is reviewed according to three topics: (1) reported experimental studies, (2) qualitative and quantitative correlations of desulfurization performance, and (3) conceptual process design and cost estimation.

A. Reported Experimental Studies

Cyclic Electromagnetic HGMS Tests. Table 4.1 summarizes the scopes of reported experimental studies of cyclic electromagnetic HGMS for

Table 4.1 Scopes of Reported Experimental Studies on Cyclic Electromagnetic HGMS for Desulfurization of Coal-Water Slurry

	Massachusetts Institute of Technology (MIT) [93-96]	General Electric Company and MIT [57]	Auburn University [43,50,68]
1. Pulverized coals used	Brazilian Siderpolis coal	Pennsylvania Upper Freeport (Delmont) coal	Illinois No. 6 coal; Pennsylvania Upper Freeport (Jefferson), Upper Kittanning (Cambria), Lower Freeport (Elk) and Lower Kittanning (Jefferson) coals
2. Scale of experiments	Bench scale	Bench scale	Bench and pilot scale
3. Separation variables and conditions investigated			
Feed coal particle size	Varied	Varied	Varied
Feed amount or matrix loading	Varied	Fixed	Varied
Field intensity	Varied	Varied	Varied
Residence time (slurry velocity)	Varied	Varied	Varied
Slurry concentration	Varied	Fixed	Varied
Type of Dispersant	Fixed	Fixed	Fixed
Type of matrix packing material	Varied	Fixed	Varied
Packing density	Varied	Fixed	Varied
Matrix backwashing tested	Yes	Yes	Yes
Recycling cleaned product or varying matrix length tested	Yes	No	Yes
4. Quantitative correlation and/or modeling of experimental performance examined	Yes	Yes	Yes
5. Preliminary process design and/or cost estimation done	Yes	No	Yes

Indiana University [66]	Department of Energy [25]	Michigan Technological University [20]	Sala Magnetics, Inc. [106]
Indiana Nos. 5 and 6 coals	Pennsylvania Upper Freeport (Delmont) coal	Illinois No. 6 and Pittsburgh No. 8 coals	Illinois No. 6 (Old Ben) coal; Pennsylvania Lower Kittanning (Canterbury) Sewickley and Upper Freeport (Delmont) coals; West Virginia Pittsburgh No. 8 coal
Pilot scale	Bench scale	Pilot scale	Bench and pilot scale
Fixed	Fixed	Fixed	Varied
Fixed	Fixed	Varied	Varied
Fixed	Fixed	Fixed	Varied
Fixed	Fixed	Varied	Varied
Fixed	Fixed	Varied	Varied
Fixed	Varied	Fixed	Fixed
Fixed	Fixed	Fixed	Fixed
Fixed	Fixed	Fixed	Fixed
Yes	Yes	Yes	Yes
Yes	No	No	No
No	No	No	No
Yes	No	Yes	Yes

desulfurization of water slurries of pulverized coals. In what follows, the effects of major separation variables and conditions on the desulfurization performance are illustrated from reported results. These effects are discussed in some detail, since similar effects are observed for HGMS in other coal desulfurization applications to be described in subsequent sections.

Evidence of Magnetic Separation of Sulfur and Ash. One simple way of demonstrating the magnetic separation of sulfur and ash from water slurries of pulverized coals is to compare magnetic tests at identical conditions with zero-field runs [43,93-96]. In addition, the comparison of magnetization measurements as well as sulfur and ash contents of the original feed, the cleaned coal (nonmags, tails, or product) and the magnetic refuse (mags), provides clear evidence of magnetic separation. In a typical experiment, samples of these components can be obtained by the standard low-temperature ashing (LTA) procedure which removes the hydrocarbon content but leaves the mineral content of the three components essentially unchanged. For example, Fig. 4.11 shows a clear differentiation in the magnetization of the low-temperature ashed products (coal minerals) derived from the feed, product, and magnetic fractions of a pulverized Brazilian Siderpolis coal [61].

Effect of Field Intensity. Figure 4.12 [106] shows the effect of field intensity on the grade and recovery of the magnetically cleaned product obtained from tests on water slurries of a pulverized Pennsylvania Lower Kittanning (Canterbery) coal at identical experimental conditions. In general, increasing the field intensity increases the sulfur and ash reductions from the feed coal, thus improving the grade of the magnetically cleaned coal; but it also decreases the recovery of the cleaned coal due to its loss to the magnetic refuse (mags) attracted and retained inside the separator matrix. Thus, the specification of a proper field intensity for HGMS applied to coal desulfurization involves a trade-off between grade (quality) and recovery (yield) of the nonmagnetic cleaned coal. Available results from HGMS for desulfurization of a pulverized Brazilian Siderpolis coal over a wide range of field intensities

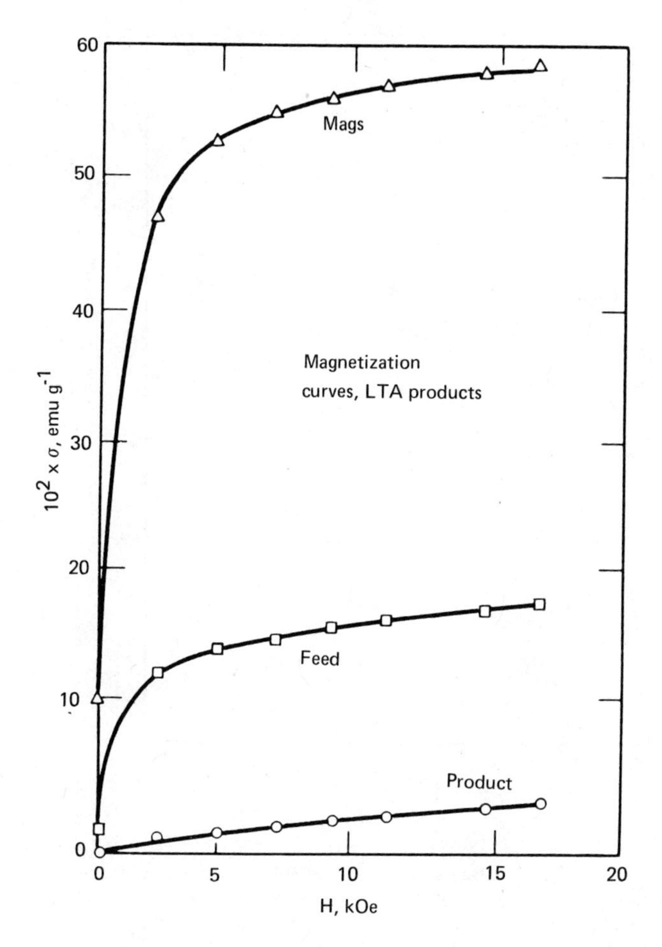

Figure 4.11 Magnetizations of the low-temperature ashed products (coal minerals) derived from the feed, product, and magnetic refuse (mags) fractions in HGMS of a pulverized Brazilian Siderpolis coal suspended in water. (From Ref. 61.)

(0 to 100 kOe) have also shown that the curves of recoveries of both magnetic and nonmagnetic products versus field intensity exhibited a trend toward saturation, especially at high field intensities [93-96]. Thus, raising the field intensity beyond a certain value, defined for each specific separation, would not substantially improve the desulfurization performance.

Effect of Slurry Flow Velocity (Residence Time). The effect of slurry flow velocity (residence time) on the sulfur and ash reduc-

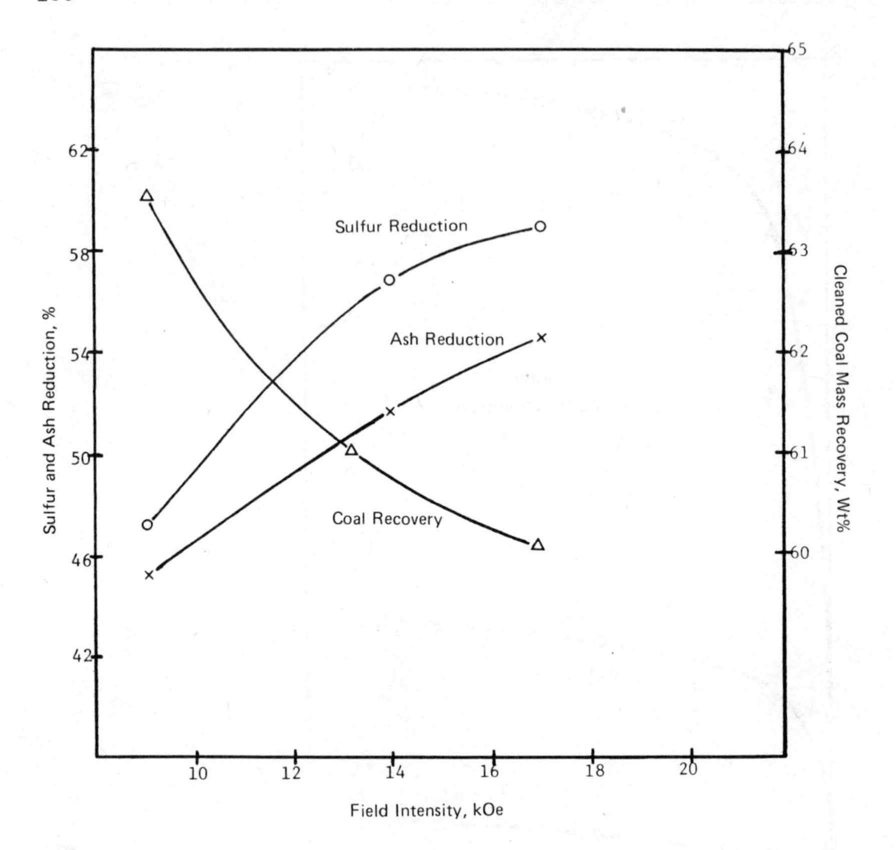

Figure 4.12 Effect of field intensity on sulfur and ash reductions and cleaned coal mass recovery obtained from HGMS of water slurries of a pulverized Pennsylvania Lower Kittanning (Canterbery) coal at identical experimental conditions. (From Ref. 106.)

tions and on the mass recovery of the magnetically cleaned coal may be illustrated by the results obtained for water slurries of a pulverized Pennsylvania Lower Kittanning (Canterbery) coal at identical experimental conditions, as shown in Fig. 4.13 [106]. The lowest slurry velocity resulted in the highest sulfur and ash reductions; however, it also produced the lowest cleaned coal recovery of just over 62 wt %. The highest slurry velocity produced the highest cleaned coal recovery of over 82 wt %, but it also gave a coal product of the poorest quality. These trends are further examples of the general trade-off relationship between grade and recovery in HGMS

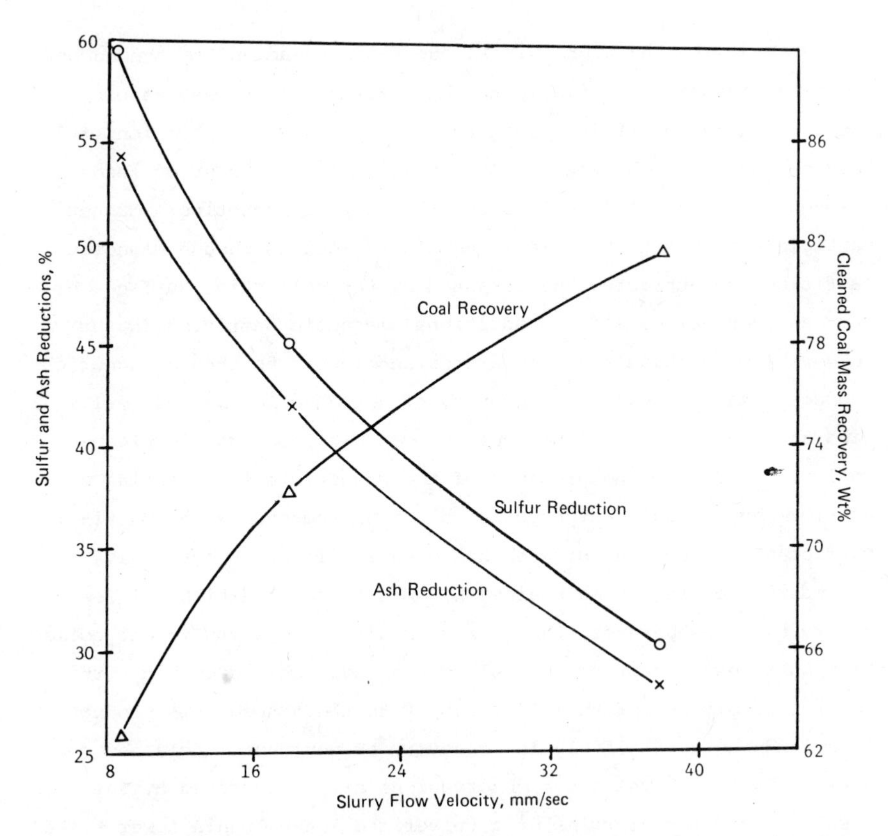

Figure 4.13 Effect of slurry velocity on sulfur and ash reductions and cleaned coal mass recovery obtained from HGMS of water slurries of a pulverized Pennsylvania Lower Kittanning (Canterbery) coal at identical experimental conditions. (From Ref. 106.)

applied to coal desulfurization. An optimum slurry velocity for HGMS applied to coal desulfurization, however, cannot be specified just on the basis of the preceding trade-off between grade and recovery, because the economics of plant operations are highly affected by the volumetric capacity of the separation unit which, in turn, is directly related to the slurry velocity. It is important that an economic balance must be attained between the product grade and recovery, and the capital investment and operating costs of the processing scheme for a given separation problem [44,106].

Effect of Matrix Loading (Separator Throughput). Matrix loading is related to the amount of magnetically susceptible particles provided to the HGMS unit in the feed stream and to the number of collection sites available to the surfaces of the separator matrix packing materials. In a typical cyclic HGMS operation, full matrix loading can be reached by passing a large amount of feed through the matrix before flushing the trapped magnetics. Magnetic particle entrapment essentially ceases as soon as enough magnetic particles are attracted and retained by the collection surfaces of the matrix packing, and any additional magnetic particles in the feed will pass through the separator and report to the nonmagnetic product. As a result, the recovery of nonmagnetic cleaned coal is high at high matrix loading; but the grade of nonmagnetic cleaned coal is low due to the presence of the magnetic sulfur-bearing and ash-forming minerals which could have been removed by the matrix if sufficient collection surfaces had been available. Conversely, if low matrix loading is maintained by small batch additions of feed slurry to the separator, the matrix should possess sufficient collection surfaces to capture the bulk of the available magnetic particles and relatively few of them will report to the nonmagnetic product. This leads to a relatively low recovery of nonmagnetic cleaned coal of a high grade. The above observations are illustrated by the results for water slurries of a pulverized Pennsylvania Lower Kittanning (Canterbery) coal at different volumetric matrix loading factors (gram coal processed per unit volume of coal slurry) shown in Fig. 4.14 [106].

It is also of interest to determine the processing times after which the separator is loaded with sulfur-bearing and ash-forming minerals. This can be done by examining the instantaneous changes in product sulfur and ash contents with processing time, the so-called concentration breakthrough curves, as illustrated by Fig. 4.15 [43]. After a relatively short processing time, the separator became loaded with sulfur-bearing minerals, but not with ash-forming minerals. Also, at a long residence time of 216 sec (low slurry velocity), the effluent concentrations of sulfur and ash approached their saturated values much slower than a short residence time of 18.3 sec (high slurry velocity).

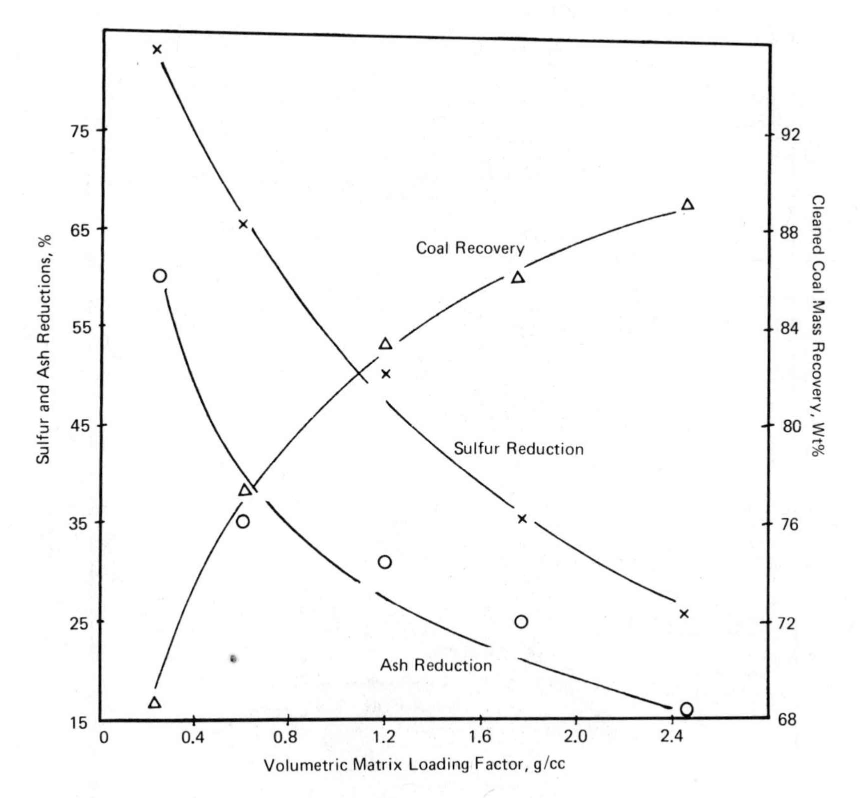

Figure 4.14 Effect of volumetric matrix loading factor on sulfur and ash reductions and cleaned coal mass recovery obtained from HGMS of water slurries of a pulverized Pennsylvania Lower Kittanning (Canterbery) coal at identical experimental conditions. (From Ref. 106.)

Effect of Type of Dispersant and of pH of Feed Slurry. The effectiveness of various dispersants in promoting the sulfur and ash removal from water slurries of pulverized coals by HGMS was evaluated in a recent report [25]. Eight different dispersants (sodium silicate, Daxad 11G, sodium pyrophosphate, bisulfite quebracho, sodium hexametaphosphate, Marasperse CBX-2 and CBAS-2, and sodium carboxymethylecellulose) were investigated. Thorough testing at equivalent experimental conditions appeared to indicate no difference in the desulfurization performance. Figure 4.16 demonstrates the independence of the pyrite reduction and magnetics recovery obtained for

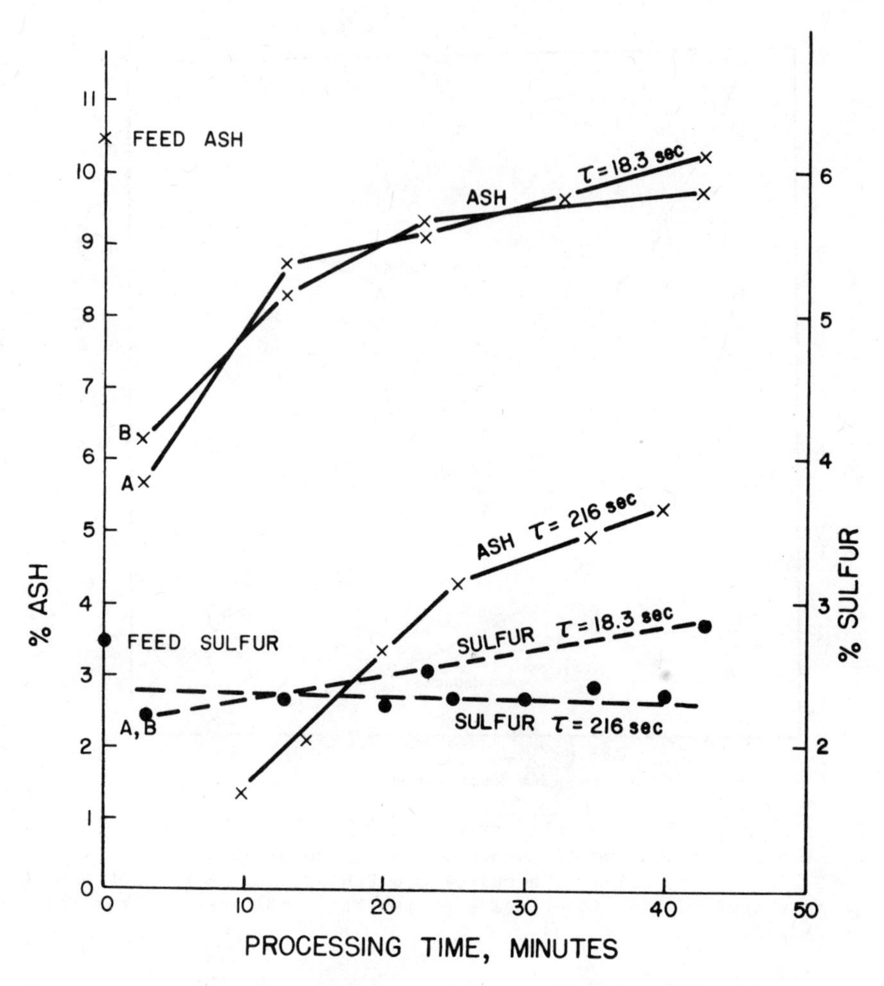

Figure 4.15 Effect of residence time on instantaneous changes in product sulfur and ash contents obtained from HGMS of water slurries of a pulverized Illinois No. 6 coal at identical experimental conditions. (From Ref. 43. © 1976 IEEE.)

water slurries of a pulverized Pennsylvania Upper Freeport (Delmont) coal on the quantity of dispersant (bisulfite quebracho) tested at a field intensity of 21.6 kOe [25]. Experimental data reported also indicated that changes in pH of the feed slurry (from pH 3 to pH 11) resulted in no appreciable change in the desulfurization performance [25].

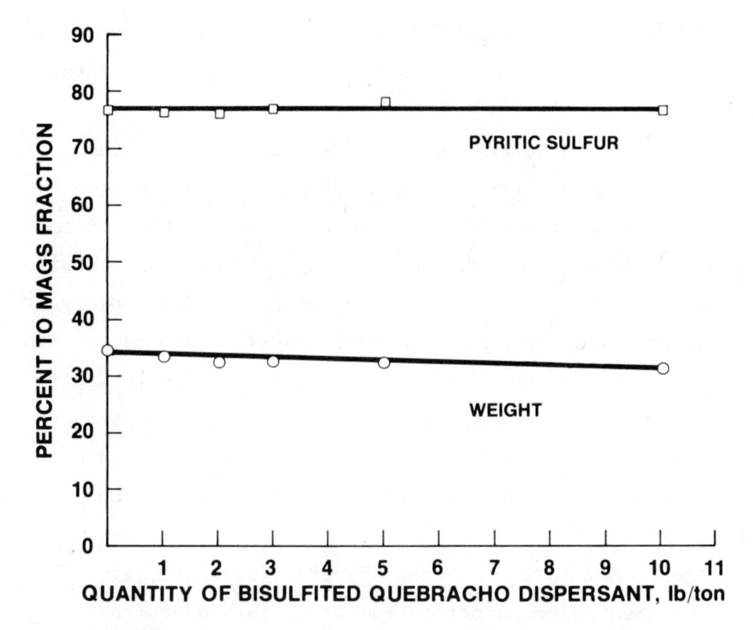

Figure 4.16 Effect of quantity of dispersant (bisulfite quebracho) on pyrite reduction and magnetics recovery obtained from HGMS of water slurries of a pulverized Pennsylvania Upper Freeport (Delmont) coal at identical experimental conditions. (From Ref. 25. © 1979 IEEE.)

Effect of Packing Material and Density. The effect of packing density on the sulfur and ash removal from water slurries of pulverized coals may be illustrated by the results obtained for an Illinois No. 6 coal (−200 mesh) [43]. With a coal slurry concentration of 2.57 wt %, a residence time of 18.3 sec and a field intensity of 20 kOe, the average percentages of ash removal obtained with a stainless steel wool packing with 94 and 97% void volume were 48.5 and 36.8%, respectively. Increasing the packing void volume or decreasing the packing density decreases both the sulfur and ash reductions. Available results [93-96] have also indicated that the use of stainless steel screens produced poor magnetic removal of sulfur and ash when compared with the use of stainless steel wools. This observation is consistent with the nature of steel wool packing which has more sharp edges and more surfaces areas so that both the field gradient and particle capture can be increased. The major advantage in using screens as packing material is its ease in allowing slurry flow and backwashing.

Effect of Coal Slurry Concentration. Experimental results obtained from studies using water slurries of several pulverized Eastern U.S. coals at 30 wt % solids concentration have shown essentially comparable sulfur reduction and coal recovery as those obtained at much lower solids concentration [43,66,93-96,106]. This observation is encouraging, since using a high coal slurry concentration can reduce the separation cost considerably. Also, the possibility of using high coal slurry concentrations suggests that HGMS may serve beneficially as an effective method for coal desulfurization in conjunction with the proposed slurry pipelines for transporting coals in the United States.

Effect of Feed Coal Particle Size. Figure 4.17 illustrates the effect of mesh of feed grind on percentage of sulfur reduction of water slurries of a pulverized Pennsylvania Lower Kittanning (Canterbery) coal under equivalent experimental conditions at 20 kOe [106]. As shown, the percentage of sulfur reduction increases as the feed coal particle size decreases. This result is related to the liberation characteristics of sulfur-bearing minerals in the coal, particularly pyrite [22,64,106]. For the Canterbery coal tested, the pyritic sulfur content was about 70% of the total sulfur. In order to fully liberate the pyritic sulfur constituents as discrete particles, available mineralogical studies indicated that grinding of the Canterbery coal to a top size smaller than 400 mesh would be required [106]. While it is not apparent that such fine grinding to −400 mesh would be practical in a commercial facility, it should be noted that using fine grinding to achieve a total liberation of pyritic sulfur is quite common in many coal desulfurization schemes and that HGMS is no worse off in this respect than many other approaches to coal desulfurization. Figure 4.17 also shows that grinding of the Canterbery coal to sizes suitable for utility boiler applications (80% −200 mesh) actually resulted in a significant sulfur reduction of over 60%.

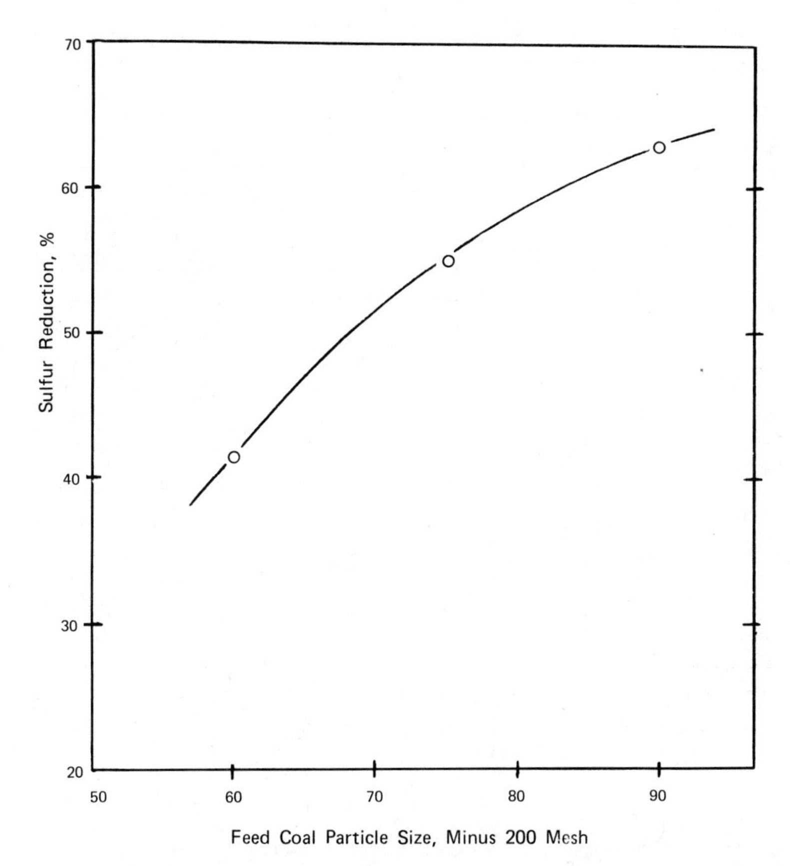

Figure 4.17 Effect of feed coal particle size of sulfur reduction obtained from HGMS of water slurries of a pulverized Pennsylvania Lower Kittanning (Canterbery) coal at identical experimental conditions. (From Ref. 106.)

Magnetic Removal of Trace Elements (Heavy Metals). Available results have shown that HGMS was effective in removing potentially harmful trace elements from water slurries of pulverized coals. Table 4.2 shows results of energy dispersive x-ray analysis of a pulverized Pennsylvania Lower Kittanning coal along with its magnetic products obtained from a wet HGMS test. As shown, HGMS could remove 55 to 74% of the compounds containing arsenic, chromium, nickel, phosphorus and sulfur elements, and 30 to 44% of the compounds containing aluminum, potassium, tin, and vanadium elements. The

Table 4.2 HGMS for Removal of Trace Elements (Heavy Metals) from Pulverized Coal Via Energy Dispersive X-ray Analysis: Pennsylvania Lower Kittanning Coal, Jefferson County (−200 +325 mesh, 19 kOe, 2.5 wt % Water Slurry, Stainless Steel Wool Packing, 96% Void Volume, and 100-g Feed Coal)

Element	Grams in feed	Percent reduction,[a] %	Mass recovery,[b] wt %
Al	1.290	30.2	48.8
As	0.0034	55.9	30.9
Cr	0.010	74.0	1.8
Fe	3.350	72.0	19.6
K	0.110	36.4	44.5
Ni	0.0007	71.4	20.0
P	0.680	69.1	21.6
S	6.250	62.6	26.2
Sn	1.930	38.9	42.8
V	0.0160	43.8	39.4

[a]Percent reduction of elemental composition of the feed coal by HGMS.
[b]Weight percent of element in the feed which remains in the magnetically cleaned coal.

effective removal of compounds of arsenic, chromium and nickel from coal by HGMS is especially noteworthy as these compounds are concentrated in the fine particulates which pass through conventional fly-ash control devices in coal-fired power plants.

Effect of Coal Type. HGMS was first tested on water slurries of a pulverized Brazilian Siderpolis coal containing 25.6 to 29.7 wt % ash and 1.30 to 2.06 wt % total sulfur, up to two-thirds of which are pyritic sulfur [93-96]. In most experiments done, feed coal particle sizes were 92% smaller than 42 μm, and slurry velocities were between 0.44 and 3.92 cm/sec. Typical results obtained at a field intensity of 20 kOe showed that HGMS could reduce the total sulfur by 20 to 53%, the ash by 15 to 21% with a mass recovery of 35 to 72 wt %.

Test results on water slurries of pulverized Illinois No. 6 coals from three studies have been reported. In the first study [41], the coal used was from Perry County with 1.17 wt % pyritic sulfur, 0.03 wt % sulfate sulfur, and 1.90 wt % organic sulfur for a total of 3.10 wt %. This was ground to a mean particle size of 32 μm and 99% smaller than 65 μm. At a field intensity of 20 kOe, a slurry concentration of 2.57 wt %, a slurry velocity of 2.90 cm/sec, and with a stainless steel wool packing of 94% void volume, 28.4% of the total sulfur and 45.6% of the ash were removed with a mass recovery of 85 wt %. This could correspond to 75% removal of the pyritic sulfur. In the second study [20], the coal used contained 2.02 wt % total sulfur, 1.39 wt % pyritic sulfur, and 26.0 wt % ash. The feed was prepared by desliming water slurries of −10-mesh coal followed by grinding to 90% −325 mesh. The initial desliming removed over half of the ash. Typical results obtained at a field intensity of 20 kOe, a slurry concentration of 15 wt % and a slurry velocity of 1.13 cm/sec and with a stainless steel wool packing of 94% void volume showed that for a coal recovery of 85 wt %, 25 to 30% of the ash and pyritic sulfur were removed. In the third study [106], the coal used was from the Old Ben Coal Company containing 1.98 wt % total sulfur, 0.97 wt % pyritic sulfur, and 19.3 wt % ash. Results obtained under a wide range of separation conditions showed that the maximum percentage of pyritic sulfur removal was 60% with a mass recovery of 70 wt %.

HGMS was used for desulfurization of water slurries of pulverized Indiana No. 5 and No. 6 coals from Warrick County, which contained over 4 wt % total sulfur, about half organic and half inorganic [66]. The coals were ground to 99% −200 mesh. At a field intensity of 20 kOe, a slurry concentration of 30 wt % and a slurry velocity of 1.78 cm/sec, the percentages of inorganic sulfur removed from Indiana No. 5 and No. 6 coals by three-pass separations were 67 and 85%, respectively. No data on mass recoveries of cleaned coals were reported.

Table 4.3 shows the maximum percent reductions of total sulfur and sulfur emission level (lb S per MM Btu) and the corresponding

Table 4.3 Maximum Percent Reductions of Total Sulfur and Sulfur Emission Levels (lb S/MM Btu) and Corresponding Mass Recoveries Obtained from HGMS of Water Slurries of Five Pulverized Eastern Coals

Coal type	Maximum percent reduction, %: Total sulfur	Maximum percent reduction, %: Sulfur emission level	Mass recovery of clean coal, wt %
1. Illinois No. 6 (Old Ben)	25	28.7	69.2
2. Pennsylvania Lower Freeport (Canterbury)	51	57.9	54.4
3. Pennsylvania Sewickley	55	66.4	37.4
4. Pennsylvania Upper Freeport (Delmont)	65	68.2	55.7
5. West Virginia Pittsburgh No. 8	33	-	70.2

Source: Ref. 106.

mass recoveries of coal coals reported from recent HGMS tests on water slurries of five pulverized Eastern coals over a wide range of operating conditions [106]. In general, if the cleaned coal recovery was high, the corresponding improvement in coal quality (grade) was not outstanding. Conversely, if the quality improvement was substantial, the resulting coal yield (recovery) was not.

Cyclic Superconducting HGMS Tests. The applications of cyclic superconducting HGMS for desulfurization of water slurries of pulverized Pennsylvania Upper Freeport (Delmont) and Illinois No. 6 (Old Ben) coals were investigated recently [89]. The coals used were ground to −170 mesh, and the tests were performed on a laboratory cyclic superconducting HGMS unit similar to that illustrated previously in Fig. 4.7. The effects of field intensity (0, 20, 50 kOe), slurry velocity (1, 2, 4 cm/sec), slurry concentration (10, 20, 33 wt %), and apparent matrix lengths (30.5, 61.0, and 91.5 cm) on the grade and recovery of the separation were quantitatively examined.

At a field intensity of 50 kOe, a slurry velocity of 4 cm/sec and a slurry concentration of 20 wt %, superconducting HGMS was effective in removing 54% of the total sulfur, 88% of the pyritic sulfur, and 75% of the ash from the Pennsylvania Upper Freeport (Delmont) coal with a corresponding Btu recovery of cleaned coal of 62%. The test resulted in a cleaned coal with a sulfur emission level of 0.35 lb S per MM Btu compared with an initial value of 1.25 for the feed. Comparable desulfurization performance was achieved by the Illinois No. 6 coal under similar test conditions, with the sulfur emission level of the feed being reduced from 2.66 to 1.23 lb S per MM Btu.

The results obtained from superconducting HGMS tests confirmed the general observations of the effects of operating variables and conditions on desulfurization performance as previously discussed for electromagnetic HGMS tests. The only exception was that at a field intensity of 50 kOe, both low and high slurry velocities resulted in essentially identical sulfur reductions for the coals tested; however, higher velocities were able to achieve higher mass and Btu recoveries of the cleaned coal. This result demonstrated the significance of using higher field intensities provided by the superconducting HGMS unit (as opposed to the generally lower field intensities available to the electromagnetic HGMS unit) to allow for the use of higher slurry velocities to enhance the sulfur reduction without sacrificing the cleaned coal recovery.

Carousel Electromagnetic HGMS Tests. The technical feasibility of applying the continuous carousel electromagnetic HGMS for desulfurization of water slurries of a pulverized Pennsylvania Upper Freeport (Delmont) coal was demonstrated recently [106]. The HGMS unit used was a SALA-HGMS Model MK III pilot-scale continuous carousel separator, and the feed coal was prepared by grinding the float products obtained from washing the raw Delmont coal with a magnetite medium with a sepcific gravity of 1.5 to 1.6. All experimental tests were conducted at a field intensity of 17.1 kOe with feed coal particle size, slurry flow velocity, matrix loading factor and method of washing as independent variables. The results obtained from continuous carousel tests were also correlated with those obtained from laboratory cyclic experiments under comparable conditions.

For similar feeds and operating conditions, the data from this study showed good correlation between the performance of the carousel separator and that of the cyclic unit, especially with respect to the mass recovery of the cleaned coal. Good agreement was also observed between the sulfur and ash reductions from carousel and cyclic separators. Although no firm conclusions could be drawn regarding the parametric effects of separation variables, results from carousel tests did show similar trends on the effects of certain separation variables as discussed previously for cyclic experiments. For example, Fig. 4.18 illustrates the similar effects of slurry flow velocity on the sulfur and ash reductions achieved by both cyclic and carousel tests [106]. It can be seen that carousel performance could be as good as cyclic performance. This is encouraging in that one can conduct relatively simple tests in a small laboratory cyclic separator to produce results which are comparable to those from more complex and time-consuming pilot-plant carousel tests.

Continuous Superconducting Quadrupole HGMS Tests. A laboratory prototype unit of a continuous superconducting quadrupole separator previously illustrated in Figs. 4.9 and 4.10, with a maximum field intensity 14 kOe, was designed and tested on water slurries of two high-ash Pennsylvania coals ground to −100 mesh [10,11]. The coal slurry of 25 wt % solids concentration was circulated through the separator at a rate of 1200 kg/hr. Experimental data showed that magnetic separation failed to achieve adequate removal of pyritic sulfur from one of the coals; while over 83% of the pyrite in the second coal could be removed with a cleaned coal recovery of 81 wt %. The first coal was apparently not sufficiently pulverized to liberate the pyrite particles in the coal. Also, the field intensity of the laboratory prototype separator was too low to remove the finely disseminated pyrite locked in the coal. It was suggested that the adequate liberation of pyrite from coal by fine grinding together

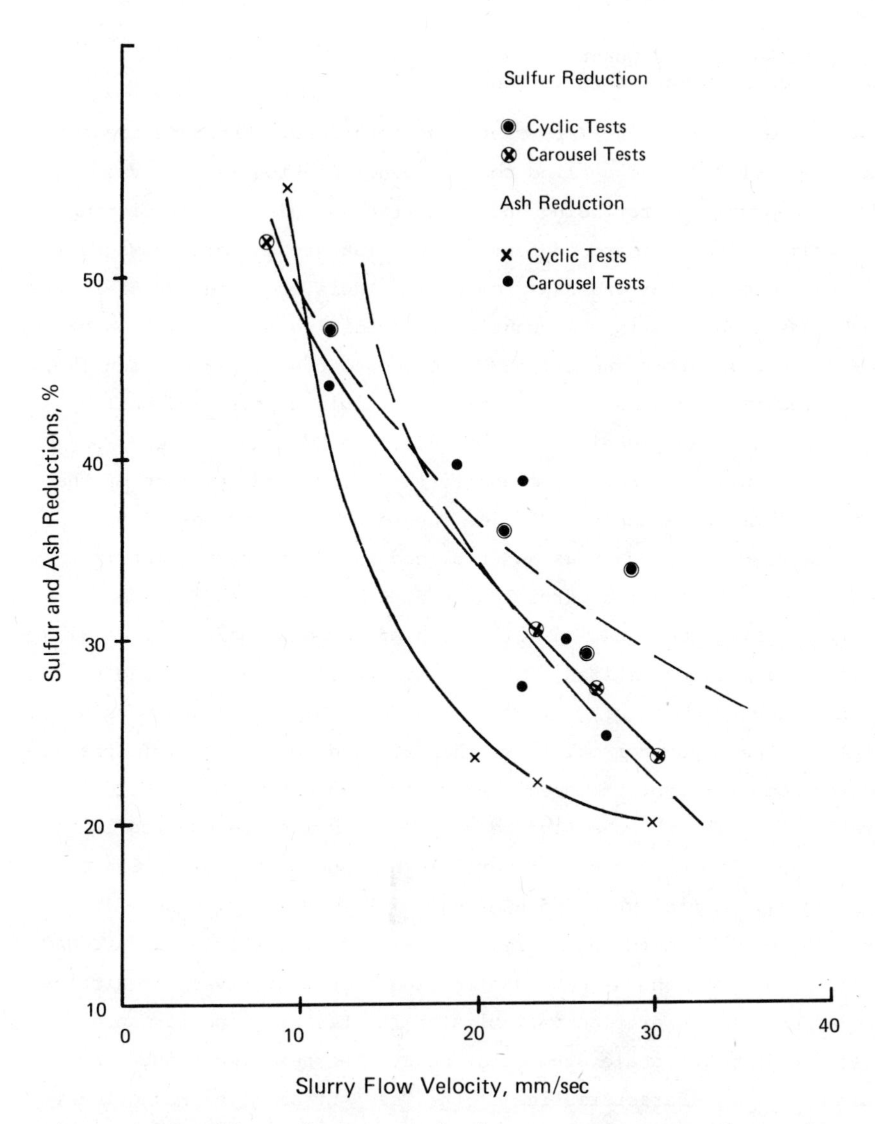

Figure 4.18 Effect of slurry velocity on sulfur reduction and cleaned coal mass recovery obtained from carousel and cyclic HGMS of water slurries of a pulverized Pennsylvania Upper Freeport (Delmont) coal at comparable experimental conditions. (From Ref. 106.)

with the development of a second generation of the new quadrupole magnetic separator with a higher field intensity would improve the desulfurization performance [10].

B. Qualitative and Quantitative Correlations of Desulfurization Performance

Modeling Objectives and Important Considerations. The objective of mathematical modeling applied to coal beneficiation is to develop the conceptual understanding and quantitative model for predicting the grade, recovery, and concentration breakthrough for the magnetic separation of sulfur and ash from coal. There are a number of important considerations in the quantitative modeling and prediction of the technical performance of HGMS applied to coal beneficiation [50, 57]. The first one is related to the particle capture and buildup. Specifically, the technical performance of a HGMS unit depends on how efficiently it captures magnetic particles, and how much of the captured particles can build up as layers on the surfaces of the magnetized collecting wires and retained in the separator matrix. The capture of magnetic particles by HGMS has been studied theoretically by using the equations of motion of magnetic particles flowing around a single magnetized, ferromagnetic collecting wire; and the performance of an ideal, unloaded HGMS unit composed of many such wires in the separator matrix is then related to the particle trajectories computed from the specified separation conditions [56,101]. Preliminary efforts have also been made to examine theoretically and experimentally the effects of particle buildup on the performance of a nonideal, partially loaded HGMS unit [57]. However, none of the existing models based on the recent analyses of particle capture and buildup has been shown to be applicable to quantitatively predicting the effects of separation variables on the technical performance observed in pilot-scale studies of HGMS. The next important consideration is the characteristics of the feed stream to be magnetically cleaned. In the literature, most of the reported modeling and experimental studies of HGMS have been limited only to the feed streams containing either pure magnetic particles, or simple mixtures of magnetic and nonmagnetic particles of approximately monodispersed or narrowly distributed sizes. However, little attention has been devoted to relating the technical performance of HGMS to the characteristics of the feed stream containing particles of a wide range of sizes, densities, and magnetic susceptibilities as found in the magnetic beneficiation of coal. The third, but relatively less important, consideration is related to the mechanical entrapment or filtration of particles at low or zero field intensity. For separations at high field intensity, however, the effect of mechanical entrapment on the technical performance of HGMS is often considered to be negligible.

Qualitative Correlation Using Force Balance Model. The first model developed for qualitatively correlating the desulfurization performance of HGMS was the force balance model due to Trindade [93-96]. The basic principles of the model may be illustrated by considering a HGMS unit which consists of a packed column inserted vertically in the bore of a solenoidal magnet. The packing, a filamentary ferromagnetic material such as stainless steel wools or a steel wire screen is the source of the field gradient and holds the magnetically captured particles. The model considers an isolated strand of steel wool taken as a cylindrical wire of uniform cross section inserted horizontally in a volume (e.g., the bore of a solenoidal magnet), where the magnetic field is uniform vertically. The pyrite particles of various sizes are carried in a water slurry flowing by gravity past the magnetized steel wool strand (see Fig. 4.19). The net vertical force, F_{net}, acting on a given particle located at a certain point close to the magnetized strand results from the balance of the following forces:

1. Vertical component of the magnetic force attracting the particle toward the strand, F_{mv}
2. Hydrodynamic drag force pushing the particle downward, F_d
3. Net weight of the particle, i.e., weight minus buoyance, also pointing downward, W

Consequently,

$$F_{net} = F_{mv} - (W + F_d)$$

$$R = \left| \frac{F_{mv}}{(W + F_d)} \right|$$

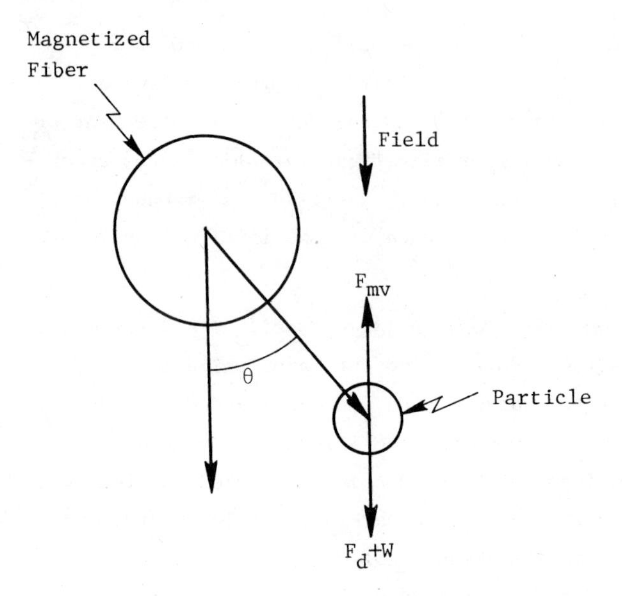

Force Ratio = $F_{mv}/(F_d+W)$

Figure 4.19 Magnetic and competing forces acting on a paramagnetic particle flowing past a magnetized steel wool strand by gravity. F_{mv}, vertical component of magnetic force; F_d, hydrodynamic force; W, net weight of particle. (From Ref. 61. © 1976 IEEE.)

where R is the ratio between the magnetic and competing forces. Here, the value of the force ratio R can be considered as a measure of the probability of capture of the particles. The force balance model postulates that for an effective separation, the ratio R must be of the order or greater than unity, i.e., $R \geq 1$.

A series of model simulations were carried out to examine the effects of major independent variables on the probability of capture of sulfur-bearing mineral particles as measured by the value of R. The magnetic field intensity was 20 kOe, the source of field gradient was a cylindrical steel wool strand of 100 μm in diameter, and only pyrite particles with a magnetic susceptibility of 2.5×10^{-6} emu/g were considered. Figure 4.20 illustrates the effect of particle size on the force ratio R for different slurry velocities [93-96]. As shown, the model predicts that there should be a certain particle

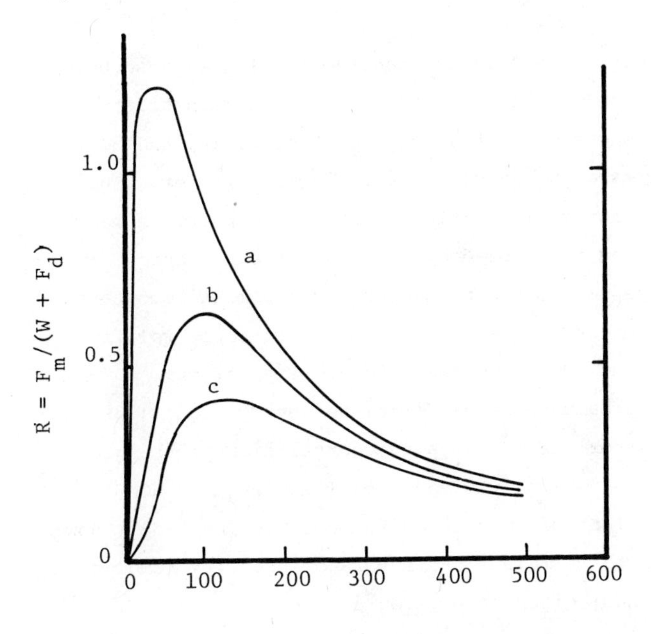

Figure 4.20 Effect of particle size on the ratio of magnetic force to competing forces on a pyrite particle at slurry velocities of 0.1 (case a), 10 (case b), and 20 (case c) cm/sec. (From Refs. 93-96.)

size for which R reaches a maximum. Consequently, one would expect that the sulfur concentration and recovery in the magnetic product would peak at a certain particle size if pyrite is the dominant form of sulfur and it is sufficiently liberated. This model prediction was confirmed by experimental data from HGMS tests on water slurries of pulverized Brazilian Siderpolis coals of narrowly distributed (approximately monodispersed) particle sizes [93-96]. A comparison of the curves with different slurry velocities shown in Fig. 4.20 suggests that the force ratio R decreases as the slurry velocity increases. Therefore, the recovery of magnetic product should decrease. Also, the sulfur concentration and recovery in the magnetic product should decrease accordingly. In other words, as R diminishes, the magnetic force becomes less able to retain the particles inside the separator. As the slurry velocity increases, the first particles to get loose from the packing should be those held marginally, i.e., those with the smallest resultant magnetizations. Figure 4.20 further shows that at a small particle size, the decrease in R due to the increase in the slurry velocity is steeper. The latter suggests that for small particles, where the particle weight is negligible, the drag force predominates over the magnetic force. In general, for large sizes, the net weight is the most important force; and for intermediate sizes, the magnetic force is relatively more important than the competing drag and gravity forces.

The simple force balance model formulated by Trindade [93-96] has provided a basis for the quantitative correlation of the grade and recovery of the magnetically cleaned coal achieved by HGMS for a variety of combinations of separation variables. The model, however, cannot be used to correlate the separator capacity (concentration breakthrough) under given separation conditions.

Quantitative Correlations Using Particle Trajectory and Buildup Models. A number of recently developed models based on the theoretical analysis of particle trajectory and buildup have been applied to quantitatively correlate the grade and recovery of the magnetically clean coal from HGMS of water slurries of pulverized coals [6,43,89, 91,101]. These models essentially analyze the capture of a magnetic particle by the magnetized ferromagnetic collecting wire such as stainless steel wool strand in the separator matrix of a HGMS unit as two separate, but related, problems. The first problem involves the analysis of the motion of the particle by finding its trajectory in the magnetic field and determining under what conditions a particle will strike the wire. The second problem is concerned with examining whether, after the impact of a particle on the wire, the particle will stick to the wire and continue to build up as a layer on the surface of the wire, or the particle will be washed away from the wire by the carrier fluid [112].

The first model for the capture of magnetic particles in a HGMS unit was presented by Bean [6]. His model was intended as an approximation of the gross characteristics of the ideal or unloaded magnetic separator performance without requiring sophisticated calculations. In the model, the slurry with an inlet velocity V_o (which is directly proportional to the reciprocal of residence time τ) and a viscosity η is assumed to be moving along the magnetic field of an intensity H_o at a high Reynolds number relative to an array of ferromagnetic collecting cylinders of a radius a and a saturation magnetization M. The cylinders occupy a void volume fraction ε (unity minus the packing density factor F) of the solenoidal volume (magnetic separator matrix) of a length L. For particles of a magnetic susceptibility χ and a radius R, the particle concentrations at the inlet and outlet of the separator matrix, C_{in} and C_{out}, respectively, can be related by the following expression:

$$\frac{C_{out}}{C_{in}} = \exp\left(-\frac{4\pi}{3}\frac{\mu MH(\chi R^2)}{\eta V_o/\varepsilon}\frac{(1-\varepsilon)L}{\pi a^2}\right) \tag{4.1}$$

In the equation, μ is the permeability of the free space which has the value of unity in the CGS system of electromagnetic units. One sees that a plot of the logarithm of the ratio C_{in}/C_{out} versus the parameter $HL\varepsilon(1-\varepsilon)/V_o$ or $H\tau\varepsilon(1-\varepsilon)$ should yield a straight line if other parameters included in this model are all kept constant and the experimental data can be satisfactorily correlated by the model. Figure 4.21 illustrates the application of Bean's model to quantitatively correlate the data from HGMS of water slurries of a pulverized Illinois No. 6 coal over a wide range of field intensity (5 to 20 kOe), residence time (7.5 to 210 sec) and packing void volume fraction (0.03 to 0.17) at a slurry concentration of 2.57 wt % [43]. The agreement between theoretical prediction and experimental data shown in the figure is quite significant considering the simplifying assumptions included in the model and the complexity of the problem of magnetic removal of sulfur and ash by HGMS. The successful experimental verification of the model allows one to quantitatively identify the trade-off of operating variables [76], such as field intensity H, residence time τ, and packing void volume fraction ε, so as to optimize the magnetic separation of sulfur and ash. Note that the Bean's model predicts

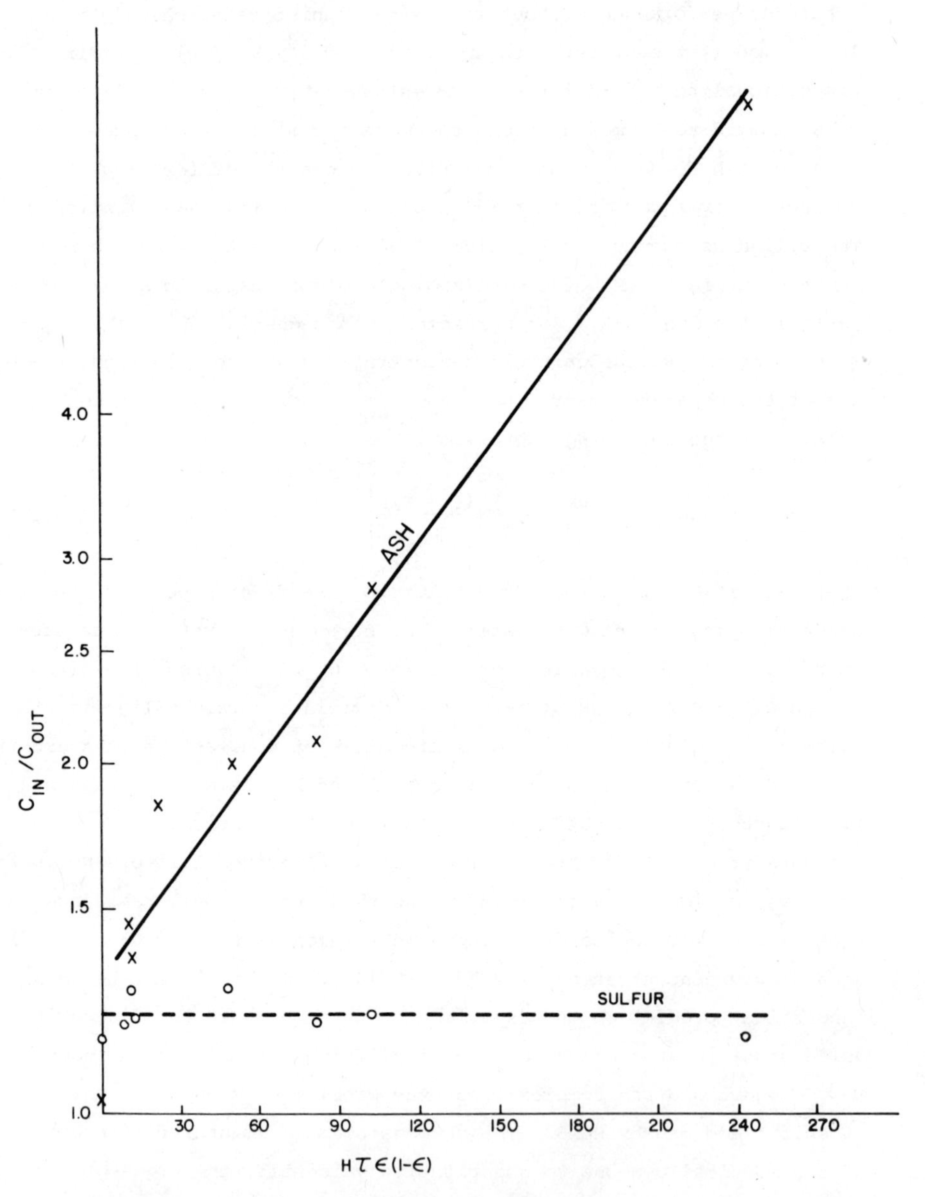

Figure 4.21 Application of Bean's model to correlate sulfur and ash concentration ratios obtained from HGMS of water slurries of a pulverized Illinois No. 6 coal over a wide range of separation conditions. (From Ref. 43. © 1976 IEEE.)

no removal of sulfur and ash when a coal slurry is processed by a HGMS unit at zero field intensity. In practice, there is always some removal due to mechanical entrapment. In addition, it is not possible to use the model to predict the separator capacity or concentration breakthrough.

Available analyses have shown that both particle capture and buildup depend highly on the ratio of the magnetic velocity V_m of the particle to the free stream fluid velocity V_∞ is defined by the inlet fluid velocity V_o divided by the packing void volume fraction ε and the magnetic velocity V_m is essentially a terminal velocity of a particle in a magnetic field. The latter is defined such that [50,101]:

$$\frac{V_m}{V_\infty} = \frac{2}{9}\,\frac{\mu MH(\chi R^2)}{\eta V_o/\varepsilon}\,\frac{1}{a} \tag{4.2}$$

One sees that the ratio V_m/V_∞ contains almost all the major independent variables in HGMS, namely (1) magnetic field terms μMH, magnetization of the particle collecting wire M and field intensity H (μ is the permeability of the free space); (2) particle property term, χR^2-particle magnetic susceptibility χ and radius R; (3) flow field term $\eta V_o/\varepsilon$, fluid viscosity η, inlet velocity V_o, and packing void volume fraction ε; and (4) matrix packing term $1/a$, where a is the wire radius. It is of interest to note that the first three terms appear explicitly in the exponent of Bean's model, and the fourth (matrix packing) term $1/a$ is replaced by $(1 - \varepsilon)L/\pi a^2$ in the exponent of Bean's model.

A particle trajectory model developed by Watson [101] and extended by Stekly and Minervini [91] has been applied to quantitatively correlate the experimental data obtained from superconducting HGMS of water slurries of pulverized Pennsylvania Upper Freeport (Delmont) and Illinois No. 6 (Old Ben) coals [89]. This model aslo relates the particle concentrations at the inlet and outlet of the separator matrix, C_{in} and C_{out}, respectively, by an exponential function of major independent variables. Depending upon the magnitude of the ratio V_m/V_∞, the model has the following two forms:

$$\frac{C_{out}}{C_{in}} = \begin{cases} \exp\left(\dfrac{-2}{9}\,\dfrac{\mu MH(\chi R^2)}{\eta V_o/\varepsilon}\,\dfrac{(1-\varepsilon)L}{\pi a^2}\right) & \text{for small } V_m/V_\infty \\ \exp\left(\dfrac{-3^{5/6}}{2^{2/3}}\,\dfrac{(\mu MH)^{1/3}(\chi R^2)^{1/3}a^{2/3}}{(\eta V_o/\varepsilon)^{1/3}}\,\dfrac{(1-\varepsilon)L}{\pi a^2}\,G\right) & \text{for large } V_m/V_\infty \end{cases} \tag{4.3}$$

In the model, G is an empirical constant representing the degree of matrix loading which must be fitted by using experimental data. This model suggests that for large values of V_m/V_∞, a plot of logarithm of the ratio C_{out}/C_{in} versus the parameter $(H/V_o)^{1/3}L$ should yield a straight line if other parameters included in the model are kept constant and the experimental data can be satisfactorily correlated by the model. Figure 4.22 illustrates the application of the model to quantitatively correlate the data obtained from superconducting HGMS of water slurries of a pulverized Pennsylvania Upper Freeport (Delmont) coal over a wide range of field intensity (20 to 50 kOe), slurry velocity (1 to 4 cm/sec), and matrix length (30.5 to 91.5 cm) at a slurry concentration of 20 wt % [89].

The preceding model suggests that the level of sulfur reduction as represented by the concentration ratio C_{out}/C_{in} is exponentially dependent on $(H/V_o)L$ for small values of V_m/V_∞, and on $(H/V_o)^{1/3}L$ for large values of V_m/V_∞. This dependence implies that when other operating parameters are maintained constant, the use of higher field intensities provided by a superconducting HGMS unit (which leads to large values of V_m/V_∞) allows one to use higher slurry velocities. In contrast, when using the generally lower field intensities available to an electromagnetic HGMS unit (which leads to small values of V_m/V_∞), it is necessary to use lower slurry velocities in order to achieve the same level of sulfur reduction. This trade-off between field intensity and slurry velocity is consistent with the experimental results obtained from superconducting HGMS of water slurries of pulverized coals as discussed in Sec. 4.2B.

Another particle trajectory and buildup model developed by Luborsky and Drummond has also been applied to quantitatively correlate the experimental data obtained from HGMS of water slurries of

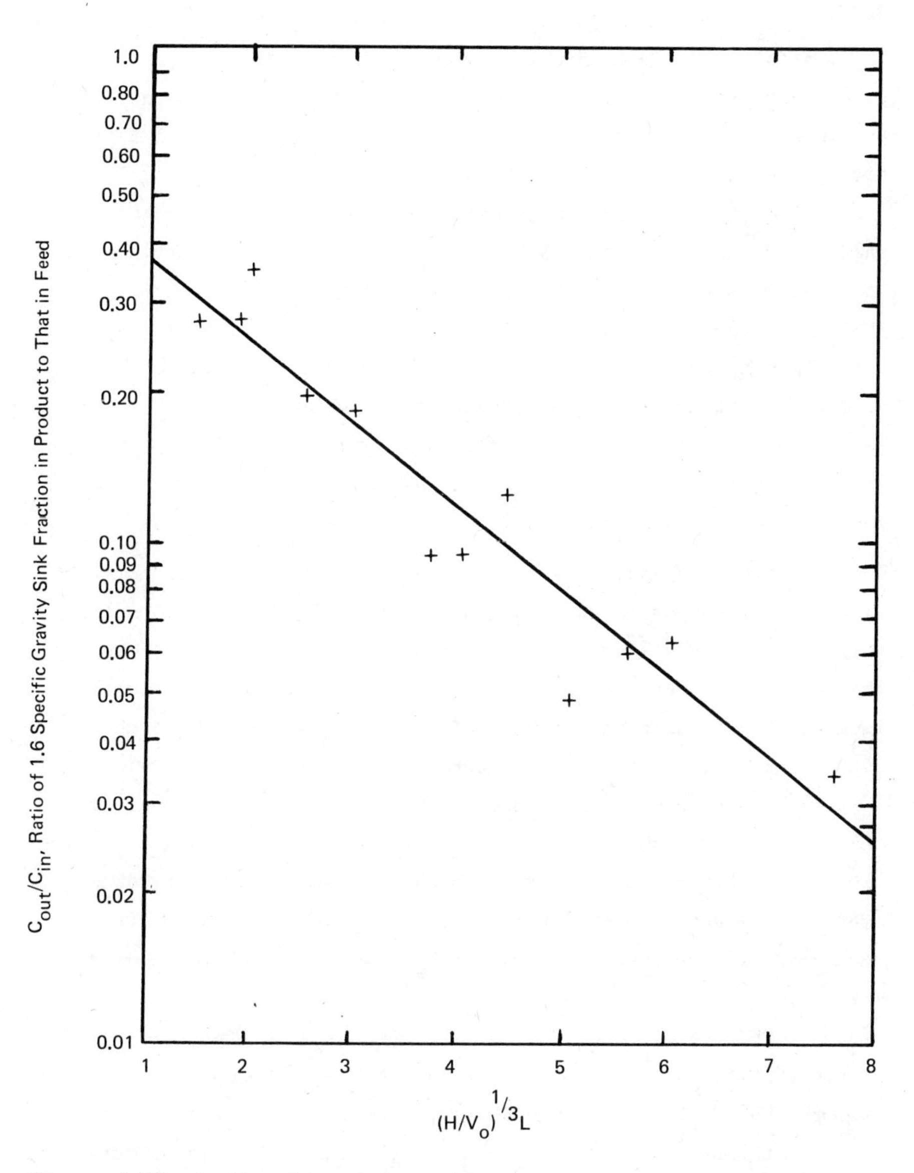

Figure 4.22 Application of Watson's model to correlate product concentrate ratios obtained from superconducting HGMS of water slurries of a pulverized Pennsylvania Upper Freeport (Delmont) coal over a wide range of separation conditions. (From Ref. 89.)

a pulverized Pennsylvania Upper Freeport (Delmont) coal [57]. This model assumes the use of ribbonlike particle collecting wires, instead of cylindrical ones as in the previous models. It considers the possibility of mechanical entrapment of particles in the separator matrix. The model also relates the particle concentrations at the inlet and outlet of the separator matrix, C_{in} and C_{out}, respectively, by an exponential function. The exponent of the model includes two empirical constants which represent the fraction of the particle collecting wires that are active in capturing particles (i.e., the capture efficiency), and the fraction of the particles that are mechanically entrapped. Both empirical constants must be determined by properly fitting the experimental data. A detailed description of the model and its experimental application to HGMS of water slurries of pulverized coals can be found in Ref. 57, in which it is shown that the model calculations can be made to fit the experimental data only approximately.

Quantitative Modeling and Prediction of Desulfurization Performance. In a recent study done at Auburn University [50,68], the technical performance of HGMS for desulfurization of water slurries of pulverized coals has been quantitatively examined with reference to the important considerations of particle capture and buildup, mechanical entrapment of particles and characteristics of the feed stream. Particular emphasis of the study has been placed on the development of a simple model which can satisfactorily predict the grade, recovery, and concentration breakthrough observed in pilot-scale studies of HGMS applied to water slurries of pulverized coals. Several practical observations resulted from this study can be summarized as follows.

1. The technical performance of HGMS is quantitatively related to the particle capture and buildup. Under the presently used or proposed conditions for wet HGMS processes, however, the capability of the wire matrix to capture magnetic particles remains practically unchanged; and essentially all magnetic particles are captured before the matrix is saturated or loaded with the buildup of particles. As

a result, there is practically no need to examine the capture of particles by calculating their trajectories in the separator matrix under the specified separation conditions. The main factor in determining the technical performance of a pilot-scale or an industrial HGMS unit appears to be the particle buildup, but *not* the particle capture.

2. Both the particle capture and buildup are highly dependent upon the ratio of the magnetic velocity V_m of the particle to the free stream fluid velocity V_∞. In particular, the maximum amount of particle buildup on the wire increases with increasing value of V_m/V_∞, and there exists a minimum value of V_m/V_∞ for the captured particles to remain sticking to the wire matrix.

3. Under conditions of constant magnetic field and flow conditions as well as fixed separator matrix packing characteristics, the magnetic velocity is mainly a function of particle properties. In order to quantitatively relate the technical performance of HGMS to the characteristics of the feed stream containing particles of a wide range of sizes, densities, and magnetic susceptibilities, it is necessary to properly determine the magnetic velocities of particles in the feed stream and the distribution of such velocities.

Based on the above observations, a simple particle buildup model incorporating the feed characteristics for predicting the technical performance of HGMS under the presently used or proposed conditions for wet separation has been developed and its details can be found in the literature [50,68]. Figures 4.23 and 4.24 illustrate the typical comparison between the theoretical and experimental concentration breakthrough curves for a pulverized Pennsylvania Upper Freeport coal and its ash, total sulfur and pyritic sulfur contents obtained in the same pilot-scale study at a field intensity of 19 kOe and a superficial velocity of 4.33 cm/sec. The closed agreement between theory and experiment demonstrated in the above and the successful application of the model to several other Eastern coals described elsewhere [50,68] suggest that there is now an experimentally verified, simple model available for quantitatively predicting the grade, recovery, and concentration breakthrough, etc.,

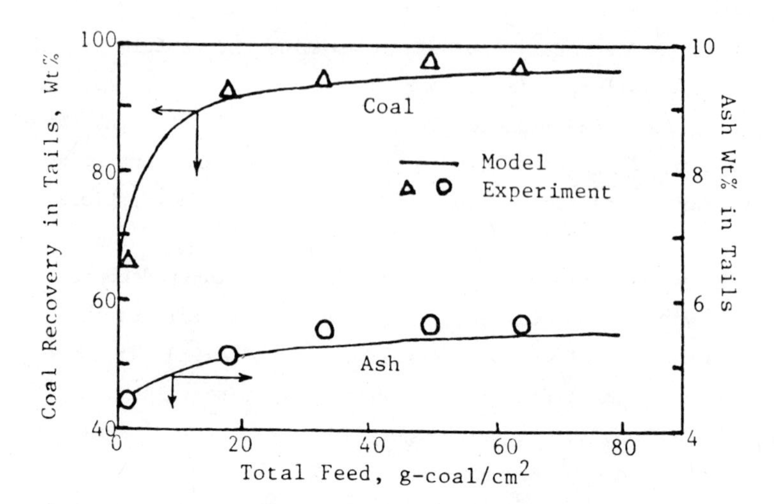

Figure 4.23 Theoretical and experimental coal and ash concentration breakthrough curves obtained from pilot-scale HGMS of water slurries of a pulverized Pennsylvania Upper Freeport coal. (From Ref. 50, 68.)

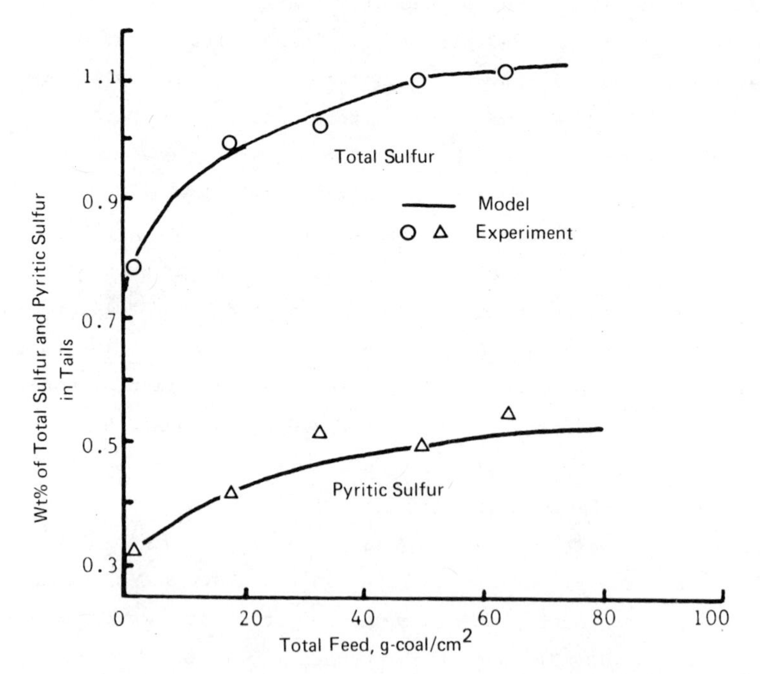

Figure 4.24 Theoretical experimental concentration breakthrough curves for total sulfur and pyritic sulfur obtained from pilot-scale HGMS of a pulverized Pennsylvania Upper Freeport coal. (From Ref. 50, 68.)

in the magnetic cleaning of water slurries of pulverized coal without actual HGMS testing. By using the new model developed, it is possible to quantitatively determine the optimum separation conditions so as to optimize the magnetic removal of sulfur and ash, while achieving an economically acceptable recovery of the magnetically cleaned coal. The new model can also be used to provide the needed information such as the proper separation duty cycle under the selected process conditions for the engineering design and cost estimation of wet HGMS processes applied to coal beneficiation. Further discussion on the practical applications of the modeling results can be found in the literature [50,68].

C. Conceptual Process Design and Cost Estimation

A generalized flowsheet of a HGMS process for desulfurization of water slurries of pulverized coals is shown schematically in Fig. 4.25 [32]. This flowsheet illustrates the major unit operations that would be required in an independent and self-sufficient process which does not share any facilities with a coal mine, power plant or any other plant. The unit operations included are: (1) coal pulverizing (grinding) and slurry, (2) wet HGMS, (3) physical dewatering of clean coal, (4) thermal drying of clean coal, (5) water treatment and supply, and (6) refuse dewatering. In a recent report [32], the key engineering problems, available equipment alternatives and suggested plant facilities for these operations are described in some details. In Chapter 8 of this book, the new developments and practical aspects of filtration and dewatering (thickening) of fine coal and fine coal refuse of importance to the design and operation of wet HGMS processes for desulfurization of pulverized coals are discussed.

Preliminary design and economics of cyclic electromagnetics HGMS processes for desulfurization of water slurries of pulverized coals have been reported in the literature [32,44,87,106]. Figure 4.26 illustrates one of such conceptual cyclic electromagnetic HGMS processes [44]. In the process, a coal slurry of a fixed solids concentration is prepared by mixing known amounts of pulverized

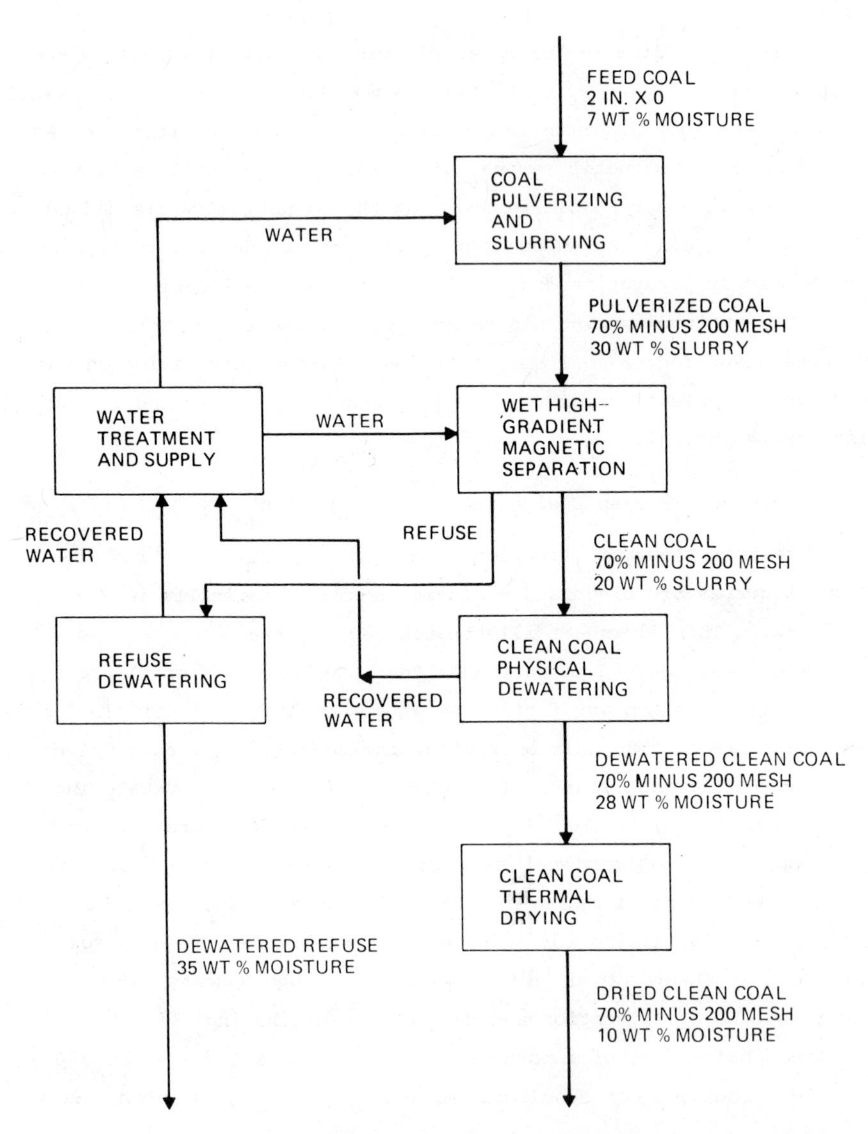

Figure 4.25 Major unit operations required for desulfurization of water slurries of pulverized coals. (From Ref. 32. © 1979 IEEE.)

coal, water, and a dispersant like Alconox. The HGMS unit employed is the largest commercial unit now in use for kaolin beneficiation. It is operated at a fixed intensity of 20 kOe generated in an open volume of 2.13 m (7 ft) in diameter and 0.508 m (20 in) long. A

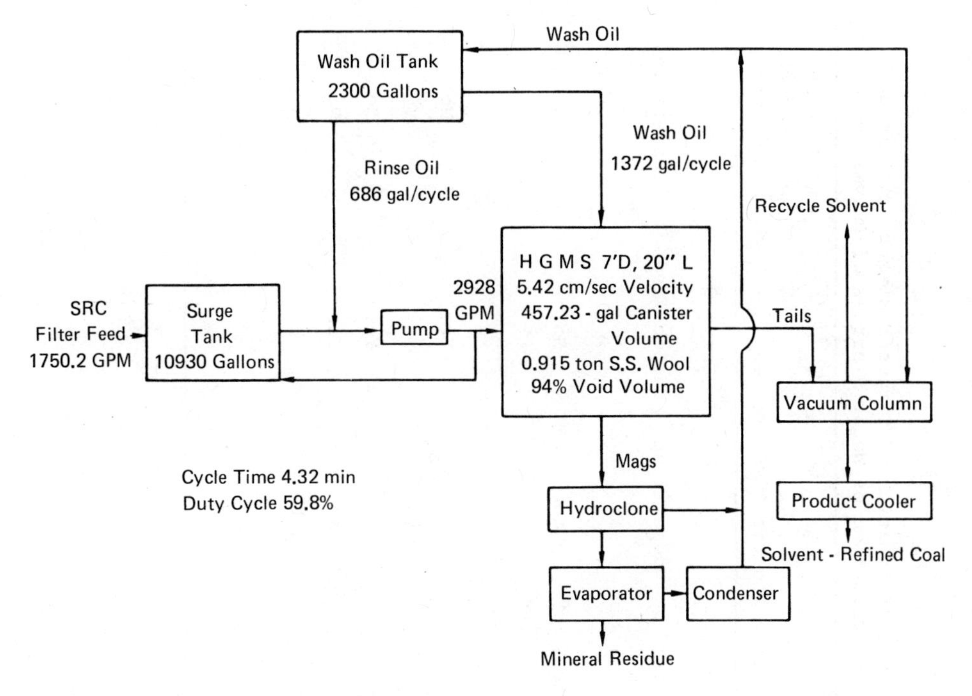

Figure 4.26 A conceptual process flowsheet for desulfurization of water slurries of pulverized coals. (From Ref. 32. © 1979 IEEE.)

separator matrix packed with compressed pads of magnetic stainless steel wool of 94% void volume percentage is placed in the open volume. The coal slurry is pumped through the energized separator matrix at a fixed flow velocity (residence time) until the matrix reaches its loading capacity. After rinsing with water, the magnetic refuse stream (mags) washed from the separator matrix is sent to a settling pond or a clarifier to recover water for reuse. The magnetically cleaned coal product (tails) is collected, dewatered, and dried. The typical specifications of process streams are also included in the figure, based on the actual data obtained from pilot-scale experimental studies of HGMS for desulfurization of a pulverized Illinois No. 6 coal [43,50,68].

Table 4.4 summarizes the scopes of the reported conceptual design and cost estimation of both carousel electromagnetic and reciprocating superconducting HGMS processes for desulfurization of

Table 4.4 Scopes of Reported Conceptual Design and Cost Estimation of Carousel Electromagnetic and Reciprocating Superconducting HGMS Processes for Desulfurization of Water Slurries of Pulverized Coals

	Carousel electromagnetic HGMS process				Reciprocating superconducting HGMS process
1. Reference	[93]	[106]	[32]	Chap. 9	[89]
2. Coals	Brazilian	Eastern	Unspecified	Unspecified	Unspecified
3. Experimentally-tested separation conditions used in design	Yes	Yes	No	No	No
4. Operations included in cost estimates					
a. Coal pulverizing	No	No	Yes	Yes	No
b. Magnetic cleaning	Yes	Yes	Yes	Yes	Yes
c. Dewatering/drying	No	No	Yes	Yes	No
d. Refuse disposal	No	No	Yes	Yes	No
5. Nature of cost estimates	a	a	b	b	a
6. Sensitivity analysis of costs examined	No	No	No	Yes	No

[a] Order-of-magnitude estimates.

[b] More detailed cost estimates.

pulverized coals. This table suggests that the most reasonable costs for magnetic separation alone by carousel electromagnetic HGMS units are those presented in Ref. 106, based on the actual experimental results for several pulverized Eastern coals (see, for example, Table 4.3). It also shows that the most complete economic analyses of the carousel electromagnetic HGMS process reported thus far are those given in Ref. 32 and in Chap. 9. The reader is referred to Chap. 9 for the detailed estimates of magnetic desulfurization characteristics and conceptual process requirements as well as installation and processing costs for the carousel electromagnetic HGMS process.

It is important to note that in most of the reported design studies of HGMS processes for desulfurization of water slurries of pulverized coals, the specific types of coals to be cleaned as well as the specific models of cyclic or continuous separators and their major operating conditions to be used have not been fully specified. These unspecified items can, however, significantly affect the desulfurization performance, and projected processing capacities and cost estimates. As an illustration, Table 4.5 shows the estimated processing capacities per magnetic head (magnetic station) for several eastern coals, considering such items as the coal type, size consist, flow velocity, solids concentration, matrix loading, and separator type [106]. The specific experimental data for the selected coals can be found in Refs. 57, 82, and 106 and are discussed in Sects. 4.2B, 4.5B, and 4.7B. It can be seen from Table 4.5 that the projected processing capacities for the selected HGMS units are highly dependent on the items which are often unspecified in most reported design studies. It is thus suggested that one should examine the results of most reported design studies of wet HGMS processes for coal desulfurization with caution.

4.3 HIGH-GRADIENT MAGNETIC SEPARATION FOR DESULFURIZATION OF DRY PULVERIZED COAL

The application of HGMS to desulfurization of dry pulverized coal was initiated in 1976. Much of the earlier work reported [57,66]

Table 4.5 Estimated Processing Capacities of Typical Cyclic and Carousel Electromagnetic HGMS Units for Desulfurization of Water Slurries of Selected Eastern Coals

Coal type (data reference)	Flow velocity, cm/sec	Solids concentration, g/cm^3	Volumetric matrix loading, g/cm^3	Capacity per magnetic head, metric ton/hr		
				Cyclic[a] SALA/HEMF No. 214-30-19	Linear[a] SALA/HGMS No. 160	Carousel[a] SALA/HGMS No. 480
1. Illinois No. 6 (Old Ben), 70% −200 mesh [106]	0.5	0.30	0.37	5.9	13.4	8.1
2. Pennsylvania Lower Freeport (Canterbury) 70% −200 mesh [106]	0.5	0.30	1.75	12.8	19.3	11.7
3. Pennsylvania Sewickley, 70% −200 mesh [106]	0.9	0.30	1.50	20.4	33.0	20.0
4. Pennsylvania Upper Freeport (Delmont) −60 mesh [57]	1.36	0.30	0.50[b]	15.3	29.0	17.6
5. West Virginia Pittsburgh No. 8, −14 mesh [82]	4.0[b]	0.30	0.50[b]	27.0	85.4	51.8

[a]Manufactured by Sala Magnetics, Inc.

[b]Estimated values.

Source: Ref. 106.

had been limited to investigating the technical feasibility of HGMS for desulfurization of dry pulverized coal by conventional or vibration-assisted gravity feeding, or by downward air-entrained separation. Although some degree of magnetic removal of inorganic sulfur and ash was observed in earlier investigations [57,66], the reported results had not achieved desulfurization performance comparable to those obtained from coal-water slurries. Since late 1977, new effective approaches to HGMS for desulfurization of dry pulverized coal based on gravity feeding assisted by a slow-velocity air flow and recirculating air fluidization have been developed. Table 4.6 summarizes the different cyclic HGMS tests which have been reported for desulfurization of dry pulverized coal. In what follows, the available experimental results and their important implications are discussed according to three different approaches: (1) gravity feeding, (2) air-entrained separation, and (3) recirculating air fluidization.

A. HGMS for Desulfurization of Dry Pulverized Coal by Gravity Feeding

The ORNL/Sala Electromagnetic HGMS Process for Desulfurization of Selectively-Sized Powdered Coals by Gravity Feeding Assisted by a Slow-Velocity Air Flow. The simplest way of feeding dry pulverized coal to a HGMS unit is by gravity feeding. Preliminary results reported using either conventional [66] or vibration-assisted [57] gravity feeding were not encouraging. An experimental study which demonstrated the technical feasibility of applying HGMS by gravity feeding assisted by a slow-velocity air flow for desulfurization of selectively-sized powdered coals was carried out by Sala Magnetics, Inc. (Sala) for the Oak Ridge National Laboratory (ORNL) [15,24]. Figure 4.27 shows a schematic diagram of the HGMS system used, in which the matrix was of 2.54-cm inside diameter and 0.153 m in length and packed with stainless steel expanded-metal screens. The coals used were Kentucky No. 9 from Muhlenberg County and Pennsylvania Lower Freeport from Lancaster County. The Kentucky coal was ground to —24- and 100-mesh levels; while the Pennsylvania coal was pulverized to —100- and 200-mesh levels before being classified into

Table 4.6 Reported Cyclic HGMS Tests for Desulfurization of Dry Pulverized Coal

Process number	Experimental approach	Typical field intensity, kOe	Typical feed coal particle sizes	Coals used	Refs.
1	Conventional gravity feeding	20		Indiana No. 5 and No. 6	[66]
1A	Vibration-assisted gravity feeding	20 and 64.5	60 × 0 mesh, −200 +200 mesh	Pennsylvania Upper Freeport	[57]
1B	Gravity feeding assisted by slow-velocity air flow (ORNL/SALA process)	20	CSF/PSFR[a]	Kentucky No. 9 and Pennsylvania Lower Freeport	[24]
2	Upward air-entrained flow	20	CSF/PSFR	Same as above	[24]
2A	Upward air-entrained flow	20	UBF[b]	Pennsylvania Upper Kittanning	[65]
2B	Downward air-entrained flow	20	−60 mesh	Pennsylvania Upper Freeport	[57]
3	Auburn/ORNL electromagnetic fluidized-bed[c]	20	CSF/PSFR	Pennsylvania Upper Freeport, Lower Freeport, Upper Kittanning and Lower Kittanning	[14,46, 51,53]
4A	Auburn/NEPSCO electromagnetic fluidized-bed[c]	20	UBF	Pennsylvania Upper Kittanning	[54,65]
4B	Auburn/NEPSCO superconducting fluidized-bed[c]	40 to 50	UBF	Pennsylvania Upper Kittanning and Upper Freeport, Ohio/ Pittsburgh No. 8	[99]

[a]CSF/PSFR: classified size fractions per preseparation fines removal (e.g., −100 +200 mesh, −200 +325 mesh, −100 mesh +10 μm, −200 mesh +10 μm).

[b]UBF: utility boiler feed coal, typically 80% −200 mesh.

[c]Process 3 uses a novel expanded fluidized-bed separator matrix, while processes 4A and 4B employ new straight-bed separator matrices with unique partial product recycle techniques.

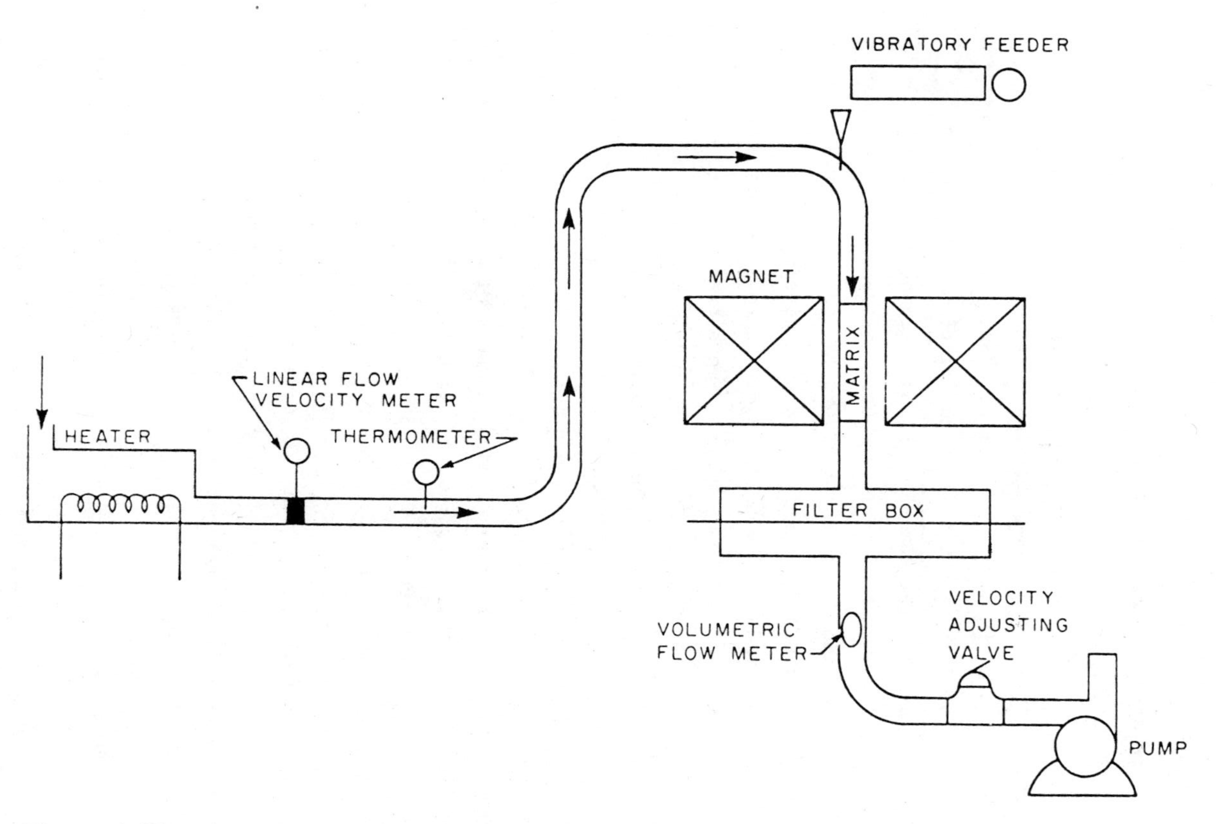

Figure 4.27 A laboratory-scale HGMS system for desulfurization of dry pulverized coal by gravity feeding assisted by a slow-velocity air flow [24].

several size fractions. During each experiment, a selected size fraction was fed to the HGMS unit by gravity feeding assisted by a slow-velocity, heated air stream. Ultrafine coal particles (below 10 μm) were separated from the feed stream and not subjected to magnetic desulfurization.

Table 4.7 illustrates the results obtained for several size fractions classified from coal samples originally crushed to –100-mesh level with a field intensity of 20 kOe and an air velocity of 25.4 cm/sec (50 ft/min). The separator matrix used for all the tabulated tests were the same one, being designated as matrix 57 in Ref. 24. The amount of coal processed per experiment was also essentially the same, being about 195 to 198 g. The tabulated results show that HGMS by gravity feeding assisted by a slow-velocity air stream could reduce 23.6 to 71.2% of the total sulfur and 60.1 to 64.7% of the ash from the –100- +200-mesh size fraction of both coals and achieve 92.0 to 94.7% Btu recovery. HGMS also removed 22.4 to 43.0% of the total sulfur and 46.5 to 49.8% of the ash from the –200- +325-mesh size fraction of both coals with a Btu recovery of 83.8 to 84.4%. Although only 9.8 to 32.4% of the total sulfur and 13.3 to 38.2% of the ash were removed from fine size fractions (–325 mesh +10 μm), the corresponding amounts of magnetic refuse were relatively large (20.2 to 39.8 wt %), resulting in relatively small amounts of mass recovery (60.2 to 69.8 wt %) of the clean coal. The latter result might be caused by the presence of ultrafine particles in the fine size fractions, which could promote the agglomeration of the pulverized coal and its mineral impurities. Consequently, certain coal particles could be mechanically entrapped together with the magnetically attracted mineral particles, and retained inside the separator matrix, leading to a relatively large amount of magnetic refuse. It can thus be concluded that for both coals tested, significant sulfur reduction and Btu recovery of coarse size fractions (e.g., –100 +200 mesh) were achieved by HGMS using gravity feeding assisted by a slow-velocity air flow. However, the process was not able to produce practically acceptable desulfurization performance for fine size fractions (e.g., –200 mesh +10 μm, –325 mesh +10 μm).

Table 4.7 Magnetic Desulfurization of Classified Size Fractions of Dry Pulverized Kentucky No. 9 and Pennsylvania Lower Freeport Coals by Gravity Feeding Assisted by a Slow-Velocity Air Flow

Coal type	Size fraction, mesh	Weight percent,[a] wt %	Mass recovery, wt %	Btu recovery, wt %	Sulfur reduction, wt %	Ash reduction, wt %	Desulfurization performance index, P[c]
Kentucky No. 9	−100 +200 mesh	52.6	80.8	92.0	23.6	60.1	1.229
	−200 +325 mesh	21.9	72.1	84.4	22.4	49.8	0.802
	−325 mesh +10 μm	15.3	60.2	71.5	9.8	13.3	0.246
	−10 μm[b]	10.2					
Pennsylvania Lower Freeport	−100 +200 mesh	21.6	81.2	94.7	71.2	64.7	3.787
	−200 +325 mesh	21.0	75.5	83.8	43.0	46.5	1.755
	−100 mesh +10 μm	77.1	65.3	73.8	32.4	38.2	0.933
	−200 mesh +10 μm	55.5	69.8	72.8	18.1	23.3	0.599
	−10 μm[b]						

[a]Weight percent of the classified size fraction in the original −100-mesh coal sample.

[b]The −10 μm fraction was not subjected to magnetic desulfurization.

[c]Defined as % total sulfur removed per % magnetic refuse.

Source: Ref. 24.

Despite its disadvantage of having to remove the ultrafine particles from the feed coal stream prior to magnetic separation (see further discussion below), the ORNL/Sala electromagnetic HGMS process by gravity feeding assisted by a slow-velocity air stream appears to be an effective approach to magnetic desulfurization of selectively sized powdered coals. A pilot-scale carousel electromagnetic HGMS system of a capacity of 1 ton/hr (see Fig. 4.4) is being used by the ORNL project to investigate the commercial feasibility of magnetic separation of sulfur and ash from classified size fractions of several dry pulverized coals, and encouraging test results have already been reported [97].

Technical Problems Associated with HGMS for Desulfurization of Dry Pulverized Coal by Gravity Feeding. Based on the preceding discussion of experimental results, possible technical problems associated with HGMS for desulfurization of dry pulverized coal by gravity feeding are suggested as follows. First, it should be evident from the available experimental results for desulfurization of dry pulverized coals of different particle size ranges (see, for example, Table 4.7) that the presence of ultrafine particles in the feed coal would often promote the agglomeration of the pulverized coal and its mineral impurities. The latter would lead to relatively small amounts of sulfur reduction and mass recovery. Thus, it has been suggested that the effective desulfurization of dry pulverized coal by HGMS by gravity feeding would require the use of crushed coal with the ultrafine particles (typically smaller than 400 mesh or 37 μm) being removed from the feed coal before magnetic separation [24,51,57,98]. Indeed, experimental data illustrated in Table 4.7 indicate that by removing the ultrafine particles smaller than 10 μm from the Pennsylvania Lower Freeport coal ground to −100 mesh, up to 71.2% total sulfur removal and 94.7% Btu recovery can be achieved when the −100- +200-mesh size fraction is desulfurized by HGMS by gravity feeding assisted by a slow-velocity air flow.

One problem in using such moderately ground, classified size fractions of pulverized coal in desulfurization, however, is that the mineral impurities may not be sufficiently liberated from the coal; and there is a trade-off between avoiding fine coal agglomeration and grinding to liberate the mineral impurities in coal. Another problem is related to the removal of the ultrafine particles from the feed coal before magnetic separation. This problem is particularly undesirable when HGMS is intended for use for desulfurization by utility coal users. This follows because typical particle size distributions of pulverized feed coals to utility boilers often contain up to 30 wt % of particles smaller than 10 to 15 μm. Also, sulfur-bearing and ash-forming minerals in many types of coals are inherently micrometer sized [22,64]. Thus, if coal fines smaller than 10 μm are removed, for example, from the Kentucky No. 9 and Pennsylvania Lower Freeport coals pulverized to −100-mesh level and only the remaining coarse size fraction is subjected to magnetic separation, then 10.2 to 25.5 wt % (see Table 4.7) of the pulverized coal will not be magnetically cleaned.

The above discussion clearly suggests that in order to effectively apply HGMS to desulfurization of dry pulverized coal by gravity feeding, a proper means must be developed to handle the ultrafine particles, which tend to promote the agglomeration of the pulverized coal and its mineral impurities. Also, it is necessary to closely examine the grinding level for a specific coal at given separation conditions, while taking into account both the inherent size distributions and liberation characteristics of its sulfur-bearing and ash-forming minerals.

B. HGMS for Desulfurization of Dry Pulverized Coal by Air-Entrained Separation

Several recent publications have suggested or described the use of high-velocity fluidization techniques such as air-entrained flow approach for feeding dry pulverized coal to HGMS devices [27,52,57, 77,109]. For example, in Ref. 77, the conceptual design and cost estimation of an air-entrained flow HGMS process have been described for desulfurization of the dry pulverized coal chemically pretreated by iron pentacarbonyl, $Fe(CO)_5$, by the Magnex process [36] (see Sec.

4.7B). Another conceptual fluidized-bed HGMS process has also been proposed in Ref. 109 for desulfurization of the dry pulverized coal pretreated by heated air or inert gas to convert and oxidize the pyrite in coal to strongly paramagnetic compounds.

Available experimental results from air-entrained flow HGMS tests [24,51,57,65] have revealed the same problem of fine coal agglomeration which can impede the desulfurization performance as observed in dry separations with gravity feeding. In order to minimize the occurrence of fine coal agglomeration, it will be necessary to remove the ultrafine particles from the feed coal to be air-entrained into a HGMS unit and to subject only the moderately ground, classified size fractions of the remaining feed coal to magnetic separation. Experimental data which demonstrated the technical feasibility of the latter processing concept have been reported [51, 65]. However, there are certain undesirable technical drawbacks associated with this approach of external size classification and ultrafines removal prior to magnetic separation. The details have already been discussed previously in connection with the gravity feeding approach to HGMS for dry coal desulfurization.

Another problem associated with the air-entrained flow approach is its limited matrix loading capacity. In other words, at a practically acceptable desulfurization performance level, the amount of coal processed per unit amount of matrix packing is relatively small. In particular, due to the high flow velocity of the air stream required to achieve an entrained flow of the pulverized feed coal, the residence time of magnetic minerals in coal in the separator matrix is very short (typically less than 0.5 sec). This short residence time will reduce the capability of the separator matrix to capture and retain magnetic mineral particles in coal. The resulting limited matrix loading capacity may be illustrated by the comparison between typical performance of dry and wet magnetic desulfurization of several classified size fractions of a Pennsylvania Lower Freeport coal (Lancaster County) shown in Table 4.8 [24]. It can be seen that for similar levels of Btu recovery and amounts of feed coal processed, the percentages of sulfur reduction achieved by air-entrained separation

Table 4.8 Typical Performance of Dry and Wet Magnetic Desulfurization of Classified Size Fractions of Dry Pulverized Pennsylvania Lower Freeport Coal

Size fraction (grind level), mesh	Desulfurization method	Feed amount, g	Mass recovery, wt %	Btu recovery, %	Sulfur reduction	Ash reduction
−100 +200 mesh[a] (−100 mesh)	Gravity feeding[c]	197.3	81.2	94.7	71.2	64.7
	Air-entrained flow[d]	199.5	90.7	97.7	14.2	38.2
	Wet separation[e]	48.8	76.5	91.7	72.1	72.7
−200 +325 mesh[b] (−100 mesh)	Gravity feeding[c]	196.8	73.6	83.5	56.2	64.7
	Air-entrained flow[d]	193.8	92.8	95.9	14.9	18.9
	Wet separation[e]	48.6	82.0	90.6	56.6	64.3
−200 +325 mesh[b] (−200 mesh)	Gravity feeding[c]	196.3	70.6	80.5	48.7	55.0
	Air-entrained flow[d]	199.0	92.7	96.6	17.8	27.2
	Wet separation[e]	49.3	89.6	94.6	61.2	67.4

[a]Matrix 57 described in Ref. 40 was used (3.80-cm inside diameter, 0.153 m long).

[b]Matrix 51 described in Ref. 40 (3.80-cm inside diameter and 0.153 m long).

[c]Assisted by a slow-velocity air flow (25.4 cm/sec).

[d]Superficial air velocity = 210 to 251 cm/sec.

[e]Velocity of coal/water slurry = 30 mm/sec.

Source: Ref. 24.

(14.2 to 17.8%) were much smaller than those achieved by dry separation with gravity feeding (48.7 to 71.2%). The data also show that although both dry separation with gravity feeding and wet separation could achieve comparable levels of Btu recovery and sulfur reduction, the amount of feed coal processed by the gravity feeding approach was approximately four times higher than that by wet separation.

C. HGMS for Desulfurization of Dry Pulverized Coal by Recirculating Air Fluidization

The Auburn/ORNL Electromagnetic Fluidized-Bed HGMS Process for Desulfurization of Selectively Sized Powdered Coal. Based on the preceding discussion of available experimental results on HGMS for desulfurization of dry pulverized coal by gravity feeding and air-entrained separation, it should be evident that an effective dry HGMS process to be developed for coal desulfurization must include at least two desirable features. First, it must have a simple means to reduce the presence of ultrafines in the pulverized coal stream to avoid their promotion of the agglomeration of the pulverized coal and its mineral impurities. Second, it must provide a sufficient residence time to the pulverized coal stream to promote the contact between the magnetic particles in coal and the active surface area available to the magnetized packing material in the separator matrix. The latter is important in increasing the capacity of the separator matrix for capturing and retaining the magnetic particles in the pulverized coal stream. As a result of research work done at Auburn University with support from the Oak Ridge National Laboratory (ORNL), a recirculating air-fluidization approach to HGMS possessing the above desirable features was successfully developed, and several novel fluidized-bed separator matrices were constructed and tested. By using one of the simplest fluidized-bed separator matrices designed, the performance of HGMS of sulfur and ash from the selectively sized fractions of several dry pulverized Pennsylvania coals were repeatedly found to be roughly equivalent to coal-water slurries. The new fluidized-bed HGMS process was effective in reducing the sulfur contents in the selectively sized fractions of the Pennsylvania coals tested to environmentally acceptable levels.

The design of fluidized-bed separator matrices can be illustrated by one of the simplest configurations described as follows [14,51,53]. Briefly, this matrix was made of three primary sections of increasing diameters arranged from the bottom to the top. These primary sections were, for example, of 5-in. inside diameter and 3 in. in length (section A), 3.5-in. inside diameter and 10 in. in length (section B), and 0.75-in. inside diameter and 7 in. in length (section C), respectively. Connecting sections A and B, and sections B and C were two expanded sections of, for example, 3 and 4 in. in length, respectively. Stainless steel expanded-metal screens along with proper spacers, for example, were used as matrix packing materials in section B.

In a typical experiment, the pulverized feed coal was charged into an auxiliary fluidized bed located outside of the separator, and then air-fluidized through the separator matrix. By properly controlling the flow velocity of the air stream through the whole separation period, it was possible to elutriate and beneficiate most of the fines in the pulverized coal stream out of the top of the bed and to collect it as a top clean coal product. At the same time, because of the unique expanded sections from the bottom to the top of the separator matrix and the resulting gradual decrease in the upward fluidization velocity of the pulverized coal stream, the majority of the pulverized coal particles of medium and large sizes would tend to recirculate inside the central separatory section containing the packing material (section B). As a result, a sufficient residence time inside the separator matrix could be provided to the bulk of the fluidized coal stream without the presence of fines, allowing the magnetic particles in coal to be captured and retained by the matrix. Toward the end of the desired separation period, the flow velocity of the air stream was reduced, and the magnetically cleaned coal of low sulfur and ash contents was collected as a bottom clean coal product. Following this, the magnetic field was turned off, and the high-sulfur and high-ash magnetic residue retained inside the separator matrix was removed as a magnetic refuse stream.

Table 4.9 Magnetic Desulfurization of Selectively Sized Pulverized Pennsylvania Upper Freeport Coal by the Auburn/ORNL Electromagnetic Fluidized-Bed HGMS Process (—100 +200 mesh, Air velocity = 17.7 cm/sec, Fluidization Time = 5 min)

	Wt %	Sulfur, wt %	Pyrite, wt %	Ash, wt %
1. Feed coal	100.0	2.123	1.519	6.320
2. Magnetic refuse (20 kOe)				
1-Pass[c]	9.05	12.80	11.79	24.87
2-Pass	5.99	4.68	4.18	13.26
3-Pass	5.12	2.25	1.93	10.00
Total	20.16	7.71	7.02	17.65
3. Clean coal (20 kOe)	79.84	0.68	0.130	3.45
4. Feed separated as mags	20.16	68.16[a]	86.80[a,b]	54.43[a,b]

[a]Weight percent of sulfur, pyrite, or ash separated from feed as mags.
[b]Weight percent of pyrite and ash separated from feed as mags by Frantz isodynamic separator are 87.7.
[c]Apparent matrix length per pass = 20 in.
Source: Ref. 51.

Table 4.9 illustrates the typical experimental results obtained with the —100- +200 mesh size fraction of a pulverized Upper Freeport coal. It can be seen that by using the recirculating air-fluidization approach, HGMS could reduce the total sulfur of the Upper Freeport coal by 68.16%, the pyrite by 86.8%, and the ash by 52.43%. Essentially comparable desulfurization performance was easily reproduced in repeated runs under the same experimental conditions.

The experimental results illustrated in Table 4.9 have clearly shown that the recirculating air-fluidization approach holds much promise as an effective method for the desulfurization of selectively sized fractions of dry pulverized coal by HGMS. A quantitative study of the effects of major variables and conditions on the performance of the Auburn/ORNL electromagnetic fluidized-bed HGMS process has also been conducted and reported elsewhere [51,53].

The Auburn/NEPSCO Superconducting Fluidized-Bed HGMS Process for Desulfurization of Utility Boiler Feed Coal. In the preceding work done by Auburn University and ORNL, the technical feasibility of the electromagnetic fluidized-bed HGMS process for desulfurization of selectively sized powdered coal was experimentally demonstrated. In addition to the disadvantages in using a selectively sized powdered coal (with external size classification and ultra-fines removal prior to magnetic separation), there are also several technical limitations of the Auburn/ORNL process. As originally proposed, the Auburn/ORNL process employs a specially designed, expanded-bed separator matrix in an electromagnetic HGMS unit. Due to the use of such an expanded-bed matrix configuration (not a straight-bed matrix configuration), not all of the available cylindrical working volume within the electromagnet structure is packed with magnetizable materials to be utilized as a separatory (packed) section for magnetic beneficiation of the fluidized coal-air stream. Further, as originally demonstrated, the Auburn/ORNL process requires a relatively long fluidization time of 5 to 10 min. Obviously, much work is needed to further improve and develop the Auburn/ORNL fluidized-bed HGMS process so that it can be used to effectively and inexpensively reduce the sulfur and ash contents in many pulverized coals to very low levels.

Beginning in January 1978, a research program to continue the research and development of novel fluidized-bed HGMS processes for desulfurization of utility boiler feed coals was initiated by Auburn University and the New England Power Service Company (NEPSCO). Pilot-scale HGMS units of both electromagnetic (phase I) and superconducting (phase II) types were used for desulfurization of dry pulverized coal suitable for utility applications (70 to 80% −200 mesh). The first phase of the Auburn/NEPSCO project was completed in 1979, which resulted in the successful experimental development and demonstration of a new, cyclic electromagnetic fluidized-bed HGMS process [54,65]. The test results from phase I indicated that by incorporating a number of novel processing concepts and equipment innovations into the fluidized-bed approach, a magnetically cleaned coal with a very low sulfur emission level (0.520 to 0.505 g SO_2 per MM j or 1.2 to 1.4 lb SO_2 per MM Btu) could be produced from a high-sulfur Pennsylvania Upper Kittanning coal of a wide range of particle sizes suitable for utility applications. Based on the experimental results obtained in phase I, several preliminary dry HGMS processes were designed for use as part of the fuel preparation system for the coal-oil mixture combustion demonstration program located at the Salem Harbor Generating Station of the New England Electric System, Salem, Massachusetts [5]. Cost analyses of the proposed processes showed that a continuous superconducting HGMS process utilizing a reciprocating fluidized-bed moving-matrix system would be most economically attractive [65].

As a result of phase II of the Auburn/NEPSCO project, an improved device and method for adapting the fluidized-bed approach to HGMS for desulfurization of dry pulverized coal was proposed and demonstrated. Briefly, the new Auburn/NEPSCO process includes the following three significant equipment innovations compared to the preceding Auburn/ORNL process.

1. The use of a *novel straight-bed separator matrix* consisting completely of a packed, fluidized-bed separatory section which provides a sufficient residence time to the fluidized coal-air stream and maximizes the active surface area available to the separator matrix packing material for enhancing the removal of magnetic sulfur-bearing and ash-forming minerals for the fluidized coal-air stream and achieving the desirable function of the fine elutriation and beneficiation as a top clean coal product.
2. The use of an *auxiliary external vibration system* which minimizes the mechanical entrapment of the nonmagnetic clean coal inside the separator matrix and reduces the gravity feeding of pulverized coal in the piping for coal feeding and product withdrawal.
3. The use of an *automatic internal vibration/washing system* to apply appropriate vibrations to the separator matrix packing material and to remove (backwash) by air the nonmagnetic coal particles mechanically entrapped or the mineral impurity particles magnetically captured by the matrix when the magnet is on or off, respectively.

In addition, the new Auburn/NEPSCO process incorporates the following four novel processing methods compared to the preceding Auburn/ORNL process:

1. The use of *higher field intensities* provided preferably by a superconducting HGMS unit to maximize the sulfur reduction, Btu recovery and processing throughout.
2. The use of *a hot air stream,* which is readily available as a hot flue gas stream in most utility facilities, as a fluidizing medium to enhance both the sulfur reduction and Btu recovery, and to eliminate the need of drying the pulverized coal prior to its feeding into the HGMS unit.
3. The use of *a partial product recycle technique* in which the top clean coal (elutriated fines) is not recycled and only the bottom clean coal is recycled, if necessary, back to the separator matrix system for further beneficiation.
4. The use of *an optimum combination of operating variables and conditions* such as matrix loading factor (gram coal processed per gram matrix packing), initial matrix void volume fraction, fluidization time per pass, and superficial fluidization velocity, so as to maximize the sulfur reduction, Btu recovery and processing throughput without any unnecessary external size classification, ultrafines removal and coal drying prior to magnetic separation.

Available results obtained from applying the Auburn/NEPSCO superconducting fluidized-bed HGMS process for desulfurization of several dry pulverized Eastern coals have shown that the new process could reduce the sulfur emission levels (lb S per MM Btu or g S per MM j) by 55 to 70% and achieve an average Btu recovery of 90 to 95%. The effects of major process variables and conditions, including field intensity, fluidizing medium, fluidization time and velocity, external and internal matrix configurations, matrix loading factor, feed coal type and particle size and types of recycle modes, on the grade and recovery of the magnetically cleaned coal and on the capacity of the process operations have been quantitatively investigated; and the detailed results have been reported elsewhere [47,99].

It is worthwhile to briefly mention the significance of the new Auburn/NEPSCO process, particularly with reference to the technical problems reported for other approaches to HGMS for coal desulfurization. For example, based on results from comprehensive technical and economic evaluations of HGMS techniques recently conducted by the Electric Power Research Institute [32,89,106], two key problems associated with desulfurization of coal-water slurry have been identified. These are (1) too much coal (and hence Btu) is lost when removing the pyrite and (2) costs for preparing the coal for the magnetic separator, and subsequently handling the product, are much higher than the magnetic separation step [32,89,106]. Since the Auburn/NEPSCO process has demonstrated fairly high Btu recoveries (about 90%) for the coals tested thus far and there is no need for drying and dewatering the product coal, it is evident that the new dry magnetic process provides an effective means to resolve the preceding two key problems and also holds much promise as an economical method for dry coal desulfurization. As such, research efforts to further develop and demonstrate the new Auburn/NEPSCO process for dry coal desulfurization should be justified.

4.4 HIGH-GRADIENT MAGNETIC SEPARATION FOR DESULFURIZATION OF SOLVENT-REFINED COAL (SRC)

A. Coal Liquefaction and HGMS

Recent experimental studies have suggested that coal desulfurization by HGMS may serve as a potential adjunct to coal liquefaction processes [2,26,43,44,61-63,74]. In particular, a close examination of the inherent physical and chemical characteristics of the hydrogenated products prior to the filtration step in the SRC and other related liquefaction processes indicates that HGMS may be developed as a practical mineral residue separation method.

At liquefaction conditions of the SRC and other related processes, namely, 425 to 475°C and 1000 to 2500 psig of hydrogen pressure, it is now established that significant amounts of pyrite (FeS_2) are converted to highly magnetic pyrrhotites (Fe_7S_8), particularly the hexagonal pyrrhotites. Further, upon cooling from the liquefaction temperature to the filtration temperature (about 150 to 200°C), some of the hexagonal pyrrhotites in the hydrogenated product are transformed into the ferrimagnetic monoclinic pyrrhotite [31,63,84,100]. This follows because the latter is apparently the stable state form of iron-sulfur compounds with atomic percentages of iron between 33.33 (for FeS_2) and 46.67 (for Fe_7S_8) at temperatures below 254°C. Indeed, results obtained by thermomagnetic and Mössbauer spectroscopic analyses of filter-cake solids in the SRC process seem to confirm this observation [30]. Further, the sulfur-bearing and ash-forming minerals tend to be liberated more easily from the dissolved organic components in the SRC filter feed slurry than, for example, from the pulverized coal suspended in water. Also, the typical mean particle size of the solid particles in the SRC filter feed slurry is often less than 5 μm [43], which dictates the use of separation methods like HGMS capable of handling micrometer-sized materials. All of the above factors seem to suggest the potential for utilizing HGMS to remove mineral residue from liquefied coal. For certain types of coal, even without further improvement in the magnetic removal of ash, the magnetically cleaned SRC would be acceptable for use as a feed to boilers which already have electrostatic precipitators. This follows because the cost of solid-liquid separation in coal liquefaction is generally substantial [4], and the use of moderately low-ash SRC plus particulate control could be less expensive. Indeed, preliminary cost estimates for the magnetic desulfurization of liquefied coal based on bench-scale experimental data reported by Hydrocarbon Research, Inc. seem to support this observation [75]. In order to provide the experimental data needed for a more complete technical and economical evaluation of the magnetic approach, bench- and pilot-scale tests to quantitatively examine the effects of key separation variables on the magnetic desulfurization of liquefied coal have been conducted, and the scopes and results of the reported studies are reviewed below.

B. Reported Experimental Studies

An Overview. The technical feasibility of applying HGMS to desulfurization of solvent-refined coal (SRC) was first demonstrated in an exploratory, bench-scale study done at Hydrocarbon Research, Inc. (HRI) in 1975 [26]. In that study, the effects of slurry flow velocity (residence time) and matrix packing density on the magnetic removal of sulfur and ash from the liquefied SRC filter feed slurry of an Illinois No. 6 coal were investigated. The results showed that HGMS effectively removed up to 90% of the inorganic sulfur; and about half of the experimental runs conducted by HRI indicated over 87% inorganic sulfur removal. HGMS was found to be less effective in ash removal, but it did remove 25 to 35% of the ash. Recently, two other bench-scale studies of HGMS for desulfurization of SRC were reported by Massachusetts Institute of Technology (MIT) and Gulf Research and Development Company. In the MIT study [2,61], the successful removal of inorganic sulfur from the liquefied SRC filter feed slurry of an Illinois No. 6 (Monterey) coal was demonstrated for a wide range of separation conditions, including field intensity, residence time, and separation temperature. At the optimum temperature of about 221°C (430°F), 80 to 95% of the inorganic sulfur and up to 40% of the ash were removed. Also, some preliminary tests done by MIT indicated that 90 to 98% of the magnetic mineral residue retained in the separator matrix could be successfully backwashed. Experimental results similar to those obtained by MIT were reported by Gulf, using the liquefied SRC filter feed slurry of a Kentucky No. 9 coal [79].

Recently, the results from an experimental study which successfully demonstrated the technical feasibility of high-gradient magnetic desulfurization of SRC on a pilot scale were reported by Auburn University [13,55]. In that study, SRC filter feed slurries of Illinois No. 6 (Monterey) and Kentucky No. 9/14 coal were magnetically filtered over a wide range of separation conditions. The effects of major separation variables on the removal of inorganic sulfur and extracted solids from SRC filter feed slurries were quantitatively investigated. Under separation conditions used in that study, HGMS removed up to 93 wt % of the inorganic sulfur. A quantitative correlation between the magnetic desulfurization performance and separator capacity applicable over a wide range of separation conditions was found experimentally; and a new SRC processing scheme incorporating HGMS potentially capable of achieving over 90 wt % total sulfur removal from pulverized feed coal was also suggested.

Table 4.10 summarizes the scopes of the four reported studies on HGMS for desulfurization of SRC.

Effects of Major Separation Variables and Conditions on the Performance on HGMS for Desulfurization of SRC. In the four experimental studies reported thus far [2,13,26,55,61,79], two types of analyses were generally performed on each feed and magnetically cleaned coal sample. The weight percentage of solids in each sample was determined by extracting the sample with tetrahydrofuran (THF) in a Soxhlet apparatus. The tetrahydrofuran would dissolve the liquefied coal, leaving only the insoluble solids in the extraction apparatus. The sulfur content in the solids was used as a measure of inorganic sulfur component in the sample. The percentages of removal of inorganic sulfur (THF insoluble sulfur) and extracted solids were used as the primary and secondary performance indexes, respectively, of the magnetic separation tests.

The quantitative effects of (1) field intensity, (2) residence time (slurry velocity), (3) packing material, and (4) packing density on the performance of HGMS for desulfurization of SRC are similar to those previously described for desulfurization of coal-water slurries [55]. In what follows, the effects of other operating variables and conditions on the magnetic desulfurization of SRC are briefly reviewed.

Effect of Separation Temperature. Changing the separation temperature has a significant effect on the percentages of removal of inorganic sulfur and extracted solids. Figure 4.28 illustrates that after 40 kg of SRC filter feed are processed, inorganic sulfur removal from SRC filter feed slurries of a Kentucky No. 9/14 coal is 85% at 176.7°C (300°F), but it drops to 55% at 148.9°C (200°F). In general, the higher viscosity of the SRC filter feed slurry at the lower temperature increases the drag force on the magnetic particles in the slurry, thus reducing the efficiency of the magnetic removal of sulfur and ash from liquefied coal. Experimental results further show that changing separation temperature has a more significant effect on the magnetic removal of inorganic sulfur as compared to that of extracted solids [13]. This observation is to be expected owing to the unique

Table 4.10 Scopes of Reported Experimental Studies on HGMS for Desulfurization of Solvent Refined Coal

	Massachusetts Institute of Technology [2,61]	Hydrocarbon Research Institute [26]	Auburn University [13,55]	Gulf Research and Development Company [79]
1. Liquefied coal used	Illinois No. 6	Illinois No. 6	Illinois No. 6 and Kentucky No. 9/14	Kentucky No. 9
2. Separator matrix used				
Cross section, cm^2	4.5	3.4	8.4	12.6
Length, cm	10.2 and 17.8	52.1	48.3	15.2
Volume, cm^3	45.9 and 80.1	177.1	405.7	192.0
3. Separation variables and conditions investigated				
Field intensity	Varied	Fixed	Varied	Varied
Residence time	Varied	Varied	Varied	Varied
Type of packing material	Fixed	Fixed	Varied	Fixed
Packing density	Fixed	Fixed	Varied	Varied
Temperature, °C	80.4 to 301.6	150 to 175	93.3 to 176.7	150 to 245
Separator feed throughput, kg	<1	<4.5	34.9 to 61.6	<5
Matrix backwashing tested	Yes	No	No	Yes
4. Time-cut samples of cleaned product taken in experiments	No	Yes	Yes	Yes
5. Feed nonuniformity or velocity fluctuation observed in experiments	Yes	Yes	No	No

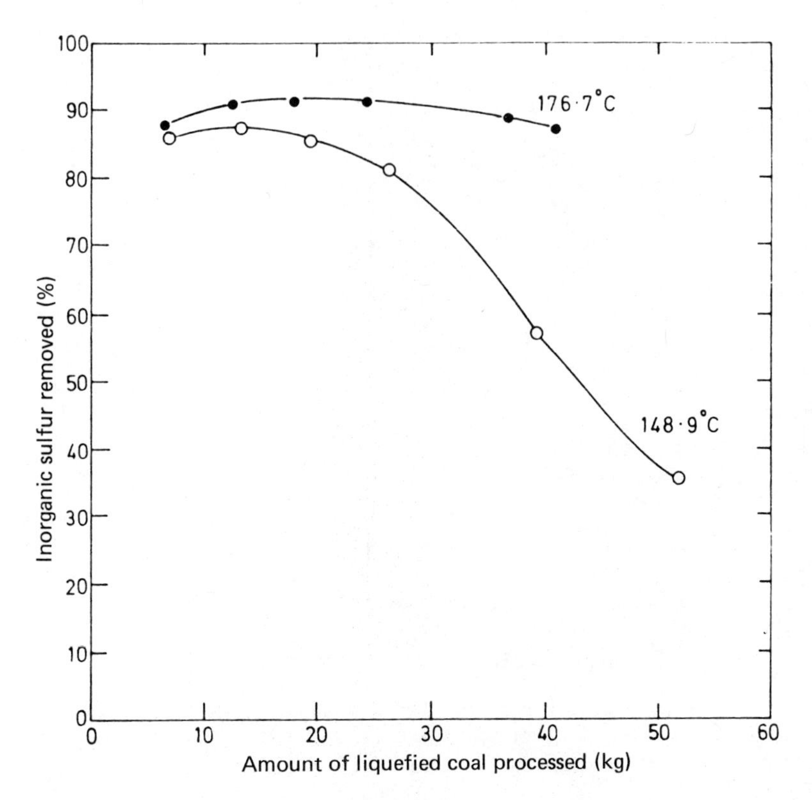

Figure 4.28 Effect of temperature on inorganic sulfur removal from SRC filter feed slurries of a Kentucky No. 9/14 coal by HGMS. (From Ref. 55.)

phase-equilibrium relationship and thermomagnetic properties of the pyrite-pyrrhotite system related to the magnetic desulfurization of liquefied coal. A detailed discussion of the latter aspect can be found in Refs. 49 and 63, in which several possible approaches to establishing the optimum separation temperatures have been described.

Effect of Thermal Treatment in Inert or Sulfidation Atmospheres. Recent work reported by General Electric [30,31] has shown that thermal treatment of the SRC filter-cake solids in inert (helium) or sulfidation (H_2S-rich gas) atmospheres can facilitate the rapid transformation of the weakly magnetic hexagonal pyrrhotites to the more strongly magnetic, monoclinic form at a temperature intermediate to

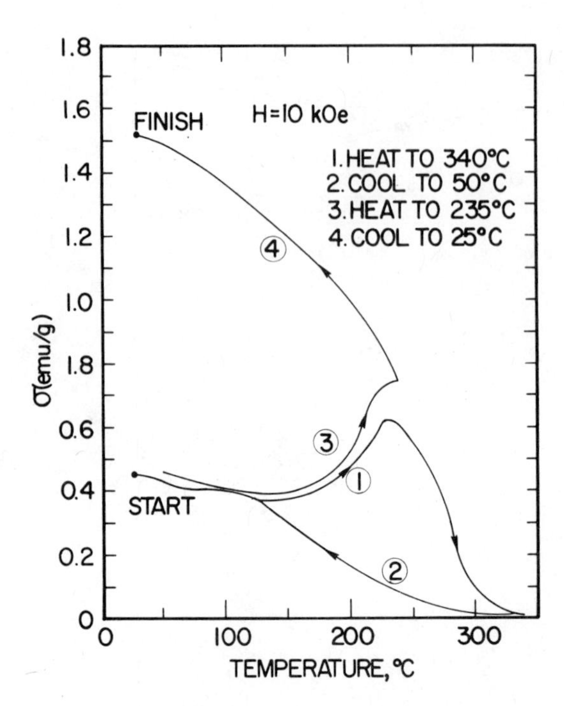

Figure 4.29 Magnetization of a SRC filter-cake sample of a West Kentucky No. 9/14 coal at a field intensity of 10 kOe as a function of thermal treatment temperature in an inert (helium) atmosphere. (From Ref. 31.)

those of the SRC hydrogenation reactor and the magnetic separation stage. The treatment yields a magnetization enhanced by a factor of 5 or 10 with respect to untreated samples. The kinetics of the transformation to a high magnetization state are strongly influenced by prior thermal and atmospheric history.

As an illustration, Fig. 4.29 shows the magnetization of a SRC filter-cake sample at a field intensity of 10 kOe as a function of thermal treatment temperature in an inert (helium) atmosphere [31]. This figure exhibits a modest initial magnetization, a peak in magnetization in the vicinity of 220°C, a marked thermal hysteresis during cooling from about 350°C, and finally, the establishment of a high-magnetization state by heating from 50°C or lower, up to but not much beyond the peak, followed by a recooling to room temperature.

This state may also be readily achieved by an initial heating of the as-received SRC filter-cake sample up to the peak followed by recooling. The high-magnetization state is identified as monoclinic pyrrhotite by the presence of resolved hyperfine splittings of the Mössbauer spectrum [30]. On the other hand, rapid cooling from 350°C yields a low-magnetization and Mössbauer spectrum associated with hexagonal pyrrhotite. The latter state, however, can be upgraded magnetically by heating to the peak followed by recooling. These results clearly demonstrate that the high-magnetization state favorable to magnetic desulfurization of liquefied coal is accessible by a thermal treatment cycle.

Another approach toward transformation of the state of the pyrrhotite in SRC filter-cake solids is to increase the sulfur content from the as-received Fe_9S_{10} (hexagonal form) up to Fe_7S_8 (monoclinic form) with the goal of achieving the high-magnetization state. This dry gas-solids sulfidation technique uses the thermal decomposition of H_2S $(2H_2S \rightarrow 2H_2 + S_2)$ to provide the desired sulfur. It has been shown to enhance the magnetic removal of sulfur and ash from the SRC filter feed samples of an Indiana No. 5 coal [63].

Quantitative Correlations of Desulfurization Performance and Separator Capacity. A quantitative measure correlating the desulfurization performance and separator capacity is the amount of liquefied coal which can be processed per unit amount of matrix packing material at a practically acceptable percentage of inorganic sulfur removal such as 90%. This measure is illustrated in Fig. 4.30, in which the specific experimental conditions are also shown [55]. It can be seen that a fairly smooth breakthrough curve passing through all experimental data points can be drawn. In particular, this common breakthrough curve coincides with the one corresponding to run 9 made at the largest packing void volume of 96.20%. The latter observation is significant in that it allows one to predict the breakthrough curves for inorganic sulfur removal at different matrix loading capacities corresponding to different packing void volumes by using only the experimental breakthrough curve obtained at one packing

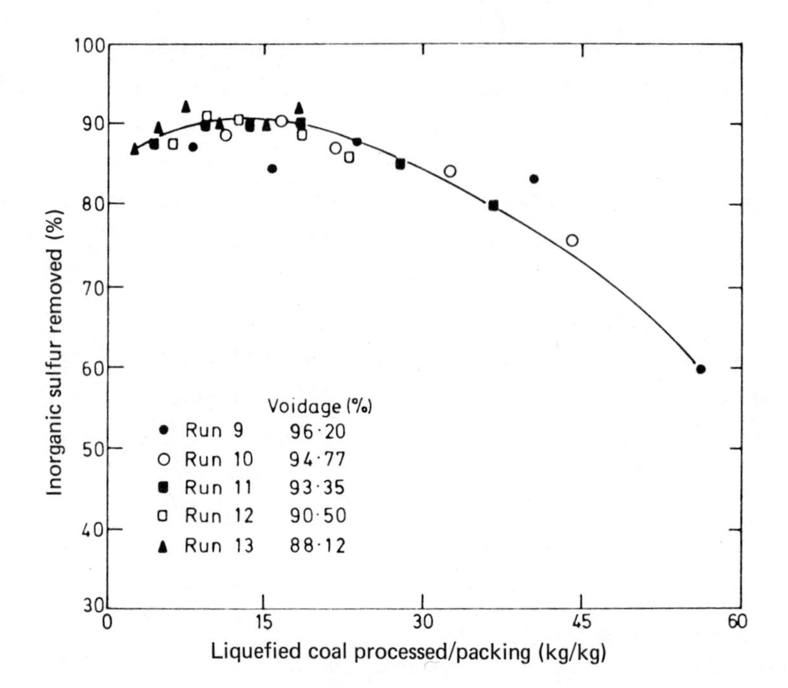

Figure 4.30 Correlation of desulfurization performance and separator capacity. (From Ref. 55.)

void volume. This predicted breakthrough curve readily provides a quantitative correlation between the desulfurization performance and separator capacity for a wide range of separation conditions without conducting additional magnetic separation tests. Figure 4.30 also shows that under the experimental conditions of runs 9 through 13, approximately 20 kg of liquefied coal can be processed per kilogram of medium steel wool packing at a weight percentage of inorganic sulfur removal of 93%.

C. Potential of Utilizing HGMS in the SRC and Related Coal Liquefaction Processes

HGMS holds much promise as a practical method for separating the mineral residue and reclaiming the mineral catalysts from the reacted coal slurry in the SRC and related coal liquefaction processes, particularly in view of the revised New Point Source Emission Standards

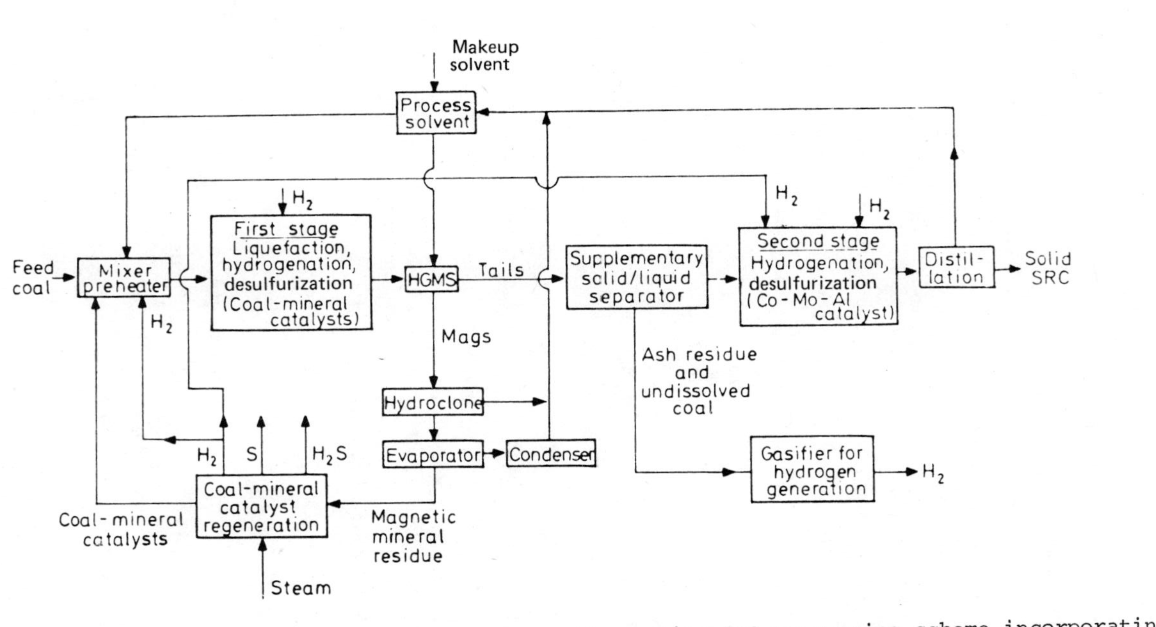

Figure 4.31 A proposed two-stage short-residence-time SRC processing scheme incorporating HGMS potentially capable of achieving a 90% total sulfur removal from pulverized feed coal. (From Ref. 55.)

(NPSES) currently being proposed by the Environmental Protection Agency (EPA). These new standards are expected to require such desulfurization methods as solvent refining to achieve a total sulfur removal of about 85 to 90 wt % from the pulverized feed coal. Unfortunately, among the seven different feed coals tested recently at the SRC pilot plant at Wilsonville, Alabama, only the Kentucky No. 9/14 coal could be processed with a high hydrogen consumption to achieve a 90 wt % total sulfur removal. Thus, in the advent that these NPSES are adopted, the scope of application of the current SRC technology for producing a solid fuel will become severly limited. Based on the results from the present work and ongoing liquefaction research at Auburn University [21], an alternative SRC processing scheme incorporating HGMS which is potentially capable of meeting the revised NPSES can be proposed as shown in Fig. 4.31 [55].

The new process has two short residence-time reaction stages along with an intermediate step for mineral residue separation and catalyst recovery consisting mainly of a HGMS device, a supplementary solid-liquid separator and a catalyst regeneration unit. A detailed discussion of this new two-stage, short-residence-time SRC scheme incorporating HGMS has been given in Ref. 55.

4.5 HIGH-GRADIENT MAGNETIC SEPARATION FOR DESULFURIZATION OF COAL-OIL MIXTURE (COM) AND COAL-METHANOL SLURRY

A. HGMS for Desulfurization of COM

As a result of the unavailability and increased cost of fuel oil for power plants, the United States faces large capital outlays in converting the existing oil-fired steam generating facilities to coal burning. At some plants, however, it may not be practical to convert to coal burning due to the lack of coal storage and handling facilities or the high capital cost and short remaining life of the plants. An alternative to the complete conversion from oil to coal burning could be to burn a pulverized coal-oil mixture (COM), which is prepared by simply mixing finely ground coal with No. 6 fuel oil [5].

An apparent disadvantage of COM combustion is the difficulty of maintaining the pulverized coal in a stable suspension, which is essential for good flame stability and also to keep coal from settling in plant facilities. Continuous agitation of COM and addition of stablizing chemicals to COM have been used to improve the settling of COM. Also, work done by Auburn University and the New England Power Service Company (NEPSCO) has shown that the use of HGMS to remove heavier compounds of iron and sulfur and other trace elements from a pulverized Pennsylvania Upper Freeport coal could improve the settling stability of the resulting COM. In this application, HGMS also reduces the sulfur dioxide and ash particulate emissions generated during the combustion of COM.

The technical feasibility of using HGMS for desulfurization of four pulverized coals dispersed in No. 2 fuel oil was investigated in a recent study [12]. Fuel oil was chosen as a carrier because the interfacial tension between fuel oil and pulverized coal is low so that the pulverized coal could be effectively dispersed for physical separation. Furthermore, residual oil remaining with the coal after filtration could be burnt with the coal to avoid drying and handling finely divided coal that would be required when water was used as a suspension medium. The coals used were an Ohio Pittsburgh coal (Harrison county), a Maryland Upper Freeport coal (Garrett county), an Ohio Lower Kittanning coal (Tuscarawas county), and an Illinois No. 6 coal (Marion county). Experiments were conducted on three grinds of each coal, including −10-μm, −325-mesh, and −20-mesh size fractions. The magnetic separator used was a Frantz 48V Ferro Filter, a low-intensity HGMS unit equipped with special high-gradient collection grids. In each experiment, 0.52 kg of pulverized coal were dispersed in 3.49 kg of No. 2 fuel oil to form a 13 wt % slurry. It was found [12] that the Upper Freeport coal exhibited a considerable reduction in sulfur content, while Illinois No. 6 coal showed a small change in sulfur content. Results with Pittsburgh and Lower Kittanning coals were between these extremes. In particular, the greatest sulfur reduction in sulfur content was achieved with coal ground to −325 mesh at higher level of field intensity (1700 Oe). The low-intensity HGMS unit was effective in reducing the total sulfur content of the –325-mesh size fraction of the Upper Freeport coal by 60 to 63% with a mass recovery of cleaned coal of over 80 wt %. Variation of either field intensity from 1300 to 1700 Oe or in flow velocity from 0.25 to 0.61 cm/sec did not appear to make any significant difference in the degree of beneficiation achieved.

B. HGMS for Desulfurization of Coal-Methanol Slurry

In a related study attempting to avoid the necessity of drying the product coal after magnetic beneficiation [57], HGMS was applied to a pulverized Pennsylvania Upper Freeport (Delmont) coal suspended in methanol or in methanol-water mixture. The feed coal used contained 1.55 wt % total sulfur, 1.06 wt % inorganic sulfur and 14.6 wt % ash. The tests were conducted with about 86 g of feed coal dispersed in methanol or methanol-water mixture to form slurries of 10, 25, and 50 wt % solid concentrations. Figure 4.32 illustrates the percent sulfur and ash reductions versus the weight percent of cleaned coal recovery obtained from experiments done with different field intensities and carrier fluids [57]. HGMS effectively removed 26 to 36% of the total sulfur from the coal sample tested with a cleaned coal recovery of 68 to 76 wt %. Also included in the figure are the data from tests on water slurries of the +200-mesh size fraction, which correspond to higher sulfur and ash reductions, but lower coal recoveries. At the same field intensity, higher sulfur and ash reductions were achieved with a +200-mesh size fraction than with the 60 mesh by 0 size fraction. Figure 4.32 also shows that the separations in methanol yielded better results than those in water; in methanol-water slurries, the results fell in between. Note that the pulverized coal particles are hydrophilic in methanol and in methanol-water mixtures, in contrast to its hydrophobic behavior in water. Thus, methanol wets coal easily and no dispersant is needed even in 10 wt % coal-methanol slurries.

A separation scheme for the above application would involve a circulating load of methanol which could carry the pulverized coal through a HGMS unit. Provided that the methanol losses could be kept small, this should be a workable scheme which could combine

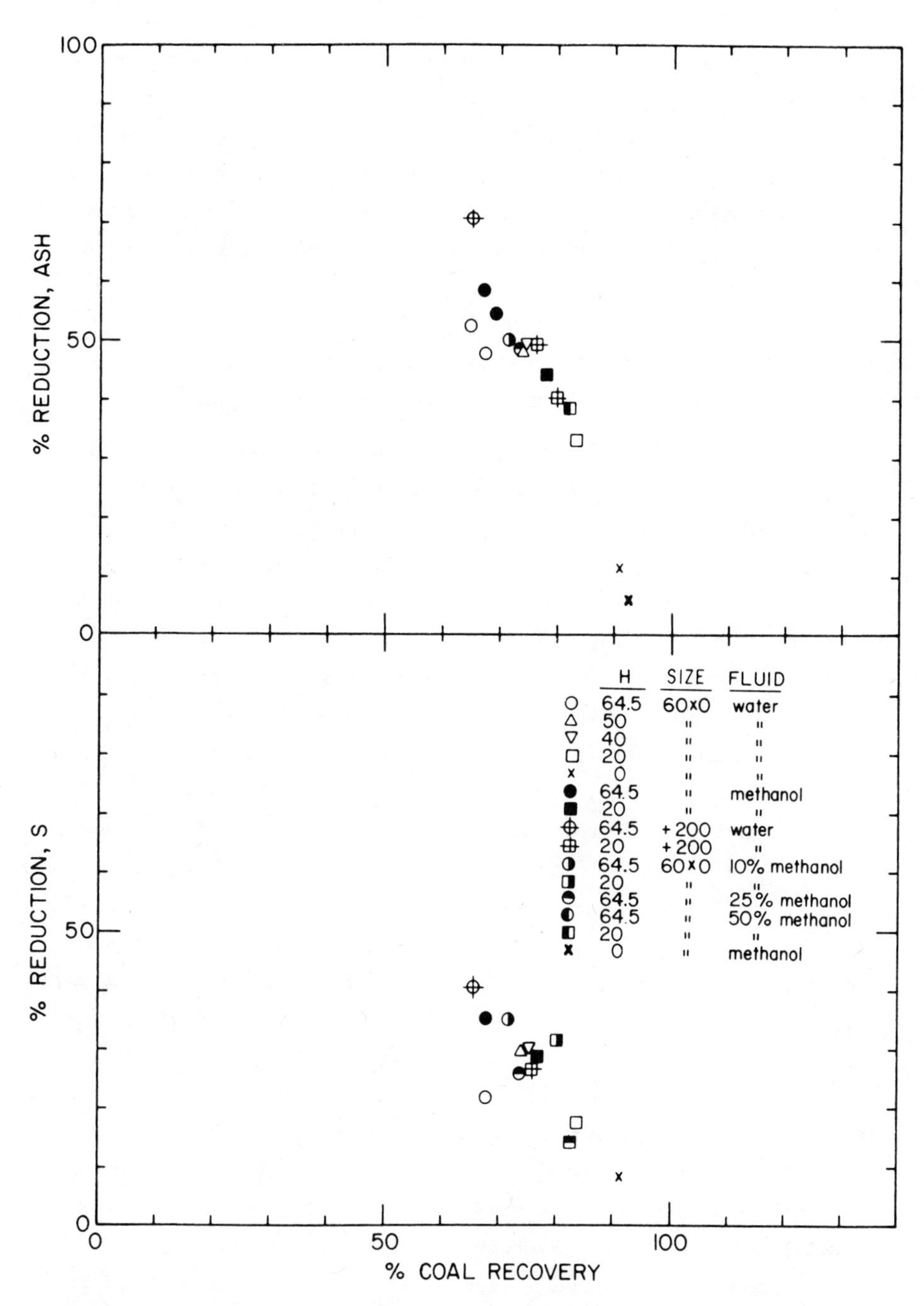

Figure 4.32 Percent sulfur and ash reductions versus weight percent cleaned coal recovery obtained from HGMS of a pulverized Pennsylvania Upper Freeport (Delmont) coal suspended in water, methanol, and water-methanol mixtures. (From Ref. 61.)

advantages of both dry and wet separations. A methanol-water mixture could also be used if it turned out that drying would be significantly easier for some methanol-water proportion than for water alone. Since methanol is a by-product of some coal gasification processes, it has been suggested that the use of methanol slurries would be particularly appropriate for coal desulfurization by HGMS prior to gasification [61].

4.6 PHYSICAL PRETREATMENT AND MAGNETIC DESULFURIZATION

A. Gravity Separation and HGMS for Coal Desulfurization

A conventional coal washing circuit is based on gravity separation methods which utilize the difference in specific gravities to separate lighter pulverized coal from heavier mineral impurities. One of the common gravity separation methods is the heavy-medium separation which uses a salt solution or a water slurry of heavy minerals of an intermediate specific gravity as a separating medium to separate two or more phases of different gravities, some above and the others below that of the medium [42].

A potential application of HGMS to coal beneficiation is to serve as an adjunct to a conventional coal-washing circuit. In this application, a HGMS unit would be installed downstream from a heavy-medium vessel to deep clean the rougher product from the washing circuit. Presumably, the rougher product would be further ground to liberate additional pyrite and ash before being fed to the magnetic separator. The feasibility of combining gravity separation with HGMS was investigated in a recent study, and several benefits of the combined approach were suggested [106]. First, fine grinding for mineral impurity liberation could be delayed until the midpoint of the beneficiation system, and the washing unit could be provided with its most efficient feed, namely, coarse coal. Secondly, the washing unit would pretreat the feed to the magnetic separator by removing a substantial amount of the nonmagnetic ash-forming minerals which could not be removed by the magnetic separator. Thirdly, the washing unit would also remove some of the very weakly magnetic sulfur-bearing and ash-forming minerals which could not be efficiently removed by the magnetic separator. It should also be mentioned that there are some potential disadvantages to this combined processing concept. One such disadvantage would be the possible interference with magnetic separation of the residual medium such as magnetite retained by the products from the washing unit.

Table 4.11 illustrates the results from cyclic electromagnetic HGMS of water slurries of a pulverized Pennsylvania Sewickley coal with and without pretreatment using gravity separation [106]. The cleaned coal recoveries as well as sulfur and ash reductions listed in the table were based on the respective feeds to the magnetic separator and not to the washing operation which preceded it. If the washing operation were included in the analysis, overall cleaned coal recoveries would be lower than those shown, but sulfur and ash reductions would be higher. The data show that, for equivalent feed and similar operating conditions in the magnetic separator, combining gravity separation with HGMS produced a much cleaner coal product than was possible with magnetic separation alone. The data also show that HGMS produced fairly consistent reductions in total sulfur and ash, regardless of whether the feed was raw coal or washed coal. In addition, the data indicate higher cleaned coal recoveries for washed coal than for raw coal. As mentioned above, these recoveries should be multiplied by the recovery percentages for the washing step to obtain the overall recoveries for the two-stage beneficiation operation. Note that it is possible to calculate the minimum float-product recovery that would be required to maintain an overall cleaned coal recovery at a level equal to that attained by feeding the raw coal to the magnetic separator alone. It can be easily shown that this minimum float-product recovery percentage is the ratio of the recovery obtained for HGMS of raw coal to that obtained for the washed coal. Using the recovery data in the table, one can find that the minimum float-product yield percentage varies between 69 and 80%. Thus, the float-product recovery could be as low as 69% without sacrificing the overall cleaned coal yield (recovery). Based on the known operational

Table 4.11 Results of Cyclic Electromagnetic HGMS for Beneficiation of Water Slurries of Pulverized Pennsylvania Sewickley Coal with and without Pretreatment Using Gravity Separation (Heavy-Medium Washing) (90% −200 mesh)

Feed coal type	Total sulfur			Ash			Coal recovery,[d] wt %
	Feed, wt %	Product, wt %	Percent reduction	Feed, wt %	Product, wt %	Percent reduction	
1.[a] Washed[c]	1.66	1.07	35.5	12.9	7.4	42.6	66.7
Raw	2.19	2.32	39.7	33.1	17.6	46.8	48.5
2.[a] Washed	1.63	1.03	36.8	12.1	8.1	33.1	83.3
Raw	1.98	1.33	32.8	34.7	23.9	31.1	66.1
3.[b] Washed	1.77	1.11	37.3	12.2	7.2	41.0	65.5
Raw	2.16	1.44	33.3	33.2	19.9	40.1	52.2
4.[b] Washed	1.68	1.07	36.3	13.8	9.8	29.0	81.5
Raw	2.07	1.36	34.3	34.9	22.7	35.0	56.6

[a]Field intensity = 17.1 kOe, slurry velocity = 5.0 to 5.3 mm/sec (case 1) and 18 to 20 mm/sec (case 2).
[b]Field intensity = 13.9 kOe, slurry velocity = 5.0 to 5.3 mm/sec (case 3) and 9 to 10 mm/sec (case 4).
[c]Float product obtained from washing in a $ZnCl_2$ solution of 1.6 specific gravity.
[d]Based on the separator feed.

Source: Ref. 106.

experience with heavy-medium washing, these recovery percentages (69 to 80%) are reasonable and can be achieved easily in commercial operations.

B. Microwave Treatment (Dielectric Heating) and Magnetic Desulfurization

The potential application of microwaves to coal desulfurization is suggested by the physical characteristics of microwave energy absorption and by the inherent inhomogeneous nature of different hydrocarbon constituents and mineral impurities in coal. It is known that the microwave power dissipation by a medium is proportional to the imaginary part of the so-called complex dielectric constant, which is a measure of the dissipated power per unit electric field (squared) according to the equation

$$P = 55.63 \times 10^{-12} fE^2 \varepsilon'' \tag{4.4}$$

In Eq. (4.4), P is the power absorbed per unit volume of material (w/m^3), f is the frequency of microwave radiation (Hz), ε'' is the imaginary part of the complex dielectric constant (dimensionless) [16,111]. If the imaginary dielectric constant ε'' at a microwave radiation frequency for a material is negligible, the material is said to be transparent, nonabsorbing, or nonconducting. When an inhomogeneous mixture made up of transparent and absorbing components is subjected to microwave treatment, the electromagnetic radiation traversing it would be absorbed only by the conducting component, resulting in a selective heating of the component. In a classic report on magnetic separation of pyrite from coals [16], it was demonstrated that pyrite could indeed be selectively heated to very high temperatures in the presence of hydrocarbon constituents of coal by using a proper range of radiation frequency, at which the imaginary dielectric constant or conductivity of pyrite was much higher than that of coal. This selective heating partially converts the pyrite to strongly magnetic compounds of iron. This conversion, which would take place in the absence of any significant loss of coal volatiles, was considered beneficial in increasing the efficiency of the subsequent use of magnetic separation for removing pyrite from coal. An apparent advantage of using microwave treatment for enhancing the magnetic properties of pyrite is that it is not necessary to crush the coal in order for the pyrite to be responsive to the electromagnetic radiation and to be selectively heated. Instead, the coarser the pyrite, the more readily it will be selectively heated. Crushing process necessary to liberate pyrite from coal could be done after dielectric heating.

Recent studies directed toward gaining further understanding of using microwave radiation as a pretreatment step prior to magnetic desulfurization have been conducted at General Electric Company (GE) [111] and Iowa State University (ISU) [8]. In the GE study, measurements on the dielectric properties of coal and some of its significant sulfur compounds such as pyrite (FeS_2), thianthrane ($C_{12}H_8S$), and dibenzothiaphene ($C_{12}H_8S_2$) were made at different microwave frequencies. It was found that coal low in pyrite was relatively transparent (nonabsorbing) to microwave at a frequency of 8.3 GHz, while pyrite would greatly enhance the level of absorption. Experiments using microwave heating were also done with 0.25-in. particles of two high-pyrite coals, namely, Pennsylvania No. 6 (Clarion County) and Lower Kittanning coal (Pennsylvania State University, Coal Research Section, sample no. 255). After relatively short periods of exposure, it was found that the treated samples could be further cleaned with ease by a low-intensity bar magnet due to the extensive conversion of pyrite (FeS_2) to more magnetic forms of iron-sulfur compounds, FeS_x (where $x = 1.14$). In the ISU study, the emphasis was placed on dielectric property measurements and dielectric heating experiments on raw coal, cleaned coal, and pyrite derived from coal, mixed in varying proportions. In particular, by measuring the dielectric properties of cleaned coal and of the pyrite derived from the coal, and by measuring the dielectric constant and loss-factor properties, the heating rates of pyrite and coal were predicted. Results reported have shown that pyrite could be heated 2.6 times faster than coal. The predicted rate was 3.2 times as fast as coal, with the difference being attributed to the heat loss to the environment. Work is currently in progress at ISU on the continued development and demonstration of combining both microwave treatment and magnetic separation for coal desulfurization.

C. Thermal Treatment and Magnetic Desulfurization

Prior to the successful development of the high-intensity, high-gradient magnetic separation technology in 1969, many of the earlier investigations on the magnetic desulfurization of coal were concentrated around developing various methods for enhancing the magnetic susceptibilities of sulfur-bearing minerals in coal, particularly pyrite. Thermal treatment was one of the methods used to possibly convert the surfaces of pyrite and marcasite (both are represented by FeS_2) grains to other highly paramagnetic and ferrimagnetic compounds such as pyrrhotite (Fe_7S_8) and magnetite (Fe_3S_4), thereby rendering such grains more amenable to magnetic separation by using the conventional low-intensity or low-gradient magnetic separator. Available reviews of many earlier studies [16,49] indicate that thermal treatment prior to magnetic desulfurization was tested with a few favorable results followed by some adverse observations. The conflicting results appeared to be mainly due to a lack of clear understanding of the thermomagnetic properties of the iron-sulfur compounds existing in coals and of an adequate liberation of sulfur-bearing minerals embedded in coals. A detailed discussion of the currently available knowledge of the thermomagnetic properties of iron-sulfur compounds of importance to magnetic desulfurization of coal can be found in Refs. 7, 30, 49, 78, 83, and 85.

Fine et al. [18] investigated the use of a low-temperature thermal treatment (flash roasting) in an inert atmosphere together with low-intensity or low-gradient magnetic separation for desulfurization of dry pulverized coal. Based on an engineering analysis of available experimental data, it was claimed that 50 to 70% of the total sulfur contained in the coal could be removed by the combined approach. However, much lower sulfur reduction was observed in the preliminary experimental studies. Further work needed for improving the efficiency of the combined approach for coal desulfurization was suggested by these investigators. In a related study, Trindade et al. [96] proposed a two-stage dry fluidized-bed process for the magnetic recovery of pyrite and coal from mine rejects. This process involves a thermal decomposition of coal mine rejects rich in pyrite at about 700°C in a fluidized bed to produce a magnetic pyrrhotite and liberate the sulfur minerals, followed by an air transport of the thermally decomposed solid product into a HGMS unit to separate the coal from magnetic minerals. Analysis was made of results reported separately on the kinetics of thermal decomposition of pyrite carried out in a thermobalance and on the HGMS tests with water slurries of pulverized coals. The technical and economic attractiveness of using the proposed two-stage process for recovering pyrite and coal from Brazilian coal mine rejects was suggested in this reported study. No experimental data for demonstrating the feasibility of the second stage (dry HGMS) of the process were presented.

D. Froth Flotation and HGMS for Coal Desulfurization

Froth flotation utilizes the difference in surface properties of pulverized coal and its mineral impurities in an aqueous pulp (coal slurry) to affect their separation. Air-avid or water-repellent (hydrophobic) coal particles are floated to the surface by finely dispersed air bubbles to be collected as a froth concentrate. Mineral impurities, which are readily wetted by water (hydrophilic), do not stick to air bubbles and remain submerged in the pulp to be carried off as an underflow. Generally, for coal flotation to be effective, reagents must be added to the pulp. One type of reagents, called *collector*, adsorbs on surfaces of coal particles; another type of reagents, called *frother*, facilitates the production of a transient froth capable of carrying the hydrophobic particle load until it can be removed from the flotation cell. Although froth flotation has been used effectively to remove high-ash materials from coal, it does little in its normal mode of operation to remove sulfur. Therefore, much effort has been devoted recently to developing new processes for sulfur removal by froth flotation. One of such new processes is a two-stage process for floating pyrite from coal with xanthate, called *coal-pyrite flotation* developed by the U.S. Bureau of Mines (now the Department of Energy) as described in detail in Chap. 5. This process involves a first-stage conventional coal flotation to reject most of the high-ash refuse and some of the coarser or liberated pyrite as tailings. The coal froth concentrate, with some dilution water, then goes to a second-stage froth flotation where a hydrophilic colloid is added to change the surface characteristics of coal particles so that they become less hydrophobic, followed by the use of a sulfhydryl collector (xanthate) to float the pyrite.

Results from laboratory HGMS tests for comparing the pyrite reduction potential of HGMS with that of the coal-pyrite flotation process were reported [25]. Tests were run on three size fractions of a Pennsylvania Upper Freeport (Helvetia Mine) coal, namely, −35 mesh (typical flotation-size material), −200 mesh, and −325 mesh. These fractions were beneficiated by flotation and HGMS and by coal-pyrite flotation. In order to make a meaningful comparison of the chosen approaches, all coal samples were first subjected to conventional coal flotation. Then the froth product was repulped and upgraded by either HGMS or coal-pyrite flotation. For the HGMS tests, a field intensity of 21.6 kOe was used along with several slurry velocities. Results reported are illustrated in Figs. 4.33 and 4.34 [25,92]. In the figures, the yield of clean coal refers to the "final" clean coal, i.e., the clean coal resulting from the beneficiation step following the conventional floth flotation. Also, each data point representing a "flotation and HGMS" test is the mean of at least eight tests conducted at equivalent experimental conditions. Figure 4.33 shows that for five of the eight tests compared, both HGMS and coal-pyrite flotation gave comparable pyrite reductions; and for the other three tests, coal-pyrite flotation results were inferior. Thus, combining flotation with HGMS was more selective than or at least equivalent to coal-pyrite flotation in removing pyrite. Figure 4.34 shows that coal-pyrite flotation did not achieve the same ash reduction as did flotation and HGMS. This observation is to be expected because in the coal-pyrite flotation process, the xanthiate collector

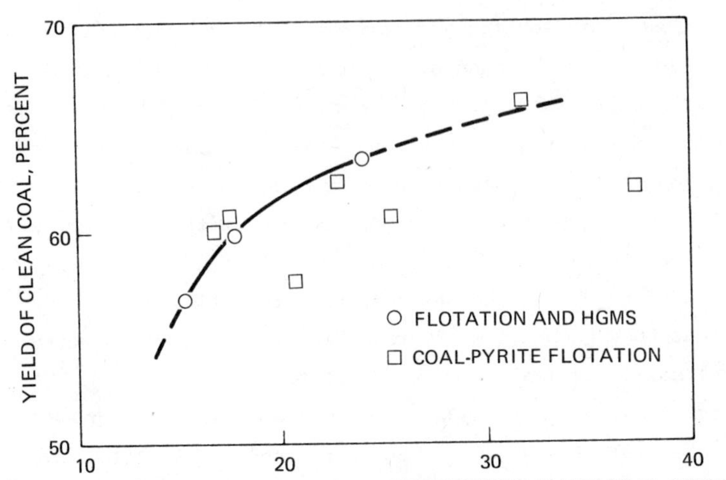

Figure 4.33 Pyritic sulfur remaining in the clean coal obtained from beneficiation of a pulverized Pennsylvania Upper Freeport (Delmont) coal (−35 mesh) by flotation and HGMS and by two-stage coal-pyrite flotation. (From Ref. 25. © 1979 IEEE.)

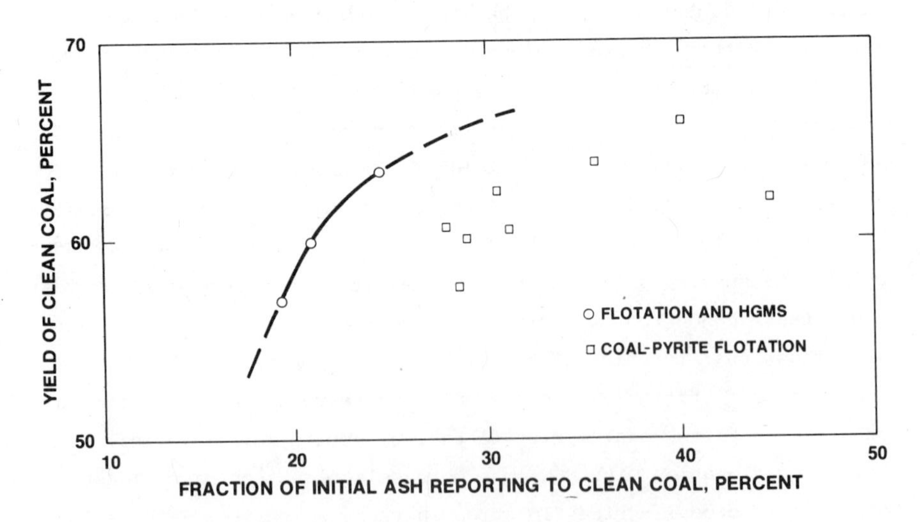

Figure 4.34 Ash remaining in the clean coal obtained from beneficiation of a pulverized Pennsylvania Upper Freeport (Delmont) coal (−35 mesh) by flotation and HGMS and by two-stage coal-pyrite flotation. (From Ref. 25. © 1979 IEEE.)

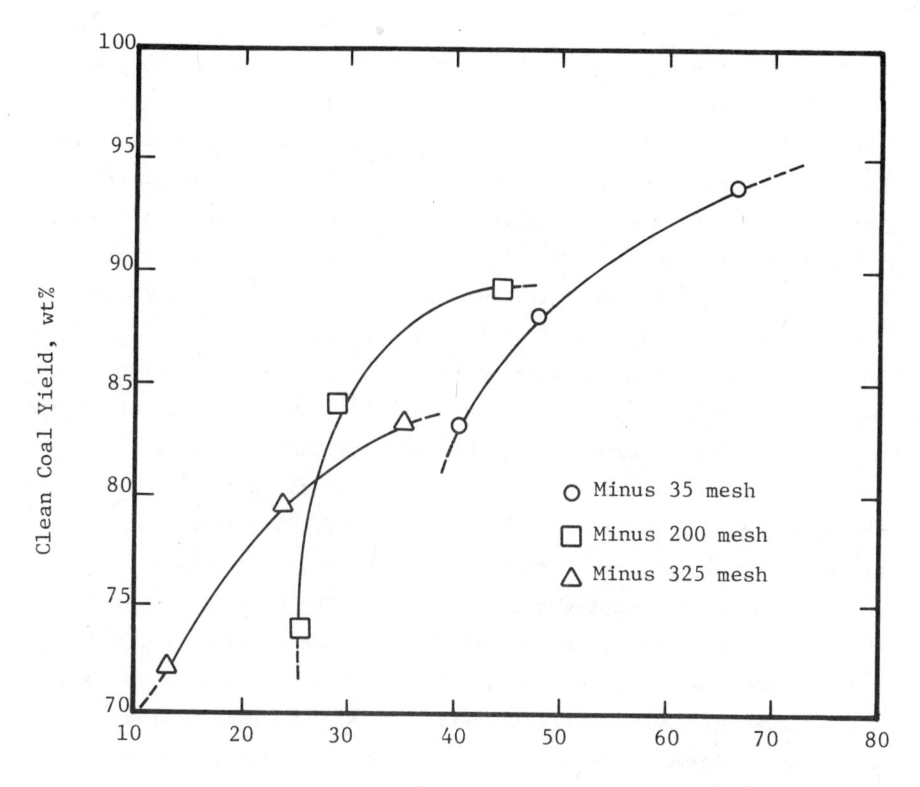

Figure 4.35 Weight percent clean coal yield versus percent of pyritic sulfur reporting to clean coal obtained from beneficiation of different size fractions of a pulverized Pennsylvania Upper Freeport (Delmont) coal by froth flotation and HGMS. (From Ref. 92.)

is highly selective for sulfide minerals. In contrast, the HGMS unit would attract much of the other mineral impurities in the separator matrix as the magnetic refuse (mags). Results reported for both −200- and −325-mesh size fractions showed that combining froth flotation with HGMS gave better performance than coal-pyrite flotation in terms of both pyrite and ash reductions; although the coal-pyrite flotation process was not optimized for these finer size feeds. Figure 4.35 compares the ability of HGMS to remove pyrite from three samples of the same coal crushed to −35-, −200-, and −325-mesh levels after an initial froth flotation [92]. The grade-recovery curve representing the —325-mesh size fraction crosses that for the —200-mesh size fraction; likewise, the latter curve appears to extrapolate through the curve for the —35-mesh size fraction. This figure demonstrates an important observation of HGMS applied to water slurries of pulverized coals, namely, equivalent desulfurization performance in terms of the grade-recovery relationship can be achieved for feeds of different top sizes. In other words, although crushing (grinding) for greater liberation can lead to less pyrite in the clean coal, the actual reduction required must be the guiding criterion so that optimal conditions of crushing and beneficiation can be merged. Moreover, exactly the same observation can be made for the ash [92].

In Ref. 20, the results from pilot-plant tests of combining froth flotation and HGMS for beneficiation of pulverized Illinois No. 6 and Pittsburgh No. 8 coals were reported. Figure 4.36 shows a flowsheet of the pilot-plant operations. The pilot-plant grinding circuit employed two stages of grinding, an open-circuit rod mill followed by a ball mill operating in closed circuit with a cyclone. The cyclone overflow was the flotation feed. The HGMS unit used was a cyclic electromagnetic model equipped with a separator matrix of 5-in. diameter and 20 in. in length packed with stainless steel wools. The operation of the unit, including feeding, rinsing and backwashing, was completely automated. Except for periodic recording of operating conditions, the unit was left unattended. In order to prevent plugging of the steel wool matrix, the magnet feed slurry had to be free of 200-mesh material. This was done by incorporating a 200-mesh screen to treat the froth concentrate prior to its processing by HGMS. The screen oversize was combined with the nonmagnetics from HGMS as the final cleaned coal.

An objective of the pilot-plant tests was to demonstrate the technical feasibility of removing up to 15% of the pyrite and ash and recovering 85 wt % of the feed coal during a continued operation of 15 consecutive 8-hr shifts, with samples being collected during each shift. The overall results of these shifts obtained for the Illinois No. 6 coal are illustrated in Fig. 4.37. At 85 wt % coal recovery, the pyrite reduction was about 24% and ash reduction about 16%. Thus,

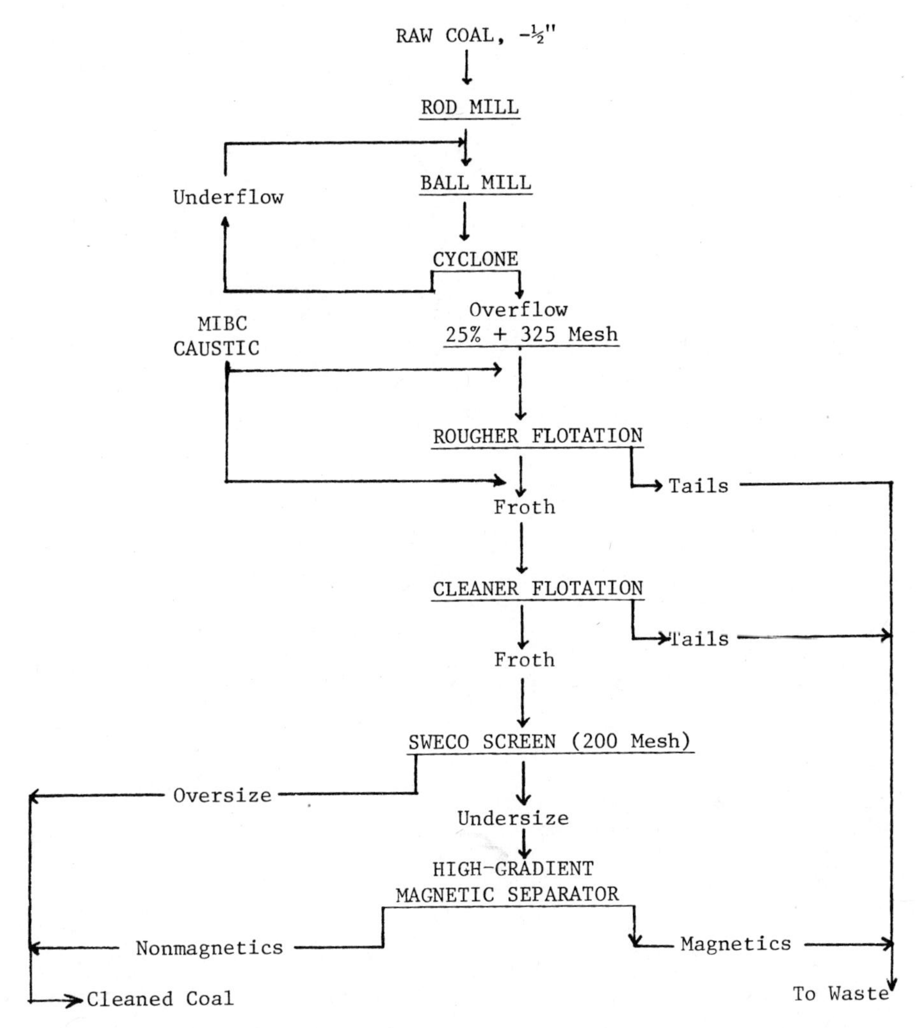

Figure 4.36 A flowsheet of a pilot plant for beneficiation of water slurries of a pulverized coal by froth flotation and HGMS. (From Ref. 20.)

the pilot-scale coal beneficiation circuit employing fine grinding, froth flotation, fine screening and HGMS of the flotation concentrate appeared to be effective in attaining the initial processing objective.

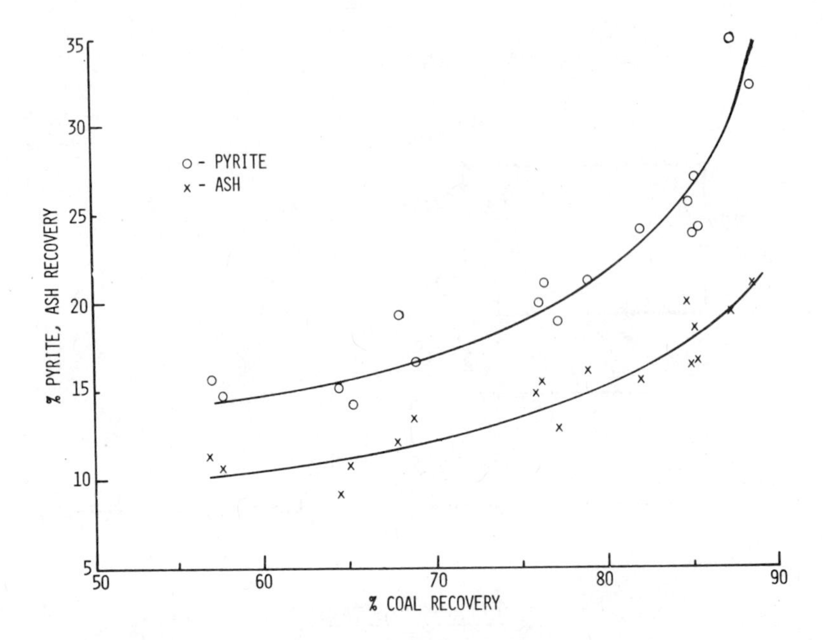

Figure 4.37 Percent pyrite and ash reductions versus cleaned coal mass recovery obtained from pilot-plant beneficiation of water slurries of a pulverized Illinois No. 6 coal by froth flotation and HGMS. (From Ref. 20.)

E. Magnetic Fluids and Magnetic Separation for Coal Desulfurization

In a recent report [86], the results from a feasibility study of coal desulfurization by using magnetic fluids together with magnetic separation conducted at the Colorado School of Mines Research Institute (CSMRI) were presented. Basically, a magnetic fluid [35] consists of a dispersed phase of colloidal particles of magnetically susceptible material (e.g., iron or magnetite), each of which is coated with a sheath of dispersing agent (e.g., oleic acid), distributed throughout a continuous phase of carrier liquid (e.g., kerosine). Thus, magnetic particles, dispersing agent, and carrier liquid constitute three essential ingredients of a magnetic fluid. Such fluids are stable, reasonably inexpensive, responsive to magnetic forces, and able to impart magnetic susceptibility to normally nonmagnetic particles. These unique characteristics of magnetic fluids could be utilized in many coal preparation and mineral processing applications. The focus of the CSMRI study was on a selective-wetting beneficiation method, in which a magnetic fluid is added to a coal feed and the fluid is selectively adsorbed to the surfaces of either organic or inorganic constituents of coal. The high magnetic susceptibility of the dispersed particles increases the magnetic susceptibility of the adsorbed constituents and prepares them for efficient magnetic separation.

Figure 4.38 is a suggested process flowsheet for desulfurization of water slurries of pulverized coals by magnetic separation and aqueous magnetic fluids (magnetic fluids with aqueous liquid carriers) [86]. As shown, run-of-mine coal or a rougher product is wet ground to a desired size range and wet screened to remove extremely fine particles which would interfere with magnetic separation. The undersize from the screens is sent to disposal or to other treatment facilities. The oversize product is contacted with a surface modifier or conditioning agent (e.g., sodium silicate or fuel oil) which makes the particle surfaces more amenable to wetting by a magnetic fluid. The conditioned feed is then contacted with an aqueous magnetic fluid and sent to a magnetic separator. Because the surfaces of most inorganic mineral impurities in coal are hydrophilic, the aqueous magnetic fluid tends to wet these surfaces, and thus increases the natural susceptibilities of some inorganics and imparts magnetic susceptibilities to those inorganics which are normally nonmagnetic. On the other hand, since the surfaces of organic coal constituents are hydrophobic, they tend to reject the aqueous magnetic fluid, and their magnetic susceptibilities will not appreciably be affected. As a result, the inorganic mineral impurities are captured in the separator matrix and depart the separator as a magnetic refuse. The unaffected coal constituents pass through the matrix and exit in the nonmagnetic product as a cleaned coal. The magnetic refuse is dewatered and sent to a disposal area such as a tailings pond. The nonmagnetic cleaned coal is dewatered and possibly agglomerated to reduce transportation problems. It is then transported to a utility for use as a boiler fuel.

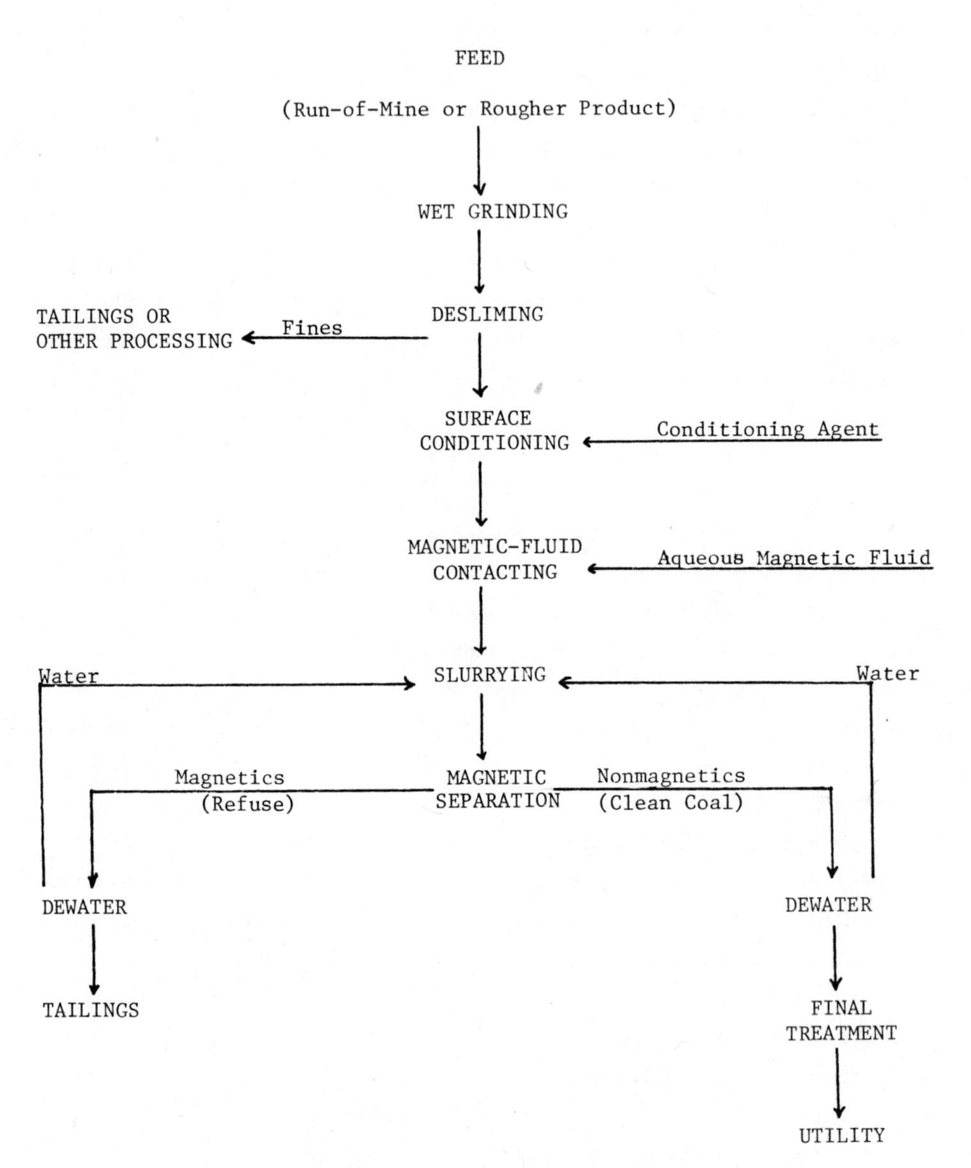

Figure 4.38 A process flowsheet for desulfurization of water slurries of pulverized coals by aqueous magnetic fluids and magnetic separation. (From Ref. 86.)

In the above flowsheet using an aqueous magnetic fluid, it is recommended that the magnetic separator be a high-intensity, HGMS unit to take advantage of the natural paramagnetism of pyrite and certain inorganic minerals. However, the magnetic fluid would increase the susceptibilities of these minerals, in addition to magnetizing normally nonmagnetic minerals. It is likely that the HGMS unit could be designed and operated with a lower field intensity than would be required for processing the untreated coal. This would result in reduced capital and operating costs for the process.

Figure 4.39 shows a conceptual process flowsheet for desulfurization of water slurries of pulverized coals which uses organic magnetic fluids and magnetic separation [86]. This process is similar to that shown in Fig. 4.38, except that the organic magnetic fluid preferentially wets the organic coal constituents and is rejected by the inorganic mineral impurities contained in coal. Consequently, it is the magnetic product from the separator which contains the cleaned coal, and the nonmagnetic product contains most of the pyrite and ash. In this application, it is likely that the magnetic separator would not be a high-intensity, HGMS device. Such a device would be unsuitable for the organic-fluid flowsheet because the separator matrix would capture both the naturally magnetic pyrite and ash *and* the organic coal constituents with high magnetic susceptibilities induced by the action of the magnetic fluid. Thus, the nonmagnetic refuse would contain only those ash constituents with extremely low magnetic susceptibilities. Although some coal beneficiation could be accomplished, the cleaned coal product would still contain most of the pyrite and ash. Low-intensity HGMS units or high-intensity, low-gradient separators, which could capture only the artificially magnetic organic coal constituents, would be more suitable.

To experimentally test the technical feasibility of the suggested process flowsheets, three magnetic fluids prepared by the Twin Cities

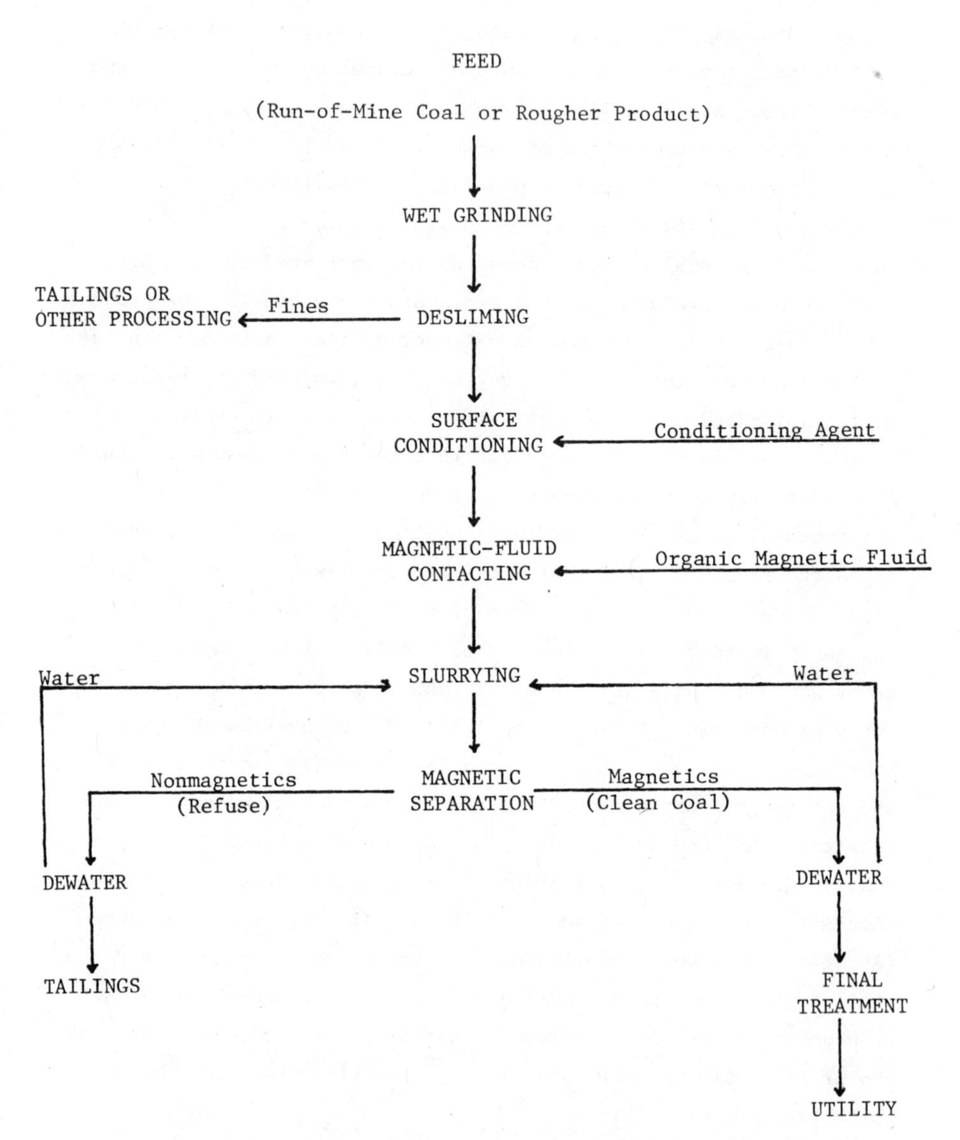

Figure 4.39 A process flowsheet for desulfurization of water slurries of pulverized coals by organic magnetic fluids and magnetic separation. (From Ref. 86.)

Metallurgy Research Center of the U.S. Bureau of Mines were utilized in the CSMRI study. These include an organic fluid using kerosene as the carrier and oleic acid as the dispersant, and two aqueous fluids using lauric acid and dodecylamine as dispersants. These magnetic fluids were used to treat seven pulverized coals, including a Pennsylvania Lower Kittanning (Canterbury) coal, a Pennsylvania Upper Freeport coal, a West Virginia Pittsburgh No. 8 coal, and a North Dakota lignite. The magnetic separators used included a Colburn HGMS unit using magnetically soft stainless steel spheres as separator matrix packing and a Carpco high-intensity induced-roll separator. Two specific objectives of the CSMRI study were (1) to determine the wetting mechanisms by which magnetic fluids would selectively attach to the surfaces of organic and inorganic coal constituents and (2) to determine the extent of coal desulfurization which could be achieved by combining magnetic fluid with magnetic separation.

With respect to the selective wetting tendencies of magnetic fluids, it was found that all three fluids (organic and aqueous) were strongly attracted to the surfaces of organic coal constituents (Fig. 4.39). This finding conflicted with the prediction of the classic flotation theory which would suggest the selective wetting of the surfaces of inorganic mineral impurities by an aqueous magnetic fluid (Fig. 4.38). This odd behavior might be due to the chemical composition or structure of the aqueous magnetic fluids used. In any case, the selective wetting of magnetic fluids to the surfaces of organic coal constituents would yield a magnetic cleaned coal product by using a process similar to that shown in Fig. 4.39 with a low-intensity HGMS unit or a high-intensity, low-gradient separator (e.g., the Carpco high-intensity separator). The latter would have no discernable effect on coals untreated by magnetic fluids. In the best Carpco test conducted, the separator produced a cleaned coal with 13.3% ash and 1.17% total sulfur from a Pennsylvania Upper Freeport coal with 17.3% ash and 2.68% total sulfur, and achieved a 77 wt % cleaned coal recovery. Results indicated that the beneficiation performance was enhanced by fairly high magnetic-fluid dosages and by long conditioning times for contact between pulverized coal and magnetic fluid.

Processing of untreated coals in the Colburn wet high-intensity HGMS unit indicated that substantial coal beneficiation could be accomplished without the use of magnetic fluids. In one of the untreated coal tests, 41% of a Pennsylvania Lower Kittanning coal reported to the nonmagnetic cleaned coal. The cleaned coal had 6.81% ash and 0.98% total sulfur for a feed containing 36.7% ash and 3.42% total sulfur. Although the cleaned coal recovery was somewhat low, the HGMS unit did remove up to 81% of the ash and 71% of the total sulfur. To examine the possible improvement in the beneficiation performance that could be achieved by using magnetic fluids, a Pennsylvania Upper Freeport coal which had not yielded satisfactory performance in tests without magnetic fluids was chosen for use in high-gradient magnetic-fluid separations. Test results showed that the use of magnetic fluids improved coal beneficiation (compared to untreated separations), but reduced cleaned coal recovery. In general, organic magnetic fluids had better effects on beneficiation than aqueous fluids; and surface-conditioning reagents reduced the cleaned coal recovery, but increased the sulfur and ash reductions. In one of the tests conducted, the high-gradient magnetic-fluid separation yielded a cleaned coal with 5.30% ash and 0.85% total sulfur for a feed with 18.8% ash and 2.46% total sulfur. Thus HGMS aided by a magnetic fluid reduced 72% of the ash and 65% of the total sulfur. However, the cleaned coal recovery was reduced from 84.2 wt % for the untreated test to only 40 wt % for the test with magnetic fluid. Obviously, further work aimed at improving the cleaned coal recovery for high-gradient magnetic-fluid separations is needed.

It is also of interest to combine the magnetic-fluid treatment with dry magnetic separation for coal beneficiation. In this application, the flowsheet is somewhat simpler than the wet-separation flowsheets because both the feed and products are handled dry and the dewatering steps are eliminated. The nature of the magnetic fluid treatment is also different. In the wet process, the coal feed could be treated by immersing the coal in the magnetic fluid and decanting any excess. In the dry process, magnetic fluid may be added to the coal as an aerosol, possibly in a fluidized-bed contactor. Results obtained by CSMRI indicated that aerosol magnetic-fluid addition was less effective than immersion addition, but some desulfurization could be achieved by the aerosol technique. Fine coal in the feed would interfere with separation performance, particularly in dry magnetic-fluid separations.

4.7 CHEMICAL PRETREATMENT AND MAGNETIC DESULFURIZATION

A. Chemical Pretreatment and Magnetic Enhancement

The possibility of using chemical reactions to convert pyrite to more magnetic compounds of iron, such as pyrrhotite (Fe_7S_8) or magnetite (Fe_3O_4), so as to enhance the performance of subsequent magnetic desulfurization has been known for years. Ergun and Bean [16] examined the effect of chemical oxidation or reduction of liberated pyrite in a pulverized Pittsburgh bed coal at elevated temperatures (250 to 350°C) in a fluidized bed on the performance of its subsequent magnetic desulfurization. Based on their experimental results and magnetic susceptibility measurements, these investigators concluded that for an effective separation of pyrite from coal, the average susceptibility of the pyrite particles embedded in coal should be raised to about 3×10^{-6} emu/g. Such a value of susceptibility could be achieved by converting less than 0.1% pyrite into ferromagnetic compounds of iron.

Recently, a novel magnetokinetic technique was proposed by Marusak et al. [59] to quantitatively follow the kinetics of oxidation of pyrite to form strongly magnetic compounds of iron. By using this technique, it was found that the oxidation of pyrite would tend to yield only α-Fe_2O_3 and γ-Fe_2O_3. The latter is a strongly magnetic or ferrimagnetic iron compound, which can be easily separated from coal by magnetic separation.

The kinetics of the transformation of pyrite to pyrrhotite of relevance to magnetic desulfurization have been studied by many investigators, and a review of representative literature can be found in Refs. 49, 61, and 78. Due to the inherent complexity of such kinetic and thermomagnetic relationships, past attempts to develop a practical chemical method as an effective pretreatment step prior to magnetic desulfurization have not been successful. A notable exception is the Magnex process described below.

B. Magnex Process for Desulfurization of Dry Pulverized Coal

The Magnex process is a dry physiochemical method for coal beneficiation which involves a chemical treatment of moderately crushed coal by iron carbonyl [$Fe(CO)_5$] vapor followed by the removal of pyrite and ash from the treated coal by dry magnetic separation. The process works because iron carbonyl selectively decomposes on the surfaces of ash-forming minerals to form strongly magnetic crystallites of iron on the ash, Fe•ash; and it also selectively reacts with pyrite to form a strongly magnetic, pyrrhotite-like material, FeS_x ($x \simeq 1.14$). However, iron carbonyl will not react or deposit on the surfaces of organic coal constituents. Both the ash with iron crystallites and the pyrrhotite-like material could be easily removed from coal by magnetic separation.

The Magnex process was originally invented at Hazen Research, Inc. in 1974 and described in a patent [36]. Figure 4.40 illustrates the four basic steps of the process [28]. These steps are as follows:

1. Crushing. In this step, raw coal is comminuted to a certain top size, typically −14 mesh (1.14 mm). The size consist of the crushed coal usually contains less than 8% −200 mesh. Note that the main goal of the crushing step is to facilitate the exposure of pyrite and ash-forming minerals to the iron carbonyl vapor in the subsequent treatment step, rather than to achieve the liberation of pyrite and ash. Thus, the role of crushing in the Magnex process with iron carbonyl treatment prior to magnetic separation is similar to that in the approach of microwave treatment followed by magnetic separation (Sec. 4.6B).

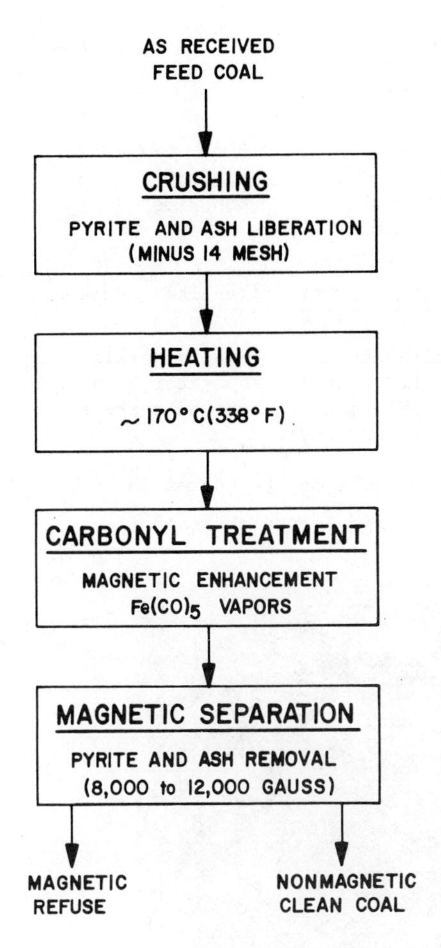

Figure 4.40 Basic steps in the Magnex process. (From Ref. 38. © 1979 IEEE.)

2. Heating. The crushed coal is heated to about 170°C (338°F). When certain coals are heated in the presence of steam, the subsequent carbonyl treatment becomes more selective. Improved selectivity results in increased yield (recovery) and reduced sulfur content of the cleaned coal.
3. Carbonyl treatment. The iron carbonyl treatment provides the magnetic enhancement of the pyrite and ash-forming minerals. The chemical reactions in the Magnex process can be summarized by three equations:

$$Fe(CO)_5 \rightarrow Fe^{o} + 5CO$$
Iron carbonyl → Iron + Carbon monoxide

$$Fe(CO)_5 + Ash \rightarrow Fe{\bullet}ash + 5CO$$
Iron carbonyl + Ash → Iron crystallites on ash + Carbon monoxide

$$(2-x)Fe(CO)_5 + xFeS_2 \rightarrow 2FeS_x + 5(2-x)CO$$
Iron carbonyl + Pyrite → Pyrrhotite-like material + Carbon monoxide

4. Magnetic separation. Pyrite and ash-forming minerals are removed by a medium-intensity magnetic separator as a magnetic refuse while the organic coal constituents are recovered as a nonmagnetic cleaned coal.

Figures 4.41 to 4.43 show photographs of the untreated feed coal, nonmagnetic clean coal and magnetic refuse obtained from

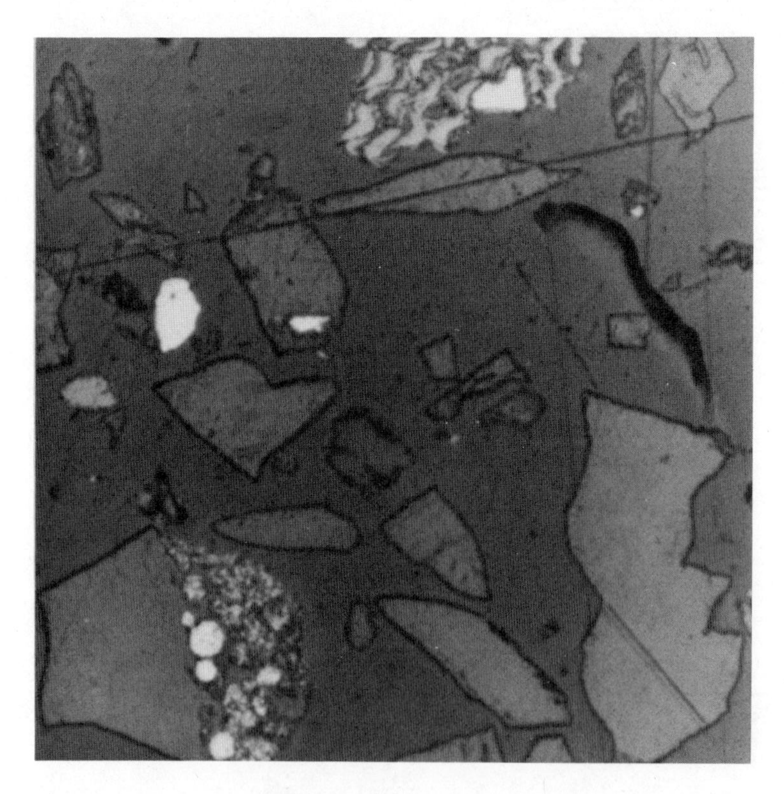

Figure 4.41 A photograph of untreated coal: −65 mesh (200 x). (From Ref. 38. © 1979 IEEE.)

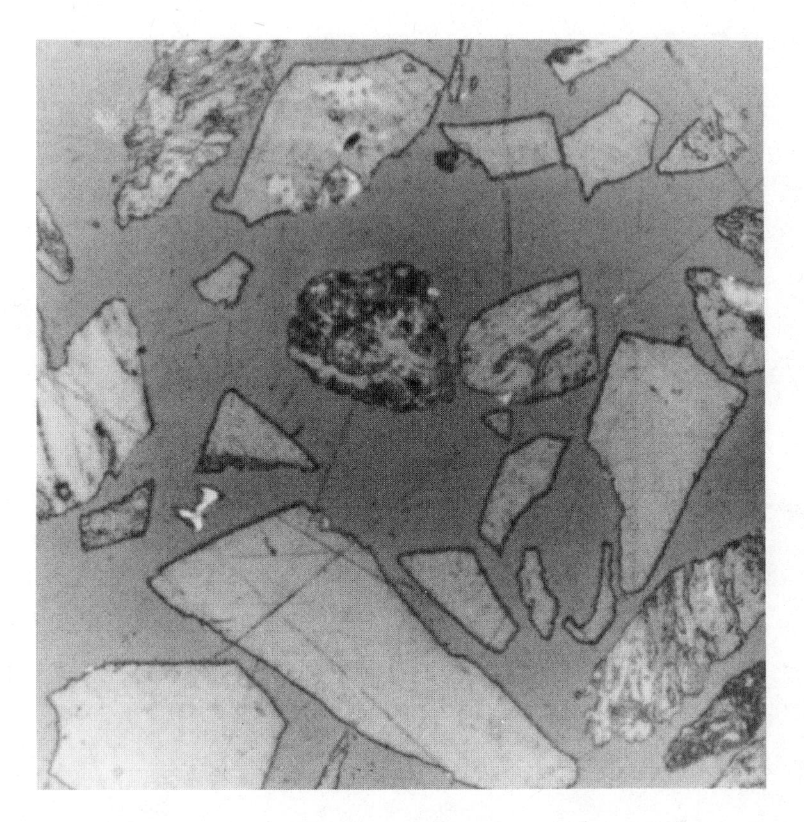

Figure 4.42 A photograph of nonmagnetic clean coal: −65 mesh (200 x). From Ref. 38. © 1979 IEEE.)

Magnex-processing of a −65 mesh coal sample [38]. The formation of the pyrrhotite-like material can be seen on the periphery and in the cracks in Fig. 4.43, the lighter and brighter material being the unconverted pyrite. Note that only a slight conversion of the pyrite to a pyrrhotite-like material and only a small deposition of iron on the ash to form crystallites are required to permit the pyrite and ash to be magnetically separated from coal. This follows because the approximate inherent magnetic susceptibility of pyrrhotite is 7000 to 9000 times greater than that of pyrite [16] and the iron crystallites on the ash are ferromagnetic, and they are many times more magnetic than pyrrhotite.

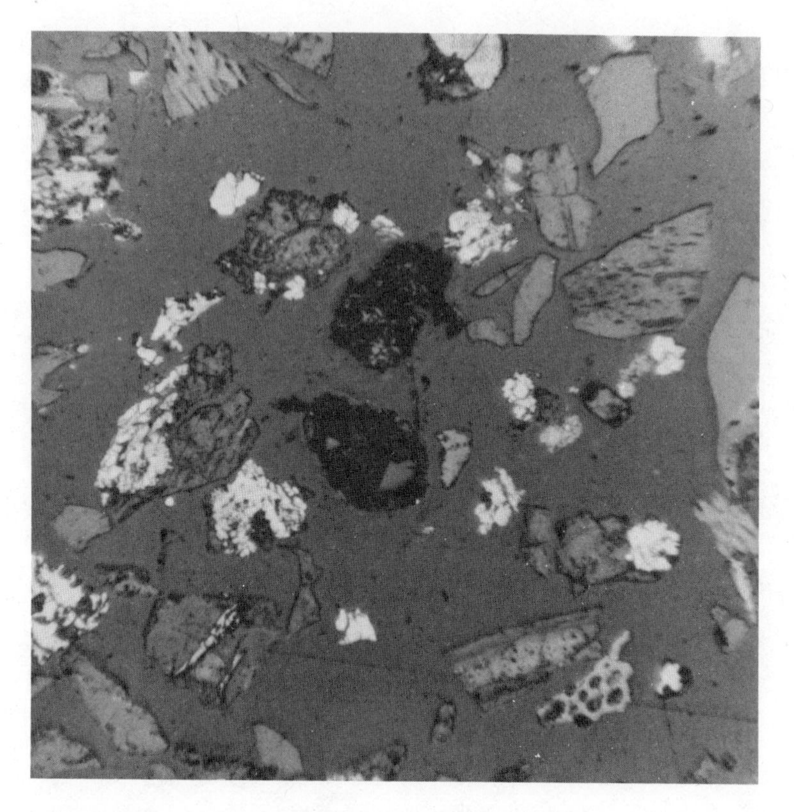

Figure 4.43 A photograph of magnetic refuse: −65 mesh (200 x). (From Ref. 38. © 1979 IEEE.)

Initial laboratory development of the Magnex process was concentrated on the chemistry of the process and not on the magnetic separation [36,37]. Through the laboratory tests, the major independent variables of the first three steps of the process were identified as follows:

1. Crushing: top size, and size consist
2. Heating: time, temperature, and conditioning atmosphere
3. Carbonyl treatment: time, temperature, iron carbonyl concentration, and conditioning with other gases

Recent developmental work involved an evaluation of the use of several commercially available magnetic separators to remove pyrite and ash from the Magnex-processed coal [80]. As expected, the performance of the magnetic separation step was found to be dependent upon the type of separator, field intensity, and field gradient.

Based on the successful laboratory development and a favorable economic study, a pilot plant of 200 lb/hr was designed and constructed to test the process on a continuous basis. Figure 4.44 illustrates a schematic diagram of the pilot-plant operations [38]. In the pilot plant, crushing was accomplished with jaw and impactor crushers. Indirect heating and pretreatment with steam at atmospheric pressure were done while the coal passed through a screw conveyor. Carbonyl treatment was accomplished in a gas phase also at atmospheric pressure. Magnetic separation of the reactor product was done on an induced-roll high-intensity magnetic separator. A detailed description of the pilot-plant facilities and operations has been given in Ref. 38.

It has been reported that approximately 60 different coals have been evaluated for their potential for beneficiation by the Magnex process. Table 4.12 summarizes the typical results obtained for beneficiation of six selected coals [39]. The Magnex process was effective in removing the total sulfur by 22.4 to 66.5%, the pyritic sulfur by 56.9 to 91.6%, while achieving a Btu recovery of 86.5 to 99.0%. The percent reduction of sulfur dioxide emission level varied from 23.9 to 73.7%, with an average of 52.7%. The sulfur dioxide emission levels of two of the listed coals were reduced to 1.20 lb SO_2 per MM Btu or less.

The effects of feed coal properties and major independent variables in the heating/pretreatment and carbonyl treatment steps on the removal of pyrite and ash from several dry pulverized coals by the Magnex process were reported in a recent paper [39]. It was found that an improved Magnex process performance could be correlated with the following feed coal properties: (1) a small amount of low-temperature volatiles, (2) a tendency to produce acid (SO_4^{-2}), (3) lower levels of copper and perhaps higher levels of zinc, and (4) lower pyritic and organic sulfur contents. In the same paper, some

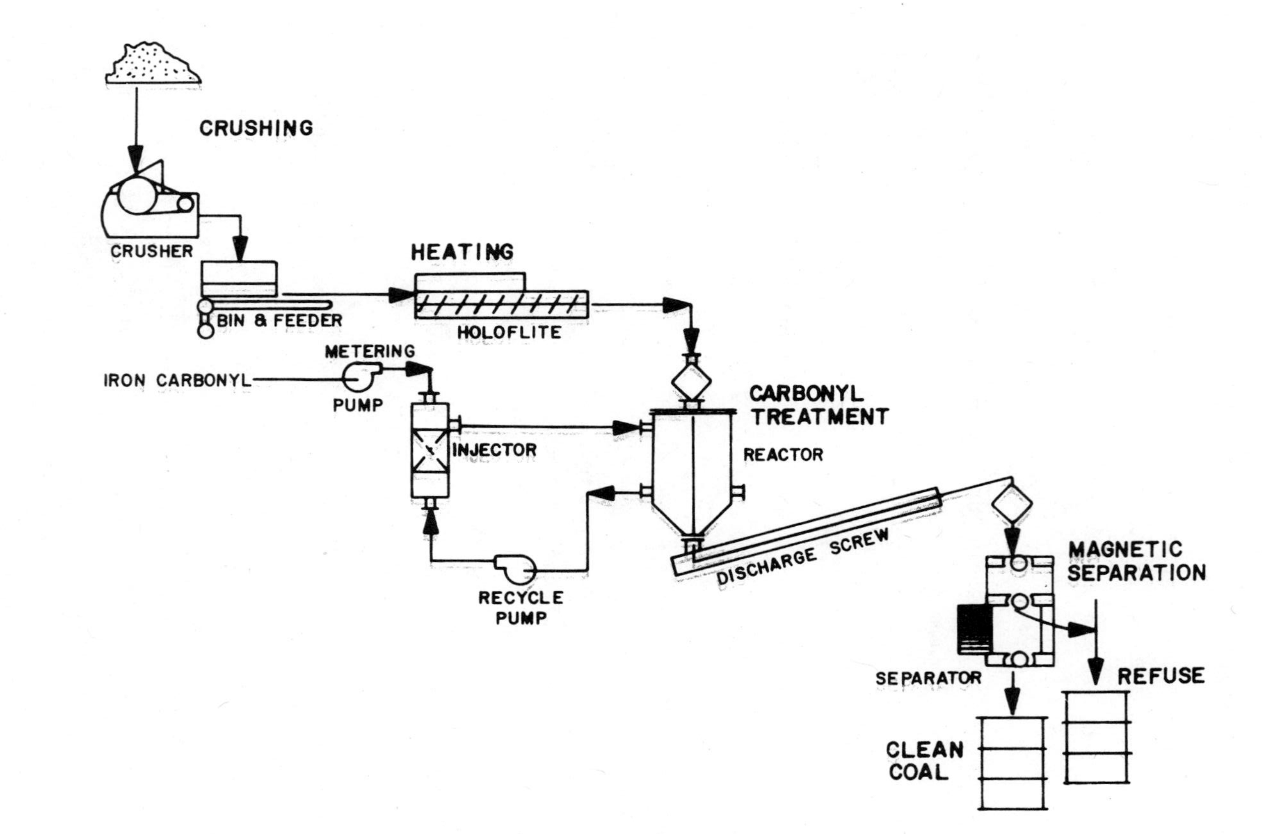

Figure 4.44 A schematic diagram of the Magnex process pilot plant. (From Ref. 38. © 1979 IEEE.)

Table 4.12 Typical Results Obtained from the Magnex Process for Beneficiation of Dry Pulverized Coal

1. Coal type	Pittsburgh No. 8	Allegheny Group	Lower Freeport	Kentucky No. 11	Lower Freeport	Rosebud McKay
2. County	Green	Mercer	Jefferson	Muhlenberg	Cambria	Big Horn
3. State	PA	PA	OH	KY	PA	MT
4. Total sulfur removal, %	38.7	57.9	43.3	22.4	66.5	59.6
5. Pyritic sulfur removal, %	68.3	91.6	62.1	56.9	84.4	73.5
6. Ash removal, %	23.3	16.3	71.2	20.2	65.2	20.7
7. Btu recovery, %	98.8	93.0	87.1	85.5	92.5	99.0
8. Pounds SO_2 per MM Btu						
Feed	4.50	2.30	6.54	5.17	4.79	1.58
Product	2.50	1.20	2.53	3.93	1.26	0.69
% Reduction	44.4	56.5	61.3	23.9	73.7	56.3

Source: Ref. 39.

of the recent work on the Magnex process have also been described. These included (1) an intensive study of iron carbonyl generation, (2) a systematic search for alternate magnetic separators capable of processing large volumes of dry solids efficiently and inexpensively, and (3) engineering studies of a coal preparation process flowsheet which employs the best blend of the advantages of conventional coal processing and the Magnex technology. Continued developmental work on the process is presently being pursued by the Nedlog Technology Group, Arrado, Colorado; and engineering studies for the construction of a demonstration plant of 60 tons/hr have been completed. A search is now underway to determine the best installation site [82]. Finally, it is worthwhile to mention that the conceptual design and cost estimation of an air-entrained flow HGMS process using a superconducting quadrupole magnet system for desulfurization of dry pulverized coal chemically treated by the Magnex process has been given in Ref. 77.

REFERENCES

1. I. Y. Akoto, High Gradient Magnetic Separation: Effect of Temperature on Performance, *IEEE Trans. Magn., MAG-12,* 511 (1976).
2. I. Y. Akoto, Magnetically Induced Desulfurization of Liquid Coal, Ph.D. dissertation, Massachusetts Institute of Technology, Cambridge, Mass., 1977.
3. W. Aubrey, Jr., J. Carpinski, D. Cahn, and C. Rauch, Magnetic Separator Method and Apparatus, U.S. Patent 3,608,718, filed Dec. 20, 1968.
4. J. D. Batchelor and C. Shih, *Solid-Liquid Separation in Coal Liquefaction Processes,* AIChE Annual Meeting, Los Angeles, Calif., Nov., 1975.
5. S. K. Batra, Coal-Oil Mixture Combustion in Boilers: An Update, in *Proceedings of Engineering Foundation Conference on Clean Combustion of Coal* (V. S. Engleman, ed.), Rindge, New Hampshire, July 31-August 4, EPA-60077-78-073, Environmental Protection Agency, Washington, D.C., Apr. 1970, pp. 193-204.
6. C. P. Bean, Theory of Magnetic Filtration, *Bull. Amer. Phys. Soc., 16,* 350 (1971).
7. M. J. Besnus and A. J. P. Meyer, New Experimental Data on the Magnetism of Natural Pyrrhotite, *Proceedings of the International Conference on Magnetism,* Nottingham, England, published by the Institute of Physics and Physical Society, London, 1964, pp. 507-511.

8. D. Bluhm, private communication, Ames Laboratory, U.S. Department of Energy, Iowa State University, Ames, Iowa, 1979.

9. N. O. Clark, Applications of Superconducting Magnetic Separation to Clay and Mineral Processing, in *Industrial Applications of Magnetic Separation* (Y. A. Liu, ed.), IEEE Publication No. 78CH1447-2MAG, Institute of Electrical and Electronic Engineers, New York, 1979, pp. 62-63.

10. E. Cohen and J. A. Good, The Application of a Superconducting Magnet System to the Cleaning and Desulfurization of Coal, *IEEE Trans. Magn., MAG-12,* 503 (1976).

11. E. Cohen and J. A. Good, The Principles and Operation of a Very High Intensity Magnetic Mineral Separator, *IEEE Trans. Magn., MAG-12,* 522 (1976).

12. A. J. Colli, Pyrite Separation from Coal for Pollution Abatement, final report, issued by Naval Ordinance Station, Indian Head, Maryland, to the U.S. Department of Energy, under contract no. ET-77-I-01-9138 (1978).

13. G. E. Crow, A Pilot-Scale Study of High Gradient Magnetic Desulfurization of Solvent Refined Coal, Master's thesis, Auburn University, Auburn, Ala., Mar. 1978.

14. D. M. Eissenberg and Y. A. Liu, High-Gradient Magnetic Beneficiation of Dry Pulverized Coal via Recirculating Fluidization, U.S. Patent 4,212,551, filed May 30, 1979 and issued July 15, 1980.

15. D. M. Eissenberg, E. C. Hise, and M. S. Silverman, ORNL Program for Development of Magnetic Beneficiation of Dry Pulverized Coal, in *Industrial Applications of Magnetic Separation* (Y. A. Liu, ed.), IEEE Publ. No. 78CH1447-2MAG, Institute of Electrical and Electronic Engineers, New York, 1979, pp. 91-94.

16. S. Ergun and E. H. Bean, Magnetic Separation of Pyrite from Coals, U.S. Bureau of Mines, Report of Investigation No. 7181, Sept. 1968.

17. Y. Eyssa, A Feasibility Study of Superconducting Magnetic Separators for Weakly-Magnetic Ores, Ph.D. dissertation, University of Wisconsin, Madison, Wis., 1975.

18. H. A. Fine, M. Lowry, L. F. Power, and G. H. Geiger, A Proposed Process for the Desulfurization of Finely Divided Coal by Flush Roasting and Magnetic Separation, *IEEE Trans. Magn., MAG-12,* 523 (1976).

19. S. Foner and B. B. Schwartz (eds.), *Superconducting Machines and Devices: Large Systems Applications,* Plenum Press, New York, 1974.

20. W. L. Freyberger, J. W. Keck, D. J. Spottiswood, N. D. Solem, and V. L. Doane, Cleaning of Eastern Bituminous Coals by Fine Grinding, Froth Flotation and High Gradient Magnetic Separation, in *Proceedings: Symposium on Coal Cleaning to Achieve Energy*

and Environmental Goals, Hollywood, Fla., Sept. 1978 (S. E. Rodgers and A. W. Lemmon, Jr., eds.), EPA-600/7-79-098a, Environmental Protection Agency, Washington, D.C., Apr., 1979, pp. 534-567.

21. D. Garg, A. R. Tarrer, J. A. Guin, J. M. Lee, and C. W. Curtis, Selectivity Improvement in the Solvent Refined Coal Process, *Fuel Processing Technology, 2,* 189 (1978).

22. R. A. Glenn and R. D. Harris, Liberation of Pyrite from Steam Coals, *J. Air Pollution Control Ass., 12*(8), 388 (1962).

23. J. A. Good and E. Cohen, A Superconducting Magnet System for a Very High Intensity Magnetic Mineral Separator, *IEEE Trans. Magn., MAG-12,* 493 (1976).

24. E. C. Hise, I. Wechsler, and J. M. Doulin, Separation of Dry Crushed Coals by High Gradient Magnetic Separation, Report No. ORNL-4471, Oak Ridge National Laboratory, U.S. Department of Energy, Oak Ridge, Tenn., Oct. 1979.

25. R. E. Hucko, DOE Research in High Gradient Magnetic Separation Applied to Coal Beneficiation, in *Industrial Applications of Magnetic Separation* (Y. A. Liu, ed.), IEEE Publication No. 78CH1447-2MAG, Institute of Electrical and Electronic Engineers, New York, 1979, pp. 77-82.

26. Hydrocarbon Research, Inc., Magnetic Separation of Coal Liquids, Report Nos. L-12-61-501 to 505, issued to the Electric Power Research Institute, under Contract No. RP-340, Trenton, N.J., 1975.

27. J. Iannicelli, New Developments in Magnetic Separation, *IEEE Trans. Magn., MAG-12,* 436 (1976).

28. J. Iannicelli, Coal Desulfurization, U.S. Patent 3,999,958, Dec. 28, 1976.

29. J. Iannicelli, N. Millman, and W. J. D. Stone, Process for Improving the Brightness of Clays, U.S. Patent 3,471,011, Oct. 7, 1969.

30. I. S. Jacobs, L. M. Levinson, and H. R. Hart, Jr., Magnetic and Mössbauer Spectroscopic Characterization of Coal, *J. Appl. Phys., 49,* 1775 (1978).

31. I. S. Jacobs and L. M. Levinson, Enhancement of Magnetic Separability in Coal Liquefaction Residual Solids, *J. Appl. Phys., 50,* 2422 (1979).

32. F. V. Karlson, H. Heuttenhain, M. Epstein, O. N. Degiovanni, and P. J. Chassagne, Coal Preparation Using Magnetic Separation: Volume 5--Evaluation of Magnetic Coal Desulfurization Concepts, EPRI Report No. CS-1517, Vol. 5, Electric Power Research Institute, Palo Alto, Calif., July 1980.

33. W. M. Kester, Magnetic Demineralization of Pulverized Coal, *Min. Eng., 17,* 72-76 (1965).

34. W. M. Kester, J. W. Leonard, and E. B. Wilson, Reduction of Sulfur from Steam Coals by Magnetic Methods, *Min. Congr. J.*, *53*,70 (1967).

35. S. E. Khalafalla, Magnetic Fluids, *Chem. Tech.*, *5*, 540 (1975).

36. J. K. Kindig and R. L. Turner, Process for Improving Coal, U.S. Patent 3,938,966, Feb. 17, 1976.

37. J. K. Kindig and R. L. Turner, Dry Chemical Process to Magnetize Pyrite and Ash Removal from Coal, Preprint No. 76-F-366, SME-AIME Fall Meeting, Denver, Sept. 1976.

38. J. K. Kindig, The Magnex Process: Review and Current Status, in *Industrial Applications of Magnetic Separation* (Y. A. Liu, ed.), IEEE Publ. No. 78CH447-2 MAG, Institute of Electrical and Electronic Engineering, New York, 1979, pp. 99-104.

39. J. K. Kindig and D. N. Geons, The Dry Removal of Pyrite and Ash from Coal by the Magnex Process: Coal Properties and Process Variables, in *Proceedings: Symposium on Coal Cleaning to Achieve Energy and Environmental Goals, Volume II* (S. E. Rogers and A. W. Lemmon, Jr., eds.), EPA-600/7/79-098b, Environmental Protection Agency, Washington, D.C., 1979, pp. 1165-1195.

40. H. Kolm, J. A. Oberteuffer, and D. R. Kelland, High-Gradient Magnetic Separation, *Scientific American*, *223*, 45 (1975).

41. J. E. Lawver and D. M. Hopstock, Wet Magnetic Separation of Weakly Magnetic Minerals, *Min. Sci. Eng.*, *6*, 154 (1974).

42. J. W. Leonard (ed.), *Coal Preparation*, 4th ed., AIME, New York, 1979.

43. C. J. Lin, Y. A. Liu, D. L. Vives, M. J. Oak, G. E. Crow, and E. L. Huffman, Pilot-Scale Studies of Sulfur and Ash Removal from Coals by High Gradient Magnetic Separation, *IEEE Trans. Magn.*, *MAG-12*, 513 (1976).

44. C. J. Lin and Y. A. Liu, Desulfurization of Coals by High-Intensity High-Gradient Magnetic Separation: Conceptual Process Design and Cost Estimation, in *Coal Desulfurization: Chemical and Physical Methods, ACS Symposium Series*, No. 64 (T. D. Wheelock, ed.), American Chemical Society, Washington, D.C., 1977, pp. 221-249.

45. Y. A. Liu (ed.), *Proceedings of Magnetic Desulfurization of Coal Symposium, IEEE Trans. Magn.*, *MAG-12*(5), 423-551 (1976).

46. Y. A. Liu (ed.), *Industrial Applications of Magnetic Separation*, IEEE Publ. No. 78CH1447-2 MAG, Institute of Electrical and Electronic Engineers, New York, 1979.

47. Y. A. Liu, Novel High Gradient Magnetic Separation (HGMS) Processes for Desulfurization of Dry Pulverized Coal, in *Recent Developments in Separation Science: Vol. VI* (Norman N. Li, ed.), CRC Press, West Palm Beach, Fla., 1981, Chap. 9.

48. Y. A. Liu, Recent Developments in High Gradient Magnetic Separation: I. General Features and Equipment Design and Development, Invited paper, *Separation Science and Technology*, in press (1982).

49. Y. A. Liu and C. J. Lin, Assessment of Sulfur and Ash Removal from Coals by Magnetic Separation, *IEEE Trans. Magn., MAG-12*, 538 (1976).

50. Y. A. Liu, M. J. Oak, and C. J. Lin, Modeling and Experimental Study of High Gradient Magnetic Separation Applied to Coal Beneficiation, *Symposium on Novel Separation Technique*, AIChE Annual Meeting, New York, Nov. 1977.

51. Y. A. Liu and C. J. Lin, Status and Problems in the Development of High Gradient Magnetic Separation Processes Applied to Coal Beneficiation, Invited paper, in *Proceedings of Engineering Foundation Conference on Clean Combustion of Coal*, Rindge, New Hampshire, July 31-Aug. 5, 1977 (V. S. Engleman, ed.), EPA-60077-78-073, Environmental Protection Agency, Washington, D.C., Apr. 1978, pp. 109-130.

52. Y. A. Liu and C. J. Lin, Research Needs and Opportunities in High Gradient Magnetic Separation of Particulate-Gas Systems, Invited paper, *Proceedings of NSF-EPA Workshop on Novel Concepts, Methods and Advanced Technology for Particulate-Gas Separation* (T. Ariman, ed.), National Science Foundation and Environmental Protection Agency, Washington, D.C., Aug. 1978, pp. 170-179.

53. Y. A. Liu, C. J. Lin, and D. M. Eissenberg, A Novel Fluidized-Bed Dry Magnetic Separation Process with Applications to Coal Beneficiation, *Digest of the 1978 International Magnetics Conference*, IEEE Publ. No. 78CH1341-7MAG, May 1978, p. 10-2.

54. Y. A. Liu, T. H. McCord, S. K. Batra, and T. Tsai, Novel Fluidized-Bed High Gradient Magnetic Separation (HGMS) Processes for Desulfurization of Dry Pulverized Coal for Utility Applications, presented at International Magnetics Conference, New York, July 1979.

55. Y. A. Liu and G. E. Crow, Studies in Magnetochemical Engineering. I. A Pilot-Scale Study of High-Gradient Magnetic Desulfurization of Solvent-Refined Coal (SRC), *Fuel, 58*, 345 (1979).

56. F. E. Luborsky, High-Field, High-Gradient Magnetic Separation: A Review, 21st Annual Conference on Magnetism and Magnetic Materials, Philadelphia, Pa., Nov. 1977.

57. F. E. Luborsky, High Gradient Magnetic Separation for Removal of Sulfur from Coal, Final report, issued by General Electric Company, Corporate Research and Development to U.S. Bureau of Mines under contract no. HO366008 (1977); published as report no. FE-8969-1 (EPA-600/7-78-208), by U.S. Department of Energy and Environmental Protection Agency, Washington, D.C., Nov. 1978.

58. P. G. Marston and J. J. Nolan, Moving Matrix Magnetic Separators, U.S. Patent 3,920,543, issued on Nov. 18, 1975.

59. L. A. Marusak, P. L. Walker, Jr., and L. N. Mulay, The Magnetokinetics of Oxidation of Pyrite (FeS_2), *IEEE Trans. Magn., MAG-12,* 889 (1976).

60. E. Maxwell, Magnetic Separation: The Prospects for Superconductivity, *Cryogenics,* 179 (1975).

61. E. Maxwell, High Gradient Magnetic Separation in Coal Desulfurization, in *Scientific Problems of Coal Utilization* (B. R. Cooper, ed.), *DOE Symp. Ser.,* No. 46, pp. 97-112, Technical Information Center, U.S. Department of Energy, Oak Ridge, Tenn., 1978; also published in *IEEE Trans. Magn., MAG-14,* 482 (1978).

62. E. Maxwell, D. R. Kelland, and I. Y. Akota, High Gradient Magnetic Separation of Mineral Particulates from Solvent Refined Coal, *IEEE Trans. Magn., MAG-12,* 507 (1976).

63. E. Maxwell, I. S. Jacobs, and L. M. Levinson, Magnetic Separation of Mineral Matter from Coal Liquids, EPRI Report No. AF-508, Aug. 1977; and Report No. AF-875, Electric Power Research Institute, Palo Alto, Calif., Dec. 1978.

64. J. T. McCartney, H. J. O'Donnell, and S. Ergun, Pyrite Size Distribution and Coal-Pyrite Particle Association in Steam Coals, U.S. Bureau of Mines, Report of Investigation No. 7231 (1969).

65. T. H. McCord, A Feasibility Study of Novel Fluidized-Bed High Gradient Magnetic Separation Processes for Desulfurization of Dry Pulverized Coal for Utility Applications, Master's thesis Auburn University, Auburn, Ala., Oct. 1979.

66. H. H. Murray, Beneficiation of Selected Industrial Minerals Using High Intensity Magnetic Separation, *IEEE Trans. Magn., MAG-12,* 498 (1976).

67. J. J. Nolan, Multiple Matrix Magnetic Separation Device and Method, U.S. Patent 3,770,629, Nov. 6, 1973.

68. M. J. Oak, Modeling and Experimental Studies of High Gradient Magnetic Separation (HGMS) with Applications to Coal Beneficiation, Master's thesis, Auburn University, Auburn, Ala., Dec. 1977.

69. J. A. Oberteuffer, Characteristics of HGMS Devices, in *Proceedings of HGMS Symposium* (J. A. Oberteuffer and D. R. Kelland, eds.), Massachusetts Institute of Technology, Cambridge, Mass., May 1973, pp. 86-101.

70. J. A. Oberteuffer, Magnetic Separation: A Review of Principles, Devices and Applications, *IEEE Trans. Magn., MAG-10,* 223 (1974).

71. J. A. Oberteuffer, Engineering Development of High Gradient Magnetic Separators, *IEEE Trans. Magn., MAG-12,* 444 (1976).

72. J. A. Oberteuffer, High Gradient Magnetic Separation: Basic Principles, Devices and Applications, in *Industrial Applications of Magnetic Separation* (Y. A. Liu, ed.), IEEE Publ. No. 78CH1447-2 MAG, Institute of Electrical and Electronic Engineers, New York, 1979, pp. 3-7.

73. J. A. Oberteuffer and B. R. Arvidson, General Design Features of Industrial High Gradient Magnetic Filters and Separators, in *Industrial Applications of Magnetic Separation* (Y. A. Liu, ed.), IEEE Publ. No. 78CH1447-2 MAG, Institute of Electrical and Electronic Engineers, New York, 1979, pp. 17-21.

74. R. R. Oder, High Gradient Magnetic Separation: Theory and Applications, *IEEE Trans. Magn., MAG-12,* 428 (1976).

75. R. R. Oder, Magnetic Desulfurization of Liquefied Coals: Conceptual Process Design and Cost Estimation, *IEEE Trans. Magn., MAG-12,* 532 (1976).

76. R. R. Oder, Method of Magnetic Beneficiation of Particle Dispersions, U.S. Patent 3,985,646, Oct. 12, 1976.

77. R. R. Oder, Engineering Aspects of Dry Magnetic Separation, *Proceedings of NSF-EPA Workshop on Novel Concepts, Methods and Advanced Technology for Particulate-Gas Separation* (T. Ariman, ed.), National Science Foundation and Environmental Protection Agency, Washington, D.C., Aug. 1978, pp. 170-179.

78. J. V. O'Gorman, Studies on Mineral Matter and Trace Elements in North American Coals, Ph.D. thesis, The Pennsylvania State University, University Park, Pa., 1972.

79. L. Petrakis, P. F. Ahner, and F. E. Kiviat, High Gradient Magnetic Desulfurization of Solvent Refined Coal, in *Industrial Applications of Magnetic Separation* (Y. A. Liu, ed.), IEEE Publ. No. 78CH1447-2 MAG, Institute of Electrical and Electronic Engineers, New York, 1979, pp. 95-96.

80. C. R. Porter and D. N. Goens, Coal Preparation Using Magnetic Separation: Vol. 1, Magnetic Separation Study of Magnex Processed Coal, EPRI Report No. CS-1517, Vol. 1, Electric Power Research Institute, Palo Alto, Calif., July 1980.

81. C. R. Porter and D. N. Goens, Magnex Pilot Plant Evaluation: A Dry Chemical Process for the Removal of Pyrite and Ash from Coal, paper presented at Society of Mining Engineers Fall Meeting and Exhibit, St. Louis, Mo., Oct. 19-21, 1977.

82. C. R. Porter, The Magnex Process, paper presented at Coal Conference and Expo V, Louisville, Ky., Oct. 23-25, 1979.

83. L. F. Powers and H. A. Fine, The Iron-Sulfur System: I. The Structure and Physical Properties of the Compounds of the Low Temperature Phase Fields, *Min. Sci. Eng., 8,* 106 (1976).

84. J. T. Richardson, Thermo-Magnetic Studies of Iron Compounds in Coal Char, *Fuel, 51,* 150 (1972).

85. E. J. Schwartz and O. J. Vaughan, Magnetic Phase Relations of Pyrrhotite, *J. Geomag. Geoelectr., 24,* 441 (1972).

86. T. A. Sladek and C. H. Cox, Coal Preparation Using Magnetic Separation: Vol. 4, Evaluation of Magnetic Fluids for Coal Beneficiation, EPRI Report No. CS-1517, Vol. 4, Electric Power Research Institute, Palo Alto, Calif., July 1980.

87. W. W. Slaughter and F. V. Karlson, Process Design Considerations for the Cleaning of Wet Pulverized Coal by High Gradient Magnetic Separation, in *Industrial Applications of Magnetic Separation* (Y. A. Liu, ed.), IEEE Publ. No. 78CH1447-2 MAG, Institute of Electrical and Electronic Engineers, New York, 1979, pp. 83-90.

88. Z. J. J. Stekly, A Superconducting High Intensity Magnetic Separator, presented at International Magnetics Conference, London, England, Apr. 1975.

89. Z. J. J. Stekly, M. L. Mallary, and H. R. Segal, Coal Preparation Using Magnetic Separation, EPRI Report No. CS-1517, Vol. 2, Electric Power Research Institute, Palo Alto, Calif., July 1980.

90. Z. J. J. Stekly, G. Y. Robinson, Jr., and G. J. Powers, Magnetic Processing Commercial Supercon? *Cryogen. Ind. Gasses,* July/Aug. (1973).

91. Z. J. J. Stekly and J. V. Minervini, Shape Effect of the Matrix on the Capture Cross-Section of Particles in High Gradient Magnetic Separation, *IEEE Trans. Magn., MAG-12,* 474 (1976).

92. S. R. Taylor, K. J. Miller, R. E. Hucko, and A. W. Deurbrouck, New Methods for Coal Desulfurization, presented at Eighth International Coal Preparation Congress, Donetsk, USSR, May 21-26, 1979.

93. S. C. Trindade, Studies on the Magnetic Demineralization of Coal, Ph.D. thesis, Department of Chemical Engineering, Massachusetts Institute of Technology, Cambridge, Mass., 1973.

94. S. C. Trindade and H. H. Kolm, Magnetic Desulfurization of Coal, *IEEE Trans. Magn., MAG-9,* 310 (1973).

95. S. C. Trindade, J. B. Howard, and G. J. Powers, Magnetic Desulfurization of Coal, *Fuel, 53,* 178 (1974).

96. S. C. Trindade, M. Suddy, and J. L. F. Monteiro, Magnetic Recovery of Sulfur from Coal in Brazil, *IEEE Trans. Magn., MAG-12,* 355 (1976).

97. U.S. Department of Energy, Contaminants in Dry Coal Separated by Magnetism in Oak Ridge Test, *Energy Insider,* May 26 (1980).

98. S. Venkatesan, High Gradient Magnetic Separation and Its Application in the Coal Industry, paper presented at Coal Conference and Expo V, Louisville, Ky., Oct. 23-29, 1979.

99. R. G. Wagner, Novel Superconducting Fluidized-Bed High Gradient Magnetic Separation (HGMS) Processes for Desulfurization of Dry Pulverized Coal for Utility Applications, Master's thesis, Auburn University, Auburn, Ala., Aug. 1980.

100. P. L. Walker, Characterization of Mineral Matter in Coals and Coal Liquefaction Residues, technical reports issued to Electric Power Research Institute, Palo Alto, California, under contract no. EPRI 366-1, Dec. 1975 to Feb. 1976.

101. J. H. P. Watson, Magnetic Filtration, *J. Appl. Phys., 44,* 209-213 (1973).

102. J. H. P. Watson, Magnetic Separation at High Magnetic Fields, International Cryogenic Engineering Conference 6, Grenoble, May 1976.

103. J. H. P. Watson, Applications of and Improvements in High Gradient Magnetic Separation, *Proceedings of Filtration Society's Conference on Filtration, Productivity and Profits at Filt-tech/79,* Olympia, London, Sept. 1977, pp. 1-5.

104. J. H. P. Watson, High-Capacity, High-Intensity Magnetic Separators Using Low Magnetic Field, *Proceedings Sixth International Conference on Magnet Technology,* Bratislava, Czechoslovakia, Sept. 1977, pp. 308-314

105. J. H. P. Watson, Improvements of a Low-Field, High-Intensity Matrix Separator, *IEEE Trans. Magn., MAG-14,* 392 (1978).

106. I. Weschler, J. Doulin, and R. Eddy, Coal Preparation Using Magnetic Separation, EPRI Report No. CS-1517, Vol. 3, Electric Power Research Institute, Palo Alto, Calif., July 1980.

107. W. Windle, Magnetic Separation, Method and Apparatus, U.S. Patent 4,054,513, Oct. 18, 1977.

108. W. A. Wooster and N. Wooster, The Magnetic Properties of Coal, in *Proceedings of a Conference of the Ultra-fine Structure of Coals and Coke,* The British Coal Utilization Research Association, London, 1943, p. 332.

109. T. F. Yan, Magnetic Desulfurization of Airborne Pulverized Coal, U.S. Patent 4,052,170, Oct. 4, 1977.

110. A. Z. Yurovskii and I. D. Remesnikow, Thermomagnetic Method of Concentrating and Desulfurizating Coal, *Koks i Khim, 12,* 8 (1958).

111. P. D. Zavitsanos, J. A. Golden, K. W. Bleiler, and W. K. Kinkard, Coal Desulfurization by Microwave Energy, Report No. EPA-600/7-78-089, Environmental Protection Agency, Washington, D.C., June 1978.

112. G. Zebel, Deposition of Aerosol Flowing Around a Cylinderical Fiber in a Uniform Electric Field, *J. Colloid Sci., 20,* 522 (1965).

5

Froth Flotation to Desulfurize Coal

KENNETH J. MILLER
A. W. DEURBROUCK

U.S. Department of Energy
Pittsburgh Mining Technology Center
Pittsburgh, Pennsylvania

5.1 INTRODUCTION

Froth flotation is a physicochemical process for beneficiating fine-sized minerals. The process utilizes a difference in the surface properties of particles in an aqueous pulp to effect a separation. Air-avid or water-repellent (hydrophobic) particles are floated to the surface by finely dispersed air bubbles to be collected as a froth concentrate. Other particles which are readily wetted by water (hydrophilic) do not stick to air bubbles and remain in suspension in the body of the pulp to be carried off as underflow.

Generally, for flotation to be effective, reagents must be added to the pulp. One type of reagent, the collector, adsorbs on the particle surfaces, rendering them water repellent or hydrophobic.

Another type of reagent, the frother, facilitates the production of a transient froth capable of carrying the hydrophobic mineral load until it can be removed from the flotation cell.

Although a variety of flotation processes were in use as early as the late 19th century [1], it was not until 1918 [2] that froth flotation was applied to coal beneficiation. Since then, the process has grown in popularity until today there is hardly a new preparation plant being designed or constructed that does not include froth flotation. Approximately 350 of the 600 or so preparation plants in the United States utilize froth flotation, with most of these in the Appalachian coal field.

Much of the popularity of froth flotation in coal preparation was brought about by increasingly stringent water pollution control legislation; i.e., froth flotation is an effective method for removing fine coal particles from preparation plant water before it is discarded into rivers or streams. In addition to this legal or environmental problem, preparation plant operators were finding that because of the increased quantity of fines produced by modern mining and preparation methods, it was to their economic advantage to recover as much fine coal as possible.

Compared with the preferential flotation of complex minerals or ores, the flotation of coal is easy. Coal particles, unless oxidized or of low rank, are generally much more water repellent than the associated clay or shale and can be floated readily with only a frother and perhaps a small amount of fuel oil.

However, while froth flotation is an excellent method for removing high-ash material from coal, it does little in its normal mode of operation to remove sulfur. Therefore, most of the research effort over the past decade has been in the area of sulfur removal. Out of this research have come new processes which borrow heavily from the more sophisticated metalliferous ore flotation techniques. It is the purpose of this chapter to outline general coal flotation procedures and to discuss in some detail one of these new techniques, a process for floating pyrite from coal with xanthate.

5.2 SUMMARY OF CURRENT PRACTICE

An important difference between the froth flotation of coal and that of other minerals is that with coal, flotation usually plays a secondary role to specific gravity separation processes for coarse coal such as dense media, jigging, and tabling. With minerals, flotation is the primary separation process and the mineral is ground to a fine size, both to liberate it and to make it fine enough for flotation.

Because of the secondary or supplementary role played by flotation in coal washing, the portion of a coal preparation plant feed going to the flotation cells is comparatively small, ranging from perhaps 4 to 20%. The quantity depends on such factors as mining methods, the degree of breaking or crushing to liberate impurities, and the tendency toward breakage during handling (friability) of the particular coal. It must be understood that coal preparation plants today do not grind coal to flotation size; they treat by flotation the fine coal (usually −28 mesh) that is produced by mining, by crushing or breaking of lump coal, and by attrition during processing.

A. Pulp Density and Particle Size

The solids concentration to coal flotation cells might range from two or three percent to 10% or more, depending on the quantity of water used in the screening or classification step and the rate of coal feed to the plant. In current U.S. practice, the solids concentration is quite variable but probably averages about 6%. Generally no attempt is made to monitor or control the feed solids content.

The solids concentration for optimum coal flotation is much lower than that used for metallic ore flotation. This can be readily appreciated when one considers that the specific gravity of ores may be over 3.0 while that of coal is about 1.5. Another factor is that in coal flotation, the bulk of the material is normally floated while in ore flotation usually only a small fraction is floated.

Typically, feed particle size in coal flotation is −28 mesh although some newer coal preparation plants classify or screen at 65 and even 100 mesh prior to flotation. In any case, compared to other methods of coal cleaning, the particle size range in the feed to flotation is very broad, ranging from 0.5-mm particles to submicrometer material. It is not unusual for a flotation feed to contain over 25% —325-mesh material.

B. Flotation Machines

Froth flotation cells come in a variety of sizes from less than 1 ft^3 to over 1000 ft^3 in capacity and are of two basic types, mechanical and pneumatic. Only a few years ago, the most popular mechanical cells in the coal industry were those of 60 to 100 ft^3 in capacity. But there is a trend in flotation toward larger and larger cells with the most popular mechanical type today being 300 to 500 ft^3 in capacity. The cells are usually arranged in banks or are of an adequate size to allow 3 to 4 min of slurry retention time.

WEMCO Fagergren Cell. The most widely used flotation machine in the United States today for coal flotation is the WEMCO Fagergren flotation cell. There are about 250 coal preparation plants in the United States using these machines [3]. The WEMCO flotation cell employs the Fagergren rotor-stator principle for agitating the pulp [4].

For aeration, air is drawn downward through the stand pipe and into the pulp by the action of the impeller. No compressed air is used. The cells come in various sizes, with the most popular today being 300 ft^3 in capacity, and can be arranged in banks with or without partitions between. These banks usually contain four to six cells.

Heyl and Patterson Cyclo-cell. The Heyl and Patterson cyclo-cell can be found in about 25 U.S. preparation plants [5]. This cell has no moving or mechanical parts. Agitation is accomplished by submerged vortex chambers which impart cyclonic motion to the slurry before it discharges in the form of a jetlike spray. The vortex chambers can be placed in an open tank or chamber of any size needed. Low-pressure air is introduced into the center of the discharge spray and is sheared into fine bubbles which are dispersed through the cell.

Denver D-R Cell. Denver flotation cells are used in about 60 U.S. preparation plants [6] with most of these installations made over the past few years. The Denver D-R cells, like the WEMCO Fagergren cell, can be arranged in banks with or without partitions between. These units, too, come in various sizes with the most popular today being the 500-ft^3 model. Unlike the WEMCO Fagergren cell, the Denver D-R machine receives air from an external blower or air pump.

Galigher Agitair Cell. The Galigher Agitair machine, which is very popular in ore flotation, can be found in about five coal preparation plants in the United States [7]. This machine uses mechanical agitation and externally supplied low-pressure air.

C. Flotation Reagents

Frother. The most commonly used frothers in coal flotation are the short-chain aliphatic alcohols of which methylisobutyl carbinol (MIBC) (or *methyl amyl alcohol* as it is sometimes called) is most popular. Of the 2.7 million lb of frothers used in the United States in 1975 for coal flotation, 1.7 million lb were MIBC [8]. MIBC now costs about 60¢/lb and the average addition level in the United States is slightly under 0.2 lb/ton of feed [8].

Collector. High-rank bituminous coals are normally floatable without collectors and pulp conditioning. A frother alone added directly into the flotation machine is usually sufficient. Lower-rank bituminous coals, anthracite, and coals that are weathered or oxidized, however, may require a collector and reagent conditioner to render the particles water-repellent or hydrophobic. Ordinarily, if a collector is needed, a nonpolar oil such as kerosine or fuel oil is used along with a frother. These oils have an affinity for coal surfaces and are relatively inexpensive. The quantity of oil needed may vary from less than ½ to 3 or more lb/ton of feed; the average addition level is about 1 lb/ton [8]. In 1975, about 1/3 of the bituminous and anthracite coals cleaned by froth flotation required a collector [8].

5.3 PYRITIC SULFUR REMOVAL

A. Improving Coal Flotation Selectivity

Sometimes the quality or selectivity of separation of coal particles from ash- and sulfur-bearing impurities is not satisfactory; that is, a clean coal concentrate with a high-ash and/or high-sulfur content is produced. This may result from clay slimes or gangue material becoming entrained in the froth or adsorbed on the coal particles; or perhaps shale and pyrite particles are not adequately liberated from the coal. In any case, it becomes necessary to find ways of improving the selectivity of separation in order to obtain a high-quality clean coal product.

Control of Process Variables. Researchers have demonstrated over the years that coal flotation selectivity, with regard to sulfur as well as ash reduction, depends to a large degree on the care taken in controlling process variables. For example, by properly controlling the reagent addition level, aeration rate and retention time, the operator can prevent much liberated pyrite from floating. Table 5.1 gives results from several froth flotation tests to illustrate this point.

Table 5.1 Froth Flotation Test Results with a Lower Freeport Bed Coal Using Different Reagent Concentrations, Aeration Rates, and Retention Times

MIBC concentration, lb/ton	Aeration, cfm	Retention time, min	Analysis of product, wt %			
			Yield	Ash	Pyritic sulfur	Total sulfur
0.1	0.12	1	54.4	6.9	0.49	1.02
		3	61.0	7.5	0.54	1.06
0.1	0.33	1	60.9	7.2	0.60	1.16
		3	66.2	8.4	0.75	1.32
0.2	0.12	1	67.9	8.6	0.83	1.32
		3	71.1	9.7	1.04	1.52
0.2	0.33	1	71.5	9.9	1.16	1.68
		3	73.4	11.2	1.35	1.86

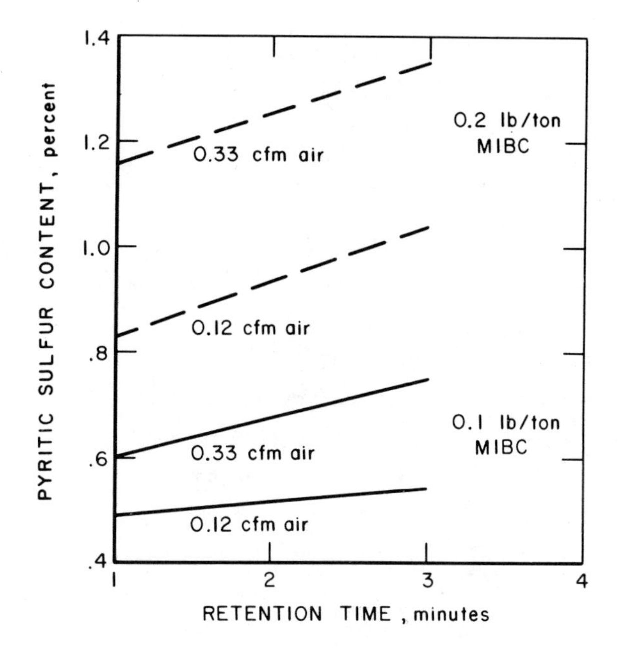

Figure 5.1 Pyritic sulfur content of clean coal froth products as a function of retention time, cell aeration, and reagent addition level.

The flotation tests in Table 5.1 were run in a laboratory WEMCO Fagergren cell with slurry of 8% solids. The Lower Freeport bed coal used in the tests was pulverized through a laboratory hammer mill to reduce it to —35 mesh. The pH of the slurry was between 7.5 and 8.0. Figure 5.1 shows that the pyritic sulfur content of the clean coal concentrate increases as retention time, reagent addition and aeration are increased. Aplan [9] and F. G. Miller [10] observed similar flotation response with other coals and have recommended gentle or careful flotation with minimal reagent, aeration, and retention time to remove the highest-quality coal prior to a second-stage flotation treatment of the underflow for maximum coal recovery.

pH Effect. It is known that pH affects the sulfur level of coal froth concentrate, but it is not clear to what degree or in what way. Zimmerman [11], for example, reported that the sulfur content of a Pittsburgh bed coal sample was reduced by flotation as pH was raised from 4 to 11 but the recovery decreased above pH seven.

Baker and K. J. Miller [12], on the other hand, showed that sulfur may either increase or decrease as pH is raised depending on what di- or trivalent metal ions are available in the water or on what base is used to adjust pH (e.g., CaOH or NaOH). And, for the same reason, recovery may increase or decrease.

Reagent Type. Besides quantity, the type of reagent or reagents used in coal flotation is of importance when sulfur reduction is desired. Kerosine and MIBC result in a clean coal product of a much higher sulfur content than would be obtained with MIBC alone and the sulfur increases with increasing kerosine addition. Also, the type of frother used has some effect on clean coal sulfur content. Bureau of Mines tests [13] showed that the collecting strength of aliphatic alcohol frothers increased with the carbon-chain length of the alcohol. That is, at equal concentrations, MIBC (6-carbon) floats more pyrite than n-amyl alcohol (5-carbon), and 2-ethyl isohexanol (8-carbon) floats more pyrite than MIBC. Also, the froth characteristics vary; generally the shorter carbon-chain aliphatic alcohols produce larger bubbles and less persistent froth.

Pretreatment and Feed Preparation. A significant factor in the desulfurization (and ash removal) capabilities of a froth flotation operation lies in the pretreatment or care taken in preparing a flotation cell feed slurry. If the material is deslimed or classified in some way according to particle size, or if the solids concentration is more uniform so that the reagent addition can be better controlled, much benefit might be gained in terms of pyrite removal. Such is not normally the practice but there are some fine coal circuits that incorporate water-only cyclones plus fine coal screens or similar classification devices to remove much of the heavier impurities and to deslime the product before froth flotation [14,15].

Rougher-Cleaner Coal Flotation. Rougher-cleaner, or two-stage, coal flotation in which the froth is recleaned, is sometimes used to reduce the ash and sulfur contents of coals that are difficult to upgrade by flotation. The technique has been used for many years and is practiced in commercial preparation plants [14]. Typical results of two-stage, rougher-cleaner coal flotation are given in Table 5.2.

Table 5.2 Two-Stage Rougher-Cleaner Coal Flotation Test Results

Product	Weight percent			
	Amount	Ash	Pyritic sulfur	Total sulfur
Lower Kittanning Coal Bed				
First stage, pH = 7.8				
Reagent addition: 0.2 lb/ton MIBC				
Clean coal	59.1	10.4	1.70	2.35
Reject	40.9	73.5	7.83	8.02
Feed	100.0	36.2	4.21	4.67
Second stage, pH = 7.9				
Reagent addition: 0.1 lb/ton MIBC				
Clean coal	92.7	6.8	1.21	1.86
Reject	7.3	56.5	7.88	8.58
Feed	100.0	10.4	1.70	2.35
Middle Kittanning Coal Bed				
First stage, pH = 4.0				
Reagent addition: 2.0 lb/ton kerosine 0.5 lb/ton MIBC				
Clean coal	85.8	9.4	1.97	3.86
Reject	14.2	76.4	5.77	6.08
Feed	100.0	18.9	2.51	4.18
Second stage, pH = 4.6				
Reagent addition: 0.1 lb/ton MIBC				
Clean coal	95.6	7.1	1.76	3.72
Reject	4.4	59.1	6.44	6.97
Feed	100.0	9.4	1.97	3.86

The test results with both the Lower Kittanning and Middle Kittanning bed coals show a significant reduction in both sulfur and ash as a result of the secondary flotation treatment. The clean coals, for example, contained 10.4 and 9.4 wt % ash, and 1.70 and 1.97 wt % pyritic sulfur, respectively, after the first-stage flotation. But after the second-stage treatment of the concentrate, the clean coals analyzed 6.8 and 7.1 wt %, and 1.21 and 1.76 wt % pyritic sulfur.

An alternative, or supplement, to rougher-cleaner coal flotation may be froth sprinkling as propounded by F. G. Miller [16]. Here, the coal froth concentrate is sprayed with water to wash free, or dislodge from the froth the less hydrophobic coal-pyrite particles and clay slimes which have been entrained.

Pyrite Depression. Pyrite depression involves changing the surface characteristics of the pyrite particles to make them less hydrophobic. This might be accomplished through chemical treatment, such as oxidation, or by the physical or chemical adsorption of a hydrophilic substance on the pyrite particles. Whatever method is used, however, it must be selective toward the coal-pyrite, allowing the coal to float.

Since the inception of coal flotation, various methods have been employed in an effort to selectively depress coal-pyrite. Numerous reagents have been tried, many of them the classical pyrite depressants used in ore flotation, such as lime and sodium cyanide.

As early as 1935, Yancey and Taylor [17] tested various organic and inorganic depressants including ferrous and ferric sulfate, the oxidation products from pyrite itself. The depressing action of ferric sulfate was attributed to the selective adsorption of a positively-charged ferric hydroxide sol on the negatively-charged pyrite particles in certain pH ranges. Similar work on the depression of pyrite with hydrolyzed metal ions derived from other multivalent salts such as $FeCl_3$, $AlCl_3$, $CrCl_3$, and $CuSO_4$ was reported in 1971 by Baker and K. J. Miller [12].

Various oxidizing agents such as potassium dichromate and potassium permanganate have been used as pyrite depressants in laboratory coal flotation [10,17]. These reagents function by oxidizing the pyrite surface to a more hydrophilic form or by producing the ferric or ferrous sulfate oxidation products of pyrite.

None of the many techniques for depressing pyrite is used in commercial coal preparation plants because even under controlled laboratory conditions, desulfurization by pyrite depression is only mediocre. It appears that some of the pyrite and marcasite particles found in coal are quite floatable and offer considerable resistance to depression. Perhaps this can be explained by looking at the physical nature of some of the pyrite or marcasite that is found in the froth concentrate from coal flotation tests.

In scanning electron microscope (SEM) photographs of the product, done at the University of Utah [18], anomalous coal-pyrite particles were distinguished by energy dispersive x-ray (EDAX) analyses. In Fig. 5.2, what appears by visual observation to be a coal particle is, in fact, an FeS_2 particle as seen by EDAX. It appears as though the FeS_2 has replaced the coal in a remnant cellular structure.

Another example of the pyrite found in froth concentrate is shown in Fig. 5.3. The top photograph, which shows an EDAX scan for iron and sulfur, reveals both an iron and a sulfur peak near the left of the photograph. When this area is magnified (bottom photograph), it reveals framboid pockets of FeS_2 on a coal particle.

The anomalous or locked pyrite or marcasite particles shown in the SEM photographs apparently have sufficient natural hydrophobicity to float during coal flotation and it is doubtful that classical pyrite depressants would help to eliminate them.

B. Coal-Pyrite Flotation

Description and Results. Because of the limited pyrite rejection achieved by coal flotation, researchers began looking at ways to remove finely disseminated or locked pyrite particles by floating them from coal. The work led to the development of a two-stage "reverse" flotation process [19]. The process, called *coal-pyrite flotation*, involves a first-stage conventional coal flotation step to reject most of the high-ash refuse and some of the coarser or liberated pyrite as tailings. The coal froth concentrate, with some dilution water, then goes to a second-stage froth flotation where a hydrophilic colloid is added to depress the coal, followed by a sulfhydryl collector to float the pyrite.

The coal-pyrite flotation process, as described in various publications [20-22], has markedly reduced the sulfur content of some coals. Table 5.3 shows results with three coals from the Appalachian

Figure 5.2 Substitution of FeS_2 in remnant cellular structure that resembles a coal particle (mag. 150 ×; particle size approximately 330 μm).

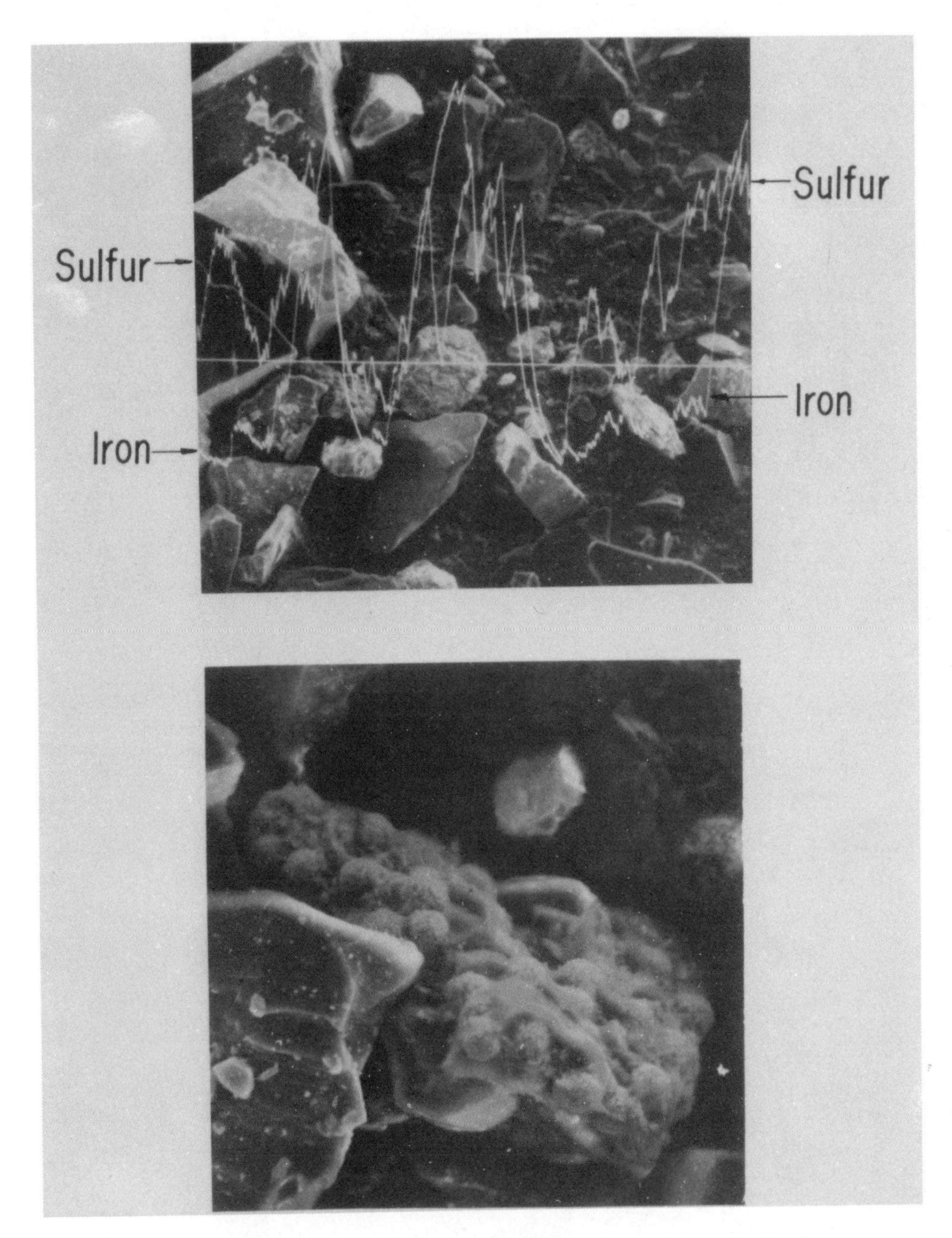

Figure 5.3 SEM photograph of particles from second-stage froth concentrate with accompanying EDAX scan for iron and sulfur. (top mag. 56 ×; bottom mag. 310 ×)

Table 5.3 Coal-Pyrite Flotation Process Test Results

Product	Weight percent: Amount	Ash	Pyritic sulfur	Total sulfur
Lower Freeport Coal Bed				
First stage, pH = 8.2				
Reagent addition: 0.1 lb/ton MIBC				
Clean coal	62.3	7.9	0.86	1.43
Reject	37.7	71.4	4.36	4.50
Feed	100.0	31.8	2.18	2.59
Second stage, pH = 6.0[a]				
Reagent addition: 0.6 lb/ton Aero Depressant 633; 0.6 lb/ton potassium amyl xanthate; 0.1 lb/ton MIBC				
Clean coal	95.1	7.5	0.55	1.12
Reject	4.9	13.3	6.85	7.41
Feed	100.0	7.8	0.86	1.43
Pittsburgh Coal Bed				
First stage, pH = 8.6				
Reagent addition: 0.2 lb/ton MIBC				
Clean coal	64.9	10.0	0.88	1.52
Reject	35.1	71.4	1.33	1.36
Feed	100.0	31.6	1.03	1.46
Second stage, pH = 5.6[a]				
Reagent addition: 0.5 lb/ton Aero Depressant 633; 0.6 lb/ton potassium amyl xanthate; 0.1 lb/ton MIBC				
Clean coal	94.1	9.7	0.56	1.18
Reject	5.9	15.7	6.02	6.89
Feed	100.0	10.0	0.88	1.52

Table 5.3 (cont'd)

Product	Weight percent Amount	Ash	Pyritic sulfur	Total sulfur
Lower Kittanning Coal Bed				
First stage, pH = 7.8				
Reagent addition: 0.25 lb/ton MIBC				
Clean coal	90.0	7.8	1.09	1.66
Reject	10.0	52.4	5.46	5.53
Feed	100.0	12.3	1.53	2.05
Second stage, pH = 6.0[a]				
Reagent addition: 0.5 lb/ton Aero Depressant 633; 0.4 lb/ton potassium amyl xanthate; 0.1 lb/ton MIBC				
Clean coal	92.6	7.1	0.55	1.16
Reject	7.4	16.1	7.81	7.84
Feed	100.0	7.8	1.09	1.65
West Kentucky No. 9 Coal Bed				
First stage, pH = 7.6				
Reagent addition: 1.0 lb/ton kerosene; 0.25 lb/ton MIBC				
Clean coal	84.7	11.2	2.55	4.15
Reject	15.3	67.1	5.81	5.96
Feed	100.0	19.8	3.05	4.43
Reagent addition: 0.7 lb/ton Aero Depressant 633; 0.5 lb/ton potassium amyl xanthate; 0.1 lb/ton MIBC				
Clean coal	85.7	9.7	1.23	2.95
Reject	14.3	20.4	10.48	11.37
Feed	100.0	11.2	2.55	4.15

[a]Second-stage pulp pH reduced with sulfuric acid.

Source: Refs. 13, 21, and 23.

region and one from the Midwest region. The selectivity of the process with the three Appalachian region coals is clearly shown by the high sulfur content of the second-stage froth reject and the low sulfur content of the final clean coal product. The West Kentucky No. 9 seam coal from the Midwest region also shows selective pyrite flotation, as the 11.37 wt % sulfur content of the second-stage froth reject reveals, but the clean coal product still contains almost 3 wt % sulfur. As reflected in this and many other flotation tests [23], Midwest region coals are typically more difficult to desulfurize than Appalachian region coals.

Reagents. The coal depressant and the pyrite collector used in most of the pyrite flotation work reported to date are Aero Depressant 633 and potassium amyl xanthate. These are reportedly the most effective reagents.

1. Coal Depressant. Aero Depressant 633 is a proprietary reagent from American Cyanamid Company and is recommended by the manufacturer for the depression of carbonaceous gangue in gold ore flotation. In Bureau of Mines tests [20], it proved more effective than quebracho and various other natural colloids. Other coal depressants, such as canary dextrins, are being looked at by various researchers but no definite results have been reported to date.

2. Pyrite Collector. Potassium amyl xanthate was selected after testing the various other homologues of xanthate. The 2-carbon ethyl xanthate was ineffective for floating coal-pyrite. Reagent addition levels of two pounds per ton of dry feed and more failed to promote adequate pyrite flotation. But with the testing of each higher homologue, pyrite flotation improved until it reached a maximum with the 5-carbon amyl xanthate.

Coal-Pyrite Flotation Response. Coal-pyrite flotation response with a sulfhydryl collector differs significantly from that of ore-pyrite. As Fig. 5.4 shows, potassium amyl xanthate (KAX) consumption is about one order of magnitude greater for coal-pyrite than for ore-pyrite. It has been suggested [18] that the reason for the high amyl xanthate requirement for coal-pyrite flotation is related to surface

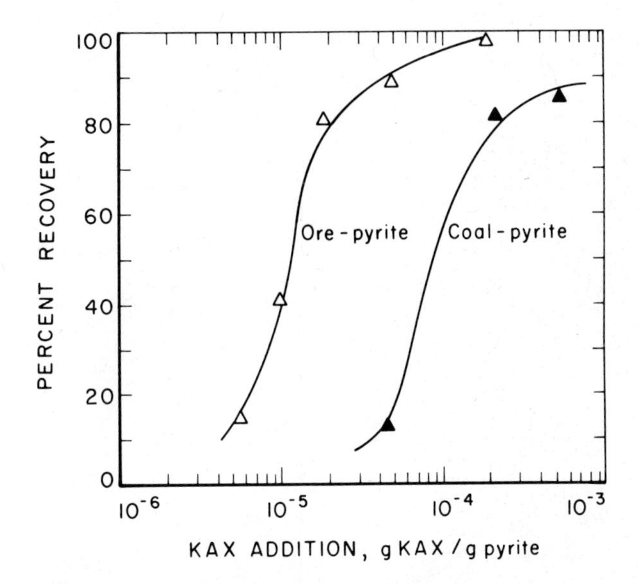

Figure 5.4 Flotation recoveries for ore-pyrite and coal-pyrite. Conditions: Hallimond flotation cell, 2 g of 65 × 100 mesh material, 140 cm^3 water, pH = 6.5.

heterogeneities, such as clay inclusions in the marcasite component, which contribute to the hydrophilic character of the coal-pyrite.

Coal-pyrite flotation response is also sensitive to pH as shown in Fig. 5.5. The graph illustrates that maximum pyrite flotation selectivity and maximum clean coal sulfur reduction occur between pH 4 and pH 6. Chernosky and Lyon [24] in a comparison of ore- and coal-pyrite flotation, and Sun and Savage [25] who floated pyrite from coal refuse have attested to this need for an acid circuit for optimum coal-pyrite flotation with xanthate. Thus, since many fresh coal slurries maintain a slightly basic pH level (pH 7 to 8), it is often necessary to acidify the second-stage circuit. This would most likely be accomplished with sulfuric acid.

Pilot-Plant Scale-up. After development and optimization of the coal-pyrite flotation process in the laboratory, efforts were directed toward scaling up the technique to the continuous pilot-plant level. For this work a two-stage flotation pilot plant was designed and

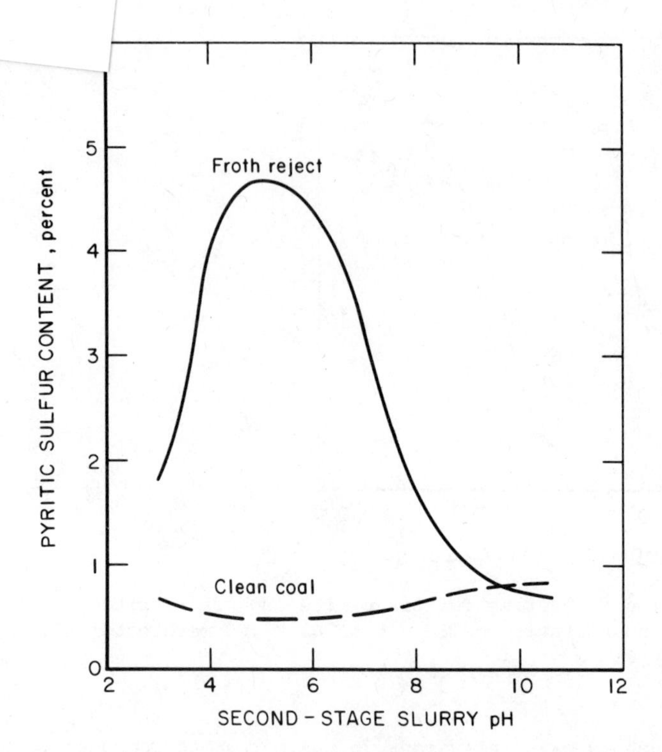

Figure 5.5 Pyritic sulfur content of products from second-stage coal-pyrite flotation as a function of slurry pH.

constructed. The initial pilot-plant flow diagram is shown in Fig. 5.6 [21].

For the pilot-plant work, a raw run-of-mine coal sample was fed from a 25-ton capacity coal storage bunker through a cone-and-ring crusher and a hammer mill to produce a —35-mesh feed material. The fine coal was mixed with water in a feed sump at a controlled rate to produce a slurry of about 8.0 wt % solids. The slurry was pumped to a 1000-gal capacity slurry holding tank from which it was fed to the flotation cells. The slurry depth in the holding tank was automatically controlled by high- and low-level probes. A frother, MIBC, was added to the feed box of the first-stage flotation machine.

For second-stage flotation, the first-stage froth concentrate was diluted with makeup water to its original volume and the coal

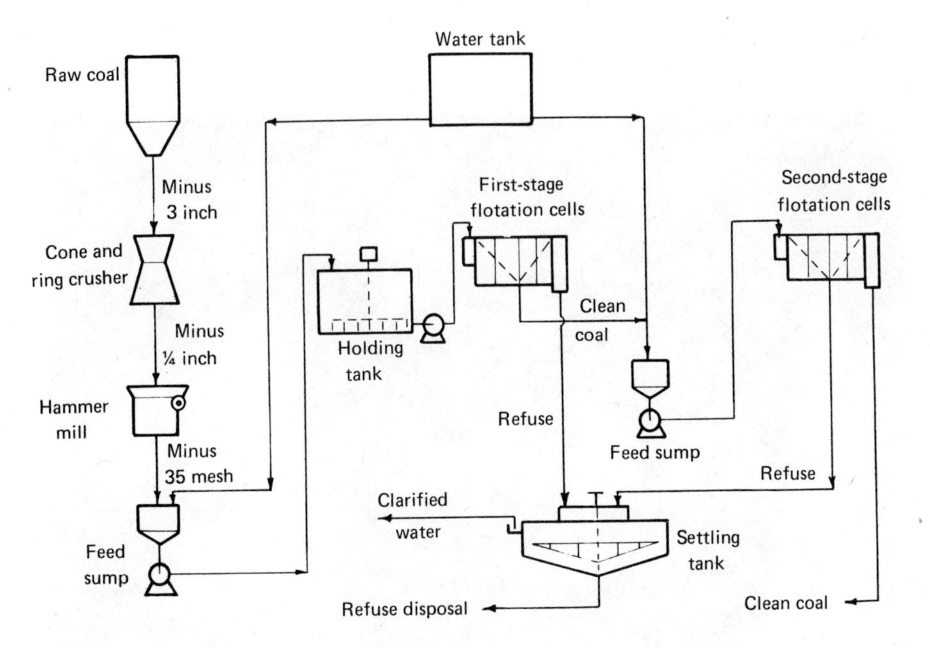

Figure 5.6 Two-stage flotation pilot plant flow diagram. (From Ref. 21.)

depressant, Aero Depressant 633, was added along with an acid to reduce pH to a slightly acidic level. The slurry was then pumped from the feed sump to the second-stage flotation cells where the pyrite collector, potassium amyl xanthate, and additional MIBC were added.

Both the first- and second-stage flotation machines in the pilot plant were composed of four, 0.6-ft^3-capacity, WEMCO Fagergren flotation cells. Figure 5.7 shows these flotation cells in operation.

Modified Circuit. Although excellent pyrite removal was achieved using the above technique, the final clean coal product was recovered as a rather dilute underflow containing from 6 to 8 wt % solids. This presented some formidable dewatering problems. Also, it was found that only half to one-fourth as much slurry retention time was needed in the second stage.

In an effort to minimize the clean coal dewatering problem and to make more efficient use of the flotation cells, the two-stage

Figure 5.7 Two-stage flotation pilot plant in operation. (From Ref. 21.)

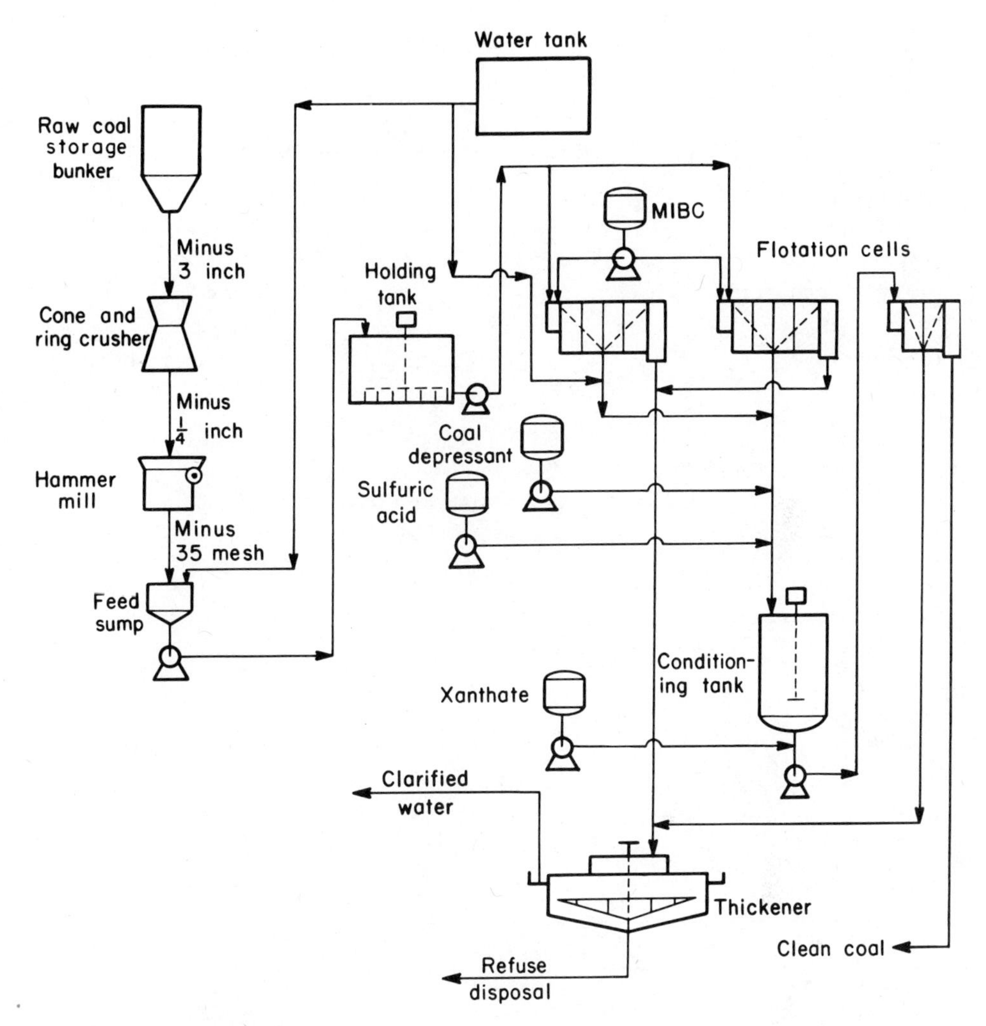

Figure 5.8 Modified two-stage flotation pilot-plant flow diagram. (From Ref. 22.)

pilot plant was modified as shown in the flow diagram in Fig. 5.8 [22].

The first-stage coal flotation was done in two parallel banks of WEMCO Fagergren flotation machines, each containing four cells. The froth concentrate was then diluted as before except the xanthate

Table 5.4 Coal-Pyrite Flotation Process Test Results with Lower Freeport Bed Coal in Modified Circuit

Product	Weight percent				
	Slurry solids	Amount	Ash	Pyritic sulfur	Total sulfur
First Stage, pH = 7.8					
Reagent addition: 0.25 lb/ton MIBC					
Clean coal	26.7	74.3	8.4	0.82	1.41
Reject	2.5	25.7	69.1	3.93	4.02
Feed	7.9	100.0	24.0	1.62	2.08
Second stage, pH = 5.9[a]					
Reagent addition: 0.9 lb/ton Aero Depressant 633; 0.8 lb/ton potassium amyl xanthate					
Clean coal	12.8	95.2	7.8	0.54	1.15
Reject	10.0	4.8	15.1	6.71	7.41
Feed	12.7	100.0	8.2	0.84	1.45
Reagent addition: 0.9 lb/ton Aero Depressant 633; 0.9 lb/ton potassium amyl xanthate					
Clean coal	12.8	92.6	7.6	0.47	1.00
Reject	10.0	7.4	14.2	5.90	6.80
Feed	12.6	100.0	8.1	0.87	1.43
Reagent addition: 0.9 lb/ton Aero Depressant 633; 1.0 lb/ton potassium amyl xanthate					
Clean coal	12.8	91.1	7.7	0.40	0.96
Reject	10.0	8.9	13.1	5.60	6.45
Feed	12.6	100.0	8.2	0.86	1.45

[a]Second-stage pulp pH reduced with sulfuric acid.

Source: Ref. 22.

was added ahead of the second-stage feed pump to provide extra conditioning time. Also, no MIBC was needed in the second-stage because of the residual in the more concentrated pulp.

Typical results (Table 5.4) with this modified circuit, seen operating in Fig. 5.9, were about the same as those obtained in earlier laboratory and pilot-plant tests, but the final clean coal slurry was in more concentrated form.

Figure 5.9 Modified two-stage flotation pilot plant in operation. (From Ref. 22.)

Full Scale Demonstration [15]

Preliminary Batch Tests. After completion of laboratory and pilot-plant work, bench-scale pyrite flotation tests were run at a coal preparation plant with a froth flotation feed of a high sulfur content. Although the feed to the flotation cell was between 15 and 20 wt % ash, only the second-stage pyrite flotation step was done. Ash reduction was not considered to be of prime importance because the feed to this flotation circuit represented only about 4% of the total plant feed and would not increase total product ash significantly.

As Table 5.5 shows, pyrite flotation selectivity was excellent with 11.8 to 13.6 wt % pyritic sulfur in the froth and only 0.79 to 0.95 wt % in the underflow. This represents a 68 to 83% reduction in pyritic sulfur content.

Full-Scale Test Circuit. Following the bench-scale work, a full-scale test circuit was designed for a coal preparation plant (Fig. 5.10). The circuit utilizes the second-stage or pyrite flotation step only, as did the bench-scale tests.

Table 5.5 Bench-Scale Coal-Pyrite Flotation Process Test Results at a Preparation Plant

	Weight percent			
Product	Amount	Ash	Pyritic sulfur	Total sulfur
pH = 5.5[a]				
Reagent addition: 0.68 lb/ton Aero Depressant 633 0.51 lb/ton potassium amyl xanthate				
Middling underflow	80.0	16.1	0.79	1.27
Reject froth concentrate	20.0	24.0	11.77	11.77
Feed	100.0	17.7	2.99	3.37
pH = 6.0[a]				
Reagent addition: 0.77 lb/ton Aero Depressant 633 0.58 lb/ton potassium amyl xanthate				
Middling underflow	82.8	18.1	0.95	1.38
Reject froth concentrate	17.2	26.8	13.61	13.66
Feed	100.0	19.6	3.13	3.50

[a]Pulp pH reduced with sulfuric acid.

Source: Ref. 15.

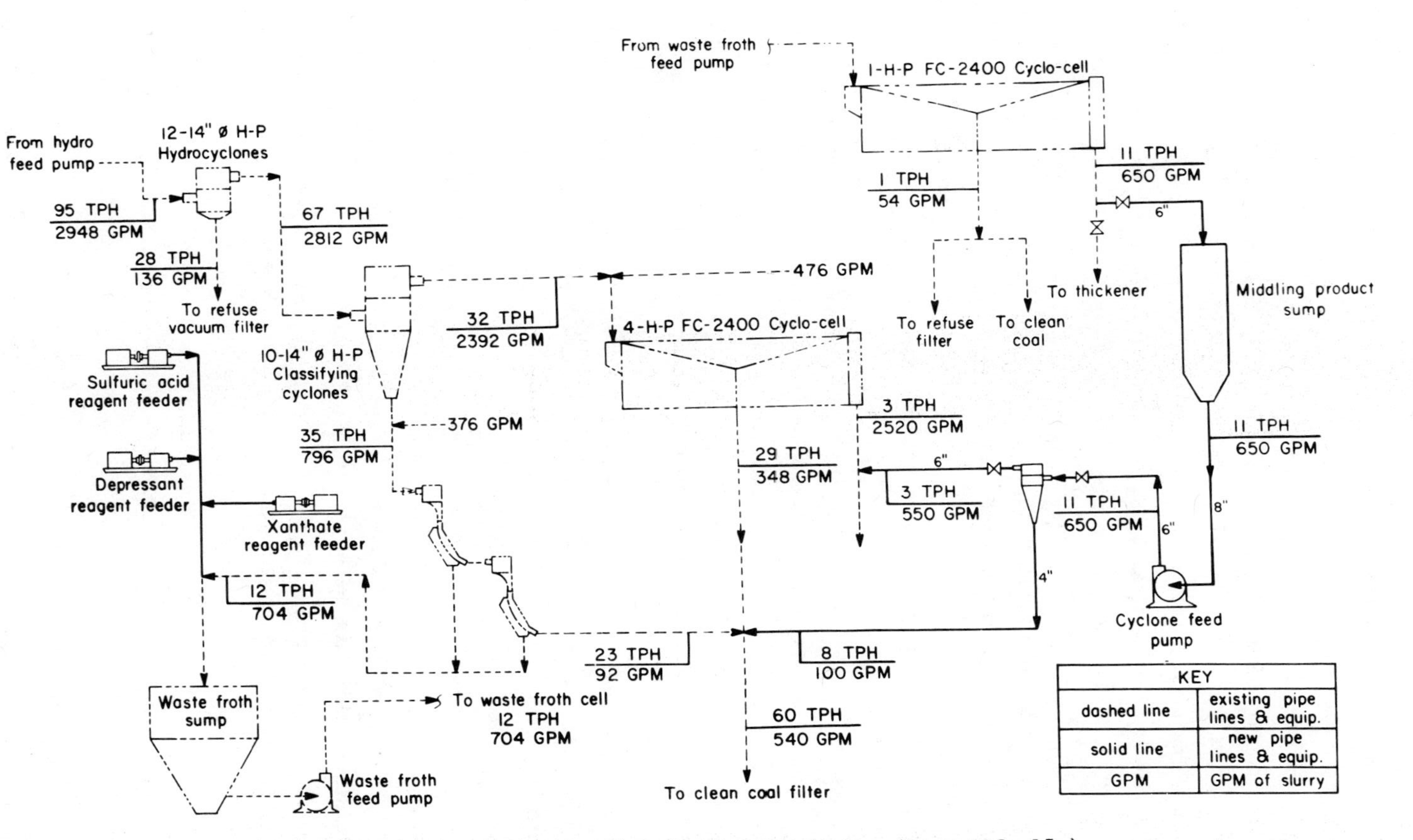

Figure 5.10 Full-scale coal-pyrite flotation circuit flow diagram. (From Ref. 15.)

The broken lines on the flowsheet represent the preparation plant's normal fine coal washing circuit, while the solid lines represent the new piping and equipment for the pyrite flotation circuit. The reagents, sulfuric acid, Aero Depressant 633, and potassium amyl xanthate, are added into the feed to the froth cell sump. The slurry goes to the secondary froth flotation cell where the refuse of high pyritic sulfur content is floated. The low-sulfur middling underflow product then passes through a sump from which it is pumped to a bank of five 8-in. classifying cyclones for thickening and desliming prior to filtration. Some ash reduction occurs in the classification/thickening step due to clay slimes removal. The overflow from the classifying or thickening cyclones goes to the plant static thickener for flocculation and water clarification.

To date, test results in the full-scale pyrite flotation circuit are few. The coal company is now mining a low-sulfur coal and, therefore, the feed to the flotation circuit is too low in pyritic sulfur for a truly meaningful evaluation of the process. Preliminary tests, however, have helped solve some of the mechanical problems in the circuit and have demonstrated the potential desulfurization capability of the coal-pyrite flotation process. For example, Table 5.6 shows that the feed to the flotation cell contained only 0.81 wt % pyritic sulfur. Despite this, the froth concentrate contained 3.07 wt % and the underflow 0.64 wt % pyritic sulfur.

Effect of Residual Reagent. One of the problems to be considered in this first commercial-scale continuous coal-pyrite flotation test circuit is the effect of residual coal depressant on subsequent coal flotation. Research at the University of Utah [18] has provided information concerning the method of adsorption of various organic colloids on coal and has shown that this adsorption is virtually irreversible. In addition, laboratory flotation tests have demonstrated that coal flotation response is sensitive to the residual concentration of Aero Depressant 633 in recycled water. However, it was shown that repeated contact with fresh coal removes much of the

Table 5.6 Full-Scale Coal-Pyrite Flotation Process Test Results

Product	Weight percent			
	Amount[a]	Ash	Pyritic sulfur	Total sulfur
pH = 6.7[b]				
Reagent addition: 0.8 lb/ton Aero Depressant 633; 0.7 lb/ton potassium amyl xanthate				
Underflow	93	19.0	0.64	1.05
Froth concentrate	7	9.7	3.07	3.49
Feed	100	18.3	0.81	1.22

[a]Weight percent computed by sulfur balance.

[b]Pulp pH reduced with sulfuric acid.

Source: Ref. 15.

residual depressant from the water. This suggests that the contact of recirculated preparation plant water with coal and clay particles in the thickener, and conditioning and wet screening of freshly mined coal entering the plant will remove most of the residual reagent prior to flotation.

Economic Analysis (Based on Late-1978 Dollars)

Incremental Cost Concept. Although there continues to be a strong emphasis on fine coal cleaning to maximize pyritic sulfur reduction, it is unlikely that there will be any immediate large-scale emergence of preparation plants built exclusively around froth flotation. This seems logical, not only because of the significantly higher costs associated with flotation (grinding, dewatering, etc.) but also because of the problem and expense of transporting and handling a plant output consisting of all fine coal. Perhaps when there is a slurry pipeline available or a power plant adjacent to the preparation plant, these impediments will become less important in the overall economic scheme. However, given the current realities of the market and the proven efficiency and economy of coarse coal washing processes, it is reasonable to assume that contemporary preparation plants will continue to use both coarse and fine coal cleaning circuits with their various clean coal products being combined as a composite output. Therefore, when addressing the economics of the coal-pyrite flotation process, it seems appropriate to assume that it would be implemented as an adjunct to the fine coal cleaning circuit of a new or an existing plant. Thus, the following analysis [26] is designed to assess the incremental cost of having this second-stage capability in a plant circuit.

1. Plant configuration. For the purpose of illustrating the incremental cost concept, a previously evaluated 600 ton/hr preparation plant configuration was selected in which the 28-mesh by 0 size fraction, representing 125 tons/hr or 21% of the plant feed, is cleaned by the conventional single-stage froth flotation. The flowsheet for this plant, with the flotation circuit highlighted, is shown in Fig. 5.11. It is assumed that to this plant is added the second-stage coal-pyrite flotation process in which the froth concentrate from the first-stage is depressed and the pyrite is floated. Since all capital, and operating and maintenance costs of the plant are known [27], only the costs associated with adding and operating this second-stage flotation circuit are assessed. This assessment is based upon amending the flotation portion of the circuit as shown in Fig. 5.12 to include a conditioning tank, a second-stage bank of flotation cells, a 60-ft-diameter static thickener, and piping to carry the products to the thickener and to the clean coal and refuse filters.

In the revised flowsheet, the first-stage flotation cells, consisting of two 4-cell banks, operate according to their original design, treating 125 tons/hr of 28-mesh by 0 coal and producing 99 tons/hr of clean coal concentrate with the 26 tons/hr of tailings product reporting to the refuse thickener. However, the 99 tons/hr of froth product, as 25-30 wt % solids slurry, now goes into a conditioning tank where water, a coal depressant (Aero Depressant 633), and a pyrite collector (potassium amyl xanthate) are added. Also, at this point concentrated sulfuric acid may be added to bring the

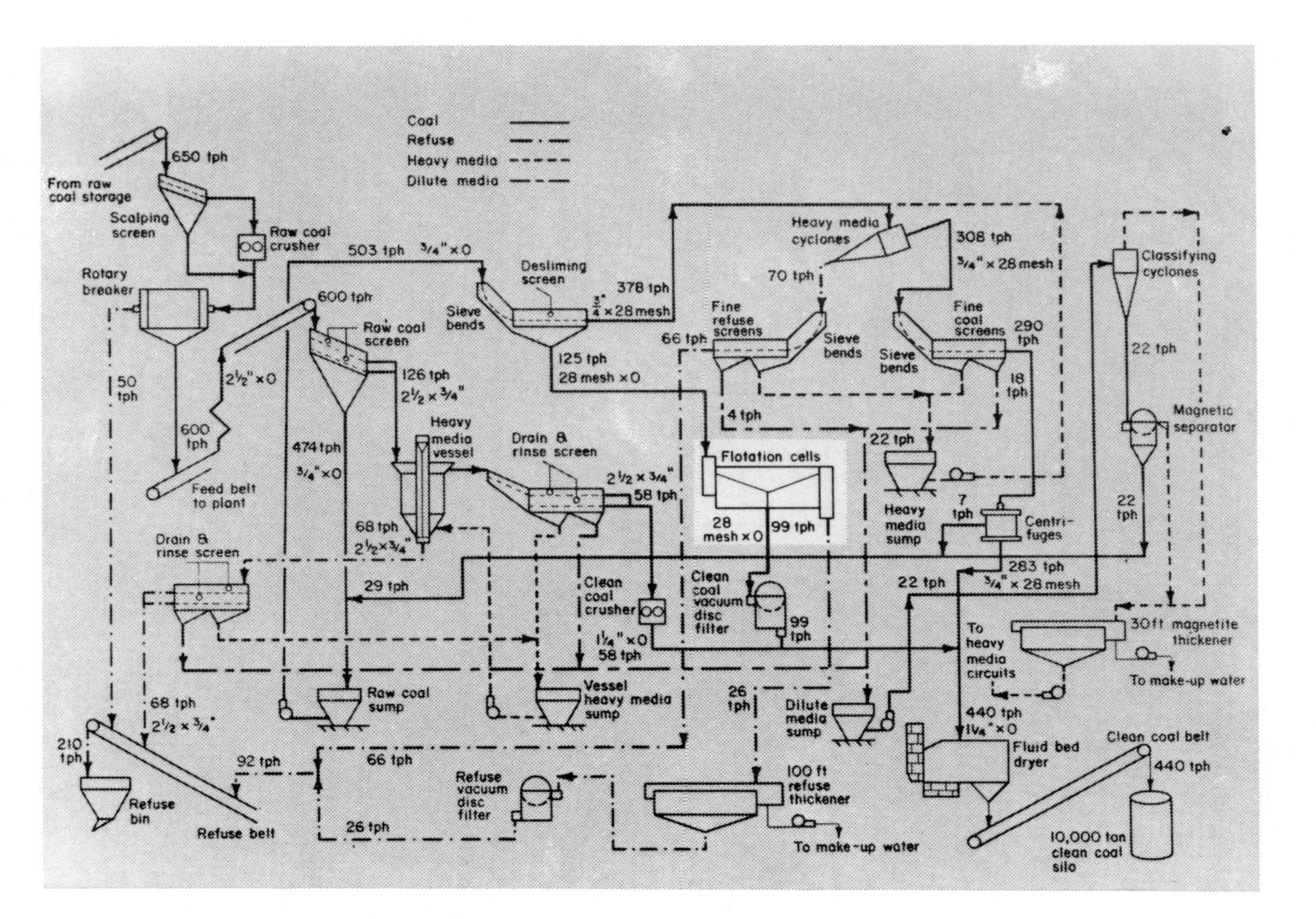

Figure 5.11 Six hundred ton per hour preparation plant flow sheet.

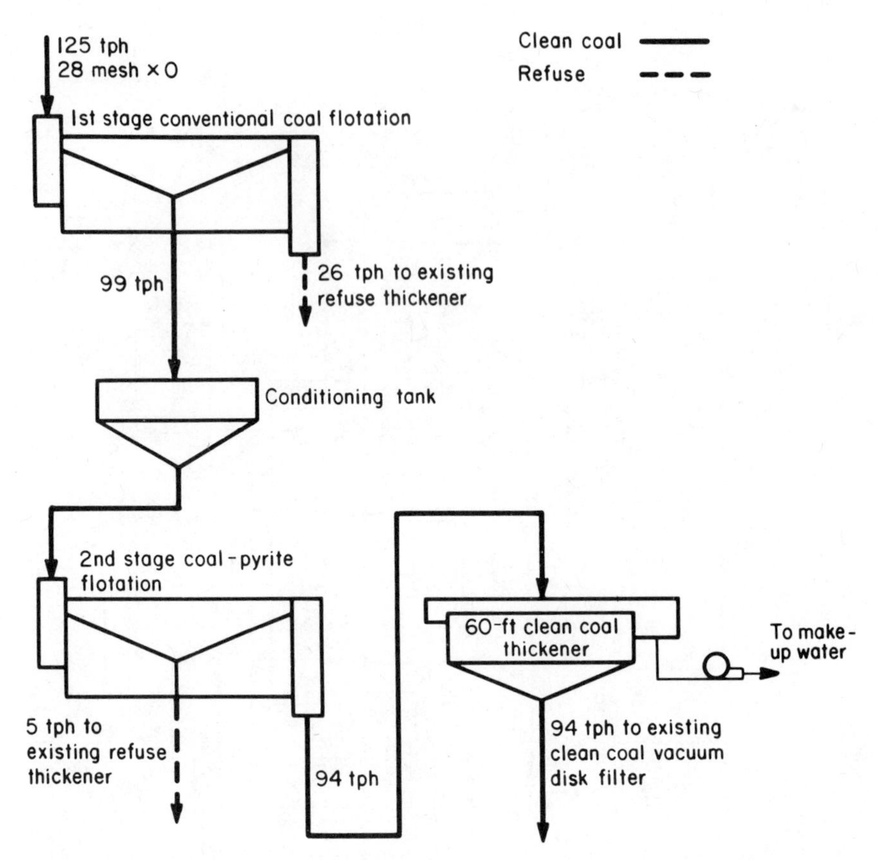

Figure 5.12 Two-stage flotation adjunct to the 600 ton/hr preparation plant.

pH of the slurry to a mildly acidic level. From the conditioning tank, the slurry, now containing 10-12 wt % solids, moves to a single 3-cell bank of second-stage flotation cells. At the feed box of the second-stage flotation cells, a frother may be added if necessary.

Of the 99 tons/hr entering the second-stage flotation cells, about 94 tons/hr is recovered as clean coal underflow with the high-pyritic-sulfur refuse being skimmed off at the rate of 5 tons/hr and reporting to the refuse thickener or directly to the refuse filter.

Since the underflow, or clean coal product, from the second-stage flotation cells contains only about 10 wt % solids, it is too dilute to be sent directly to a conventional vacuum disk filter. It must be thickened to a 20-30 wt % solids. This can be accomplished in a number of ways but for this analysis a 60-ft-diameter static thickener is considered. By this approach, essentially no material is lost, with all 94 tons/hr being pumped to the existing clean coal vacuum disk filter for additional dewatering before thermal drying.

Although additional sulfur reduction is achieved by adding the second-stage flotation cells, clean coal output from the plant is reduced by 5 tons/hr. This reduces the total plant clean coal product from 440 to 435 tons/hr. Even though this additional 5 ton/hr material discarded is high in sulfur, it is also fairly high in Btu heat value. This has the effect of reducing total Btu recovery by approximately one percent. However, because the flotation clean coal product is normally slightly higher in ash content than the coarse clean coal, the Btu content of the composite clean coal product increases from 13,619 to 13,625 Btu/lb, as noted in Table 5.7. The forementioned Btu increase is based on the assumption that the second-stage froth reject is equal in Btu value to the flotation clean coal product. It is, therefore, a conservative number because normally the froth reject has a higher ash content than the clean coal underflow, as Tables 5.3 and 5.4 show. The impact of the change in total plant output indirectly influences all costs calculated on a per-ton-of-clean-coal basis. This factor has been taken into consideration in the summary of costs presented in Table 5.7.

2. Capital Cost. Assuming the second-stage flotation cells are installed to permit gravity feed from the first stage, the following additional major equipment would be required:

One bank of three flotation cells	\$34,000
Conditioning tank with mixer	8,000
	\$42,000

If it was necessary to route the first-stage froth to a sump from which it would have to be pumped up into the second-stage feed box, there would be an additional \$10,000 associated with the difference between the conditioning tank and the sump with a heavy-duty

Table 5.7 Operating and Maintenance Costs (Late-1978 Dollars) for a 600-ton/hr Preparation Plant with and without the Second-Stage Coal-Pyrite Flotation Process

Cost category	With single-stage flotation		With two-stage flotation	
	Per ton raw coal	Per ton clean coal	Per ton raw coal	Per ton clean coal
Labor				
Supervisory (nonunion)	$0.099	$0.135	$0.099	$0.137
Operating and maintenance (union)	0.638	0.870	0.638	0.880
Overhead				
Fringe benefits: 25% nonunion	0.025	0.034	0.025	0.034
20% union	0.128	0.174	0.128	0.177
Other: includes workmens' compensation insurance, payroll taxes, welfare fund, etc.	0.213	0.291	0.213	0.294
Supplies				
Operating	0.169	0.23	0.317	0.437
Maintenance and other	0.242	0.33	0.242	0.334
Thermal dryer fuel based upon 7.0 ton/hr coal consumption and cost of coal, $20/ton	0.233	0.318	0.233	0.321
Cleaning plant repair parts	0.147	0.20	0.147	0.203
Electricity (large thermal dryer)	0.334	0.456	0.334	0.461
Operating and maintenance cost, not including capital amortization	$2.23	$3.04	$2.38	$3.28
Capital amortization (10 years, 30% utilization)	1.31	1.79	1.34	1.85
Total operating and maintenance cost	$3.54	$5.83	$3.72	$5.13
Btu/lb of clean coal product	13,619		13,625	
Cost per million Btu		$0.177		$0.188

(30-hp) open, impeller-type pump. However, for this analysis it is assumed that a gravity feed arrangement is possible.

Following the generally accepted industry rule of thumb on installed equipment prices, the $42,000 of purchased parts is multiplied by 3 to account for shipment, construction, electrical service, piping, etc. This amounts to $126,000. To this is added $120,000 to cover the installed cost of a 60-ft-diameter static thickener outside the plant. Adding this $120,000 to the previously developed $126,000 gives $246,000. To this is added a 15% contingency factor to account for such items as interest during construction to give an incremental capital requirement of $282,900 or $283,000. Using the same method of capital amortization and utilization as that applied to the basic plant configuration gives an increase in capital cost per ton of raw coal of $0.028. However, because of the decrease in total plant output from 440 to 435 tons/hr of cleaned product, this increases the capital cost per ton of cleaned coal by $0.06.

3. Supplies Cost. As pointed out earlier, various additives are applied to the coal slurry entering the second-stage froth cells to create the conditions necessary for the selective flotation of pyrite from coal. Although greater amounts of water and perhaps MIBC are required over that consumed by the existing first stage, as well as flocculants in the additional thickener, their cost is not considered significant.

For each ton of feed to the first-stage flotation cells, various quantities of the following reagents are required:

Reagent	Addition, lb/ton of first-stage feed	Cost per pound (late 1978)
Coal depressant (Aero 633)	0.4 to 0.8	$0.54
Pyrite collector (potassium amyl xanthate)	0.3 to 0.7	$0.73
Sulfuric acid	0.5 to 1.5	$0.02

The ranges of reagent addition given above are representative of the quantities required to achieve satisfactory results with a variety of coals tested in the laboratory and pilot plant. Coals vary in floatability and, therefore, some are easier to depress than others; also, they contain different amounts of liberated and unliberated pyrite or marcasite and exhibit different pH values in slurry. This then accounts for the relatively wide ranges of reagent addition needed for coal-pyrite flotation. For the purpose of this analysis, the midpoint of each range has been selected as representative.

Based upon the above prices and representative addition levels, the additional supplies cost per ton of raw coal feed for the second-stage flotation increases by \$0.148. This is based on the following relationship:

$$\frac{(125\text{ tons/hr})[(0.6\text{ lb/ton} \times \$0.54/\text{lb}) + (0.5\text{ lb/ton} \times \$0.73/\text{lb}) + (1.0\text{ lb/ton} \times \$0.02/\text{lb})]}{600\text{ tons/hr}}$$

As noted earlier, the impact on each cleaned ton is compounded by the reduction in plant yield resulting in a \$0.207 increase.

For the purpose of this cost analysis, Aero Depressant 633 is used as the coal depressant because it was almost exclusively the reagent used in the pyrite flotation developmental work. However, recent laboratory flotation studies have indicated that canary dextrin is equally effective as a coal depressant. If canary dextrin, which costs about \$0.18/lb, could be substituted for Aero Depressant 633, the reagent or supplies cost would be reduced from \$0.148 to \$0.104/ton of plant feed or from \$0.207 to \$0.146/ton of clean coal product.

4. Labor and Other Costs. Although the addition of the coal-pyrite flotation process results in slightly more equipment to operate and maintain, it is assumed that the plant staffing requirements as well as maintenance supplies, power consumption, etc., will not be influenced to any measurable degree.

Summary. The economic analysis provided here deals with the cost of installing and operating the second-stage coal-pyrite flotation process in an existing preparation plant which already has a conventional single-stage flotation circuit. The existing single-stage flotation circuit in the 600 ton/hr preparation plant processes

125 tons/hr of 28-mesh by 0 coal, or 21% of the plant feed, and produces 99 tons/hr of low-ash clean coal. With the addition of the second-stage flotation circuit, the 99 tons/hr of product would be recleaned to remove pyritic sulfur, thus producing 94 tons/hr of low-ash, low-sulfur product.

Referring to the comparison of costs (Table 4.7) for the 600 ton/hr preparation plant with and without the second-stage flotation capability, it can be seen that the inclusion of the second stage increases the total operating and maintenance cost for the plant from $4.83 to $5.13/ton of clean coal or slightly more than $0.01/ million Btu. In a more typical preparation plant where, say, 10% or less of the feed goes to the flotation circuit, the overall operating and maintenance cost would be proportionately less than the above $0.30/ton of composite plant product.

If the total cost increase attributable to the addition of the coal-pyrite flotation process were charged to the second-stage flotation circuit alone, it would compute to about $130 more to produce 94 tons of flotation clean coal product, or about $1.40/ton.

Under conditions where a reduction in sulfur is critical to meeting a coal quality specification, the additional cost shown in this economic analysis would no doubt be an acceptable price to pay for compliance. Moreover, as is true with any incremental coal preparation step, the coal-pyrite flotation process would not be applied unless there was a positive influence upon the overall economics between mining and utilization.

REFERENCES

1. K. L. Sutherland and I. W. Wark, in *Principles of Flotation*, Australasian Institute of Mining and Metallurgy, 1955, pp. 1-6.
2. E. Hindmarch and P. L. Waters, Froth-Flotation of Coal, *Trans. Instit. Min. Eng. (U.K.), 111,* 222 (1951-1952).
3. D. L. Springston, personal communication, Coal Services Division, Envirotech Corporation, Beckley, W.V., 1978.
4. R. E. Zimmerman, Froth Flotation: Flotation Equipment, in *Coal Preparation* (J. W. Leonard and D. R. Mitchell, eds.), 3d ed., AIME, New York, 1968, pp. 10-73.

5. K. E. Harrison, personal communication, Heyl and Patterson, Inc., Pittsburgh, Pa., 1978.

6. D. R. Conte, personal communication, formerly Pittsburgh District Manager, Joy Manufacturing Co., Denver Equipment Division, Denver, Colo., 1978.

7. G. A. Lawrence, personal communication, The Galigher Co., Salt Lake City, Utah, 1978.

8. Bureau of Mines, Froth Flotation in 1975, in *Mineral Industry Surveys*, U.S. Department of the Interior, 1976.

9. F. F. Aplan, Use of Flotation Process for Desulfurization of Coal, in *Coal Desulfurization: Chemical and Physical Methods, ACS Symposium Series, 64* (T. D. Wheelock, ed.), American Chemical Society, Washington, D.C., 1977, pp. 70-82.

10. F. G. Miller, Reduction of Sulfur in Minus 28-Mesh Bituminous Coal, *Trans. AIME, 229*, 7 (1964).

11. R. E. Zimmerman, Flotation of Bituminous Coal, *Trans. AIME, 117*, 338 (1948).

12. A. F. Baker and K. J. Miller, Hydrolyzed Metal Ions as Pyrite Depressants in Coal Flotation: A Laboratory Study, Report of Investigations 7518, Bureau of Mines, U.S. Department of the Interior, Washington, D.C., 1971.

13. A. F. Baker and K. J. Miller, unpublished results, Coal Preparation Group, U.S. Department of Energy, Pittsburgh, Pa., 1978.

14. S. G. Lowman, Westmoreland Coal's Bullitt Plant Upgrades Steam Coal Quality, *Coal Age*, 70-75 (1973).

15. S. R. Taylor, K. J. Miller, R. E. Hucko, and A. W. Deurbrouck, New Methods for Coal Desulfurization, Presented at Eighth International Coal Preparation Congress, Donetsk, USSR, May 21-26, 1979; also published in proceedings of this meeting.

16. F. G. Miller, The Effect of Froth Sprinkling on Coal Flotation Efficiency, *Trans. AIME, 244*, 158 (1962).

17. H. F. Yancey and J. A. Taylor, Froth Flotation of Coal: Sulfur and Ash Reduction, Report of Investigations 3263, Bureau of Mines, U.S. Department of the Interior, Washington, D.C., 1935.

18. J. D. Miller, unpublished data, University of Utah, Salt Lake City, Utah, 1978.

19. K. J. Miller, Flotation of Pyrite from Coal, U.S. Patent 3807557, 1974.

20. A. F. Baker, K. J. Miller, and A. W. Deurbrouck, Two-Stage Flotation Selectively Floats Pyrite from Coal, *Coal Mining and Processing*, 44-46, 56 (1973).

21. K. J. Miller, Flotation of Pyrite from Coal: Pilot Plant Studies, Report of Investigations 7822, Bureau of Mines, U.S. Department of the Interior, Washington, D.C., 1973.

22. K. J. Miller, Coal-Pyrite Flotation in Concentrated Pulp: A Pilot Plant Study, Report of Investigations 8239, Bureau of Mines, U.S. Department of the Interior, Washington, D.C., 1977.

23. K. J. Miller, Desulfurization of Various Midwestern Coals by Flotation, Report of Investigations 8262, Bureau of Mines, U.S. Department of the Interior, Washington, D.C., 1978.

24. F. J. Chernosky and F. M. Lyon, Comparison of the Flotation and Adsorption Characteristics of Ore- and Coal-Pyrite with Ethyl Xanthate, *Trans. AIME, 252*, 11-14 (1972).

25. S. C. Sun and K. I. Savage, Flotation Recovery of Pyrite from Bituminous Coal Refuse, *Trans. AIME, 241*, 377-384 (1968).

26. E. C. Holt, Jr., personal communication, The Hoffman-Muntner Corp., Silver Spring, Md., 1978.

27. E. C. Holt, Jr., *An Engineering/Economic Analysis of Coal Preparation Plant Operation Cost,* Federal Interagency Energy/Environment R & D Program, EPA Report No. 600/7-78-124, July 1978, pp. 232-250.

6

Selective Oil Agglomeration in Fine Coal Beneficiation

C. EDWARD CAPES

National Research Council of Canada
Ottawa, Ontario, Canada

RENE J. GERMAIN*

STELCO, Inc.
Hamilton, Ontario, Canada

---

*Present affiliation: Coal Mining Research Centre, Edmonton, Alberta, Canada.

6.1 INTRODUCTION

Fine particles in liquid suspension can be agglomerated in a number of ways. One of the oldest procedures involves the addition of electrolytes to the suspension to cause a reduction in the zeta potential and allow colliding particles to cohere [1]. A second method involves the use of polymeric flocculants to bridge between particles [2]. A third method, which is our subject here, involves the addition of a second immiscible liquid preferentially to wet the particles and cause adhesion by capillary interfacial forces. While the bonding forces in the first two methods are small and result in rather weak and voluminous agglomerates, the third method can produce more dense and much stronger agglomerates.

In the case of fine coals, the carbonaceous constituents can be agglomerated and recovered from an aqueous suspension with many different oils as collecting liquids. Inorganic or ash-forming constituents remain in suspension and are rejected. As with froth flotation, oil agglomeration relies on differences in the surface properties of coal and dirt to effect a separation. Froth flotation, however, becomes less effective where extremely fine particles of coal must be treated or if there is considerable clay slime present. By contrast, there appears to be virtually no lower limit on the particle size suitable for oil agglomeration. For example, colloidal particles of silica have been collected by the technique [3]. If desirable, size distributions containing particles 1/8 in. in diameter or larger can also be treated. In addition to this ability to treat a broad size range of coal particles, oil agglomeration can produce a dense, coarse granular product of acceptable strength and low moisture content without the need for vacuum or thermal drying.

A. Historical Perspective

The preferential wetting of solid surfaces by immiscible liquids has been utilized, if not fully understood, for a long time. Indeed Gaudin [4], in tracing the ancestry of froth flotation, refers to a fifteenth century oil-pulp kneading procedure in which water-wetted particles were freed under water from an oiled mass. During World

War I, the bulk oil Trent process [5] was developed to clean and recover fine coal. This process consisted essentially of the churning together of a mixture of finely ground coal and water, together with oil to the extent of about 30% of the weight of the raw coal to be cleaned. After agitation up to 15 min or longer, small nodules of cleaned coal were formed which could be easily separated from the water suspension of the dirt. The separated coal nodules formed an "amalgam" suitable for a number of industrial and domestic uses. Similarly in more recent times, the Convertol process [6] was examined as a means of recovering and dewatering fine coal. In this case, high-shear mills were used to produce intimate contact between the oil and the coal over short residence times of 15 to 30 sec. The coal agglomerates formed by the mixer were then removed from the slurry by centrifugal filtration.

There is renewed interest today in oil agglomeration because of the increasing quantity of fines which must be dealt with in coal preparation. These fines are produced by natural degradation, increasingly mechanized mining methods and by the grinding necessary to liberate finely disseminated impurities from the lower-quality coals now being processed. Much of the current examination of processes based on preferential wetting by immiscible liquids is due to studies at the National Research Council of Canada. This work resulted from interest in a densified rounded form of agglomerates [7] formed when suitable amounts of bridging liquid are added under appropriate agitation to a solids suspension. A family of techniques, generally known under the original label of "spherical agglomeration processes," has subsequently been developed.

Work on the application of selective oil agglomeration to coal preparation has recently been reported in the United States [8-10], Poland [11], Australia [12], Germany [13], and other countries [14, 15].

B. Immiscible Liquid Wetting

Before giving specific details of the oil agglomeration process as applied to coal beneficiation, it is instructive to examine the

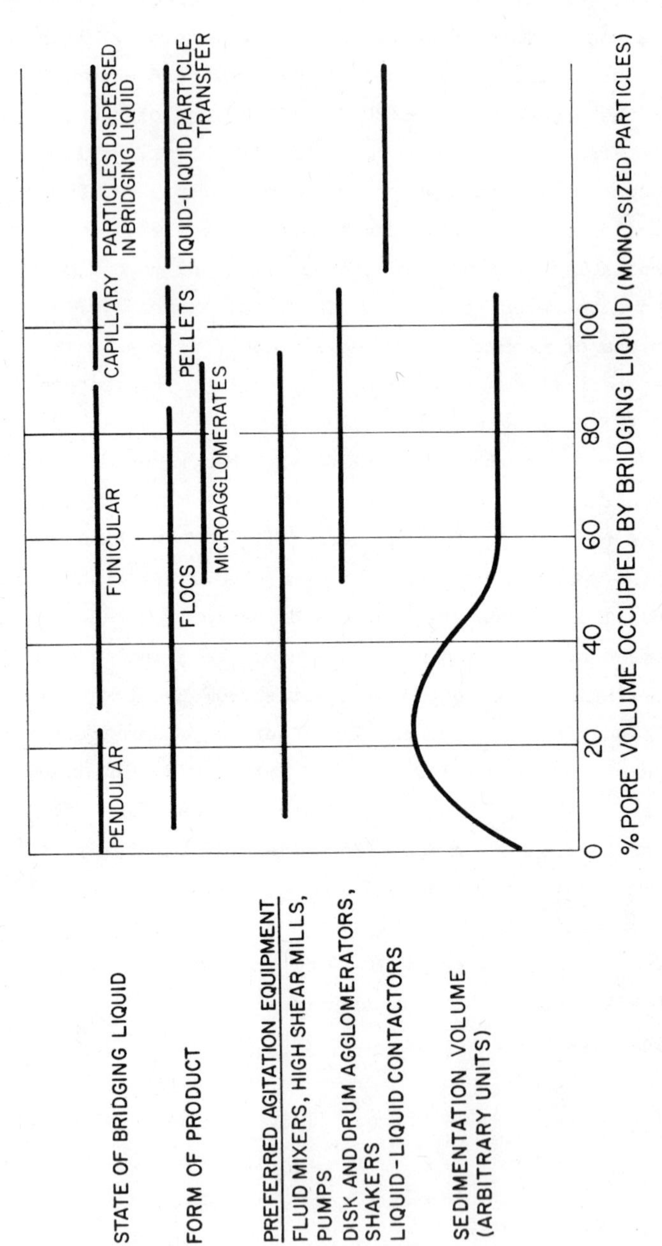

Figure 6.1 Agglomeration "phase diagram." The effect of increasing amounts of bridging liquid on the process. A typical system is fine sand suspended in organic liquid and agglomerated with water. (From Ref. 71.)

general characteristics of processes in which particles suspended in one liquid are wetted by a second immiscible liquid. The phenomena which occur as progressively larger amounts of bridging liquid are added to a solids suspension are depicted in Fig. 6.1. The general relationships shown are not specific to a given system. They apply equally well, for example, to siliceous particles suspended in oil and collected with water, or to coal particles suspended in water and agglomerated with oil.

At low levels of bridging liquid, only pendular bridges can form between the particles with the result that an unconsolidated floc structure exists. As seen in the lower part of Fig. 6.1, a loose settled mass of volume larger than that of the unflocculated particles results. As the funicular region of bridging liquid levels is reached, the flocs consolidate somewhat and lower settled volumes are recorded. Some compacted agglomerates appear and increase in number until about midway in the funicular region the whole system has been formed into "microagglomerates." As the amount of bridging liquid is increased, the agglomerates grow in size and reach a peak of strength and sphericity near the capillary region. Beyond this region the agglomerates exist as pasty lumps; the solids are then essentially dispersed in the bridging liquid. Figure 6.2 shows these different stages in the development of coal agglomerates.

Thus, depending upon the amount of bridging liquid and the type and intensity of agitation used, particles in liquid suspension may be agglomerated as flocs, microagglomerates, pellets or may be transferred from one liquid phase to the second immiscible one. Each type of process has applications particularly suited to it. These applications may range from recovery of solids and displacement of suspending liquid in the flocculation regime through a novel method of sphere formation in the capillary region to selective concentration of one or more components of complex mixtures in the region of liquid-liquid particle transfer. Examples of these applications are available in the literature. Table 6.1 gives examples of the recovery and separation of solids from liquid suspension. Table 6.2 deals with

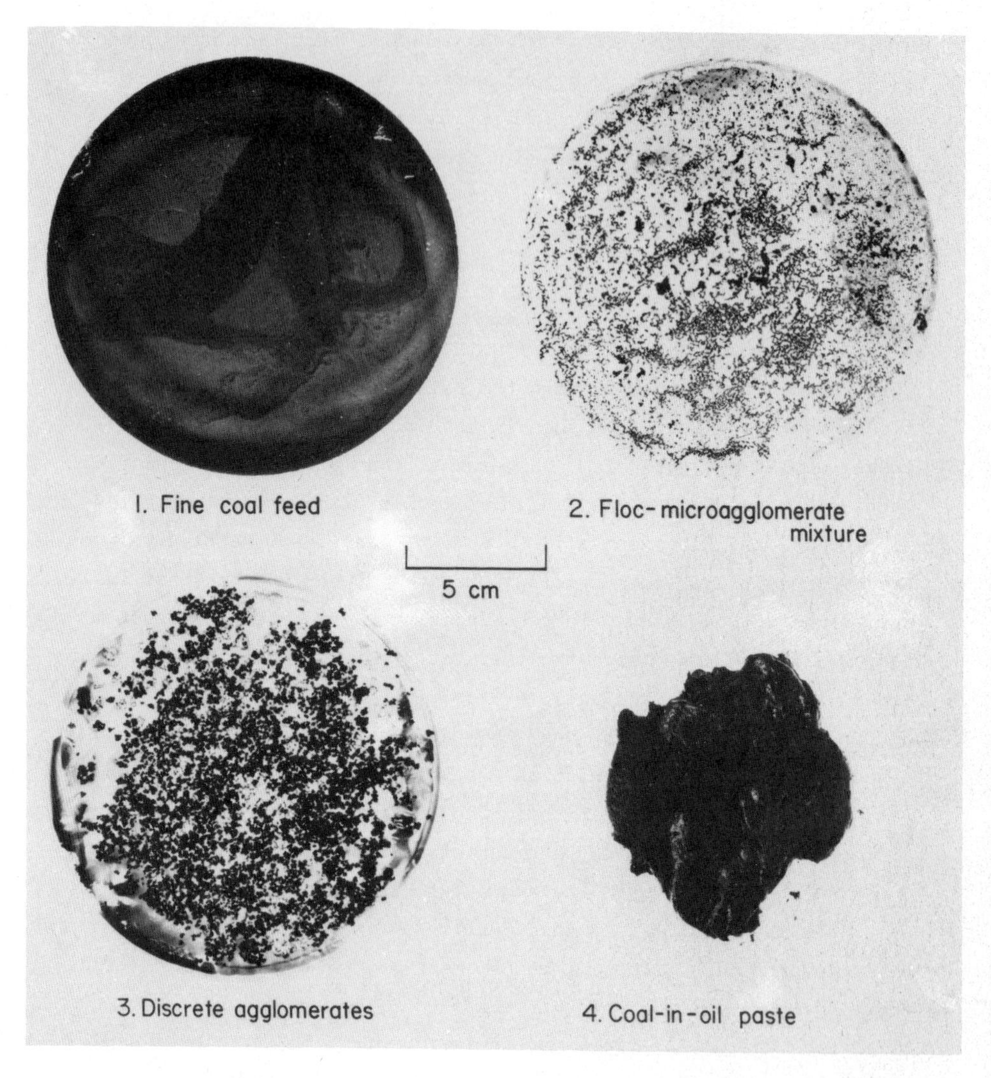

Figure 6.2 Photographs of coal agglomerates formed with increasing amounts of bridging oil.

the fractionation of complex mixtures such as ores in which surface-modifying chemicals are often used to agglomerate selectively different particles in the mixture.

Table 6.1 Recovery and Separation of Solids from Liquid Suspension

Suspension treated	Bridging liquid	Equipment	Remarks	Ref.
Calcium sulfate in phosphoric acid (wet process for phosphoric acid)	Petroleum sulfonate surfactants/hydrocarbon liquid	Conditioned in mixer, pellets formed by tumbling	Ashed $CaSO_4$ pellets contained 1.1 to 1.8% P; greater than 99% recovery of phosphoric acid from a semigelatinous precipitate.	[47]
Iron ore in water (tailings slurry)	Oleic acid/motor oil	Mixing tank	Agglomerates removed by cyclone separator which may also aid in forming agglomerates.	[48,49]
Fine coals in wash plant slurries	Light hydrocarbon oil for selective flocculation; heavy oil for balling	High intensity mixer for flocculation/beneficiation, balling on a disk	Waste fine coals upgraded to acceptable ash levels and desirable pelletized form.	[34,43]
Soot in water	Heavy oil	Rotating annular mixer/pelletizer	Wash-water stream from oil gasification process. Soot removal down to less than 5 ppm.	[22]
Phosphate ores in water	Fatty acid/fuel oil	Conditioned in mixer, separated on tables	"Agglomeration tabling" applied to phosphate potash minerals. Selective flocculation followed by separation on tables.	[50]
Oil dispersed in water	Finely-divided hydrophobic solid (e.g., coal) acts as emulsion breaker	Turbine mixer	Addition of powder which is agglomerated by dispersed oil and causes emulsion breakdown.	
Peat moss in petroleum solvent	Aqueous salt solutions		Recovery of peat moss following solvent extraction of resins.	[51]
Sodium lignate in water, pH 10	Rubber latex, acidify to pH 3	Turbine mixer	Lignin reinforcement of rubber.	[52]
Silica aquagel, preparation of dry colloidal silica	Ammonium oleate/hydrocarbons	Turbine mixer	Dewatering of gelatinous precipitate to form high surface area powder.	[3]

Source: Ref. 71.

Table 6.2 Surface Conditioning and Selective Agglomeration

Material treated and objective	Suspending liquid	Conditioning agents and/or other additives	Collecting liquid	Ref.
Barite in mill tailings; upgrading	Water	Sodium dodecyl sulfate, sodium silicate, acid wash of agglomerates	Still bottoms	[53]
Brine (NaCl) solution, removal of colloidal iron oxides	Saturated NaCl solution	Tall oil	Crude oil, sulfonated petroleum oil	[54]
Coal, bituminous, separation of pyrite and other ash constituents	Water	Iron-oxidizing bacteria, alkali	Petroleum distillates	[30]
Coal, bituminous, balling of ash constituents	Petroleum distillates	Balling nuclei added	Water	[41]
Coal, subbituminous and lignitic	Water	Coke oven tar, pitch, petroleum crudes and fractions	Hydrocarbons	[43]
Germanium in carbonaceous sandstone	Water	Coal tar, Na_2CO_3 and sodium silicate	Petroleum still bottoms	[55]
Glauber's salt ($Na_2SO_4 \cdot 10H_2O$), removal of siliceous matter	Saturated Na_2SO_4 solution	Various amines	Hydrocarbon	[56]
Gold ore for upgrading	Water	sec-Butyl xanthate	Still bottoms	[57]
Graphite-sulfur mixture	Light petroleum distillate			
1. Microagglomeration and removal of graphite		Tannic acid in collector	5% aqueous tannic acid	[58]
2. Microagglomeration and removal of sulfur		Aerofloat 15	Dilute NaOH solution	[58]

Graphite-zinc sulfide-calcium carbonate mixture	Water			
1. Microagglomeration and removal of graphite		None	Nitrobenzene	[58]
2. Microagglomeration and removal of ZnS		sec-Butyl xanthate $CuSO_4$, NaOH	Nitrobenzene	[58]
3. Microagglomeration and removal of $CaCO_3$		Oleic acid	Nitrobenzene	[58]
Ilmenite concentrate; removal of complex silicates	Water	Oleic acid, sodium silicate, pH adjuster	Petroleum distillates	[59]
Iron ores; removal of phosphatic and siliceous matter	Water, $CaCl_2$ solution	Various fatty acids, bases and acids for pH adjustment	Crude or semirefined viscous petroleum oils	[48,49, 60,61]
Marl ($CaCO_3$) deposits, removal of siliceous matter	Water	Fatty acid after heat treatment	Kerosene	[62]
Methyl methacrylate suspension; balling	Light petroleum distillate		Aqueous chloral hydrate	[63]
Shale; removal of $CaCO_3$	Water	Sodium oleate	Oxidized crude oil	[64]
Tar sands, agglomeration of bitumen	Water	Alkali	Hydrocarbons present in sands	[65]
Tar sands, extraction of bitumen	Water	Alkali	Water with alkali	[66,67]
Tin ore for upgrading	Water	Tall oil, acids and bases for pH adjustment	Viscous petroleum oil	[68,69]

Source: Ref. 71.

C. Scope of the Chapter

From these considerations of the general characteristics of immiscible liquid wetting, Sec. 6.2 moves on to discuss the details of unit operations and equipment needed to apply the oil agglomeration process to coal beneficiation. The next three sections deal with the results obtained from the process in relation to the major goals of coal cleaning. In Sec. 6.3, the operating and system variables affecting impurity reduction and fines recovery are detailed. Operating variables included are oil concentration, suspension density, and agitation characteristics, while the system variables of coal and oil characteristics are also considered. The reduction of pyrite and of trace metals in the agglomerated product are treated as separate topics. In Sec. 6.4, the factors affecting the moisture content of the product, especially the oil loading and its attendant effect on agglomerate size, are dealt with. A mathematical model to correlate these variables and their influence on moisture content is described. In Sec. 6.5 important benefits in the treatment of tailings as a result of oil agglomeration are noted. Finally, Sec. 6.6 summarizes potential benefits to be realized by the end user of oil agglomerated coal products in a number of applications while Sec. 6.7 assesses the economics of the oil agglomeration of coal fines.

6.2 COAL AGGLOMERATION PROCESS

From the background on immiscible liquid wetting presented in Sec. 6.1B and especially from the information given in Fig. 6.1, it will be appreciated that oil agglomeration of coal can be tailored to form the type of product needed in a given application. In particular, the process flowsheet will depend on the size of agglomerates required which in turn is controlled by such factors as the amount and type of oil used, the type and intensity of agitation, and the size distribution and wetting properties of the coal particles.

Although a specific flowsheet cannot be given, Fig. 6.3 presents a generalized flow diagram with suggested equipment for coal agglomeration processes. The sequence of operations includes selective

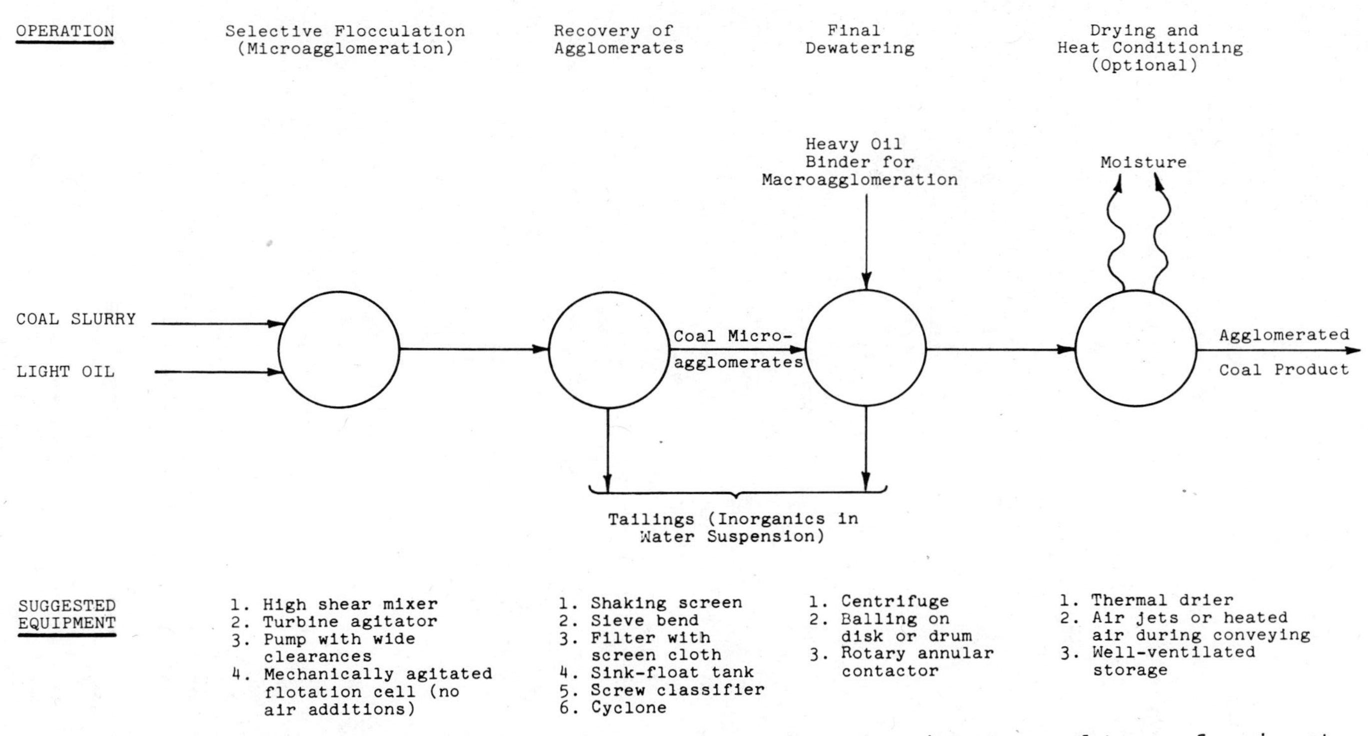

Figure 6.3 Coal agglomeration flowsheet indicating alternative processing steps and types of equipment. (From Ref. 71.)

flocculation or microagglomeration, agglomerate recovery with simultaneous impurity rejection, size enlargement of the recovered and beneficiated coal particles with simultaneous further rejection of tailings and moisture, and finally, thermal treatment of the product to remove moisture and/or harden binders. Some important features of the process should be noted:

1. To achieve initial oil dispersion and efficient wetting of the carbonaceous surfaces, some form of high-intensity mixing or prolonged contact is required. Predispersion of the oil in water, prior to mixing with the coal slurry [16] or during an upstream size-reduction operation [12] has been used to improve this initial oil-coal contacting.
2. Although two stages of agglomeration are included in Fig. 6.3, the small agglomerates produced in the first stage may be adequate for many applications such as when the product is to be transported in a slurry pipeline or fired directly in a thermal power-generating unit.
3. Where larger pellets are required for conventional storage and shipment of the coal product, the heavy binding oils used to form such larger agglomerates are generally poorly suited to good initial ash rejection. The two-stage process of Fig. 6.3 then allows the initial selective agglomeration and ash rejection to be accomplished with lighter oils to maximize impurity removal.
4. The steps indicated in Fig. 6.3 can be accomplished in conventional equipment well proven in commercial operations in the chemical and mineral industries.
5. Oil agglomeration displaces moisture effectively from even the finest coal particles. If agglomerates larger than about 2 mm are produced, thermal moisture removal is not required to obtain product with less than about 10 wt % moisture, although some thermal treatment may be necessary to harden binders.

Each of the operations depicted in Fig. 6.3 will now be discussed in detail in relation to the major objectives of coal preparation, namely,

1. Recovery of coal matter while rejecting impurities
2. Reducing the moisture content of the product
3. Dealing with rejected tailings

6.3 IMPURITY REDUCTION AND PRODUCT RECOVERY

Batch laboratory testing using high-speed blenders accounts for much of the experimental work [10,17] done on the oil agglomeration of coal. Typically coal slurries with size distributions containing 100 wt % −200-mesh particles and 40-50 wt % −325-mesh particles have been treated. A standard test uses 500 cm^3 of a 10 wt % suspension with the agglomerated product recovered on a 100-mesh screen. Product analysis is generally performed on material dried at 110°C. When high-boiling hydrocarbons are used for agglomeration, they are removed by toluene extraction prior to analysis.

Information on continuous pilot-scale operation of various coal agglomeration processes is, however, becoming more available [12,13, 18,19] and is included in the present discussion.

The major factors affecting coal agglomeration results are similar to those affecting the froth flotation process. Both methods depend on surface chemical effects and are influenced by the wetting properties of the coal, type and intensity of agitation, pulp density, and type and quantity of collecting phase.

A. Operating Parameters

Most published studies of oil agglomeration indicate that excellent ash reduction and product recovery are possible over a broad range of operating conditions. For example, Table 6.3 summarizes the results of batch kerosene agglomeration tests on a −325-mesh Lower Freeport bed coal from Cambria County, Pennsylvania. The variables tested and levels of treatment were kerosene concentration (2, 4, and 8 wt %), agitation speed (6000 and 12,000 rpm), agitation time (1 and 2 min) and slurry pH (natural pH and pH 10). The combustibles recovered in the cleaned coal ranged from 63.1 wt % at a 2 wt % kerosene dosage to 95.5 wt % at an 8 wt % kerosene dosage. In a broad laboratory study [20] of the application of oil agglomeration to six U.S. coal samples from which the data in Table 6.3 were taken, similar results were found. That is, of the variables and levels tested, kerosene concentration was found to be the most significant parameter with agitation speed, retention time, and slurry pH having lesser effects on product recovery and quality.

The data in Table 6.4, which pertain to tests with longer mixing times and generally higher oil levels than those in Table 6.3, show

Table 6.3 Laboratory Oil Agglomeration Tests for a Pennsylvania Lower Freeport Coal[a]

Process variables				Agglomerated product			
				% Recovery			
Kerosene, wt %	Agitator speed, rpm	Mixing time, min	Slurry pH	Weight	Combustibles	% Ash	% S
2	6,000	1	6.0	58.5	63.1	13.0	1.88
4				70.5	77.6	11.2	2.19
8				81.8	89.8	11.4	2.08
2	6,000	2	6.0	60.4	65.3	12.8	1.96
4				75.0	82.5	11.2	1.94
8				82.5	91.4	10.6	2.18
2	6,000	1	10.0	68.5	74.0	12.8	1.88
4				78.5	85.6	12.0	1.94
8				82.1	91.3	10.2	2.14
2	6,000	2	10.0	66.8	73.4	11.3	1.91
4				80.8	89.8	10.3	2.03
8				82.7	93.0	9.2	2.22
2	12,000	1	6.00	62.0	67.6	11.6	1.76
4				70.1	83.9	12.0	1.94
8				82.6	91.4	10.7	2.08
2	12,000	2	6.0	69.1	74.9	12.5	1.82
4				77.1	83.9	12.0	1.91
8				83.7	94.0	10.4	2.04
2	12,000	1	10.0	68.6	74.9	10.7	1.73
4				76.3	84.4	11.9	1.97
8				85.0	95.5	9.4	2.15
2	12,000	2	10.0	74.5	82.9	10.2	1.81
4				78.5	88.7	8.8	1.93
8				83.6	94.7	8.6	2.07

[a]Analysis of raw coal: ash, 19.3 wt %; pyritic sulfur, 1.77 wt %; total sulfur, 2.50 wt %.

Source: Refs. 20 and 43.

Table 6.4 Laboratory Oil Agglomeration Tests for a Variety of Coals[a]

Origin	Type of sample	Approximate size consist	Bridging oil		Feed coal		Agglomerated product		
			Type	Quantity	% Ash	% S	% Ash	% S	% Recovery combustible
European No. 1	Bituminous refuse slurry	100% −28 mesh 30% −200 mesh	Stoddard solvent	∿15%	27.12		3.9		99
European No. 2	Bituminous settling pond refuse	−200 mesh nominal	No. 2 fuel oil	12.1%	59		9.5		83
European No. 3	Bituminous flotation feed	90% −100 mesh	No. 2 fuel oil	5%	19.0		8.6		97
European No. 4	Bituminous seam coal	−200 mesh nominal	Paraffinic crude oil	∿15%	18.7	1.8	4.1	1.2	>90
European No. 5	Bituminous seam coal	−200 mesh nominal	Paraffinic crude oil	∿15%	9.7	1.1	3.9	1.0	>90
North American No. 1	Bituminous thickener underflow	100% −28 mesh 66% −325 mesh	Froth flotation test 0.25 lb/ton MIBC		41.4		28.4		64.3
North American No. 1	Bituminous thickener underflow	100% −28 mesh 66% −325 mesh	Kerosene	10%	42.6		5.4		83.0
North American No. 2	Bituminous seam coal	−325 mesh	Kerosene	8%	19.3	2.50	8.6	2.07	94.7
North American No. 3	Bituminous flotation feed	−200 mesh	Stoddard solvent	3.2%	25.7	2.0	6.7	1.4	92.5
North American No. 4	Bituminous seam coal	−400 mesh	Paraffinic crude oil	36%	10.0		2.2		>90
North American No. 5	Subbituminous seam coal	−200 mesh	Paraffinic crude oil	∿15%	8.0		4.9		>90
North American No. 6	Lignite seam coal	−200 mesh	Mixture of crude oil and tar	∿20%	15.4		8.3		95.9

[a]Tests on 500 cm^3 slurry in blender operated at 9000 rpm; mixing for 5 to 10 min; naturally occurring pH; product recovered on 100-mesh screen; all analyses are in weight percent.

Source: Ref. 43.

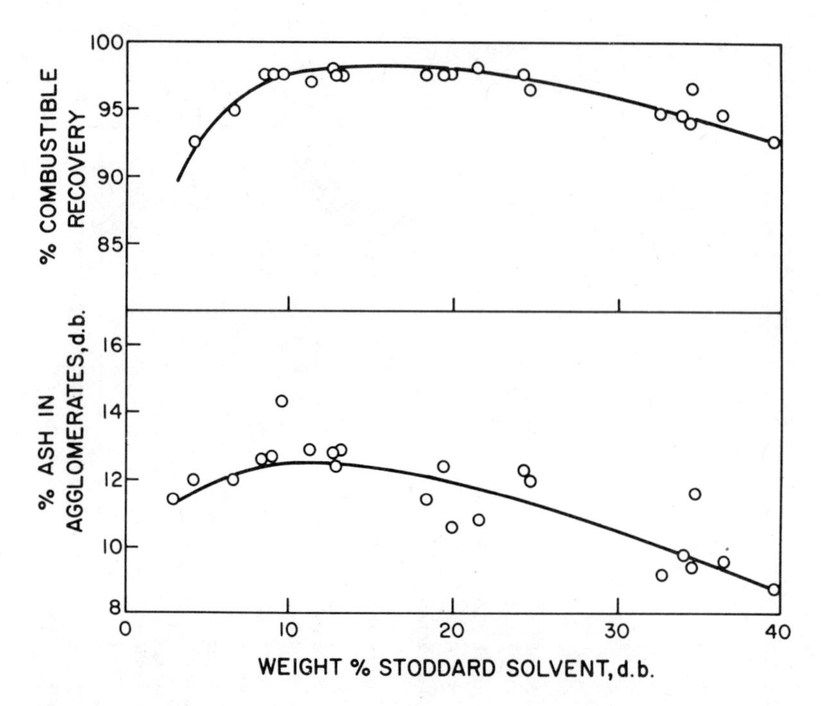

Figure 6.4 Ash content and recovery of combustibles of flotation tailings from coking coal preparation plant as a function of amount of bridging liquid (feed material: —28 mesh, 26 wt % ash, 20% w/v slurry in mixer). d.b. = dry basis.

that fine grinding followed by agglomeration can produce premium products with low ash levels for a variety of coals from many different geographical sources.

Oil Concentration. The paramount importance of the level of bridging agent was noted above. Further results in Fig. 6.4 for the recovery of coal from the flotation tailings of a metallurgical coal preparation plant demonstrate the effects of oil concentration. When the first few increments of oil are added, the recovery of combustible matter increases dramatically as the quantity of agglomerates retained by screening increases. At the same time, the recovered coal increases somewhat in ash content. There is, however, a broad oil range (typically from about 10 to 30 or 40% by weight of solids) over which product ash and recovery are relatively constant. Of course, with even

Table 6.5 Laboratory Oil Agglomeration Tests with 25 wt % Solids Slurry Concentration[a]

Stoddard solvent, wt %	Agglomerated product	
	% Ash	% Recovery combustibles
2	7.5	64.1
4	6.3	87.2
6	6.3	88.2
8	7.2	91.2
10	6.2	91.3

[a]Thickener underflow of bituminous coal; 4 wt % +48 mesh and 70 wt % −325 mesh; 42.6 wt % ash.

Source: Ref. 43.

higher oil loadings, coal recovery will again be reduced. This is caused by the loose structure of the coal-oil amalgam formed at these higher oil concentrations, some of which passes through the screen.

Pulp Density. Pulp density is not a critical factor in oil agglomeration. Slurries as high as 50% by weight of solids can be dewatered and deashed to levels comparable to those at lower pulp densities [18, 19]. Table 6.5 shows the low ash levels and high recoveries obtained with 25 wt % solids concentration. In contrast, flotation would require solids concentrations of about 3 to 5 wt % [21] to treat effectively coal as fine as that of Table 6.5 (4 wt % +48 mesh and 70 wt % −325 mesh). Indeed, very low pulp densities would negatively affect oil agglomeration since the process requires intimate mixing and contact among oil, coal, and oil-wetted coal particles. Slurries with very low pulp densities would require prolonged mixing to ensure that these contacts occur. Low pulp densities are undesirable due to the large volumes of water which must be handled.

Agitation Duration and Intensity. The data cited in Table 6.3 show that agitator speed and mixing time both affect product quality and recovery. Figure 6.5 and Table 6.6 indicate that the importance of agitation time increases as oil density and viscosity increases.

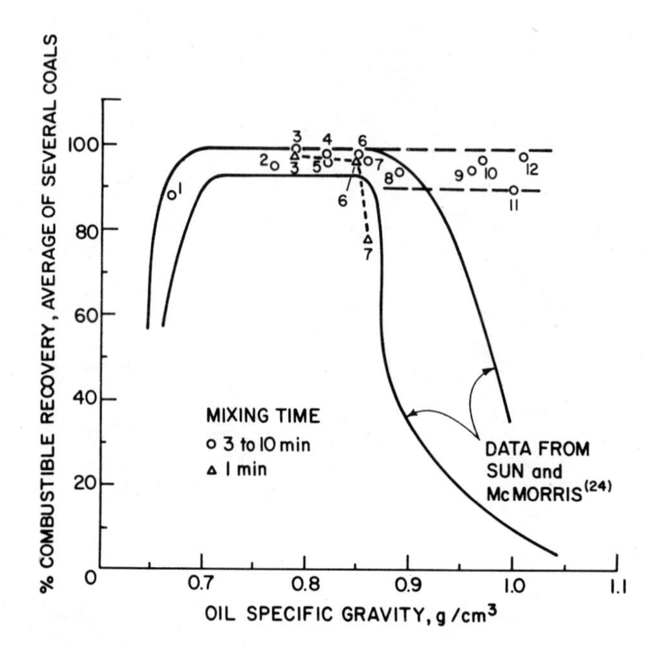

Figure 6.5 The effect of mixing time on the ability of various oils to recover bituminous coals.

In general, the mixing time required to form "good" coal agglomerates decreases as the agitation is intensified, other factors being constant.

A variety of mixing conditions have been used for coal agglomeration. In a laboratory high-speed blender, an agitator tip speed of 8 to 10 m/sec is common. For a 10 wt % coal slurry and, say 10 wt % bridging oil concentration, adequate agglomeration occurs in 2 to 5 min. These agitation conditions are similar to those in the propeller-agitated region of the plant-scale Shell Pelletizing

Table 6.6 Influence of Oil Viscosity on Agglomeration[a]

Agglomerating oil	Viscosity, cP	Agitation time, min	Product ash, wt %	Combustibles recovery, wt %
Stoddard solvent	0.9	1	8.8	98.0
Paraffin 0-119	23.9	1	8.1	97.1
Paraffin 0-120	64.4	1	8.5	77.5
Stoddard solvent	0.9	7	7.8	98.1
Paraffin 0-119	23.9	7	7.6	98.0
Paraffin 0-120	64.4	7	7.4	98.0

[a]Test conditions: 50 g of 20 wt % ash bituminous coal, 60 wt % −100 mesh. Agglomerated in 450 cm^3 of water with 10 cm^3 oil per test.

Source: Ref. 25.

Separator [22] where a peripheral speed of about 12 m/sec is used with a residence time of the order of 3 to 5 min. By contrast, the early Trent process [5,23] apparently combined low intensity agitation (a churn-type mixer was used for laboratory demonstration) with high oil loadings (about 30% of the coal weight) and prolonged agitation (up to 15 min or more, depending on the coal) to effect "amalgamation." In more recent times, work in Germany on the Convertol [19] and Olifloc [13] processes has taken an opposite tack. High-intensity phase inversion (colloid-type) mills with peripheral speeds of the order of 20 to 30 m/sec have been combined with short residence times (15 to 30 sec) and low oil contents (2 to 7% of coal weight) to accomplish agglomeration on slurries containing 40 to 50 wt % coal.

Information on power input to the mixing stage is not readily available in the literature. The authors estimate, however, that between 10 and 40 kW/m^3 of mixer volume have been used. The higher figure corresponds to processes with high pulp density and agitator speed.

Agitation serves initially to disperse the oil phase and then to contact the oil droplets and carbonaceous particles so that pendular bonds are formed between colliding oil-coated particles. Some attrition or cleaning of the coal particle surfaces may take place to expose fresh surfaces which may be readily wetted by oil. Lemke [19] notes that too much breakage of the coal may not be desirable since recovery may suffer as a given quantity of oil must wet an increasing particle surface area. Bensley et al. [16] prefer emulsification of the oil phase prior to its addition to the coal pulp. Mixing time and intensity for comparable coal recovery by oil agglomeration are reduced by this procedure and reduced oil requirements are also claimed.

B. Oil Characteristics

The type of oil used is virtually as important as its concentration in the agglomeration of coals. To be effective, the oil must not only recover the combustibles which are present, but the recovered material must also have a low ash content. That is, the inorganic or ash constituents must be rejected.

Consider the first of these factors, the recovery of coal constituents. In the case of bituminous coals, Fig. 6.5 gives some indication of the effect of oil density on combustibles recovery. In this figure, the solid lines indicate the trend found by Sun and McMorris [24] during their work on the Convertol process. Oils of medium density were required to obtain satisfactory recovery levels for the coal fines. The low-density oils were found to have insufficient viscosity to "pull" the coal particles together into strong agglomerates. On the other hand, the high-viscosity oils were not dispersed sufficiently well in the slurry to be able to wet the particles and cause agglomeration. It has been found, however, that these effects of oil density and viscosity are primarily due to the hydrodynamics of the system and not due to the chemical makeup of the bridging liquid. It should be noted that the data of Sun and McMorris are for relatively short mixing times of less than 1 min. In experiments [25] with three paraffinic oils of increasing density (see data in Table 6.6 plotted as triangles in Fig. 6.5) poor coal recovery was found with the most dense liquid when a short mixing time (of the order of 1 min) was used. This observation was in agreement with the data of Sun and McMorris. When longer mixing times were used, however, as indicated by the circles in Fig. 6.5, these paraffin oils (liquids 3, 6, and 7 in Fig. 6.5) and other more dense oils all yielded excellent coal recoveries. It may be concluded that there is nothing inherent in the coal-surface wetting properties of most oils to prevent their use as bridging agents for bituminous coals with excellent combustible-matter recoveries, provided that intimate oil-particle contact is achieved in the mixing. It should also be noted that shorter coal-oil mixing times can be used when the oil phase is predispersed prior to contact with the coal slurry [9,16].

Turning now to the effect of oil characteristics on ash rejection, Table 6.6 shows that the use of viscous paraffinic oils may have a beneficial effect on product ash. When sufficient mixing is employed viscous oils yield relatively larger agglomerates due to improved strength through viscous bonding. The larger agglomerates in turn hold less moisture, and since the liberated ash is associated with the water phase, the ash content of the agglomerates is reduced accordingly.

Figure 6.6 gives further information on the ability of bridging oils to agglomerate selectively the carbonaceous material while rejecting the inorganic or noncombustible matter. This figure shows the ratio of the ash levels obtained for a number of different coals with various oils to the ash levels obtained with Stoddard solvent for the same coals. Stoddard solvent, a highly paraffinic oil which generally gives a good level of ash rejection, is used as a reference material. Notice again in this case that the oils of lower and higher density must be considered in two separate categories. The oils with densities below about 0.9 g/cm^3 generally give good ash rejection with ash levels of agglomerates often within 10 or 20% of those obtained with the Stoddard solvent. These lighter oils which give good ash rejection include various refined paraffins, such as hexane and Stoddard solvent, and commercial oils such as B.T.X., kerosene, No. 2 fuel oil and diesel fuel oil. On the other hand, the oils with densities greater than 0.9 g/cm^3 generally show more erratic behavior with

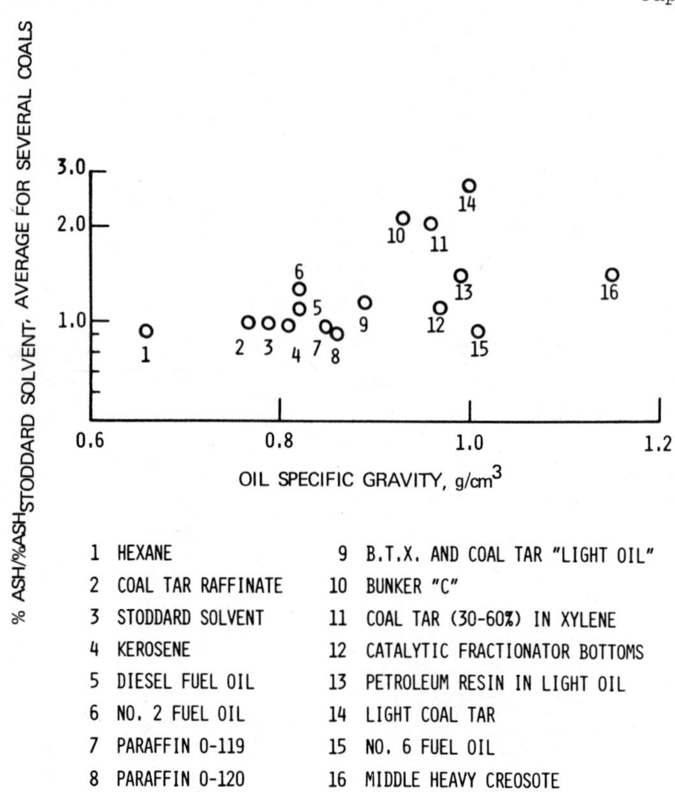

Figure 6.6 Ash content of agglomerates of bituminous coals using various oils as bridging agents. (From Refs. 43 and 71.)

substantially higher ash levels in the agglomerated products. These denser and more viscous oils include the complex mixtures such as petroleum resins and tars, the bunker fuels, heavier crudes, and coal tar and its derivatives (e.g., creosote). These complex oils contain higher levels of inorganic polar constituents such as nitrogen, oxygen, and sulfur groups which make their surface activity and wetting properties more unpredictable. These polar functional groups of the oil molecules are apparently able to attach the oil molecules to the surfaces of the ash-forming particles and render the particles hydrophobic, similar in function to that of a flotation collector. The ash-forming particles would then report with the carbonaceous particles in the agglomerates, leading to the higher ash levels observed.

Table 6.7 Beneficiation of Bituminous Coals by Agglomeration with Refined, Low-Density Oils (Specific Gravity < 0.9)

Oil	Vol % olefins plus aromatics	Ratio of wt % ash to wt % ash with Stoddard solvent[a]	
		Authors' work	Data of Sun and McMorris [24]
Paraffin O-120	0	0.94	
Hexane	0	0.94	0.90
Paraffin O-119	0	0.97	
Stoddard solvent	21	1.00	1.00
Kerosene	25	1.00	1.01
Coal tar raffinate	26-45	1.01	
Diesel fuel	30	1.09	
No. 2 fuel	37	1.29	1.21
B.T.X. and coal tar "light oil"	100	1.17	1.34

[a]Average values for a number of different coals and/or different oil sources.

Some further interesting features of the lighter, more refined oils are highlighted in Table 6.7. A number of the oils used in Fig. 6.6 were subjected to a modified ASTM procedure D1019-68* in which olefins and aromatics present in the oils were reacted with fuming sulfuric acid, leaving the unreacted paraffins to be measured. Data from the authors' work and from that of Sun and McMorris [24] indicate a similar trend to higher product ash levels as the olefin plus aromatic content becomes greater.

Table 6.8 shows a similar relationship between the ash and sulfur contents of agglomerates produced from two bituminous coals in which a number of petroleum crude oils were used to agglomerate the coal particles. The crudes originated in various parts of the

*ASTM procedure D1019-68 requires that a meniscus be formed and measured between reacted and unreacted portions of the sample. The test was not suitable for the heavier, opaque oils since the meniscus did not form and/or could not be clearly seen.

Table 6.8 Beneficiation of Bituminous Coals by Crude Oil Agglomeration[a]

Type of crude	Viscosity-gravity constant	No. of different crudes	Agglomerate analyses			
			Coal A		Coal B	
			Wt % S[b]	Wt % Ash[c]	Wt % S[b]	Wt % Ash[c]
Paraffinic	0.79-0.82	3	1.25	2.9	2.96	4.6
Relatively paraffinic	0.82-0.85	5	1.38	3.4	3.03	5.4
Naphthenic	0.85-0.90	4	1.40	3.9	3.14	5.8
Relatively aromatic	0.90-0.95	1	2.04	8.7	5.27	11.3

[a]Feed analysis: (1) coal A, 1.97 wt % S and 20.4 wt % ash; (2) coal B, 7.88 wt % S and 29.7 wt % ash.

[b]Oil-free basis, average values.

[c]Oil-in basis, average values.

United States, Canada, Africa, South America, and the Middle East. The type of crude oil indicated in Table 6.8 was inferred from their viscosity-gravity constant as measured by ASTM procedure D2501-67 and discussed by Hills and Coats [26]. As in Table 6.7, the agglomerates with the lowest ash and sulfur levels tend to be produced by the bridging oils of the highest paraffin content.

This tendency to yield higher ash levels in the agglomerates as the aromaticity of the oil increases is consistent with the greater affinity of the aromatics for water and for other inorganic materials. For example, Table 6.9 shows the increasing solubility in water and the corresponding decreasing interfacial tension between water and oils in progressing from paraffins to aromatics. It is also known that aromatics are better absorbed on silica than are paraffins. This property has been used as a method to remove small amounts of unsaturates such as olefins and aromatics from saturated hydrocarbons by percolation through silica gel [27]. These facts indicate that the aromatics would be expected to absorb more readily than the paraffins on the ash constituents and cause them to report with coal. That is, the agglomerates might be expected to have higher ash levels as the aromaticity of the oil is increased.

In summary, it is recommended that the lighter (less than 0.9 g/cm^3 density), more refined oils of higher paraffin content be used

Table 6.9 Effect of Oil Type on Oil-Water Affinities

	Solubility in water, g/l	Interfacial tension against water dyne/cm
Paraffins		
Hexane	0.052	51
Octane	0.015	51
Aromatics		
Benzene	0.82	35
Toluene	0.47	36

for selective agglomeration when the rejection of ash is an important consideration. In addition to their more desirable wetting properties discussed above, these lighter oils achieve efficient and economical coating of the coal particles during mixing. The denser, more viscous and generally more aromatic oils are suitable for agglomeration of coal in which recovery and dewatering of fines is a major objective, with ash rejection being somewhat secondary. These heavier oils are often, of course, suitable as binders to form pellets after rejecting the ash by preliminary microagglomeration (see Fig. 6.3). Some heavier hydrocarbons may achieve acceptable ash rejection in which case a single separating-pelletizing step may be used [14].

C. Coal Characteristics

Ash dissemination size primarily affects product ash content. The degree to which a coal can be beneficiated by agglomeration is limited by the extent to which the ash has been liberated from the coal. Thus, as Fig. 6.7 indicates, grinding to finer sizes produces cleaner coal. As the coal particle size decreases, the oil concentration must be increased to maintain high product recoveries since the smaller particles have a greater total surface area which must be wetted by the oil.

The lower-rank subbituminous coals and lignite are distinguished by their greater oxygen content and the hydrophilic nature of their surfaces relative to those of bituminous coals. The light oils which are used successfully with bituminous coals are not able to wet the lower-rank coals and form only emulsions with no discrete agglomerates when agitated with them in a water slurry. If heavier oils, such as coke oven tars and pitches as well as petroleum crudes and their higher boiling components, are used as conditioners, however, distinct agglomerates are formed with the lower-rank coals. Apparently the nitrogen, oxygen and sulfur functional groups of these complex oils are able to adsorb sufficiently well on the relatively hydrophilic surfaces of the lower-rank coals to form agglomerates.

As illustrated by the examples given in Table 6.4 for lower-rank coals, ash rejection does occur when the heavier, complex oils

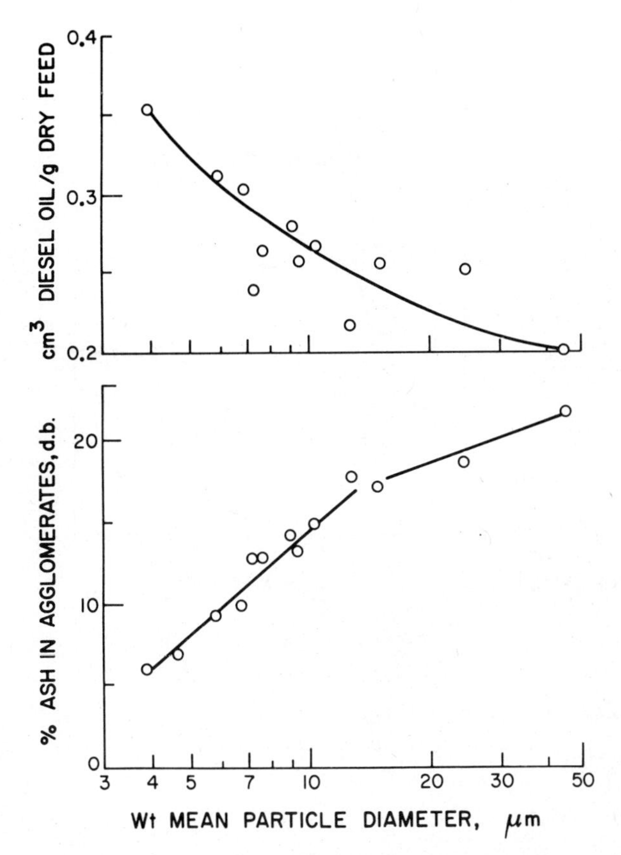

Figure 6.7 Effect of particle size on ash content and oil requirement in recovery of flotation tailings from coking coal preparation plant (feed material: 33 wt % ash, 10% w/v slurry in mixer).

containing multiple functional groups are used as conditioning agents. As with the bituminous coals in Fig. 6.6, however, the amount of ash rejection is less than that might be expected if the lighter, more refined oils alone could be used for agglomeration. The procedure also produces a granular material from which a large portion of the surface moisture has been displaced. Unfortunately, the treatment is not able to reduce the internal moisture bound within the structure of the lower-rank coals without thermal drying. The consistent granular texture of the product is well suited to rapid thermal drying and, as seen in Fig. 6.8, the absorbed oil in the agglomerates

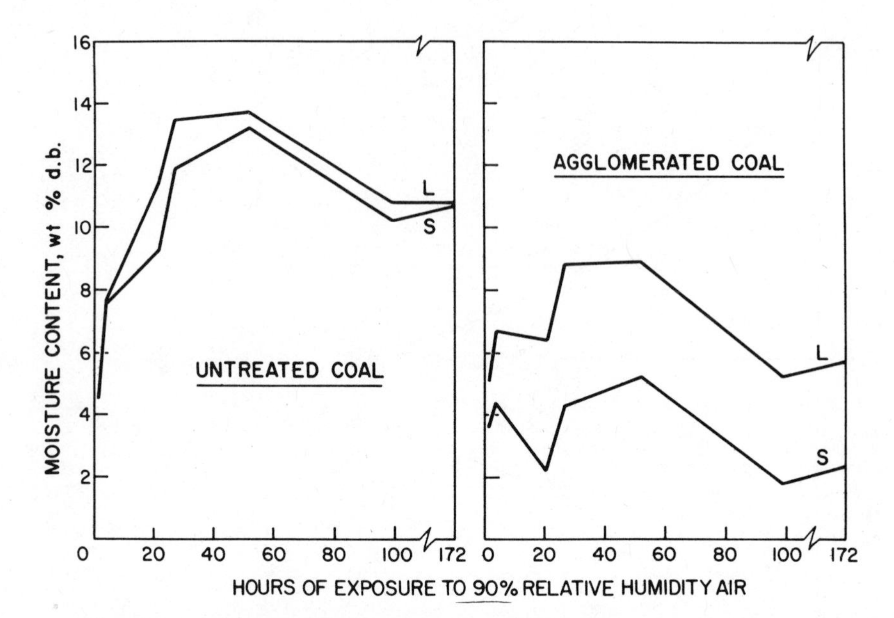

Figure 6.8 Moisture absorption by lignite and subbituminous coals and by their agglomerates after thermal drying (L refers to 14 × 28 mesh lignite coal and its agglomerates of the same size; S refers to 16 × 42 mesh subbituminous coal and its agglomerates of the same size). Agglomerates formed using coal tar-xylene conditioner and Leduc crude oil bridging agent.

reduces considerably the readsorption of moisture following thermal drying.

D. Reduction of Pyrite

Two main problems are associated with the removal of pyritic sulfur from coals. The first is due to the fine state of dissemination of pyrite, which requires grinding to a very fine size to accomplish a high degree of liberation [28]. The second problem is caused by the similar surface chemical characteristics of coal and pyrite, which complicate the use of separation methods based on this property [29].

The problem of fine particle size for pyrite liberation can be overcome by the selective agglomeration method. As noted already, even submicrometer particles of coal can be effectively recovered,

Table 6.10 Effect of Various Chemicals on Sulfur Reduction

Chemical[a]	Analyses of agglomerated products											
	Natural pH				Acidic				Basic			
	pH[c]	% Ash	% S	% Rec.[d]	pH[c]	% Ash	% S	% Rec.[d]	pH[c]	% Ash	% S	% Rec.[d]
None	8.1	6.8	5.9	97	2.8	6.4	5.9	98	10.9	7.8	6.8	97
Glycerine	8.7	6.7	6.5	97	2.5	9.1	8.8	96	10.5	7.1	7.1	95
Chlorinated lime	8.1	6.5	5.9	97	2.8	8.1	6.5	97	10.8	6.6	6.0	98
Ferric sulfate	8.3	7.5	6.9	97	2.7	7.7	6.7	96	10.0	6.3	6.4	97
Pyrogallic acid	8.1	7.2	5.9	99	3.1	9.9	8.9	96	10.4	6.1	5.8	97
Sodium cyanide	8.4	5.8	4.9						10.3	8.0	7.1	
Sodium carbonate	9.2	5.5	4.0		3.1	6.8	5.7		11.5	7.3	6.6	
Dextrine	8.0	7.3	6.3		2.8	7.1	5.9		11.5	7.3	6.6	
Potassium ferrocyanide	8.5	5.9	4.1						12.0	5.8	4.8	
Potassium ferricyanide	8.2	4.8	4.5						10.2	5.7	4.5	

[a]One-tenth gram of chemical added for 150 g coal in 400 g water prior to grinding; pH adjusted with NaOH or HNO_3.

[b]Feed coal: (1) run-of-mine high-volatile A bituminous; (2) 100 wt % −50 μm, and 70 wt % −22 μm; (3) 16 to 20 wt % ash, 8.3 to 9.0 wt % Fe_2O_3, 6.6 to 8.0 wt % total sulfur, 0.1 to 0.3 wt % sulfate sulfur, 6.0 to 6.4 wt % pyritic sulfur, and 1.4 to 1.7 wt % organic sulfur.

[c]pH during agglomeration.

[d]% Recovery is the weight percent of combustibles recovered in the agglomerated product.

Source: Ref. 30.

Table 6.11 Reduction of Trace Metals by Selective Oil Agglomeration[a]

Element	Analytical method	No. of determinations[b]	Concentrations, ppm[c]						% Element rejected/ % ash rejected
			Feed coal		Agglomerates		Tailings		
			Average	High, low	Average	High, low	Average	High, low	Average (standard deviation)
Aluminum	OES	1	15,000		10,000		100,000		0.88
Antimony	SSMS	2	1.3	2.0 0.5	0.7	0.8 0.5	7	8 5	0.68
Arsenic	OES	3	72	115 50	15	45 0	270	410 200	1.11 (0.18)
Barium	OES	4	40	100 0	38	100 0	100	300 0	0.42 (0.21)
Beryllium	OES	2	0.5	0.8 0.1	0.7	1.2 0.1	0.03	0.05 0	0.02
Boron	OES	7	21	50 5	18.8	50 3	32	60 10	0.49 (0.26)
Cadmium	SSMS	1	0.3		0		2.0		1.23
Calcium	OES	1	2200		1300		4800		0.65
Chromium	AA	10	652	1690 25	186	488 25	2837	7690 100	0.90 (0.19)
Copper	AA	9	301	800 42	171	566 21	1103	4300 165	0.79 (0.15)
Germanium	OES	1	14	23 5	14	22 5	0		0
Iron	AA	8	49,000	87,000 15,000	15,700	30,000 7,700	193,500	330,000 50,000	0.89 (0.15)
Lead	OES	5	75	160 15	33	105 10	242	410 100	0.90 (0.25)

Magnesium	OES	2	250	500 0	125	250 0	1500	3000 0	0.88
Manganese	AA	8	275	410 67	55	152 17	935	1326 400	0.97 (0.14)
Mercury	AA	10	0.49	0.86 0.07	0.41	0.91 0.05	1.2	3.4 0.4	0.51 (0.34)
Molybdenum	OES	3	183	250 150	43	100 10	667	1000 500	1.01 (0.24)
Nickel	AA	10	59	86 18	24	56 6	204	325 111	0.88 (0.17)
Selenium	SSMS	1	4		3		20		0.44
Sodium	OES	2	50	100 0	25	50 0	500	1000 0	1.48
Tin	SSMS	1	65		20		170		0.90
Titanium	OES	2	425	600 250	400	550 250	710	800 620	0.33
Vanadium	OES	5	150	500 50	61	150 10	410	1000 150	0.87 (0.28)
Zinc	AA	6	284	570 41	144	385 25	1150	4160 110	0.84 (0.29)
Zirconium	OES	1	25		25		100		0.42

[a]Six different steam and coking coals from North American sources were tested. Wet ground to −200 mesh prior to agglomeration.
[b]Not replicate analyses but the number of measurements for which calculated and measured feed concentrations agreed within ±30%
[c]All analyses are on a moisture-free basis.

Source: Reprinted with permission from C. E. Capes, A. E. McIlhinney, D. S. Russell, and A. F. Siriani, Rejection of Trace Metals from Coal During Beneficiation by Agglomeration, *Envir. Sci. Tech. 8*, 35-38 (1974). Copyright by the American Chemical Society.

if necessary. However, even with very fine grinding and the aid of a number of possible pyrite depressants under acidic, neutral and alkaline conditions as detailed in Table 6.10, it was found [30] to be difficult to remove even 50% of the pyrite present in a high-volatile A bituminous coal. This agrees with the general difficulty experienced in removing pyrite by surface wetting differentiation as is used in both oil agglomeration and froth flotation.

It will be noted that the sulfur reduction is quite variable in the results given in Table 6.4. Chemical conditioning agents were not used in this case. This variability is apparently related to the form of sulfur present, organic sulfur being unaffected by agglomeration, inorganic (sulfate) sulfur being readily rejected in the aqueous phase and pyritic sulfur residing between these two extremes depending on the state of oxidation and wetting characteristics of the pyrite particles. Consistent sulfur reductions during agglomeration (as well as in flotation) therefore appear to depend on discovery of a reliable and economic means to render the pyrite surface more hydrophilic. Naturally occurring iron-oxidizing bacteria have been found [30,31] to lead to greater than 90% pyrite rejection, but it was concluded that the long treatment times involved (1 or 2 days or longer) make this technique unattractive on a large scale. Consideration has also been given to chemical pretreatment (wet oxidation) of coal fines to enhance the difference in surface properties of coal and pyrites. Details of this approach can be found in Sec. 7.2.

E. Reduction of Trace Metals

The ability of selective oil agglomeration to remove heavy metals during coal beneficiation has been examined by Capes et al. [32]. The results of the study are summarized in Table 6.11 where the data in the last column (ratio of weight percent element rejected to weight percent ash rejected) are of particular interest. From the values of this ratio, it is evident that many of the trace elements are concentrated in the tailings and are thus removed by fine grinding and oil agglomeration. A number of elements, however, are apparently associated with the organic constituents of coal and report with the agglomerated product. Most prominent among these materials are barium, beryllium, boron, germanium, mercury, selenium, titanium, and zirconium. Oil agglomeration is hence not a satisfactory separation method for these constituents. At least three reasons why these elements might tend to report with the clean coal can be cited. First, as noted already, a major portion of the element may be organically associated with the coal matrix. Such elements cannot be removed by physical beneficiation techniques. Second, the average dissemination size of the element, as present in the fine coal, might be less than that of the ash. Third, it is possible that these elements are present as compounds with hydrophobic surfaces, as is apparently the case discussed already for pyrite, and are thus retained in the oil agglomerates.

6.4 MOISTURE REDUCTION

All wet coal beneficiation techniques create the additional problem of dewatering. Water must be removed from coal to improve its heating value and to avoid the cost of shipping inert material. Freezing of the coal in winter months must also be avoided. An important aspect of oil agglomeration for moisture reduction is its inherent dewatering capability due to the displacement of moisture by the absorbed oil.

Oil concentration is the most significant parameter in coal dewatering by agglomeration. Figure 6.9 is a generalized representation of the complicated relationships of agglomerate size, moisture content and collecting oil concentration. The concentration regions, designated as low, intermediate and high, correlate with the physical characteristics of the agglomerates, as discussed previously in connection with Fig. 6.1. Low oil concentrations in the pendular region produce flocculated coal, intermediate concentrations in the funicular and capillary regions yield microagglomerates and larger discrete agglomerates, while higher oil concentrations result in a coal-oil amalgam. The dewatering characteristics of each of these oil concentration regions is different [33].

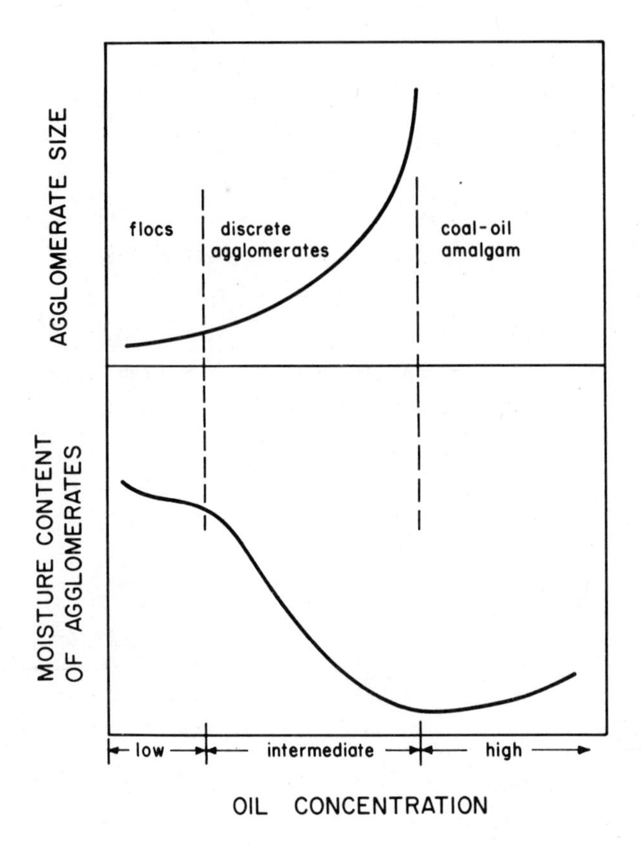

Figure 6.9 The effect of oil concentration on agglomerate size and moisture content.

A. Low Oil Concentrations

In this region, which would generally include oil concentrations less than 5% of the dry coal weight, initial oil additions flocculate the coal particles to form a two-dimensional network structure. Settled volume of the solids increases in this region and, as indicated in Fig. 6.10 especially for very fine coal, the moisture content of the recovered product remains very high as oil is added. The unconsolidated floc structure traps much suspending liquid.

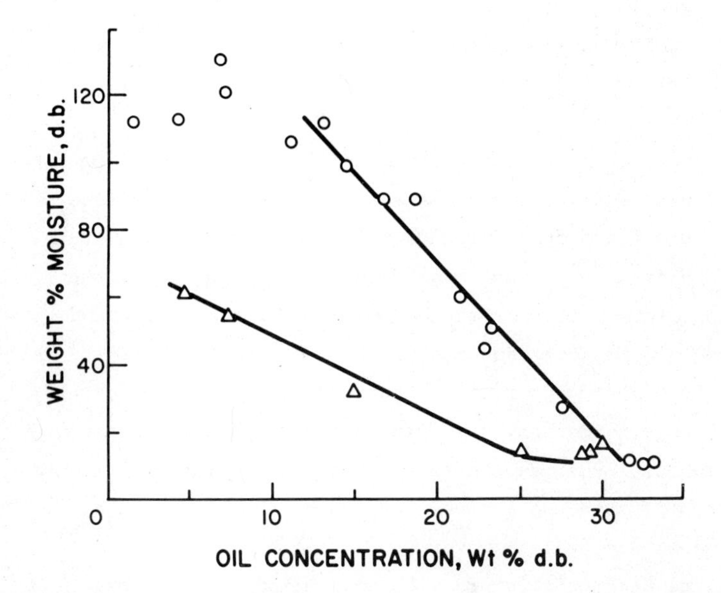

Figure 6.10 Agglomerate moisture content as a function of oil level for two coal grinds (bituminous coals, collected with paraffinic oils and dewatered on screen; O refers to −100-mesh feed, 17 wt % ash; Δ refers to −28-mesh feed, 26 wt % ash).

B. Intermediate Oil Concentrations

The transition from low to intermediate oil concentrations occurs when the voluminous two-dimensional coal flocs begin to compact into three-dimensional discrete agglomerates. The moisture content of the recovered coal begins to decrease more significantly. As illustrated in Fig. 6.10, this transition occurs at higher oil concentrations as the coal particles become finer.

As the oil concentration increases within the intermediate range, the agglomerates grow rapidly in size. With this increase in size the total surface area of the agglomerates decreases and correspondingly, so does the surface moisture. The internal moisture reduces because the oil fills in the agglomerate voids and displaces the water. The buffeting action in the agglomerating vessel tends to compact the agglomerates and squeeze out the internal moisture. On the other hand, the shearing action of the agitator may also tend to counteract this effect.

It is possible to improve dewatering at intermediate oil concentrations by using a two-stage process as in Fig. 6.3. Low oil concentrations in the first stage produce a low-ash, but high-moisture, flocculated feed for the second stage. In this second stage, the shearing action of the agitator may be replaced by the compaction forces produced by tumbling, for example in a balling disk [34]. This squeezes out the trapped internal moisture and tends to reduce the agglomerate porosity. For any specified oil concentration in the intermediate range, larger agglomerates with lower moisture contents will be produced if tumbling is used instead of the buffeting and shearing action of a propeller mixer.

To correlate product moisture with agglomerate size, Capes et al. [35] proposed the following equation:

$$W = \frac{6K_1}{\rho D} + \frac{K_2}{\rho} \tag{6.1}$$

where

W = weight fraction of moisture in agglomerate
K_1 = weight of surface moisture per unit surface area of agglomerate
K_2 = weight of internal moisture per unit volume of agglomerate
ρ = true density of agglomerate
D = agglomerate diameter

As seen in Figs. 6.11 and 6.12, the model is adequate for agglomerates which are discrete. Figure 6.12 shows, however, that the model does not correlate laboratory results for small flocculated coal. It should also be noted that these data indicate that the moisture content may be reduced to less than 10% by mechanical dewatering alone without the need for thermal drying, provided that the product diameter is larger than about 2 mm.

Given the relationship in Eq. (6.1), it is important to be able to relate the agglomerate size to the bridging liquid content so that the moisture content of the agglomerates may be predicted from the

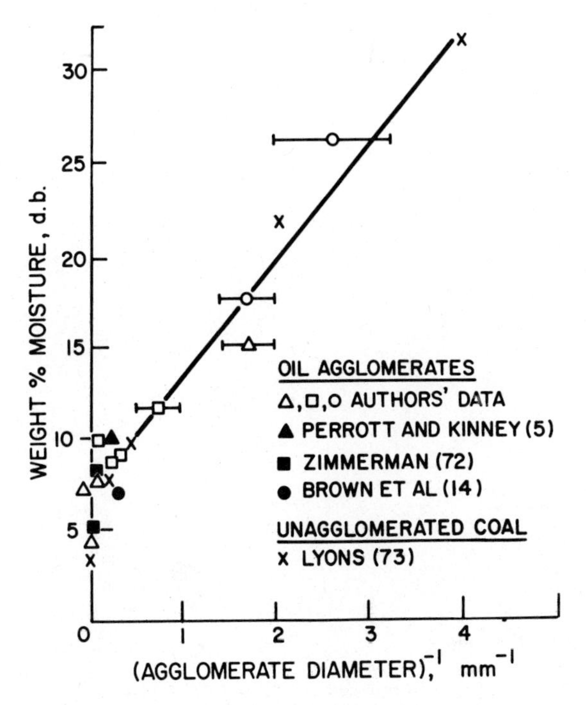

Figure 6.11 Agglomerate moisture content as a function of diameter. (From Ref. 35.)

knowledge of the oil consumption. This may be accomplished by modifying the theoretical models of Sherrington [36] and Butensky and Hyman [37]. In these models it is assumed that the bridging liquid (that is, the oil in the case of coal agglomerates) is withdrawn by capillarity to a depth td/2 into the interstices between the surface particles of the agglomerates, where t is a "withdrawal parameter" and d is the diameter of the particles. Then the amount of bridging liquid required to form the agglomerates is somewhat less than that to fill the voids within the agglomerates. The amount of this deficit is related to the agglomerate surface area through the reciprocal of agglomerate diameter. Equation (6.2) then results

$$M = \frac{k\rho_1}{\rho_s}\left(1 - 3K\,\frac{td}{D}\right) \tag{6.2}$$

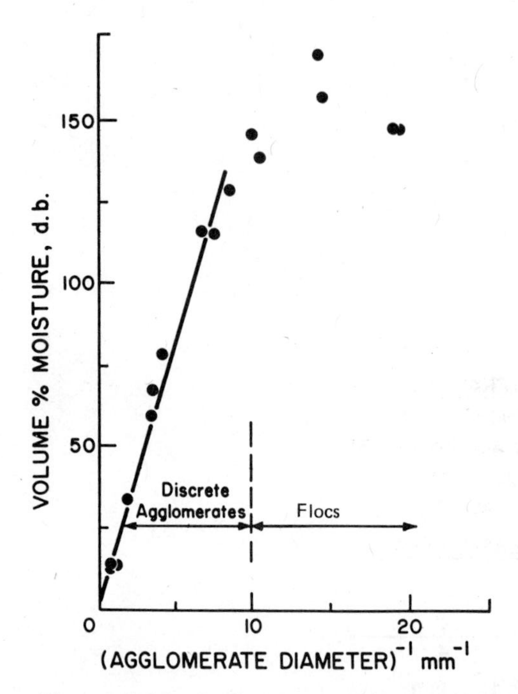

Figure 6.12 Agglomerate moisture content as a function of diameter [33]. (Bituminous feed coal, —100 mesh, 17 wt % ash, collected with paraffinic oil.)

where

M = weight fraction of bridging liquid on a dry solids basis
k = ratio of void volume to solid volume in an infinitely large agglomerate
ρ_1 = bridging liquid density
ρ_s = solid density
D = agglomerate diameter
K = agglomerate shape factor

If Eqs. (6.1) and (6.2) are combined with the elimination of D, a linear reduction in the moisture content W is predicted as the oil concentration M is increased. Figure 6.10 shows that such a relationship is indeed observed experimentally when discrete coal agglomerates are formed. This approach cannot be applied, however, to very small flocculated agglomerates, nor will it hold for products with high oil concentrations. Use of Eqs. (6.1) and (6.2) requires knowledge of several parameters which must be determined experimentally for the particular agitation and dewatering system involved.

C. High Oil Concentrations

The transition from intermediate to high oil concentrations occurs when there is sufficient oil to saturate completely the internal voids and surface pores of the agglomerates. At this point, a paste or amalgam of coal and oil forms, as in the early Trent process [5]. The agglomerates become cohesive and begin sticking to one another. At very high concentrations, the agglomerates break down and disperse in the oil to form a coal-in-oil slurry.

The moisture content of the agglomerates reaches a minimum at the transition concentration, which is usually in the range of 30 to 50% by weight of the coal (see Fig. 6.13). With higher oil

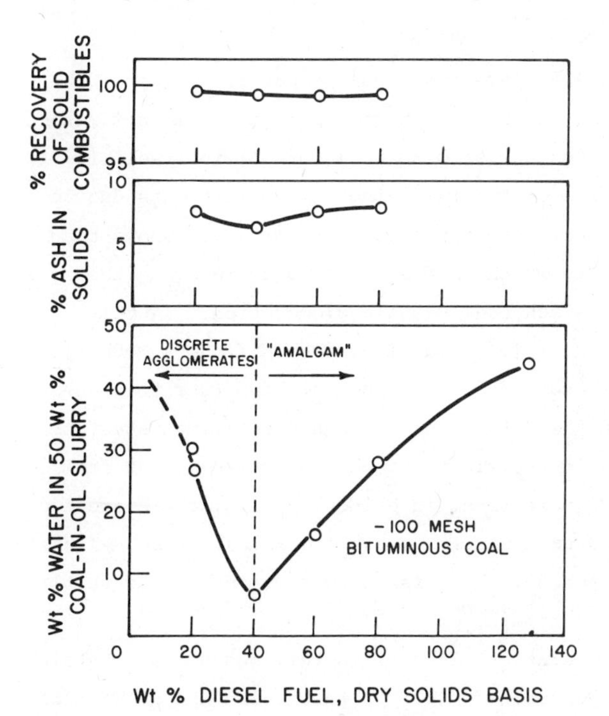

Figure 6.13 Displacement of water from coal-in-water suspension by agglomeration with hydrocarbon and screening. (Products subsequently resuspended as 50 wt % coal-in-oil slurry.) (From Ref. 70.)

concentrations, the moisture content increases as more and more water is trapped in the continuous oil network of the coal-oil amalgam. Thus, the use of excessively high oil concentrations should be avoided to minimize product moisture. The Trent process [5] provided a kneading action to work trapped moisture out of the coal-oil amalgam by mechanical means.

6.5 TREATMENT OF TAILINGS

It is not sufficient today to plan a beneficiation process simply to recover the coal; the disposal of reject tailings must also be considered. This must be done to prevent pollution problems due to fine nonsettling clays and to permit recycling of water in a closed system.

A. Settling of Tailings

Considering the fine waste clays from coal preparation plants, the removal by agglomeration of the carbon constituents from this material (which may be as high as 70 to 80 wt % carbon) in itself increases the settling rate of the remaining ash constituents due to the reduction in the slurry concentration. Figure 6.14 shows the increase in settling rate which can be realized by removing the coal particles from a 30 wt % ash coal using agglomeration. In this figure, the Richardson-Zaki [38] equation, as modified by Michaels and Bolger [39] and by Scott [40] relating settling rate to concentration, was fitted to the data. The increased settling rate, due simply to the decrease in the slurry concentration resulting from the removal of the coal particles, is evident. Thus, there are two benefits in treating waste slurries by agglomeration: first, valuable coal is recovered and second, there is an improvement in the settling rate of the solids which remain.

In an existing preparation plant, if a tailings slurry containing 70 wt % coal is agglomerated prior to thickening and 90 wt % of the coal is recovered, then the solids feed to the thickener and tailings pond decreases by 63 wt %. Thus, not only is settling rate

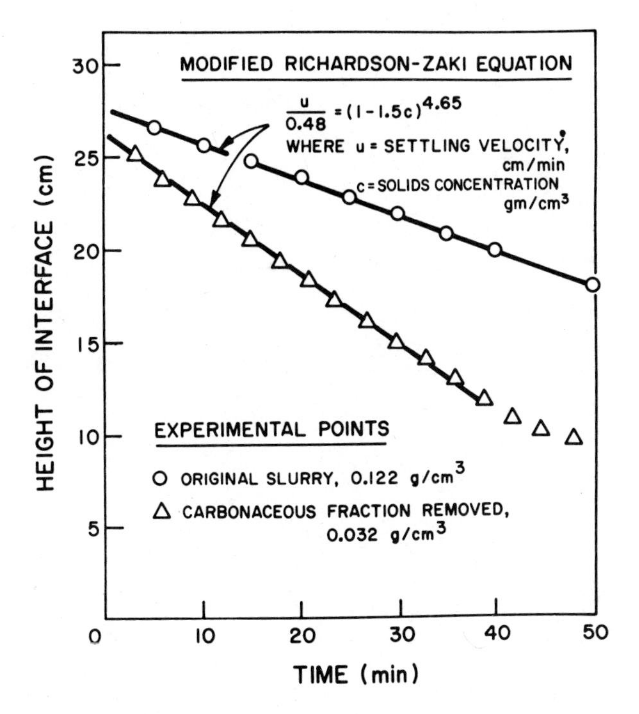

Figure 6.14 Settling rates of waste coal slurry before and after removal of carbonaceous fraction by oil agglomeration. (−200-mesh bituminous coal, 30 wt % ash initially; agglomerates formed with Stoddard solvent bridging liquid contained 6 wt % ash.) (From Ref. 70.)

improved, but tailings pond life almost triples. Agglomeration of coal in existing tailings ponds recovers lost coal values, extends pond life and in some cases, may eliminate the need for new ponds.

In Sec. 6.3B, it was shown that the more dense and viscous oils containing a variety of functional groups are less selective in recovering carbonaceous particles and rejecting inorganics. Traces of some oils are apparently able to adsorb on the ash constituents and render them less hydrophilic. These traces of adsorbed oil also influence the settling rate of the tailings. The results in Table 6.12 were obtained by measuring the maximum settling velocity of tailings produced from coal which was agglomerated with different oils. Obviously, coal tar and the other complex oils significantly

Table 6.12 Influence of Oil Type on Tailings Settling Rate[a]

Bridging liquid		Maximum tailings settling velocity, cm/min	Tailings slurry concentration, $g/cm^3 \times 10^{-2}$	Ash in tailings, wt %	Ash in agglomerates, wt %
Type	Concentration cm^3/g, %				
50% coal tar 50% xylene	33.3	12.4	1.8	47.2	11.3
50% maltenes 50% xylene	34.8	3.6	2.9	51.0	6.5
30% asphaltenes 70% xylene	33.3	2.5	1.5	76.7	5.9
50% 350°C coal tar 50% xylene	32.3	1.4	2.0	66.3	5.5
Leduc crude	30.8	1.0	1.9	79.2	4.9
Xylene	31.9	0.7	1.5	87.0	5.9
Stoddard solvent	30.1	0.5	1.7	87.2	5.6

[a]Feed coal: 20 wt % ash bituminous; settling velocities measured in a 500-cm^3 graduated cylinder.

increased the tailings settling rate. The faster settling rate of the tailings produced by the heavier oils can be accounted for in terms of their poor selectivity. Poor selectivity for carbon over inorganics means that the beneficiated coal has a higher ash content and that the inorganics in the tailings will be flocculated by traces of adsorbed oil. In addition, the higher concentration of coal in the tailings is readily flocculated by oil and, hence, settles rapidly. Visually, the tailings flocs produced by agglomeration with coal tar in the experiments reported in Table 6.12 were much larger than the flocs produced by agglomeration with xylene, Stoddard solvent, or Leduc crude.

B. Ash Agglomeration: The Reverse Process

A novel approach to the avoidance of waste slurries of fine tailings has been developed on a laboratory scale [41]. In what may be termed the *reverse* process, the inorganic ash-forming material is removed from a hydrocarbon suspension of feed coal as discrete agglomerates, while the cleaned coal remains in oil suspension. This oil slurry is suitable for pipeline transportation, for use as a coal-in-oil fuel or as feed to advanced coal conversion processes. Clean coal could also be recovered from the oil by filtration or by agglomeration using tannic acid conditioner [42]. Conventional coal agglomeration (the *forward* process) is compared with the reverse process in Fig. 6.15.

The feed to the reverse system consists of a slurry of fine coal in oil. As with the conventional process, the first agglomeration stage occurs under the influence of intensive mixing. In the reverse system, however, the shearing is provided to help the aqueous bridging liquid (water) displace the oil from the ash surfaces. The amount of aqueous bridging liquid added in this stage will depend upon the moisture content of the feed slurry. For example, if the coal were very wet prior to being mixed with the oil, then no additional bridging liquid would be required. If, on the other hand, the coal were dry, then all of the necessary bridging liquid would have to be added. Also, if dry adsorbents are used in the tumbling

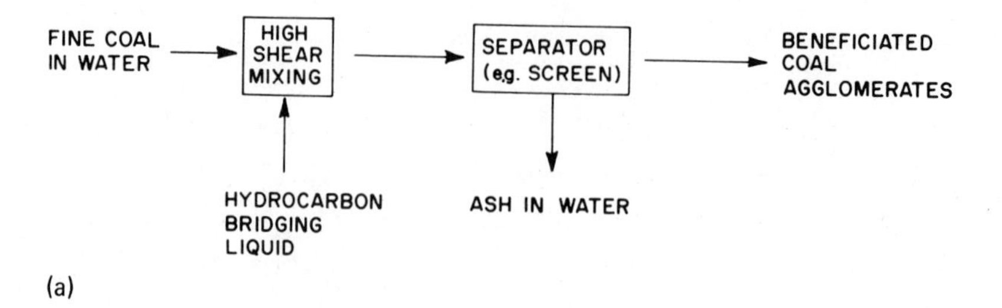

Figure 6.15 Coal agglomeration processes: (a) conventional; (b) reverse. (From Refs. 70 and 71.)

stage, then the concentration of bridging liquid must be increased accordingly. Excessive bridging liquid concentrations can be compensated for by increasing the load of adsorbents in the tumbling stage. Some bench-scale test results relating beneficiated coal ash content to bridging liquid concentration are presented in Fig. 6.16.

As seen in Fig. 6.15, coarse inert adsorbents are used to "seed" the reverse agglomeration process. In a continuous operation, these adsorbents would be recycled ash agglomerates, a portion of which would have to be discarded as their volume increased during ash agglomeration. The adsorbents have a twofold effect. Since they are much larger than the fine ash particles, their surfaces have a much lower curvature and readily pick up the ash particles by capillary adhesion. In addition, since coal particles normally form the bulk of solids in the feed slurry, the chances of water-ash and ash-ash contact are poor. The coal particles tend to mask the water and

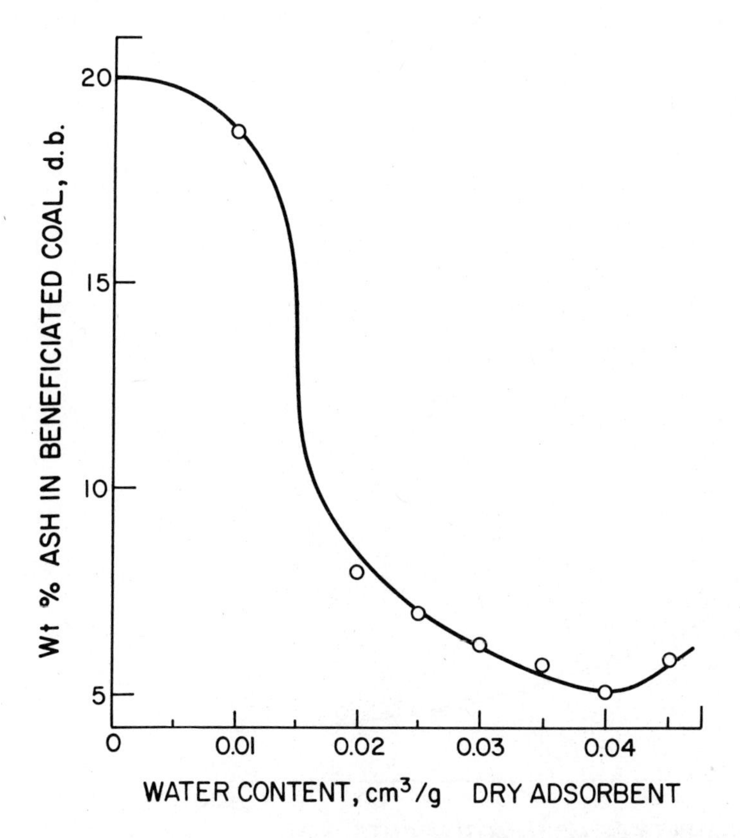

Figure 6.16 The influence of aqueous bridging liquid volume on ash reduction in the reverse process (50 g of 20 wt % ash bituminous coal suspended in hexane and tumbled with 200 g dry 10-mesh silica adsorbents). (From Ref. 41.)

ash from one another. By adding adsorbents, the total hydrophilic surface area is increased and so are the chances for favorable contact. Figure 6.17 shows the effect of increasing adsorbent loading on ash content. Another method of increasing contacts is to increase tumbling time. Higher adsorbent loadings require shorter tumbling times to achieve acceptable ash reduction. In practice, the economics of long tumbling times would be balanced with that of handling large volumes of adsorbents. In some instances, high adsorbent loadings might be required to compensate for excess water present in the feed slurry.

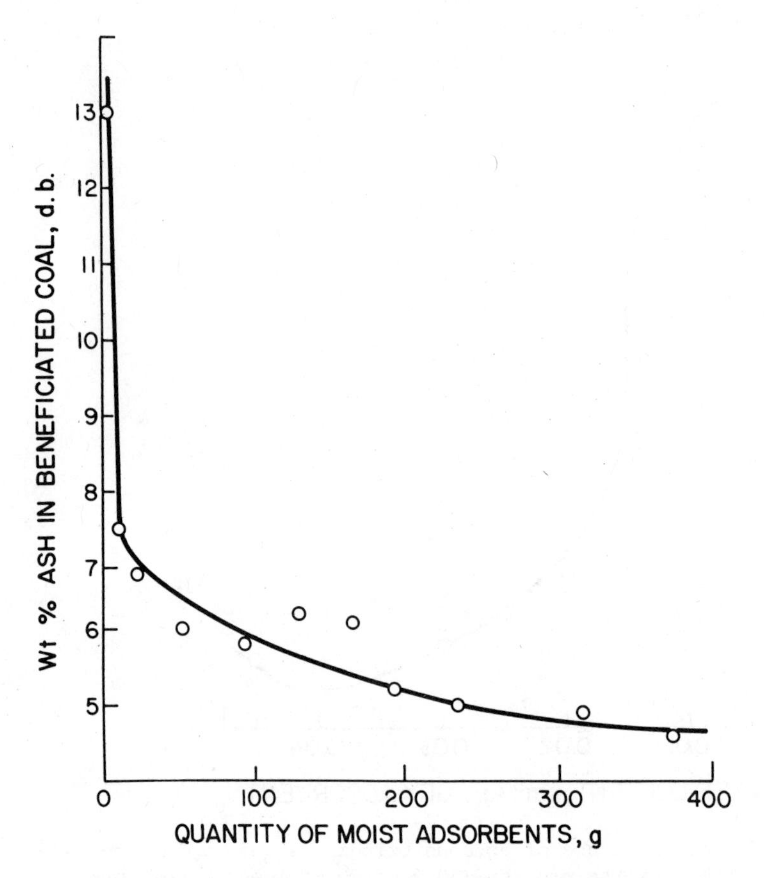

Figure 6.17 The influence of adsorbent loading on ash reduction in the reverse process (50 g of 20 wt % ash bituminous coal suspended in hexane and tumbled with moist 7-mesh ash-coated gravel adsorbents). (From Ref. 41.)

The dewatering and ash-reducing potential of reverse agglomeration is further indicated in Table 6.13. The test results were obtained from a combined forward-reverse agglomeration process. In such a process, the coal is first beneficiated by conventional oil agglomeration. Only one agglomeration stage is required in which the optimum concentration of light oil is used to produce low-ash, low-moisture agglomerates. These agglomerates are then redispersed, this time in a hydrocarbon. The coal-in-oil slurry is then tumbled with inert adsorbents. As Table 6.13 indicates, some ash reduction

Table 6.13 Combined Forward and Reverse Processes[a]

Run no.	Wt % ash (dry basis)		Wt % moisture (wet basis)		% Recovery of combustibles	
	Forward	Combined	Forward	Combined	Forward	Combined
1	4.5	3.3	22.3	6.2	96.4	94.7
2	4.7	3.3	20.8	5.8	96.1	94.7
3	4.7	3.4	19.5	7.0	95.8	94.6

[a]Test conditions: forward feed: 20 wt % ash, 95 wt % −200-mesh particles; 30 cm^3 of No. 2 fuel oil used to treat 50 g of feed suspended in 450 cm^3 of water. Forward concentrate suspended in 300 cm^3 of hexane and tumbled with 100 g of 10-mesh silica to produce combined concentrate.

Source: Refs. 70 and 71.

occurs and the overall recovery of combustibles for the combined processes is excellent. The significant effect, however, is the reduction in coal surface moisture to low levels without thermal drying. It should be noted that the high internal moistures of lignite and subbituminous coals cannot be reduced by this technique. In fact, the internal moisture and hydrophilic surface characteristics of these coals make their beneficiation by the reverse system difficult. Without special surface conditioning, recoveries of lower-rank coals are poor.

In a continuous forward-reverse process, the recycled adsorbents would soon become saturated with water. Course adsorbents can be mechanically dewatered, with the water being recycled to the forward process. This would also recycle any oil trapped with the adsorbents. Alternatively, the adsorbents could be thermally dried. This is more efficient than heating fine coal since the coarse adsorbents dry more easily. By not heating the fine coal, the chances of coal oxidation are greatly reduced. In addition, the dust load during the drying of adsorbents would be substantially reduced from that during the drying of fine coal. This would reduce air pollution and the costs associated with preventing it.

6.6 POTENTIAL APPLICATIONS

It is evident that oil agglomeration improves our ability to deal with very fine particles during coal preparation. In addition, a number of benefits are possible for the end user. These have recently been summarized [43].

A. In Power Generation

Agglomerates provide a consistent, free-flowing feed material, reducing the need for continuous adjustments during the combustion process. Low moisture and ash levels mean less inerts to ship, recover, and dispose of; less erosion in the pulverizer and the boiler would also be expected. Because there are lower levels of inerts, combustion should be more efficient with less loss of combustibles in the fly ash. The generally lower levels of inorganics in the agglomerates are useful in reducing emissions of trace metals from combustion and in reducing corrosion from, for example, alkali materials. A higher calorific value is made possible not only from the utilization of the hydrocarbon bridging agent, but also from the lower moisture content and inerts levels of the agglomerates.

Currently there is considerable interest in substituting coal-oil mixtures (COM) for oil in oil-fired furnaces. Oil agglomeration is well suited to this application since some of the fuel oil to be used in forming the COM can first be used to beneficiate the coal in the agglomeration process. The agglomerates can then be blended with additional oil to form the COM. The potential of oil agglomeration in this application is being investigated in a small power station boiler [44].

B. In Coke Making

Agglomerate size may be tailored to the traditional 80% —1/8-in. size distribution used as a feed to slot-type coke ovens. The elimination of fines reduces the problem of their release to the atmosphere or capture in the tars. In the case of the newer coking processes involving preheating and pipeline charging, agglomerates appear to be a superior feed with less hang-up and reduced chance of fires and explosions. There are indications, from carbonization tests [45] carried out in a BM/AGA apparatus and a 12-in. movable-wall coke oven, that oil agglomerating a coal charge for coke making can contribute to higher coke yields and increased ASTM coke stability factors. It is postulated that the increased coke yield is due to thermal cracking of the gases from the tar used for agglomeration in the fissures of the incandescent coke with the extent of carbon deposition increasing with increased temperature and residence time. Oil agglomeration permits the intimate blending of fine coal in a multicomponent blend conducive to uniform coke quality and, in general, higher strength properties. Lower quality coals may be upgraded to metallurgical levels by agglomeration.

A recent publication [12] notes that the coking properties of some coals may be seriously deteriorated during water slurry transportation. This deterioration was shown to result from clays present in the coal which were liberated as fine suspensions during pipelining. The clays coated the larger coal particles and resulted in weak and friable cokes. While removal of the ultrafine clay particles without loss of fine coal was impractical by conventional techniques, they were displaced from the coal particles through the selective wetting of the coal particles by the oil phase during agglomeration. Thus oil agglomeration proved to be an effective method to recover the coal after pipelining with negligible losses of coal matter and without deterioration of coking properties.

C. In the Blast Furnace

Tuyère injection of agglomerates offers the possibility of replacing coke carbon with coal carbon and substituting metallurgical coal with lower quality coals. The ease of handling and feeding of agglomerates and their abrasion resistance make them ideal for pneumatic conveying and charging. Thermal shock is expected to cause the agglomerates to explode, exposing the large surface area of the ultimate particles and improving reactivity. The low ash and moisture levels of the agglomerates are also attractive in this application.

6.7 ECONOMIC CONSIDERATIONS

A number of economic assessments [13,14,16] have been made for the oil agglomeration of coal fines. For example, based on 1972 data [46] updated to 1978, the capital and fixed costs for the 200 ton/hr plant shown in Fig. 6.18 producing a pelletized product total $3.75 per annual ton based on an 8000-hr year. If a flocculated or micro-agglomerated product is produced, using a single stage process without pelletizing, a figure of $2.50 per annual ton is estimated. In another recent analysis [13], favorable economic results were found to justify incorporation of oil agglomeration into the fines circuit of German coal preparation plants.

The operating costs for an oil agglomeration plant are very sensitive to the quantity and price of the oil used. For the plant of Fig. 6.18, to produce a pelletized product using 5 wt % agglomerating oil (No. 2 fuel oil at 5.8¢/lb) and 10 wt % pelletizing oil (Bunker C at 4.3¢/lb), the total oil cost would be $14.40/ton of feed coal. The average oil cost would be 96¢ per weight percent oil per ton of feed. To produce microagglomerates using 5 wt % agglomerating oil (No. 2), the oil cost would be $5.80/ton of feed, or $1.16 per weight percent oil per ton of feed. From these costs, it would seem that oil agglomeration is highly uneconomical; however, many situations exist wherein oil agglomeration is economical.

Obviously, if the oil cost can be reduced or eliminated the economics improve. If the coal is to be used for power generation then on the average about 1/3 to 1/2 of the oil cost will be recovered. The complete cost is not recovered because, when associated with the coal, the oil would probably be assessed at the same rate (dollars per million Btu) as the coal. If No. 2 fuel oil, at $2.97/million Btu, is used to agglomerate steam coal, at $1.20/million Btu, then a loss of $1.77 would be incurred per million Btu of oil used. Alternatively, the coal operator recovers $1.20 of the $2.97/million Btu, he paid for the oil. Thus, the oil cost of $1.16 per weight percent oil per ton of coal quoted above to produce the microagglomerated product is reduced by 40.4% to $0.69 per weight percent oil per ton of coal.

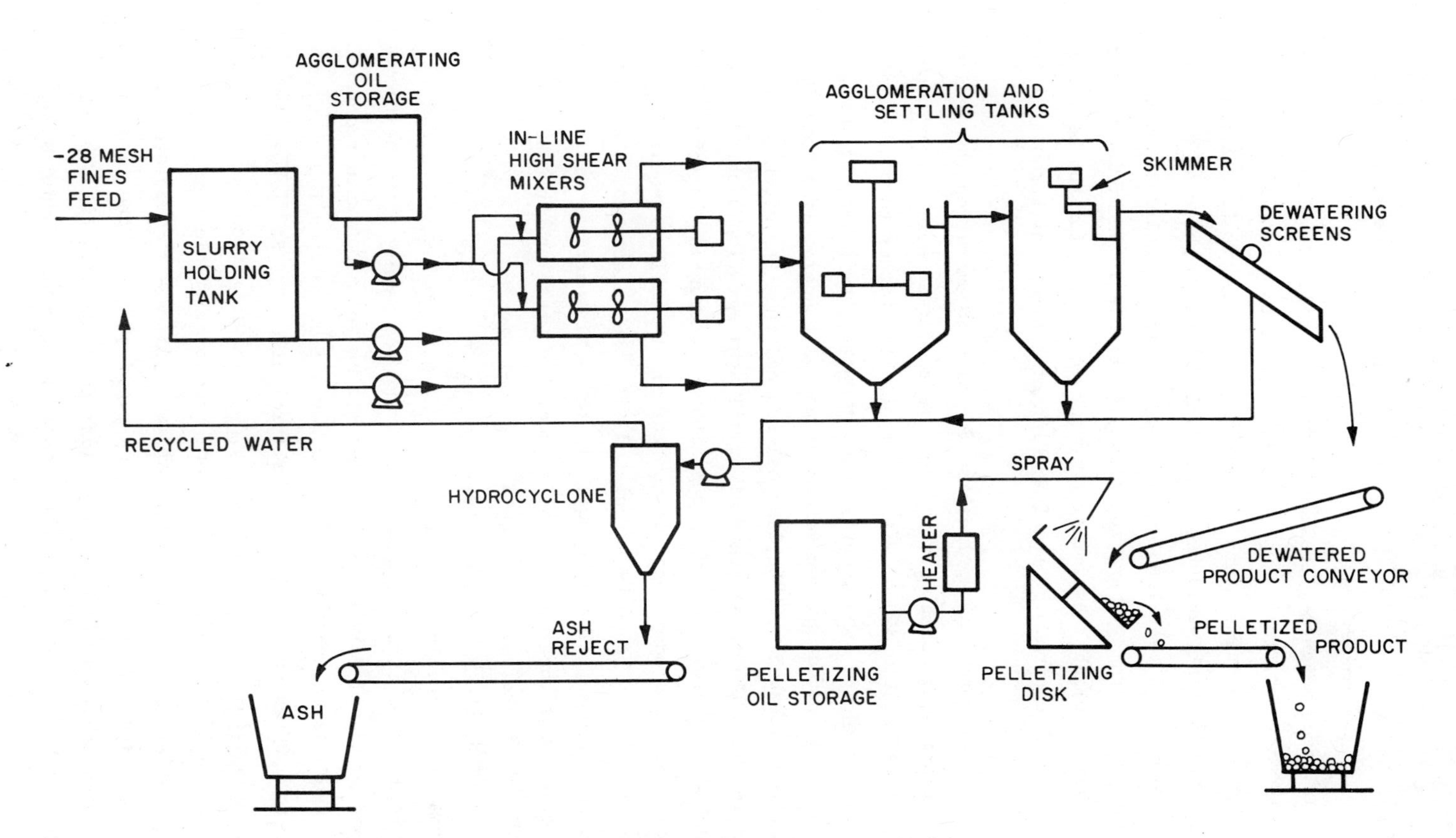

Figure 6.18 Process flow plan for economic analysis. (From Ref. 46.)

In some instances, coal is slurried in oil. In the solvent refining of coal, for example, coal is added to a process-derived solvent. Potentially, this solvent could be used to beneficiate the coal. The solvent would then be recycled back to the coal refining plant along with the clean coal. The cost of the solvent should not be assessed to the preparation plant, particularly if it is owned by the refining plant. A similar argument would hold for the beneficiation of coal upstream of the formation of coal-oil mixtures as a fuel supply.

Another potential application for oil agglomeration involves the beneficiation of coals destined for use as formed coke. Some formed-coke processes blend coal with a tar or pitch binder to produce briquettes which are subsequently coked. During the coking process, new pitch is generated and recovered for use as binder. If a formed-coke plant were located next to its coal source, then the pitch could be used in the pelletizing stage of an oil agglomeration system. There would be no charge for this pitch although, depending on the coking conditions, some additional makeup tar might still be required. There would also be costs associated with the selective light oil used in the beneficiation stage. If the coal source is distant from the coke plant the revenue from sale of the pitch at the coke plant could offset the cost of pitch at the coal source.

Perhaps the most economically favorable application of oil agglomeration involves the treatment of waste coal slurries and tailings ponds. The economic incentive comes mainly from the recovery of lost coal values. Waste coal has no value but when recovered and beneficiated, the coal could be worth from $30/ton for a steam coal to $50/ton or more for a metallurgical coal. For recovering such high value metallurgical coals, oil concentrations of 20 wt % or more could be economical. In one recent pilot-plant investigation [18], metallurgical coal from a tailings pond was beneficiated to 7 to 8 wt % ash from 40 wt % at a mass yield of 55% of the feed, using oil concentrations of 7 to 8 wt % (dry feed basis). Pilot-plant coking tests suggested that the beneficiated product would be an excellent addition to coking blends. Obviously the oil costs would be more than compensated for.

There are many benefits of oil agglomeration which enter the overall economics. The ease of dewatering large agglomerates reduces dewatering costs and can even eliminate the need for thermal drying. The handling properties of agglomerates are superior to those of unagglomerated fine coal, particularly with regard to dusting and freezing. The improved settling characteristics of the rejected tailings reduce waste-handling costs, while the reduced volume of tailings extends pond life.

Only general considerations of the economic implications of oil agglomeration can be given here. Since a variety of applications are possible, each specific application must be considered with its unique set of circumstances.

6.8 CONCLUSIONS

The most important feature of oil agglomeration as applied to wet coal preparation is its ability to recover virtually quantitatively the finest coal particles encountered in plant fines circuits and waste streams. Inorganic impurities and moisture are simultaneously rejected from the agglomerated product.

Selective oil agglomeration can be carried out over a broad range of conditions. For a given feed material, product ash and combustible recovery are essentially constant from oil concentrations of 5 or 10% to 30 or 40% by weight of solids. Pulp densities from very low levels to as high as 40 or 50 wt % solids can be treated. Intensive agitation over short periods of time (less than 1 min) or less intensive mixing over prolonged periods (e.g., 15 min) can be used.

Coal and oil properties are important system variables in selective agglomeration. The ash level of agglomerates attained with a given coal depends on the extent of liberation of the impurities and hence on the fineness of ash dissemination and the level of grinding. Grinds down to a few micrometers in diameter can be agglomerated successfully, if desirable. With bituminous coals, many different oils yield excellent coal recovery levels although the refined, light oils give the best ash rejection. Lower-rank coals require special oils for agglomeration because of their more hydrophilic character.

Moisture content in the agglomerates depends on the quantity of oil used. As larger and more compact agglomerates are produced at higher oil levels, both the internal and surface moisture held in the product is reduced. Excessive amounts of oil, however, may trap relatively large amounts of moisture in the resulting coal-oil amalgam.

Waste tailings rejected in the oil agglomeration process have good settling characteristics. Reduced waste fines loadings resulting from the technique extend tailings pond life. The treatment of waste fine coal slurries and recovery of coal from tailings ponds are probably the most attractive applications for selective oil agglomeration at the present time.

ACKNOWLEDGMENT

Many individuals have contributed to the development of oil agglomeration for fine coals. Their names are generally found in the references and the authors thank them all. We particularly acknowledge the cooperation, encouragement and advice of I. E. Puddington, A. F. Sirianni, and the late A. E. McIlhinney of the National Research Council of Canada, of L. Messer of the American Minechem Corporation and of J. H. Walsh and B. J. P. Whalley of the Department of Energy, Mines, and Resources of Canada.

REFERENCES

1. J. Gregory, Stability and Flocculation of Colloidal Particles, *Effluent Water Treat. J., 17,* 641-651 (1977).
2. M. Yusa and A. M. Gaudin, Formation of Pellet-like Flocs of Kaolinite by Polymer Chains, *Bull. Amer. Ceram. Soc., 43,* 402-406 (1964).

3. A. F. Sirianni and I. E. Puddington, Novel Preparation of Finely Divided Colloidal Silica, *Can. J. Chem. Eng., 42,* 42-43 (1964).

4. A. M. Gaudin, *Flotation*, McGraw-Hill, New York, 1957, Chap. 1.

5. G. St. J. Perrott and S. P. Kinney, Use of Oil in Cleaning Coal, *Chem. Met. Eng., 25*(5), 182-188 (1921).

6. A. H. Brisse and W. L. McMorris, Jr., Convertol Process--Efficient Method Removes Usable Coal from High Ash Slurries, *Mining Eng., 10,* 258-261 (1958).

7. H. M. Smith and I. E. Puddington, Spherical Agglomeration of Barium Sulphate, *Can. J. Chem., 38,* 1911-1916 (1960).

8. L. Messer, *LoMAg Technology: Low Moisture Agglomerates and Recycle Water from Coal Slurries,* American Minechem Corporation, Coraopolis, Pa., 1977.

9. R. H. Shubert, Method of Treatment of Coal Washery Waters, U.S. Patent 3,856,668, Dec. 24, 1974.

10. S. Min and T. D. Wheelock, A Comparison of Coal Beneficiation Methods, in *Coal Desulfurization: Chemical and Physical Methods, ACS Symposium Series,* 64 (T. D. Wheelock, ed.), American Chemical Society, Washington, D.C., 1977, pp. 83-100.

11. J. Szczypa, J. Neczaj-Hruzewicz, Z. Skimina, and Z. Koszman, Spherical Agglomeration--A Technique for Dewatering and Upgrading Coal Slurries, *Metody Fizkochem. Oczyszania Wod Sciekow, Ref. Konf. Nauk. Tech., 1,* 85-100 (1976) (in Polish).

12. G. R. Rigby and T. G. Callcott, Prospects for the Slurry Transportation of Australian Coking Coals, *Chemeca 77, 5th Australian Conf. Chem. Eng.,* 282-288 (1977).

13. W. Blankmeister, B. Bogenschneider, K. H. Kubitza, D. Leininger, L. Angerstein, and R. Kohling, Optimized Dewatering Below 10 mm, paper A1, *Proc. 7th Int. Coal Prep. Congress,* Sydney, Australia, May 1976.

14. R. B. Brown, J. H. Brookman, and C. G. Haupt, A Continuous Process for Agglomeration and Separation, *Proc. Inst. Briquet. Agglom., 11,* 61-69 (1969).

15. G. G. Sarkar, B. B. Konar, S. Sakha, and A. K. Sinha, Demineralization of Coals by Oil-Agglomeration," paper H3, *Proc. 7th Int. Coal Prep. Congress,* Sydney, Australia, May 1976.

16. C. N. Bensley, A. R. Swanson, and S. K. Nicol, The Effect of Emulsification on the Selective Agglomeration of Fine Coal, *Int. J. Miner. Process, 4,* 173-184 (1977).

17. C. E. Capes, A. E. McIlhinney, D. S. Russell, and A. F. Sirianni, Rejection of Trace Metals from Coal During Beneficiation by Agglomeration, *Environ. Sci. Technol., 8,* 35-38 (1974).

18. S. K. Nicol and A. R. Swanson, Selective Agglomeration in the Treatment of Fine Coal Refuse, *Aust. Min., 69*(2), 42-43 (1977).

19. K. Lemke, The Cleaning and Dewatering of Slurries by the Convertol Process, *Second Int. Coal Prep. Congress*, Essen, Germany, Sept. 1954.

20. R. E. McKeever and A. W. Deurbrouck, Coal Preparation and Analysis Laboratory, U.S. Department of Energy, Pittsburgh, Pa., unpublished data, 1975.

21. R. E. Zimmerman, Froth Flotation, in *Coal Preparation* (J. W. Leonard and D. R. Mitchell, eds.), 3d ed., AIME, New York, 1968, pp. 10-73.

22. F. J. Zuiderweg and N. van Lookeren Campagne, Pelletizing of Soot in Waste Water of Oil Gasification Plants--The Shell Pelletizing Separator, *Chem. Eng. (London), 220,* CE 223-CE 227 (July/August, 1968).

23. W. W. Batley, The Trent Process for Cleaning Coal, *Fuel, 2,* 236-41 (1923).

24. S. C. Sun and W. L. McMorris, Factors Affecting the Cleaning of Fine Coals by the Convertol Process, *Mining Eng., 11,* 1151-1156 (1959).

25. R. J. Germain, Coal Preparation by Agglomeration, unpublished report, National Research Council of Canada, Apr. 17, 1975.

26. J. B. Hills and H. B. Coats, The Viscosity-Gravity Constant of Petroleum Lubricating Oils, *Ind. Eng. Chem., 20,* 641-644 (1928).

27. C. L. Mantell, *Adsorption*, 2d ed., McGraw-Hill, New York, 1951.

28. R. T. Greer, Coal Microstructure and Pyrite Distribution, in *Coal Desulfurization: Chemical and Physical Methods, ACS Symposium Series, 64* (T. D. Wheelock, ed.), American Chemical Society, Washington, D.C., 1977, pp. 3-15.

29. J. W. Leonard and D. R. Mitchell (eds.), *Coal Preparation,* 3d ed., AIME, New York, 1968.

30. C. E. Capes, A. E. McIlhinney, A. F. Sirianni, and I. E. Puddington, Bacterial Oxidation in Upgrading Pyritic Coals, *Can. Inst. Mining Met. Bull., 66*(739), 88-91 (1973).

31. A. E. McIlhinney, A. F. Sirianni, C. E. Capes, and I. E. Puddington, Bacterial Oxidation in Upgrading Sulfidic Ores and Coals, U.S. Patent 3,796,308, Mar. 12, 1974.

32. C. E. Capes, A. E. McIlhinney, D. S. Russell, and A. F. Sirianni, Rejection of Trace Metals from Coal During Beneficiation by Agglomeration, *Environ. Sci. Technol., 8,* 35-38 (1974).

33. R. J. Germain, Suspending Liquid Rejection from Agglomerates Produced by Spherical Agglomeration, M.A. Sc. thesis, University of Waterloo, Ontario, Canada, 1977.

34. C. E. Capes, A. E. McIlhinney, and R. D. Coleman, Beneficiation and Balling of Coal, *Trans. AIME, 247,* 233-237 (1970).

35. C. E. Capes, A. E. McIlhinney, A. F. Sirianni, and I. E. Puddington, Agglomeration in Coal Preparation, *Proc. Inst. Briquet. Agglom., 12,* 53-65 (1971).

36. P. J. Sherrington, The Granulation of Sand as an Aid to Understanding Fertilizer Granulation, *Chem. Eng. (London), 220,* CE 201-215 (1968).

37. M. Butensky and D. Hyman, Rotary Drum Granulation. An Experimental Study of the Factors Affecting Granule Size, *Ind. Eng. Chem. Fund., 10,* 212-219 (1971).

38. J. F. Richardson and W. N. Zaki, Sedimentation and Fluidization, *Trans. Instn. Chem. Eng., 32,* 35-53 (1954).

39. A. S. Michaels and J. C. Bolger, Settling Rates and Sediment Volumes of Flocculated Kaolin Suspensions, *Ind. Eng. Chem. Fund., 1,* 24-33 (1962).

40. K. J. Scott, Thickening of Calcium Carbonate Slurries, *Ind. Eng. Chem. Fund., 7,* 484-490 (1968).

41. C. E. Capes, R. J. Germain, A. E. McIlhinney, I. E. Puddington, and A. F. Sirianni, Method of Separating Inorganic Material from Coal, U.S. Patent 4,033,729, July 5, 1977.

42. C. E. Capes, A. E. Smith, and A. E. McIlhinney, Method of Displacing Liquid Suspendant of a Particulate Material Liquid Suspendant Mixture by Micro Agglomeration, Canadian Patent 1,020,880, Nov. 15, 1977.

43. C. E. Capes, A. E. McIlhinney, R. E. McKeever, and L. Messer, Application of Spherical Agglomeration to Coal Preparation, paper H2, *Proc. 7th Int. Coal Prep. Congress,* Sydney, Australia, May 1976.

44. H. Whaley, The Canadian Coal-Oil Mixture (COM) Program, *Eng. Digest,* 15-20 (June, 1980).

45. J. H. Walsh, B. J. P. Whalley, and J. C. Botham, Upgrading Coking Coals and Coke Production, U.S. Patent 3,637,464, Jan. 25, 1972.

46. C. E. Capes, A. E. Smith, and I. E. Puddington, Economic Assessment of the Application of Oil Agglomeration to Coal Preparation, *Can. Inst. Mining Met. Bull., 67*(747), 115-19 (1974).

47. A. F. Sirianni, G. Paillard, and I. E. Puddington, Separation of Wet-Process Phosphoric Acid, *Can. J. Chem. Eng., 47,* 210-211 (1969).

48. A. L. Mular and I. E. Puddington, A Technically Feasible Agglomeration-Separation Process, *Can. Inst. Mining Met. Bull., 61* (674), 726-30 (1968).

49. A. L. Mular and I. E. Puddington, Pilot Plant Studies of Iron Ore Agglomeration, *Proc. Inst. Briquet. Agglom., 11,* 71-84 (1969).

50. F. D. DeVaney, Agglomeration Tabling, in *Chemical Engineers' Handbook* (J. H. Perry, ed.), 4th ed., McGraw-Hill, New York, 1963.

51. M. M. J. Ruel and A. F. Sirianni, Agglomeration and Extraction of Peat Moss, U.S. Patent 3,844,759, Oct. 29, 1974.

52. A. F. Sirianni and I. E. Puddington, Laundered Amorphous Reinforcing Lignin, U.S. Patent 3,817,974, June 18, 1974.

53. F. W. Meadus and I. E. Puddington, The Beneficiation of Barite by Agglomeration, *Can. Inst. Mining Met. Bull., 66*(734), 123-126 (1973).

54. I. E. Puddington and J. R. Farnand, Oil Phase Agglomeration, U.S. Patent 3,399,765, Sept. 3, 1968.

55. J. R. Farnand and I. E. Puddington, Oil-Phase Agglomeration of Germanium Bearing Vitrain Coal in a Shaly Sandstone Deposit, *Can. Inst. Mining Met. Bull., 62*(683), 267-271 (1969).

56. I. E. Puddington, H. M. Smith, and J. R. Farnand, Process for Separation of Solids by Agglomeration, U.S. Patent 3,268,071, Aug. 23, 1966.

57. J. R. Farnand, F. W. Meadus, E. C. Goodhue, and I. E. Puddington, The Beneficiation of Gold Ore by Oil-Phase Agglomeration, *Can. Inst. Mining Met. Bull., 62*(692), 1326-1329 (1969).

58. J. R. Farnand, H. M. Smith, and I. E. Puddington, Spherical Agglomeration of Solids in Liquid Suspension, *Can. J. Chem. Eng., 39,* 94-97 (1961).

59. B. D. Sparks and R. H. T. Wong, Selective Spherical Agglomeration of Ilmenite Concentrates, *Can. Inst. Mining Met. Bull., 66*(729), 73-77 (1973).

60. A. F. Sirianni, R. D. Coleman, E. C. Goodhue, and I. E. Puddington, Separation Studies of Iron Ore Bodies Containing Apatite by Spherical Agglomeration Methods, *Trans. Can. Inst. Mining Met., 71,* 149-153 (1968).

61. B. D. Sparks and A. F. Sirianni, Beneficiation of a Phosphoriferous Iron Ore by Agglomeration Methods, *Int. J. Miner. Process., 1,* 231-241 (1974).

62. B. D. Sparks and F. W. Meadus, The Separation of Silica from Sturgeon Lake Marl Deposits, *Can. Inst. Mining Met. Bull., 67*(747), 111-114 (1974).

63. C. E. Capes, J. P. Sutherland, and A. E. McIlhinney, Spherical Agglomeration Process, U.S. Patent 3,471,267, Oct. 7, 1969.

64. R. D. Coleman, J. P. Sutherland, and C. E. Capes, Reduction of the Calcite Content of Ground Shale by Liquid-Liquid Particle Transfer, *J. Appl. Chem., 17,* 89-90 (1967).

65. B. D. Sparks, F. W. Meadus, and I. E. Puddington, The Continuous Separation of Tar Sands by Oil-Phase Agglomeration, *Can. Inst. Mining Met. Bull., 64*(710), 67-72 (1971).

66. B. D. Sparks and F. W. Meadus, A Combined Solvent Extraction and Agglomeration Technique for the Recovery of Bitumen from Tar Sands, presented at 27th Canadian Chemical Engineering Conference, Calgary, Alberta, Oct. 23-27, 1977.

67. W. Campbell, Oil from Alberta's Tar Sands, *Science Dimension, 8*(1), 10-15 (1976) (National Research Council, Ottawa).

68. J. R. Farnand, F. W. Meadus, P. Tymchuk, and I. E. Puddington, The Application of Spherical Agglomeration to the Fractionation of a Tin-Containing Ore, *Can. Metal. Q., 3,* 123-135 (1964).

69. F. W. Meadus, A. Mykytiuk, I. E. Puddington, and W. D. MacLeod, The Upgrading of Tin Ore by Continuous Agglomeration, *Trans. Can. Inst. Mining Met., 69,* 303-305 (1966).

70. C. E. Capes, Basic Research in Particle Technology and Some Novel Applications, *Can. J. Chem. Eng., 54,* 3-12 (1976).

71. C. E. Capes, A. E. McIlhinney, and A. F. Sirianni, Agglomeration from Liquid Suspension--Research and Applications, in *Agglomeration 77* (K. V. S. Sastry, ed.), AIME, New York, 1977, pp. 910-930.

72. R. E. Zimmerman, Flotation of Bituminous Coal, *Amer. Inst. Mining Met. Eng., Coal Technol., 3*(2), Tech. Pub. No. 2397 (1948).

73. O. R. Lyons, Filter-cake Particle Size and Moisture Relationships, *Mining Eng., 3,* 868-870 (1951).

7

Development and Demonstration of Selected Fine Coal Beneficiation Methods

THOMAS D. WHEELOCK

Iowa State University
Ames, Iowa

7.1 INTRODUCTION

The development of improved coal beneficiation methods has been underway at Iowa State University (ISU) since 1974 when the Iowa Coal Project was established with funds provided by the Iowa Legislature and administered by the Energy and Mineral Resources Research Institute. The general purpose of the project is to solve important problems associated with mining and utilizing Iowa coal. Most of this material is classified as a high-volatile bituminous coal with high sulfur and ash contents. Much of the sulfur is present as iron pyrites, and it has been shown that an appreciable fraction of this material is present as microscopic single crystals and framboids which are finely disseminated throughout the coal [1].

When the Iowa Coal Project was launched, a dual attack was made on the high sulfur content of Iowa coal. One approach was to build a coal preparation plant to evaluate the performance and economics of selected coal cleaning methods when applied to Iowa coal. By necessity, the first methods selected for evaluation and demonstration were those in commercial use which seemed to have the lowest cost-benefit ratio. But these were gravity separation methods which removed only coarse refuse. Since Iowa coal also contains finely disseminated pyrites, a second approach was to screen a number of promising but largely undeveloped methods for removing such crystallites, select several methods for further development, and then proceed to develop these methods. The separation methods selected for further development include those based on froth flotation, oil agglomeration, hydrocyclones, heavy-media cyclones, and high-gradient magnetic separation. All of these methods are designed for cleaning fine-size coal and this is the size which must be cleaned if finely disseminated pyrites are to be liberated and removed.

Since funds and human resources are never unlimited, the development of these methods has been uneven. While methods based on froth flotation and oil agglomeration have received extensive testing and laboratory development and are being demonstrated in larger facilities at the Ames coal preparation plant, the other methods have received limited attention. However, hydrocyclones and heavy-media cyclones which do not lend themselves to small-scale systems are being incorporated into the Ames coal preparation plant and will be evaluated on a pilot-plant or demonstration-plant scale.

To improve the separation of coal and pyrites achieved by either froth flotation or oil agglomeration, research has focused on chemical pretreatment of coal fines to enhance the difference in surface properties of the two components. Also, attention has been given to the optimization of process conditions. In addition, various combinations of gravity separation, froth flotation, oil agglomeration, and comminution methods have been tested to see to what extent these methods complement each other.

Although initially this effort was devoted largely to Iowa coal, it has been broadened gradually to encompass coals from other regions with an influx of support from the U.S. Department of Energy through the Ames Laboratory.

The results of the laboratory development and application of both the oil agglomeration and froth flotation methods of cleaning coal and of a chemical pretreatment step to improve the effectiveness of these cleaning methods are reviewed below, as well as the results of applying various combinations of these methods and gravity separation to the cleaning of high-sulfur Iowa coals. In addition, the experimental facilities at the ISU coal preparation plant which are designed to demonstrate major improvements in methods for cleaning fine-size coal are described and problems encountered in the start-up of these facilities are discussed.

7.2 CHEMICAL PRETREATMENT AND PHYSICAL CLEANING

Both the oil agglomeration [2] and froth flotation [3] methods take advantage of the difference in surface properties of coal and inorganic mineral particles to effect a separation of these components in an aqueous suspension. In the first method, the hydrophobic coal particles are selectively coated and agglomerated by the addition of a small amount of fuel oil to the suspension, whereas the hydrophilic mineral particles are not coated by oil or agglomerated. The relatively large agglomerates can be recovered by screening the suspension. In the second method, numerous small air bubbles are generated within the suspension; some of the bubbles become attached to the hydrophobic coal particles and buoy them to the surface where they are recovered in a froth. The hydrophilic minerals are again left behind in the aqueous suspension. Generally, a frother is added to facilitate frothing and sometimes a collector to increase the hydrophobicity of the coal.

In the past, these methods have not generally provided a good separation between coal and microscopic crystallites of iron pyrites because of the similarity in surface properties of the two materials. Thus both unweathered bituminous coal and freshly exposed (unoxidized) pyrite tend to be naturally water repellant or hydrophobic [4,5]. Therefore, both of these materials are wetted somewhat more readily by hydrocarbon oils than by water [5,6]. Consequently, both materials are agglomerated when oil is added to a well-agitated suspension of the materials in water [6]. Also, both materials tend to float readily in an aerated and agitated aqueous suspension using only a frothing agent such as methyl isobutyl carbinol (MIBC) [5,7-9].

In order to improve the separation of coal and pyrites, consideration has been given to the use of certain chemical reagents regarded as pyrite depressants which would be adsorbed selectively by the pyrite surface to render it more hydrophilic [5,7-11]. Although this is an attractive concept, it does not appear to have reached the stage of commercial practice in coal flotation [3,12] or oil agglomeration [13]. At Iowa State University, Min [14], Le [15], and Laros [16] tested a number of potential pyrite depressants in conjunction with Iowa bituminous coals and had only slight success. These coals contain appreciable amounts of finely disseminated pyrites and are floated in the presence of MIBC or agglomerated by fuel oil.

Consideration has also been given to the alteration of the pyrite surface through chemical reaction. Thus Capes et al. [13] used iron-oxidizing bacteria to oxidize the surface of the pyrites in coal and improved the subsequent separation between coal and pyrites by means of oil agglomeration. However, since bacterial oxidation is slow, the treatment requires 1 to 3 days. A much shorter, but equally effective, chemical treatment was demonstrated by Min [14]. Thus when ground coal was pretreated for a few minutes with a warm alkaline solution containing dissolved oxygen and subsequently separated by either oil agglomeration or froth flotation, significantly more sulfur was removed than was removed by either oil agglomeration or froth flotation alone. Subsequent measurements reported by Le [15] and Patterson et al. [17] showed that the floatability of pyrite is reduced by this type of chemical treatment, whereas the floatability of bituminous coal is not affected greatly by it. The reduced floatability of pyrite is in agreement with the observation of Glembotskii et al. [5] that wet oxidation forms a highly hydrated film of ferric hydroxide on the surface of pyrite which reduces the floatability of the pyrite. Furthermore these authors noted that such oxidation is promoted by an alkaline solution.

Patterson [18] further investigated and developed the wet oxidation pretreatment step in conjunction with the oil agglomeration of coal and pyrites extracted from coal. The principal results of his work are summarized below.

A. Pretreatment and Agglomeration of Pyrites

The effectiveness of wet oxidation in an alkaline solution as a means of preventing the oil agglomeration of pyrite particles was demonstrated in a series of laboratory experiments. For these experiments, pyrite nodules were extracted from the ISU demonstration mine coal and pulverized to provide a 230 mesh × 0 size consist. These particles contained 86 wt % FeS_2 and significant amounts of other iron, sulfur and calcium minerals. In some experiments 50 g of pyrite was pretreated for 15 min at 80°C with 250 ml of an aerated solution containing dissolved air and 2.0 wt % sodium carbonate. The cooled suspension was then placed in a kitchen blender which was operated for several minutes at high speed while 5 ml of fuel oil was added. Any agglomerates which formed were recovered by pouring the suspension through a 140-mesh sieve (U.S. standard). Other experiments were performed where the pyrite was subjected to oil agglomerating conditions without being chemically pretreated. Two methods of introducing fuel oil during the oil agglomeration step were also tested. In one method the oil was added directly to the agitated suspension of pyrite particles and water, while in the other method the oil was added as an oil and water emulsion. This emulsion was prepared with an ultrasonic generator.

The percentage of charged pyrite which was recovered as agglomerates is reported in Table 7.1 for various experimental conditions.

Table 7.1 Agglomerating Chemically Pretreated and Untreated 230 mesh × 0 Pyrite Particles with Different Oils

Fuel oil no.	Method of oil introduction	Chemical treatment	Recovery, %
200 LLS	Direct	None	63.8
200 LLS	Emulsified	None	91.7
200 LLS	Emulsified	Treated	8.8
5 LLS	Direct	None	46.4
5 LLS	Emulsified	None	74.9
5 LLS	Emulsified	Treated	8.0

Source: Ref. 17.

The results indicate that the untreated pyrite was readily agglomerated and recovered by either No. 200 LLS or No. 5 LLS fuel oil with the greatest recovery being obtained in either case when the oil was emulsified before use. Thus 92% of the pyrite was agglomerated and recovered with emulsified No. 200 LLS fuel oil. But when the pyrite was chemically pretreated, the recovery was only 8 or 9% with either of these oils applied as an emulsion. Hence, the pretreatment step was very effective in preventing pyrite agglomeration. These results were fairly reproducible since the absolute difference in the percentage recovery between experiments made under the same conditions was only 4 to 8%. Each value listed in Table 7.1 is an average of the recovery obtained in two separate experiments.

Chemical pretreatment of pyrite at 50°C with an aerated alkaline solution applied for either 5 or 15 min was nearly as effective in preventing agglomeration as pretreatment with such a solution at 80°C for 15 min. On the other hand, pretreatment at room temperature for 15 min was not effective. The agglomeration step was always carried out at room temperature.

B. Pretreatment and Agglomeration of Coal

To further evaluate and develop the wet oxidation method of pretreatment, numerous laboratory experiments were conducted with a variety

Table 7.2 Typical Cumulative Size Distribution of Dry Pulverized Coal

Sieve no., U.S. std.	Coal retained, wt %
100	11
140	30
200	49
270	59
400	92

Source: Ref. 18.

of high-volatile bituminous coals from Iowa, Illinois, and Western Kentucky. In these experiments the coal was ground to a specific size consist, chemically pretreated, and oil agglomerated. Two size consists of coal were employed, 60 mesh × 0 and 400 mesh × 0. The first was prepared by dry pulverizing the coal with a Mikro-Samplmill (Pulverizing Machinery Division, American-Marietta Co.); the cumulative size distribution of this material was typically as indicated in Table 7.2.

Some of the dry pulverized coal was mixed with water and further ground in a ball mill for 24 hr to prepare the second size consist. For the chemical pretreatment step, some of the pulverized or ground coal was mixed with an alkaline solution which was subsequently heated to a specific temperature and aerated for a predetermined time in a stirred flask fitted with an air sparger. The suspension was then transferred to a high-speed kitchen blender where a small amount of fuel oil was added and the coal agglomerated. The agglomerates were recovered by screening the suspension with either a 60- or 140-mesh sieve depending on the size of the original material. The results were evaluated in terms of recovery of coal combustible matter and the reductions in both ash and inorganic sulfur contents as defined below.

$$\text{Recovery (\%)} = \frac{\text{WP}(100 - \text{AP})}{\text{WF}(100 - \text{AF})} \times 100 \tag{7.1}$$

$$\text{Ash reduction (\%)} = \frac{AF - AP}{AF} \times 100 \qquad (7.2)$$

$$\text{S reduction (\%)} = \frac{SF - SP}{SF} \times 100 \qquad (7.3)$$

where

AF = ash content of dry feed
AP = ash content of dry, oil-free product
SF = inorganic sulfur content of dry feed
SP = inorganic sulfur content of dry, oil-free product
WF = weight of dry feed
WP = weight of dry, oil-free product

The recovery is essentially the yield of dry, oil-free, ash-free coal. Values of the recovery and other parameters reported below are generally the average of two or three values since each experiment was usually repeated several times.

In one set of experiments, 60 mesh × 0 size coal from the ISU demonstration mine in Mahaska County, Iowa, was pretreated with different alkalis and then recovered by agglomeration with a mixture of No. 1 (86 vol %) and No. 6 (14 vol %) fuel oils. This particular mixture was chosen because Min [14] had obtained relatively high ash rejection with it in preliminary agglomeration tests. During the pretreatment step, the coal was exposed to oxidizing conditions for 15 min at 50°C in the presence of 2.0 wt % concentration of alkali, and for the agglomeration step 10 ml of the fuel oil mixture was used for 100 g coal. The results presented in Table 7.3 show that chemical pretreatment with almost any of the alkalis listed followed by agglomeration reduced the inorganic sulfur content of the coal more than oil agglomeration alone. However, sodium carbonate and magnesium carbonate appeared to be the most effective of the alkalis tested since they resulted in the greatest reduction in the inorganic sulfur content of the coal and provided a high recovery of combustible matter. Reduced yields were obtained with calcium carbonate or calcium hydroxide and no coal was recovered when potassium hydroxide was employed. The pH of the coal-alkali suspension before and after pretreatment is indicated; or, in the case where the coal was oil agglomerated only, the natural pH of the coal suspension is shown.

Table 7.3 Oil Agglomerating 60 mesh × 0 ISU Coal Pretreated with Different Alkalis

Treatment	Alkali	pH		Recovery, %	Composition, wt %			Reduction, %	
		Initial	Final		Inorganic S	Ash	Alkali	Inorganic S	Ash
None (ISU coal)	-	-	-	-	4.56	15.7	0.05 Na 0.17 Ca 0.02 Mg	-	-
Oil aggl. only	-	2.4	2.4	87.4	3.93	12.8	0.05 Na 0.10 Ca 0.05 Mg	18.4	13.8
Chem.-oil aggl.	Na_2CO_3	9.6	9.5	83.9	1.88	10.2	0.28 Na	58.9	35.3
Chem.-oil aggl.	$MgCO_3$	6.1	7.9	86.3	1.66	11.2	0.16 Mg	63.7	28.7
Chem.-oil aggl.	$CaCO_3$	5.3	6.6	69.4	2.56	11.4	0.31 Ca	43.8	27.7
Chem.-oil aggl.	$Ca(OH)_2$	12.1	11.6	50.7	2.04	11.5	0.43 Ca	55.3	26.6
Chem.-oil aggl.	Na_3PO_4	7.6	6.6	82.0	2.54	11.7	0.21 Na	44.3	25.6
Chem.-oil aggl.	KOH	13.1	13.1	0	-	-		-	-

Source: From Ref. 18.

There appeared to be an association between a low coal recovery and pretreatment at a high pH. A comparison of the alkali metal content of the product with that of the feed shows that the coal adsorbed a significant amount of alkali from the pretreatment solution. Thus when sodium carbonate was employed in the pretreatment step, the product contained 0.28 wt % sodium which corresponded to 2.7% of the ash. Further study is needed to determine whether this alkali concentration is sufficiently high to cause fouling of boiler tubes. While high alkali levels can cause tube fouling, the fouling mechanism is complex and depends on other ash components and the total ash content of the coal as well as the alkali content [20]. Consequently, an alkali content which corresponds to 1 or 2% of the ash may be tolerable in some Western steam coals. However, if the pretreatment step leaves too much alkali in the coal, the alkali content can be reduced by acid washing as indicated below.

Higher-rank coals which have less oxygen and, hence, fewer carboxylic acid groups and hydroxyl groups that can exchange cations should pick up less sodium than the high-volatile C bituminous coal used in the present investigation. Therefore, the alkaline pretreatment step should create less of a problem when applied to the higher-rank coals. On the other hand, the allowable sodium content of coking coal seems to be lower than that of steam coal because alkalis create serious operating problems in blast furnaces [21]. Consequently, the pickup of sodium by coking coals would need to be evaluated carefully.

Although the preceding set of experiments showed that magnesium carbonate was as effective as sodium carbonate for the chemical pretreatment step, the next set of experiments, which utilized ISU mine coal precleaned by float-sink separation in trichloroethylene (specific gravity = 1.47), produced contrary results. The second set of experiments was designed to compare different grades of fuel oil for agglomeration as well as different alkalis for pretreatment. After precleaning in trichloroethylene, the coal was reduced to 60 mesh × 0 size. For some experiments, the coal was not chemically pretreated; while for others, it was pretreated for 15 min at 80°C with an aerated solution containing 2.0 wt % alkali. In each experiment, 5 ml of oil was used to agglomerate 50 g of coal. After the coal was agglomerated, it was recovered by screening and then resuspended in either water or a dilute hydrochloric acid solution. The purpose of the acid was to remove sodium which the coal adsorbed when sodium carbonate was employed for the pretreatment step. The acidified suspension had a pH of 2.0. The suspension was returned to the blender for recleaning after which the product was rescreened. The results of these experiments (Table 7.4) show that coal treated with sodium carbonate, agglomerated, and washed with acid contained only 0.06 wt % sodium whereas in the previous set of experiments (Table 7.3) the coal treated with sodium carbonate contained 0.28 wt % sodium. Therefore, the final acid treatment was effective in reducing the sodium content of the coal. The data of Table 7.4 also indicate that the pretreatment step enhanced the removal of inorganic sulfur from coal by oil agglomeration particularly when sodium carbonate was employed as the alkali. Obviously, the results were not as good with magnesium carbonate as with sodium carbonate. The results indicate, furthermore, that the lighter fuel oils (No. 1, No. 2, and No. 200 LLS) produced a product with a lower sulfur content than the heavier residual oils (No. 5 LLS and 6) produced, whether the coal was pretreated or not. Although the data of Table 7.4 do not indicate a significant effect of oil grade on ash removal, Capes et al. [19] found that the lighter oils removed more ash than did the heavier oils.

Some of the ISU coal which had been precleaned by float-sink separation in trichloroethylene was also ground to 400 mesh × 0 size and agglomerated with different fuel oils (Table 7.5). In this case the coal was not chemically pretreated. The coal was agglomerated as before with an amount of oil equivalent to about 10% of the weight of dry solids. The amount of sulfur removed in these experiments with No. 1, No. 2, or No. 200 LLS fuel oils was between that removed before with untreated 60 mesh × 0 coal and with pretreated 60 mesh × 0 coal. On the other hand, the recovery of coal combustible matter achieved with these oils was lower for the smaller size consist than for the larger one. For both size consists, the recovery of untreated

Table 7.4 Agglomerating Untreated and Chemically Treated 60 mesh × 0 ISU Coal[a] with Different Fuel Oils

Treatment	Fuel oil no.	Recovery, %	Composition, wt %			Reduction, %	
			Inorganic S	Ash	Alkali	Inorganic S	Ash
None (ISU coal)[a]	-	-	1.71	9.65	0.03	-	-
Oil aggl. only	1	95.9	1.02	7.82	0.03	40.2	18.9
Oil aggl. only	2	98.3	1.02	7.95	0.03	40.4	17.7
Oil aggl. only	200 LLS	100.0	1.12	7.78	0.03	34.7	19.3
Oil aggl. only	5 LLS	91.7	1.20	7.93	0.03	30.0	17.8
Oil aggl. only	6	98.7	1.21	8.20	0.03	29.2	15.0
Na_2CO_3							
Chem.-oil aggl.	1	96.9	0.63	7.74	0.06	63.0	19.8
Chem.-oil aggl.	2	98.0	0.75	7.56	0.06	55.3	21.7
Chem.-oil aggl.	200 LLS	96.7	0.70	8.01	0.06	59.3	17.0
Chem.-oil aggl.	5 LLS	89.1	0.86	7.66	0.07	49.7	20.7
Chem.-oil aggl.	6	97.2	0.87	7.77	0.07	48.9	19.4
$MgCO_3$							
Chem.-oil aggl.	1	95.4	0.92	8.33	-	46.2	13.7
Chem.-oil aggl.	2	100.0	0.97	7.93	-	43.5	17.9

[a]Precleaned by float/sink separation in trichloroethylene.

Source: From Ref. 18.

Table 7.5 Agglomerating Untreated 400 mesh × 0 ISU Coal[a] with Different Fuel Oils

Treatment	Fuel oil no.	Recovery, %	Composition, wt %		Reduction, %	
			Inorganic S	Ash	Inorganic S	Ash
None (ISU coal)[a]	-	-	1.82	9.97	-	-
Oil aggl. only	1	74.8	0.96	8.56	47.5	14.1
Oil aggl. only	2	88.2	0.99	7.76	45.3	22.2
Oil aggl. only	200 LLS	94.1	1.04	8.34	42.9	16.3
Oil aggl. only	5 LLS	95.4	0.78	8.22	57.1	17.6
Oil aggl. only	6 LLS	89.7	0.62	8.21	65.9	17.7
Oil aggl. only	6 LLS	0	-	-	-	-
Oil aggl. only	6	0	-	-	-	-

[a]Precleaned by float-sink separation in trichloroethylene.

Source: From Ref. 18.

coal increased in the same order with regard to grade of fuel oils (i.e., No. 1, No. 2, and No. 200 LLS). The results achieved with No. 5 LLS fuel oil were unusual in that both the recovery of combustible matter and reduction in sulfur content were greater for the smaller size consist than for the larger one. Mixed results were achieved with No. 6 LLS fuel oil; in one experiment with 400 × 0 coal a recovery of 89.7% was achieved, while in another experiment none of the coal was recovered by oil agglomeration and screening. However, in the experiment where a high recovery was achieved, the sulfur reduction was the highest of all. None of the smaller size consist was recovered by agglomeration with No. 6 fuel oil. The failure to achieve agglomeration with these oils may be related to the high viscosity of the oils which made them difficult to disperse. But until the mechanism of oil agglomeration is better understood, a complete explanation of these anomalous results cannot be offered. In the experiments with 400 mesh × 0 coal, it was observed that the lighter oils produced large, shapeless, weak flocs which clogged the openings in the 140-mesh dewatering screen and interfered with the

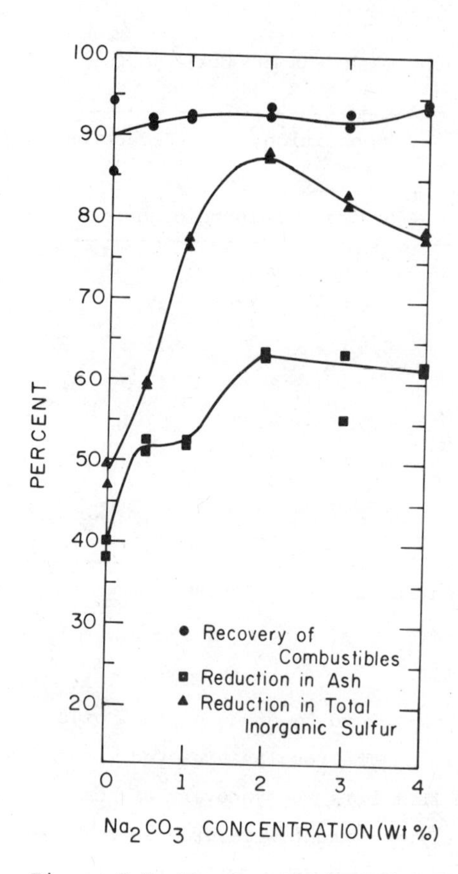

Figure 7.1 Results of oil agglomerating 400 mesh × 0 ISU Mine coal pretreated at 80°C for 15 min. (From Ref. 17.)

removal of water and mineral matter; whereas the heavier oils formed microspherical agglomerates which were more easily dewatered and freed of mineral matter.

Further experiments were conducted to establish the effects of different pretreatment conditions on the recovery of coal combustible matter and reductions in ash and sulfur contents achieved by oil agglomeration. The effect of alkali concentration is illustrated by Fig. 7.1. To obtain these results, 400 mesh × 0 ISU coal was pretreated for 15 min at 80°C and then agglomerated with a mixture of No. 200 LLS (25 vol %) and No. 6 LLS (75 vol %) fuel oils. This

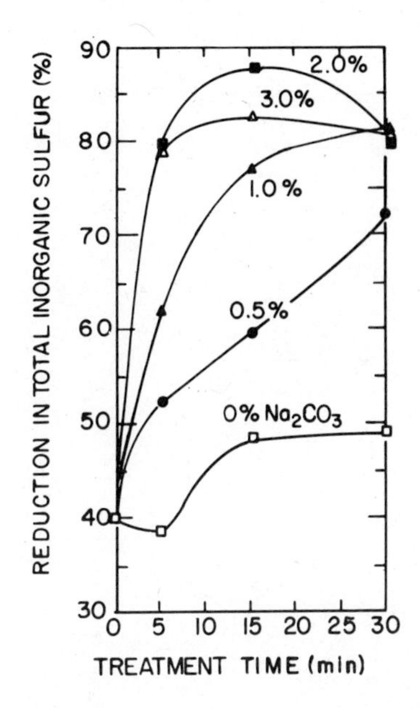

Figure 7.2 Effect of pretreatment time and alkali concentration at 80°C on oil agglomeration of 400-mesh × 0 ISU Mine coal. (From Ref. 17.)

mixture was chosen because it gave consistent results and utilized a relatively high proportion of the less costly No. 6 LLS fuel oil. Again, an amount of oil equivalent to about 10% of the weight of coal was used. The coal initially contained 4.96 wt % inorganic sulfur and 15.7 wt % ash and was not precleaned. Figure 7.1 shows that the greatest reduction in inorganic sulfur content (88%) was achieved when the pretreatment solution contained 2.0 wt % sodium carbonate. Although the removal of sulfur and ash was found to be affected greatly by the alkali concentration, the recovery of combustible matter was affected very little by this parameter. The effect of alkali concentration on sulfur removal is further illustrated by Fig. 7.2 as well as the effect of chemical treatment time. This diagram shows that a treatment time of 15 min was best when the optimum alkali concentration of 2.0 wt % was employed. However, a treatment time of 30 min gave better results when an alkali concentration of 1.0 wt % or less or over 3.0 wt % was employed. Hence,

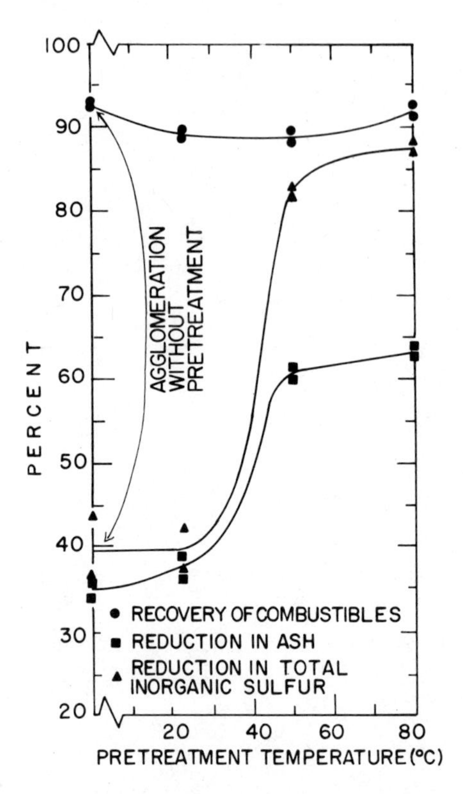

Figure 7.3 Results of oil agglomerating 400-mesh × 0 ISU Mine coal pretreated for 15 min with 2.0 wt % Na_2CO_3 solution at different temperatures. (From Ref. 17.)

a longer treatment time seemed to make up partially for a less-than-optimum alkali concentration. The effect of pretreatment temperature on the results is illustrated by Fig. 7.3. The data plotted along the left-hand side of the diagram represent the results of oil agglomeration alone at room temperature without pretreatment. Although a pretreatment temperature of 50°C or higher was quite effective in reducing both the sulfur and ash contents of the coal when it was agglomerated, a pretreatment temperature of 23°C was not effective.

Coals from several other locations also responded favorably to chemical pretreatment and oil agglomeration. These coals were ground to 400 mesh × 0 size, pretreated for 15 min at 80°C, and agglomerated

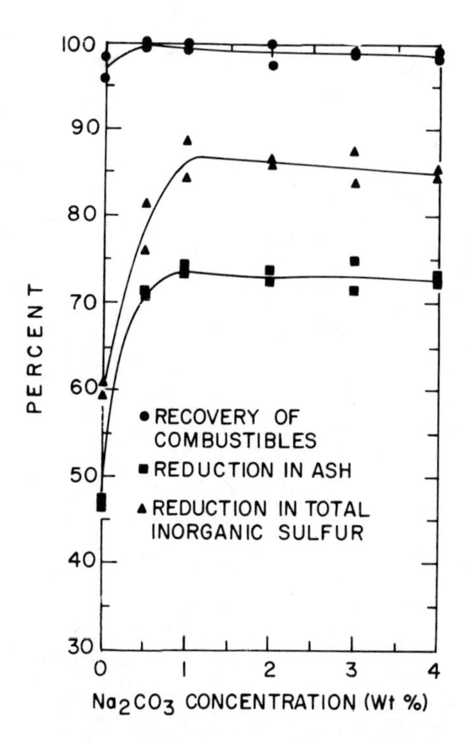

Figure 7.4 Results of oil agglomerating 400-mesh × 0 Kentucky No. 9 coal pretreated at 80°C for 15 min. (From Ref. 18.)

with a mixture of No. 200 LLS and No. 6 LLS fuel oils. The results obtained with Kentucky No. 9 seam coal from the Fies Mine in Hopkins County, Kentucky, are shown in Fig. 7.4. This coal initially contained 1.77 wt % inorganic sulfur and 20.6 wt % ash. Pretreatment with a solution containing 1 to 4 wt % sodium carbonate followed by oil agglomeration was very effective; the inorganic sulfur content was reduced by 86% and the ash content by 74%. Also the recovery of coal combustible matter was very high, being about 99%. The results obtained with coal from the Lovilia No. 4 Mine in Monroe County, Iowa, are shown in Fig. 7.5. This coal initially contained 2.99 wt % inorganic sulfur and 15.2 wt % ash. The greatest reduction in sulfur content (70%) was obtained with an alkali concentration of 1.0 wt %, but this reduction was noticeably poorer than was achieved with

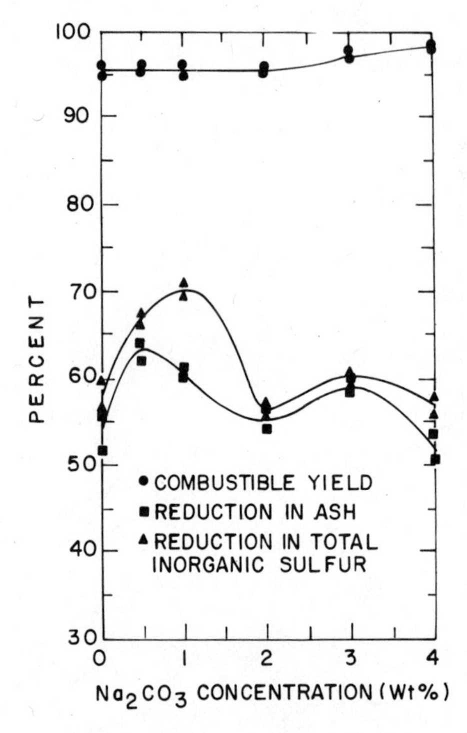

Figure 7.5 Results of oil agglomerating 400-mesh × 0 Lovilia Mine coal pretreated at 80°C for 15 min. (From Ref. 18.)

either the ISU or Western Kentucky coals. Finally, the results of pretreating and oil agglomerating an Illinois No. 5 coal from the Rapatee Mine in Middle Grove, Illinois, are shown in Fig. 7.6. This coal had been precleaned by an industrial jig washer operated by Midland Coal Co. and contained 1.39 wt % inorganic sulfur and 13.2 wt % ash. Although 50% of the inorganic sulfur was removed by pretreatment with a 1.0 wt % solution of alkali and oil agglomeration, this was considerably lower than that obtained with the other coals.

One reason why the various coals responded differently to chemical pretreatment and oil agglomeration may have been due to the difference in the proportion of sulfate sulfur among these coals. Thus, the data in Table 7.6 indicate an apparent association between the proportion of inorganic sulfur in the form of sulfates and the

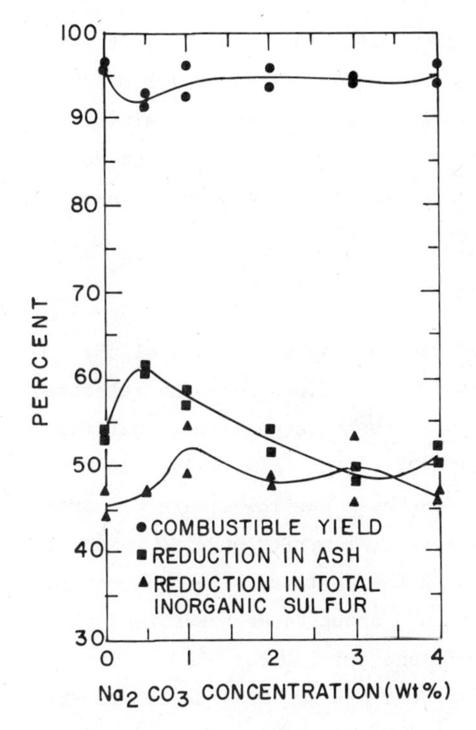

Figure 7.6 Results of oil agglomerating 400-mesh × 0 Illinois No. 5 coal pretreated at 80°C for 15 min. (From Ref. 18.)

Table 7.6 Association between Maximum Reduction of Inorganic Sulfur Achieved by Pretreatment and Oil Agglomeration and the Fraction of Inorganic Sulfur in Sulfate Form

Coal	Sulfate S/Inorganic S	Maximum reduction of inorganic S, %
ISU Mine	0.44	88
Kentucky No. 9 Seam	0.50	86
Lovilia No. 4 Mine	0.25	70
Illinois No. 5 Seam	0.09	50

Source: From Ref. 18.

maximum reduction in inorganic sulfur achieved. Moreover, it was shown that 90% of the sulfate sulfur would be extracted from 400 mesh × 0 ISU coal in 15 min by a hot (80°C) solution containing 2.0 wt % sodium carbonate. Therefore, the pretreatment step must have extracted most of the sulfate sulfur from these coals in addition to altering the surface of the pyrites.

C. Pretreatment and Floatability

To gain a better understanding of the factors which control the flotation and depression of pyrite in coal flotation, the relative floatability of pyrite particles was investigated with a modified Hallimond tube microflotation cell [22]. For this study, 124 × 104 μm size particles of pyrite were used. These particles were prepared by cleaning and dry pulverizing and screening pyrite nodules which had been recovered from coal produced at the Childers site adjacent to the ISU mine. The particles contained about 74 wt % FeS_2 and significant amounts of other iron, sulfur, and calcium compounds. The calcium content was 4.8 wt %. Some of the pyrite was cleaned with hot 5% hydrochloric acid to remove various impurities. This treatment reduced the calcium content to 0.2 wt % and increased the pyrite content to 88 wt % FeS_2.

High-volatile C bituminous coal from the Childers site was also used in this study. The coal was cleaned in the laboratory by gravity separation at 1.29, ground and wet screened to produce 177 × 149 μm size particles for use in the Hallimond cell. These particles contained 3.46 wt % ash, 0.62 wt % pyritic sulfur and 4.58 wt % organic sulfur.

The relative floatability of the coal or pyrite particles was determined by aerating a suspension consisting of 1 g solids and 100 ml of water in the Hallimond cell and measuring the weight of solids which floated and was recovered in 4 min. Nitrogen was used for aeration. In most instances, 10 μl of methyl isobutyl carbinol (MIBC) were used as a frother. Without the frother, only 2 to 3% of the pyritic material was floated regardless of whether impure or acid cleaned pyrites were used. But with 10 μl of MIBC, 11% of the impure pyrites was floated and 53% of the acid cleaned pyrites. In a subsequent experiment, the floatability of acid cleaned pyrites was observed to decline slowly over a period of several days from exposure to air in the laboratory. Thus after 5 days the floatability was about one-third that of the freshly cleaned material. Another experiment showed that calcium ions greatly reduced the floatability of the acid cleaned pyrites. Therefore, it seemed likely that the low floatability of the impure pyrites was due to the presence of both oxidation products and calcium compounds on the surface of the particles.

To determine the effectiveness of wet oxidation as a means of depressing the flotation of pyrite, numerous experiments were conducted in which pyrite particles were subjected first to oxidizing conditions and then tested in the Hallimond cell. For the oxidizing treatment, the particles were suspended in a solution which usually contained an alkali and which was heated to 80°C. Air was bubbled through the solution to create an oxidizing environment and the treatment lasted 15 min. After this treatment, the suspension was cooled to room temperature and transferred to the Hallimond cell where the floatability of the material was determined in the presence of MIBC (usually 10 μl). Typical results for acid cleaned pyrites are shown in Fig. 7.7. These results were obtained using sodium carbonate as the alkali. One set of data represents the floatability of unoxidized pyrites in various concentrations of alkali, while another set represents the floatability of the surface oxidized pyrites. Although the alkali by itself reduced the floatability of the pyrites markedly, wet oxidation in alkaline solutions reduced the floatability an additional amount.

Results are also presented in Fig. 7.7 for coal which had been subjected to the same oxidizing treatment as the pyrite and for coal which had not been. These results show that while the alkali depressed the floatability of the coal, the effect was much smaller than the effect of the alkali on the pyrite floatability. Also the results indicate that wet oxidation had very little additional effect on the floatability of the coal.

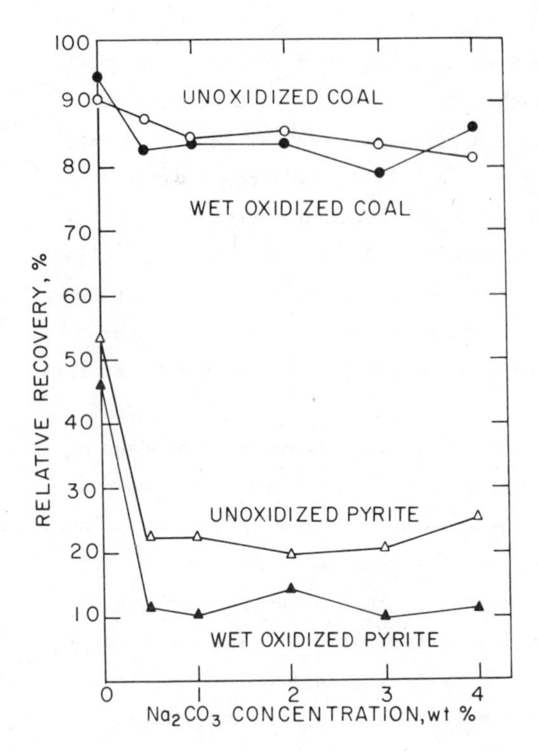

Figure 7.7 Effect of wet oxidation on the floatability of coal or pyrites. (From Ref. 22.)

Somewhat similar results were achieved using other alkalis including sodium bicarbonate, sodium hydroxide, and calcium hydroxide. Among these reagents calcium hydroxide was the most effective flotation depressant for pyrite. A concentration of only 0.1 wt % reduced the relative recovery of acid washed pyrite in the presence of MIBC to 8%. Moreover, wet oxidation of pyrite in this concentration of calcium hydroxide reduced the recovery of pyrite to 5%.

From the various results presented above, it appears that clean, unoxidized particles of pyrite should present the greatest problem in coal flotation (aside from unlocked pyrite). Such particles may correspond to the single microcrystals and framboids of pyrite which appear to constitute an appreciable fraction of the pyrite content of some coals [23]. Photomicrographs of these microcrystallites give the appearance of pure materials; and as long as the microcrystallites are imbedded in the coal, they should remain in a reduced state. Therefore, in freshly ground coal these microcrystallites may float readily. Both wet oxidation in alkaline solutions and the use of calcium hydroxide as a depressant offer considerable promise for reducing the floatability of clean pyrites. The development and application of these techniques for coal flotation is proceeding apace.

7.3 MULTISTAGE, MULTIMETHOD CLEANING

It is well known that mineral impurities are present in coal in a variety of physical and chemical forms and that these forms respond differently to different methods of cleaning. Thus coarse mineral particles which are relatively dense compared to coal are readily removed by gravity separation; whereas fine particles, though dense, are difficult to remove by this method. On the other hand, fine-size hydrophilic mineral particles can be separated from hydrophobic coal particles by froth flotation or oil agglomeration. Therefore, given a coal with a variety of impurity forms, one would anticipate that multistage cleaning utilizing several different methods of cleaning would be more effective than single-stage cleaning based on only one method. Also, one would anticipate that different methods of comminution would not be equally effective in unlocking mineral impurities. The experimental work described below was designed to test these theories.

A. First Series of Treatments: ICO and Jude Coals

The first series of laboratory treatments (Fig. 7.8) designed to compare various combinations of comminution and separation methods was applied to run of mine high volatile C bituminous coal from the ICO and Jude mines in Iowa [14,24]. The first treatment involved crushing the coal to 6 mm × 0 size with a small roll crusher, pulverizing to 35 mesh × 0 size with a Mikro-Samplmill, and oil agglomerating the pulverized coal in a kitchen blender. For the oil

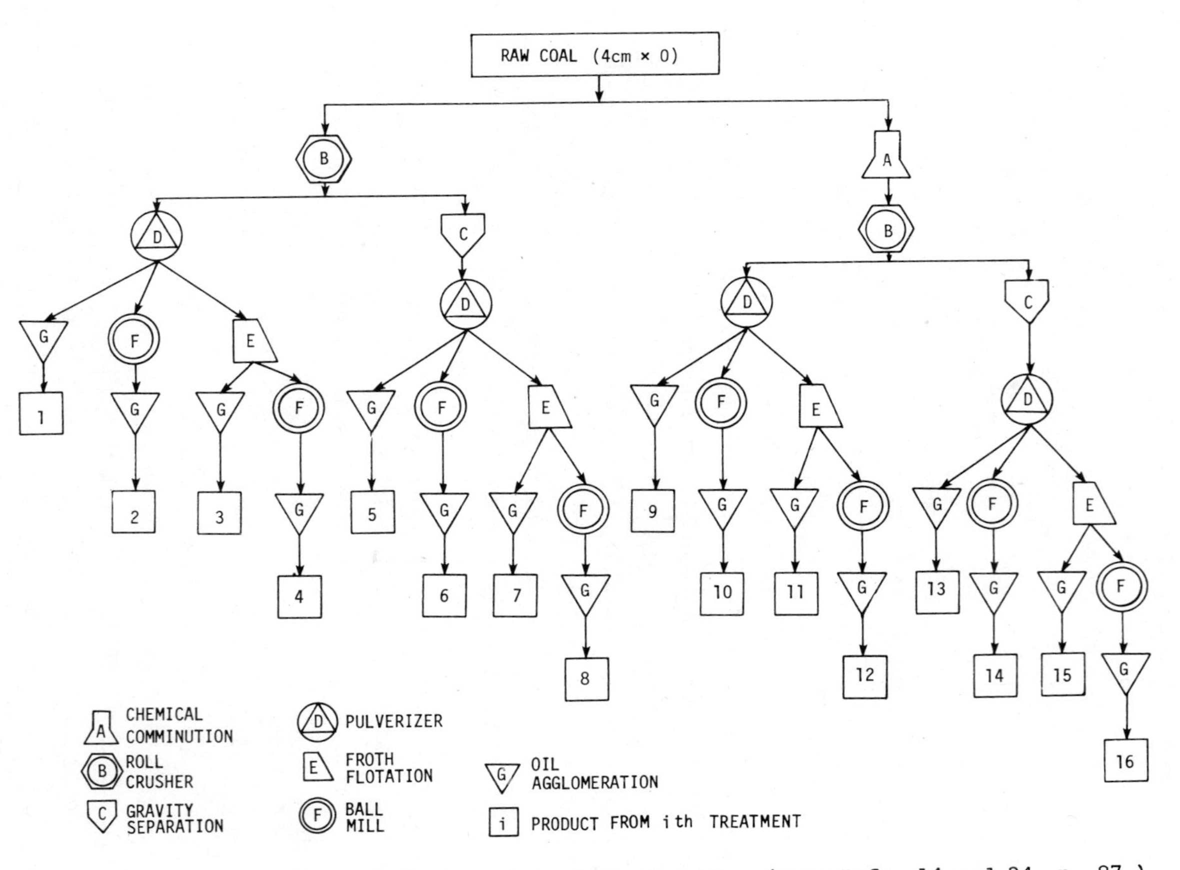

Figure 7.8 Flow diagram of first series of treatments. (From Refs. 14 and 24, p. 87.)

agglomeration step, 20 g of coal suspended in about 200 ml of water was mixed with 200 ml of an oil-water emulsion containing 2.0 ml of fuel oil (86 vol % No. 1 and 14 vol % No. 5). The final suspension was placed in a separatory funnel where the agglomerated coal floated and the refuse sank to effect a separation. This method of separation appeared to work better than screening for this series of treatments. The second treatment included a ball milling step which reduced the coal to 400 mesh × 0 size in addition to the other steps. The third and fourth treatments included a froth flotation step in which the pulverized coal was floated by a two-stage rougher-cleaner process using a small amount of kerosene and MIBC as reagents; the first stage was conducted at low pH and the second stage at high pH. In the fifth through eighth treatments, a larger portion of crushed coal was subjected to gravity separation in tetrachloroethylene (specific gravity = 1.61) before being pulverized and split into smaller portions which were then treated as in the first four treatments. In the last eight treatments, a quantity of coal was chemically comminuted by soaking it in liquid anhydrous ammonia before crushing it with the roll crusher. Following the chemical comminution step, the pattern of treatments was the same as for the first eight treatments. The final step of each treatment was an oil agglomeration step.

The results of applying this series of treatments to coal from the ICO mine, which initially contained 2.41 wt % pyritic sulfur, 0.05 wt % sulfate sulfur, 0.99 wt % organic sulfur, and 8.3 wt % ash are presented in Fig. 7.9. The cumulative percentage reductions in both the pyritic sulfur and ash contents and the cumulative weight yield of product are presented over the various separation steps in each treatment. The yield is reported on an oil-free and moisture-free basis. Coal from the ICO mine contained finely disseminated crystals of iron pyrites as well as framboids and other larger massive forms [1].

It can be seen that as the number of separation steps within treatments was increased, the greater the reductions in sulfur and

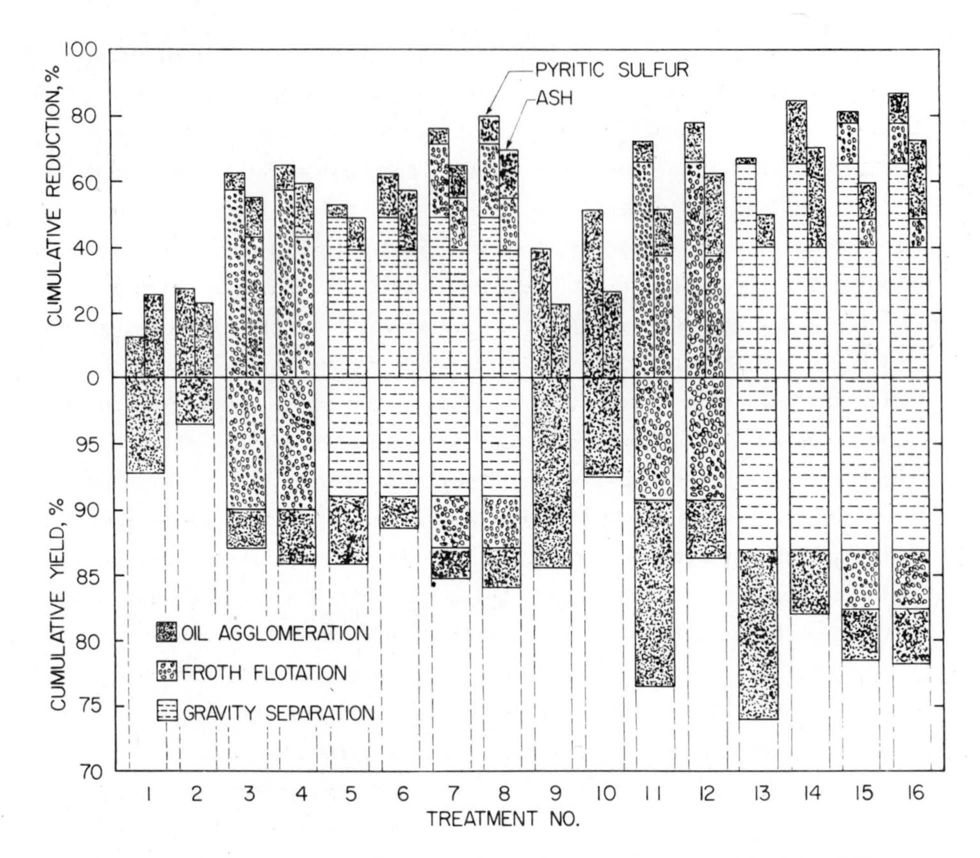

Figure 7.9 Results of first series of treatments applied to ICO Mine coal. (From Refs. 14 and 24.)

ash contents achieved. Thus an overall reduction in pyritic sulfur content of 12 to 27% was achieved in treatments 1 and 2 with one separation step, 53 to 65% in treatments 3 to 6 with two separation steps and 76 to 80% in treatments 7 and 8 with three separation steps. Similar results were obtained with coal from the Jude mine in Iowa which also contained a wide distribution of pyrite sizes. From these results, it is apparent that both froth flotation and oil agglomeration of fine-size coal complement gravity separation of somewhat coarser coal as a means of cleaning this type of coal. Also oil agglomeration complements froth flotation, particularly when it is used in conjunction with fine grinding.

Unfortunately, the data in Fig. 7.9 also indicate a trade-off between greater reductions in sulfur and ash contents and lower yields. This trade-off is also shown by Figs. 7.10 and 7.11, where the overall yield of product is plotted against the corresponding pyritic sulfur or ash content resulting from the different treatments. The upper curve in each diagram is drawn through the points representing the treatments which achieved the highest yields for the corresponding levels of pyritic sulfur or ash. Thus for ICO coal, treatments 2, 10, 12, 14, and 16 were the most efficient from

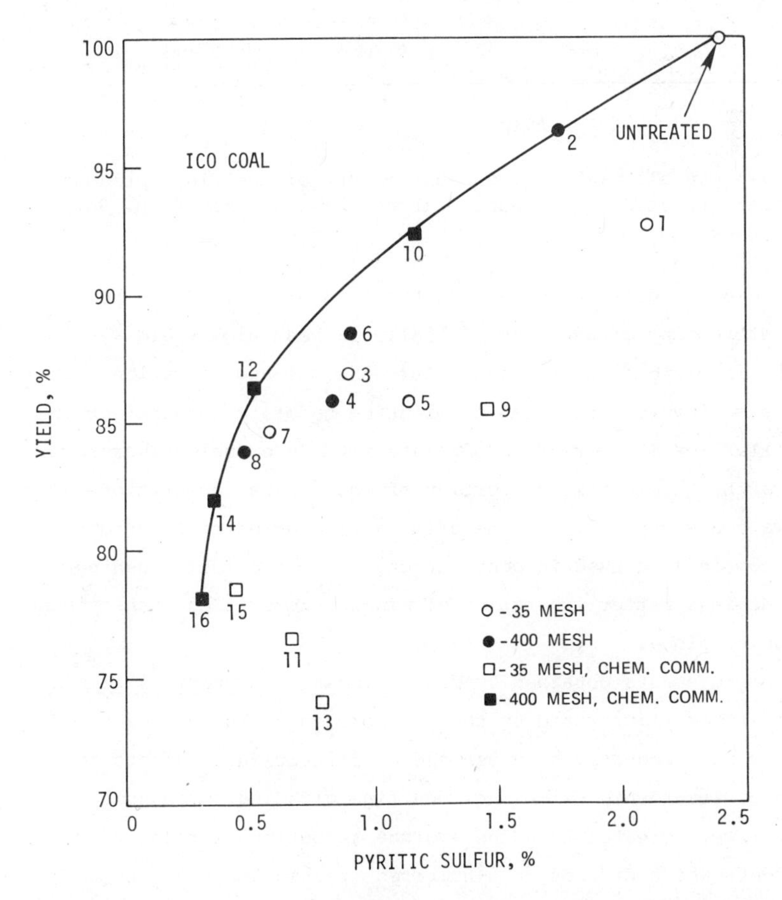

Figure 7.10 Overall yield versus pyritic sulfur content of the final product of the first series of treatments. (From Refs. 14 and 24, p. 93.)

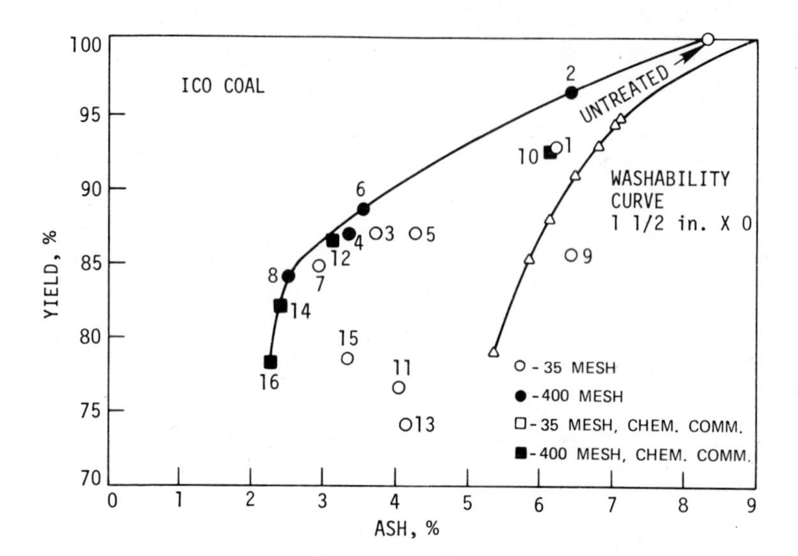

Figure 7.11 Overall yield versus ash content of the final product of the first series of treatments. (From Refs. 14 and 24, p. 94.)

the standpoint of sulfur removal and treatments 2, 6, 8, 14, and 16 from the standpoint of ash removal. All of these high-yield treatments involved fine grinding before the oil agglomeration step and most also involved chemical comminution before mechanical crushing. Similar results were also obtained with Jude coal. Therefore, it appears that a significant portion of the mineral impurities in these Iowa coals can only be liberated by fine grinding. Moreover, chemical comminution used in conjunction with mechanical crushing seems to achieve better liberation of mineral impurities than mechanical crushing alone.

The improvements obtained by fine grinding and chemical comminution are further illustrated by the data presented in Table 7.7. Thus the average reductions in ash and pyritic sulfur contents for the group of treatments which included fine grinding (even-numbered treatments) were greater than the average reductions for the group of treatments which did not (odd-numbered treatments). At the same time, the average overall yield of clean coal for the group with fine grinding was larger than that for the group without it. Similarly,

Table 7.7 Applying Different Comminution Methods to ICO Mine Coal

Comminution method	Treatment combination	
Grinding in ball mill	Odd	Even
Mean ash reduction, %	47	55
Mean pyritic sulfur reduction, %	58	67
Mean yield, %	83	87
Chemical comminution	1-8	9-16
Mean ash reduction, %	50	52
Mean pyritic sulfur reduction, %	55	70
Mean yield, %	88	82

Source: From Ref. 24, pp. 96-97.

the average reductions in ash and pyritic sulfur contents for the group of treatments which included chemical comminution were greater than the average reductions for the group of treatments which did not. However, the average overall yield of clean coal was slightly lower for the group which included chemical comminution. Similar results were also achieved with Jude coal except that the average yield of clean coal was larger for the group of treatments which included chemical comminution than for the group which did not.

Nearly all of the treatments described by Fig. 7.8 were more efficient than single-stage gravity separation of 4 cm × 0 coal. The yield and corresponding ash content provided by the gravity separation of coarse ICO coal in liquids of different specific gravity are represented by the washability curve in Fig. 7.11. Similar results were obtained with Jude coal. Thus for either kind of coal the yield fell off sharply as the ash content was reduced to lower levels by separation in lighter liquids.

B. Second Series of Treatments: ISU Coal

The second series of laboratory treatments (Fig. 7.12) designed to compare various combinations of comminution and separation methods was applied to bituminous coal from the ISU demonstration mine [18].

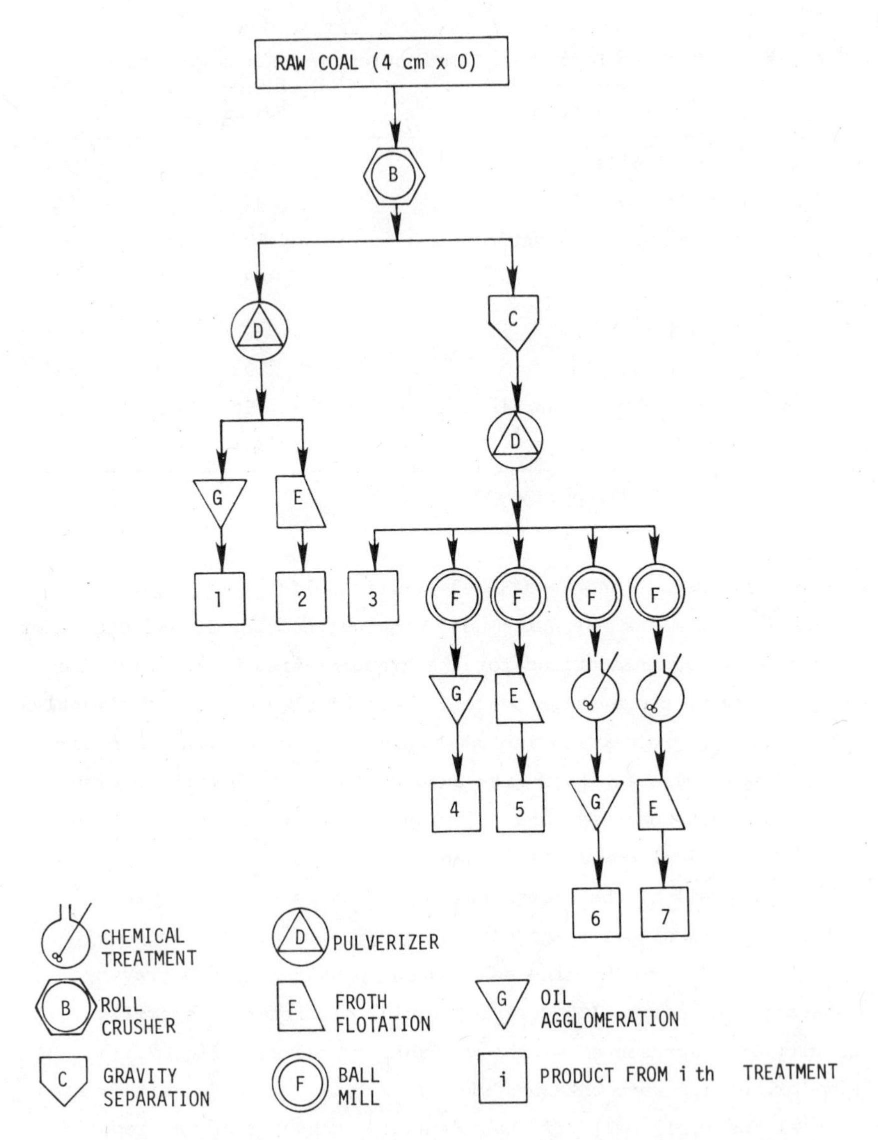

Figure 7.12 Flow diagram of second series of treatments. (From Ref. 18.)

This coal initially contained 6.42 wt % inorganic sulfur and 19.7 wt % ash. The first treatment involved crushing the coal to 6 mm × 0 size with a small roll crusher, pulverizing to 60 mesh × 0 size with a Mikro-Samplmill, oil agglomerating the pulverized coal in a kitchen blender, and recovering the product with a separatory funnel. For the oil agglomeration step, 100 g coal suspended in 500 ml deionized water was mixed with 10 ml of fuel oil (86 vol % No. 1 and 14 vol % No. 6). The second treatment substituted a froth flotation step for the oil agglomeration step. For the froth-flotation step, a suspension containing 100 g pulverized coal, 2000 ml deionized water, and 1.0 ml MIBC was placed in a laboratory flotation cell and aerated until the froth was nearly free of coal. For the third and subsequent treatments, a quantity of 6 mm × 0 size coal was cleaned by gravity separation in tetrachloroethylene (specific gravity = 1.61), pulverized, and split into smaller portions for the individual treatments. For the fourth and subsequent treatments, the pulverized coal was wet ground in a ball mill. In the fourth and fifth treatments, the ground coal was beneficiated by oil agglomeration and froth flotation, respectively. The sixth and seventh treatments were similar to the previous two treatments except that the ground coal was chemically treated by aeration in a 2.0 wt % solution of sodium carbonate at 80°C for 15 min.

The cumulative results of applying this series of treatments to ISU coal are shown in Fig. 7.13. In most cases the data shown for a specific treatment represent an average of two runs. Here again as with the first series of treatments, the overall reduction in inorganic sulfur content was greater when two separation steps were employed than when one separation step was used. On the other hand, the reduction in ash content was not always greater when two separation steps were applied. Thus the final ash content achieved in treatments 4, 6, and 7, which included both gravity separation and either oil agglomeration or froth flotation, was higher than that achieved with only a single separation step in any of the first three treatments. In treatments 6 and 7, but not in treatment 4, the coal may have adsorbed sodium from the alkaline solution employed in the chemical treatment step and this would have increased its ash content. Also the coal may have picked up mineral particles abraded from the grinding media used in the ball mill.

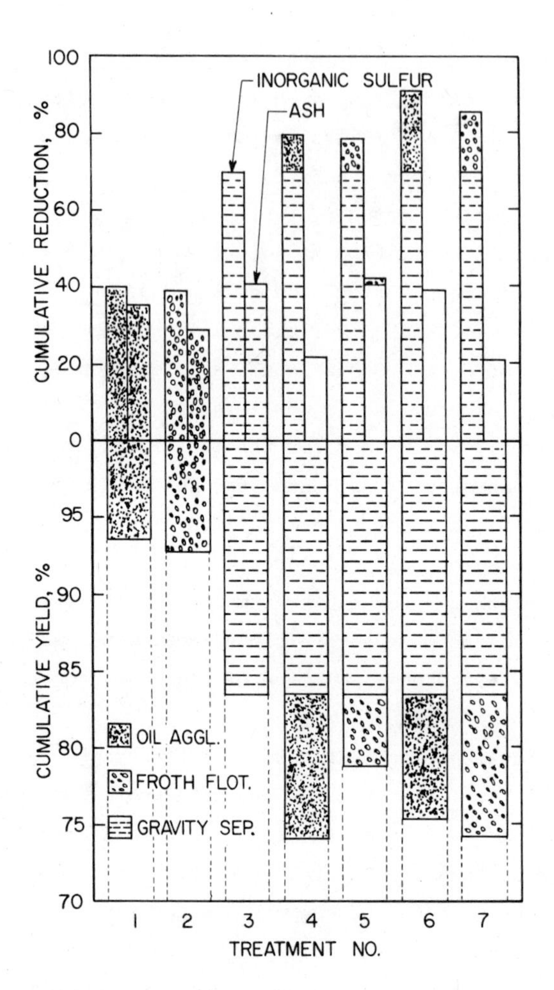

Figure 7.13 Results of second series of treatments applied to ISU Mine coal. (From Ref. 18.)

Treatments 6 and 7 demonstrated the value of the chemical treatment step, since the sulfur reduction achieved in these runs was significantly greater than that achieved in treatments 4 and 5 which did not include this step. Moreover, in treatment 6 the increase in desulfurization was achieved without sacrifice in coal recovery (compared to treatment 4). Treatment 6 was also noteworthy because it achieved the highest overall reduction in total inorganic sulfur content (91%).

7.4 DEMONSTRATION PLANT

A coal preparation plant has been built on the campus of Iowa State University to demonstrate various methods of cleaning coal on a larger scale. A building to house the plant (Fig. 7.14) and the first section of the plant to clean coarse- and medium-size coal were completed in 1976. This section included a primary crusher, heavy-media separator, wet concentration table, size separation and dewatering screens, and materials handling equipment. Hydrocyclones for cleaning fine-size coal and equipment for dewatering fines were added in 1978 and heavy-media cyclones for cleaning medium to fine-size coal were added in 1979. All of this equipment is of an

Figure 7.14 Aerial view of Iowa State University coal preparation plant.

industrial scale and was paid for by the State of Iowa. Other equipment of a smaller size to demonstrate the froth flotation and oil agglomeration methods of beneficiating fine-size coal was installed in 1978 using funds provided by the Fossil Energy Division of the U.S. Department of Energy, In the future, equipment may be added for demonstrating new methods under development for cleaning, dewatering, and consolidating fines. Additional information concerning the design, operation, and use of the plant is presented below.

A. Main Plant

A process flowsheet of the main coal preparation plant as it exists in early 1979 is shown in Fig. 7.15. Run-of-mine coal is crushed by a Cedar Rapids Model 30 impact mill which can produce up to 90 t/hr of 4 cm × 0 size coal. The crushed coal is stored in a feed hopper from which it is fed via conveyor belt to the process. This belt is equipped with a weighing scale which controls the feed rate and provides an indication of the total amount of coal fed. The belt discharges onto a vibrating screen with 9.5-mm-wide openings; the oversize material from this screen is directed to a heavy-media unit and the undersize to a wet concentrating table.

The heavy-media unit manufactured by Eagle Iron Works utilizes a cone-shaped vessel with mechanical agitation to contain the magnetite suspension which effects the separation of coarse coal and refuse. The coal and refuse from the heavy-media unit are drained and rinsed to recover the magnetite across another set of vibrating screens. The clean coal and refuse are then carried by belt conveyors to separate storage piles. The heavy-media unit can treat up to 46 t/hr. of 4 cm × 9.5 mm size coal.

The wet concentrating table is a Deister No. 88 double-deck table which has processed as much as 36 t/hr of 9.5 mm × 0 size coal. The clean coal and refuse separated by the table are dewatered by still another set of vibrating screens and then are conveyed to the clean coal and refuse storage piles.

Most of the coal fines (0.25 mm × 0) are suspended in the water recovered by the last set of dewatering screens. This suspension is

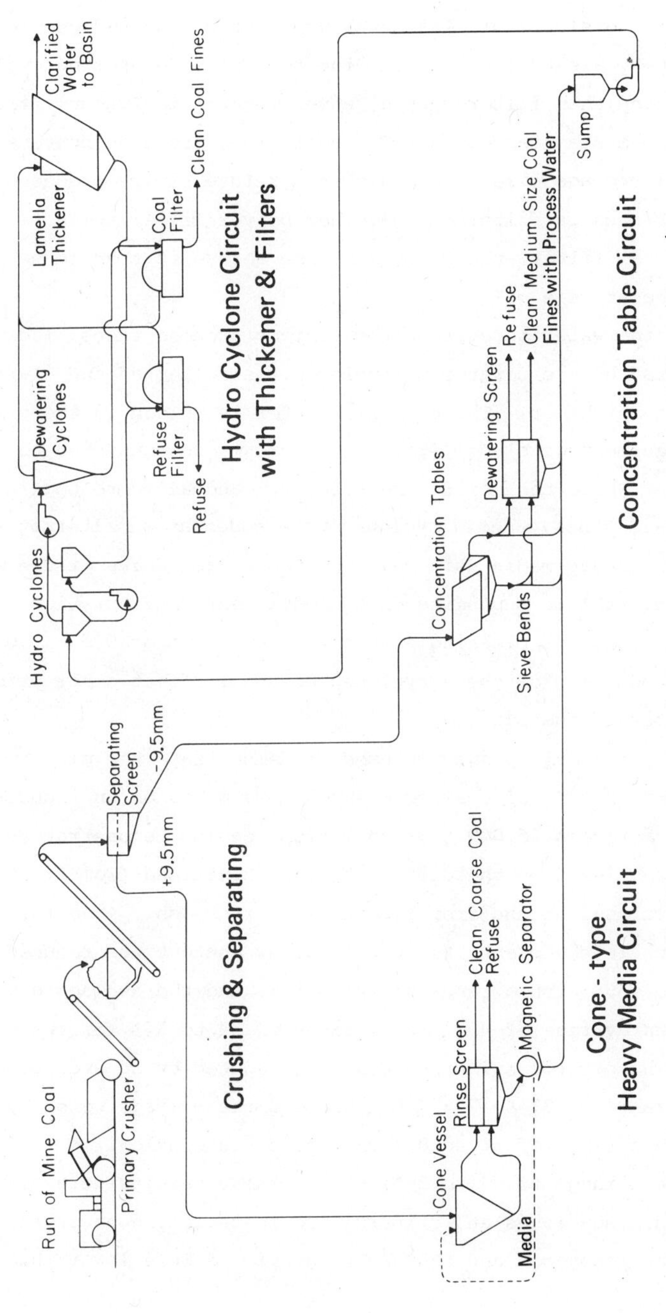

Figure 7.15 Schematic flowsheet of the main coal preparation plant at Iowa State University.

conducted to a series of two 25-cm-diameter Krebs hydrocyclones for separation. The slurry of clean coal produced by the hydrocyclones is thickened by a set of two 15-cm-diameter Krebs dewatering cyclones in parallel and then filtered by a Denver rotary disk vacuum filter with four 1.8-m-diameter disks. The coal fines are also conveyed to a clean coal storage pile. The thickened refuse slurry produced by the hydrocyclones is filtered by another Denver filter which is similar to the coal filter; the recovered fine refuse is then conveyed to the refuse storage pile.

All of the water recovered from various process steps, including that recovered by the dewatering cyclones, is collected and clarified by a Lamella thickener. The clarified water is reused in the plant, while the sediment or sludge from the thickener is conducted to the refuse filter where the solids are recovered and added to the refuse.

In mid-1979 heavy-media cyclones were added as an alternative to either the heavy-media unit built by Eagle Iron Works or the wet concentrating table. Depending on the equipment configuration, the cyclones are used for cleaning either 4 cm × 9.5 mm or 9.5 × 0.6 mm size coal. Addition of these cyclones does not affect other principal features of the plant.

In 1977, the main plant was used to demonstrate the cleaning of large samples (907 t) of coal from seven Iowa mines on an industrial scale and to process 45,000 t of coal from the ISU Demonstration Mine [25]. The samples from the different mines contained from 2.5 to 8.7 wt % total sulfur and from 11.6 to 20.0 wt % ash. As a result of processing in the plant, the total sulfur content was reduced by an average of 35% with a range of 24 to 45%, and the ash content was reduced by an average of 45% with a range of 34 to 57% for the series of coals. The pyritic sulfur content was reduced by an average of 52% with a range of 37 to 70%. Furthermore the average weight yield was 74% with a range of 66 to 80% and the average calorific yield was 84% with a range of 74 to 96%. Since these results were obtained before the hydrocyclones and filters were installed, none of the −48-mesh coal was recovered and therefore the yields were lower than would be obtained with the present layout. All of the processed coal was burned in the ISU power plant which is adjacent to the coal preparation plant.

In 1978 after the installation of the hydrocyclones and filters, the plant was used to clean more than 7000 t of coal from the ISU mine. Since this coal had been stockpiled for nearly a year, it was highly weathered and the proportion of fines which passed through the plant exceeded the design limits. Consequently the dewatering cyclones, refuse filter, and Lamella thickener were all overloaded. Therefore, less-than-optimum cleaning of fine-size coal was achieved. Late in 1978, a 500-t sample of freshly mined coal from the Dahm mine in Iowa was processed. Since this sample had a normal proportion of fines, the fine coal circuit was not overloaded and the entire plant operated normally.

At present the operation of the main plant is being integrated with the operation of the froth flotation and oil agglomeration unit described below. This will make it possible in the future to compare the effectiveness of froth flotation, oil agglomeration, and hydrocyclone for cleaning fine-size coal. It is hoped that this comparison can be extended further to include high gradient magnetic separation in the future.

B. Froth Flotation and Oil Agglomeration Unit

Construction of the unit (Fig. 7.16) for demonstrating the froth flotation and oil agglomeration methods of cleaning fine-size coal was completed in 1978. This unit includes equipment for grinding and chemically pretreating 500-kg batches of coal and for continuously beneficiating the pretreated coal by froth flotation or oil agglomeration at a rate of 50 to 150 kg/hr. It also includes means for pelletizing the beneficiated coal.

A schematic diagram of the unit is presented in Fig. 7.17. Coal fines from the Ames coal preparation facility are placed as an aqueous slurry in either of two agitated tanks which serve for both storage and chemical pretreatment. For the pretreatment step, an

Figure 7.16 Equipment for cleaning fine-size coal is supported by the structure in the foreground.

alkali is added to the coal slurry which is then heated to the required temperature. Air is introduced next to oxidize the surface of the pyrite particles. After the oxidation step, the slurry is cooled to a temperature appropriate for the subsequent separation steps. If a finer particle size is required, the coal is ground with a ball mill before applying the chemical treatment. The ball mill circuit includes cyclones for both thickening the pulp supplied to the mill and classifying the particles according to size. Consequently only the coarser particles enter the ball mill.

After the feed has been adequately ground and/or pretreated, it is pumped to either a bank of froth flotation cells or the first

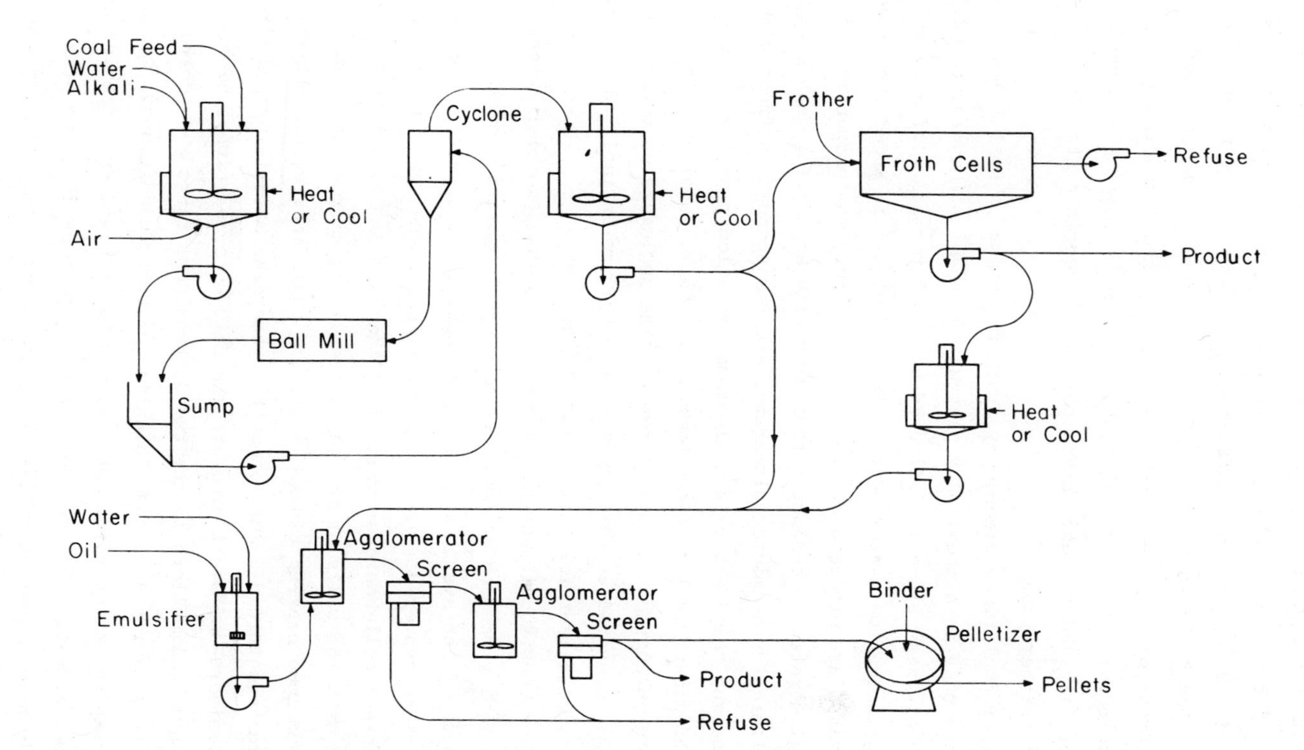

Figure 7.17 Schematic flowsheet of the froth flotation and oil agglomeration unit.

stage of an oil agglomeration system. If the feed is directed to the bank of flotation cells, a frothing agent is added and the coal is floated and removed in the froth while the refuse is removed in the underflow. In cases where a flotation collector is required, the collector may be mixed with the feed in a small conditioning tank just before it enters the flotation cells. The float product is either filtered to recover the coal or placed in a storage tank to await further treatment.

Either coal fines cleaned by froth flotation or coal fines which have only been chemically pretreated can be oil agglomerated. A slurry of these fines is delivered to the first stage of a two-stage agglomeration system (Fig. 7.18). Fuel oil is added and micro-agglomerates are produced by high-shear mixing. The suspension of microagglomerates is conducted to a vibrating screen for dewatering and desliming. The microagglomerates are resuspended in fresh water in the second stage where less vigorous agitation promotes the coalescence and growth of larger agglomerates. The suspension is then dewatered on another vibrating screen. The agglomerated coal can either be recovered at this point or conveyed to an inclined rotating disk pelletizer for further size enlargement.

After construction of the unit was completed, several batches of coal were processed to test the mechanical operation of the system. While most of the system met design expectations, several minor problems were encountered. The two pretreatment tanks were out of round; so the panel coil heaters attached to the outside did not make good contact with the walls of the tanks, resulting in poor heat transfer rates and long heat-up times. Also, the coal fines supplied by the main plant contained a small amount of oversize material which plugged pump strainers and small pipes in the unit. In addition, the carbon steel turbine impellers in the two pretreatment tanks were corroded by the coal slurry even under alkaline or basic conditions. The problem caused by the oversize material has been solved by screening the feed to the unit. The impeller corrosion problem is being attacked by applying and testing various protective coatings. Finally the problem of long heat-up times will be solved by either direct steam

Figure 7.18 Agitated tanks and screens for the oil agglomeration and recovery of fine coal.

injection or the addition of internal heating coils to the pretreatment tanks.

In the future the unit will be utilized in two ways. One way will be to treat a sidestream of fine-size coal from the main plant. Another way will be to treat small batches of coal which are introduced directly into the unit. In order to have the capability for direct introduction, an independent coal crushing and feeding system was added during 1979. This system adds flexibility and makes it possible to operate the unit when the main plant is shut down and to investigate the application of froth flotation or oil agglomeration to coals which cannot always be treated in the main plant because of cost considerations. Consequently, it is possible to treat more kinds of coal in the smaller unit than in the main plant.

C. Use of Facilities

The ISU coal preparation plant is the only large integrated plant in the public domain which is presently available for demonstrating and applying a variety of cleaning methods on an experimental basis. Therefore it represents an important and a unique national asset which can and will be used for a wide range of tasks. First, it will be used to develop and demonstrate new cleaning technology such as the use of chemical pretreatment to improve the efficiency of the froth flotation and oil agglomeration methods of cleaning fine coal. Second, it will be used to develop and demonstrate optimum cleaning methods for specific types of coal. Third, it will be used to clean and prepare sizable lots of coal for special purposes such as for conducting boiler tests or pilot plant tests of specific conversion processes.

7.5 CONCLUSIONS

It has been shown by means of small-scale laboratory experiments that the separation of coal and pyrites by oil agglomeration can be enhanced by an oxidative chemical pretreatment step. This step involves treating coal fines with an aerated solution containing 1 to 2% sodium carbonate at temperatures in the range of 50 to 80°C for about 15 min. By applying this treatment to 400 mesh × 0 size coal from the Iowa State University demonstration mine and then agglomerating the coal with a mixture of No. 200 LLS and No. 6 fuel oils, the inorganic sulfur content of the coal was reduced 88% and the ash content 63% while 93% of the combustible organic matter of the coal was recovered. When the same coal was oil agglomerated without pretreatment, the inorganic sulfur content was reduced only 40% and the ash content 35% while the recovery of combustible matter was the same. Similar results were realized with No. 9 seam coal from Western Kentucky and poorer results with an Illinois No. 5 coal.

The pretreatment step described above has also been shown to reduce the floatability of clean, unoxidized pyrite in the presence of MIBC, while not greatly affecting the floatability of an Iowa bituminous coal. In addition, various alkalis have been shown to reduce the floatability of clean pyrite more than the floatability of coal. Consequently, the combination of an oxidative surface treatment in an alkaline solution seems to hold considerable promise as a means of depressing pyrite in coal flotation.

Other laboratory experiments have shown the advantage of combining various comminution and physical separation methods to clean coal containing various forms of mineral impurities. Also, these experiments have shown the importance of chemical comminution and/or fine grinding for unlocking mineral impurities in coal containing an appreciable fraction of finely disseminated pyrites and other mineral matter.

An experimental coal preparation plant has been constructed for developing and demonstrating various methods of coal cleaning including those based on hydrocyclones, heavy-media cyclones, froth flotation, and oil agglomeration which are specifically designed for cleaning fine coal. Although this plant has been used until now to determine the technical and economic feasibility of cleaning Iowa coals, it will be used in the future for a broader range of tasks. These tasks include developing and demonstrating new cleaning technology such as that based on the laboratory work described above, developing and demonstrating optimum treatment schemes for specific coals, and preparing clean coal for special research and development applications.

ACKNOWLEDGMENT

This work was supported by the U.S. Department of Energy, Division of Fossil Energy, and the Iowa Coal Project.

REFERENCES

1. R. T. Greer, Characterization of Coal Microstructure, in *Advanced Development of Fine Coal Desulfurization and Recovery Technology*, by T. D. Wheelock, R. T. Greer, R. Markuszewski, and R. W. Fisher, Annual Technical Progress Report, Oct. 1, 1976-Sept. 30, 1977, IS-4363, Ames Laboratory, U.S. Department of Energy, Iowa State University, Ames, Iowa, March 1978.
2. C. E. Capes and R. J. Germain, Selective Oil Agglomeration in Fine Coal Beneficiation, in *Physical Cleaning of Coal* (Y. A. Liu, ed.), Marcel Dekker, New York, 1982, Chap. 6.
3. K. J. Miller and A. W. Deurbrouck, Froth Flotation to Desulfurize Coal, in *Physical Cleaning of Coal* (Y. A. Liu, ed.), Marcel Dekker, New York, 1982, Chap. 5.
4. H. J. Gluskoter, Forms of Sulfur, in *Coal Preparation* (J. W. Leonard and D. R. Mitchell, eds.), 3d ed., AIME, New York, 1968, pp. 1-50.
5. V. A. Glembotskii, V. I. Klassen, and I. N. Plaksin, *Flotation*, Primary Sources, New York, 1972, pp. 450-457.
6. G. St. J. Perrott and S. P. Kinney, The Use of Oil in Cleaning Coal, *Chem. Met. Engr., 25*(5), 18 (1921).
7. D. J. Brown, Coal Flotation, in *Froth Flotation, 50th Anniversary Volume* (D. W. Fuerstenau, ed.), AIME, New York, 1962, p. 518.
8. R. E. Zimmerman, Flotation Reagents, in *Coal Preparation* (J. W. Leonard, ed.), 4th ed., AIME, New York, 1979, pp. 10-82 to 10-84.
9. F. F. Aplan, Coal Flotation, in *Flotation, A. M. Gaudin Memorial Volume* (M. C. Fuerstenau, ed.), AIME, New York, 1976, p. 1235.
10. H. F. Yancey and J. A. Taylor, Froth Flotation of Coal; Sulphur and Ash Reduction, Report of Investigations, No. 3263, U.S. Bureau of Mines, Washington, D.C., 1935.
11. A. F. Baker and K. J. Miller, Hydrolyzed Metal Ions as Pyrite Depressants in Coal Flotation: A Laboratory Study, Report of Investigations, No. 7518, U.S. Bureau of Mines, Washington, D.C., 1971.
12. F. F. Aplan, Use of the Flotation Process for the Desulfurization of Coal, in *Coal Desulfurization: Chemical and Physical Methods* (T. D. Wheelock, ed.), *ACS Symposium Series*, No. 64, American Chemical Society, Washington, D.C., 1977, pp. 70-82.
13. C. E. Capes, A. E. McIlhinney, A. F. Sirianni, and I. E. Puddington, Bacterial Oxidation in Upgrading Pyritic Coals, *Can. Min. Metal. Bull., 66*(739), 88 (1973).
14. S. Min, *Physical Desulfurization of Iowa Coal*, Report No. IS-ICP-35, Energy and Mineral Resources Research Institute, Iowa State University, Ames, Iowa, Mar. 1977.

15. H. V. Le, *Flotability of Coal and Pyrite,* Report No. IS-T-779, Ames Laboratory, U.S. Department of Energy, Iowa State University, Ames, Iowa, July 1977.

16. T. J. Laros, *Physical Desulfurization of Iowa Coal by Flotation,* Report No. IS-ICP-47, Energy and Mineral Resources Research Institute, Iowa State University, Ames, Iowa, June 1977.

17. E. C. Patterson, H. V. Le, T. K. Ho, and T. D. Wheelock, Better Separation by Froth Flotation and Oil Agglomeration, CEP Technical Manual, *Coal Processing Technology,* AIChE, 5, pp. 171-177 (1979).

18. E. C. Patterson, *The Effect of Chemical Pretreatment on the Desulfurization of Coal by Selective Oil Agglomeration,* Report No. IS-ICP-64, Energy and Mineral Resources Research Institute, Iowa State University, Ames, Iowa, Sept. 1978.

19. C. E. Capes, A. E. McIlhinney, and R. D. Coleman, Beneficiation and Balling of Coal, *Transactions Society of Mining Engineers, AIME, 274,* 233 (1970).

20. E. A. Sondreal, P. H. Tufte, and W. Beckering, Ash Fouling in the Combustion of Low Rank Western U.S. Coals, *Combustion Science and Technology, 16,* 95 (1977).

21. R. J. Germain, The Steel Company of Canada, Ltd., Hamilton, Ontario, Canada, private communication, Oct. 12, 1978.

22. T. D. Wheelock and T. K. Ho, Modification of the Floatability of Coal Pyrites, presented at 1979 AIME Annual Meeting, New Orleans, La., Feb. 18-22, 1979.

23. R. T. Greer, Coal Microstructure and Pyrite Distribution, in *Coal Desulfurization: Chemical and Physical Methods* (T. D. Wheelock, ed.), *ACS Symposium Series,* No. 64, American Chemical Society, Washington, D.C., 1977, pp. 3-9.

24. S. Min and T. D. Wheelock, A Comparison of Coal Beneficiation Methods, in *Coal Desulfurization: Chemical and Physical Methods* (T. D. Wheelock, ed.), *ACS Symposium Series,* No. 64, American Chemical Society, Washington, D.C., 1977, pp. 83-100.

25. R. A. Grieve and R. W. Fisher, *Full Scale Coal Preparation Research on High Sulfur Iowa Coal,* Report No. IS-ICP-53, Energy and Mineral Resources Research Institute, Iowa State University, Ames, Iowa, Feb. 1978.

8

Practical Aspects of Filtration and Dewatering in Physical Cleaning of Fine Coal

DONALD A. DAHLSTROM
RONALD P. KLEPPER

Envirotech Corporation
Salt Lake City, Utah

8.1 INTRODUCTION

The use of filtration and sedimentation equipment in coal cleaning has broadened since the price of coal has increased substantially in the past few years. The material in this chapter covers the theory of predicting filtration and thickening performance as well as discussing some of the practical considerations that cause variations in coal preparation plant operations.

Different types of moisture found in coal are defined and described so that surface moisture, commonly used as a performance measurement in dewatering, can be better discussed in the test. Applied filtration theory is also discussed so that rates and performance can be predicted by bench-scale test procedures. Innovations that are used in coal dewatering or in other mineral processing industries are described.

Fine coal refuse dewatering equipment operation and design are discussed. The sizing of both conventional and high-rate thickeners is presented along with the effect of the fine coal circuit operation on the dewatering equipment performance. Several types of dewatering alternatives are evaluated for different types of fine coal refuse.

8.2 DEWATERING OF FINE COAL

A. Dewatering of Fine Coal by Continuous Filtration

Dewatering of fine coal by continuous filtration involves the filter-cake formation and removal of surface moisture by drawing air through the capillaries of the cake. This continuous dewatering operation is actually a cyclic process in that a batch process (with continuous cake discharge) is repeated every revolution or cycle of the filter operation. In fine coal filtration, the cycle will generally range from 1.5 to 5 min/rev. Thus, both cake formation and cake dewatering occur in relatively short periods of time.

Figure 8.1 shows a schematic of a vacuum Agidisc filter. This disk-type filter is very widely employed in coal filtration in probably 90 to 95% of the applications, because cakes of at least 3/8 in. (9.5 mm) thick are easily formed. This minimum thickness is required for the disk-type filters for proper cake discharge. The disk filter yields the lowest capital cost per unit area of filtration and also requires the least amount of floor space. However, agitation of the feed slurry within the filter tank is imperative, as generally the solids are −28 mesh in sizes (typically 5 to 10% +28 mesh) and will

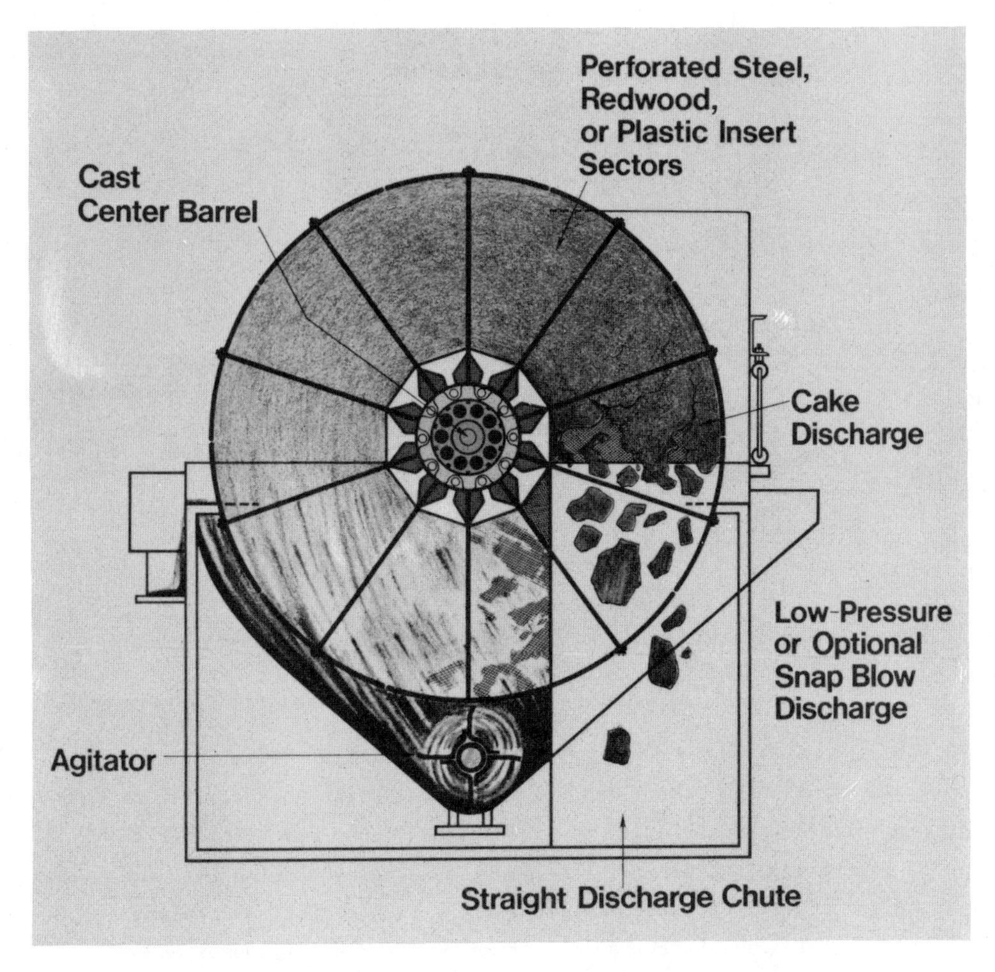

Figure 8.1 Schematic of an Agidisc filter. (Courtesy of Envirotech Corporation.)

otherwise settle out. Thus, a rotating shaft with special blades between each disk is employed at the base of the filter tank to maintain suspension of the slurry. Note that the scraper discharge drum and the string discharge drum filters have also been employed for coal filtration but to a much lesser extent.

Figure 8.2 shows a schematic of a typical Agidisc filter station. It is seen that the filter station consists not only of the filter,

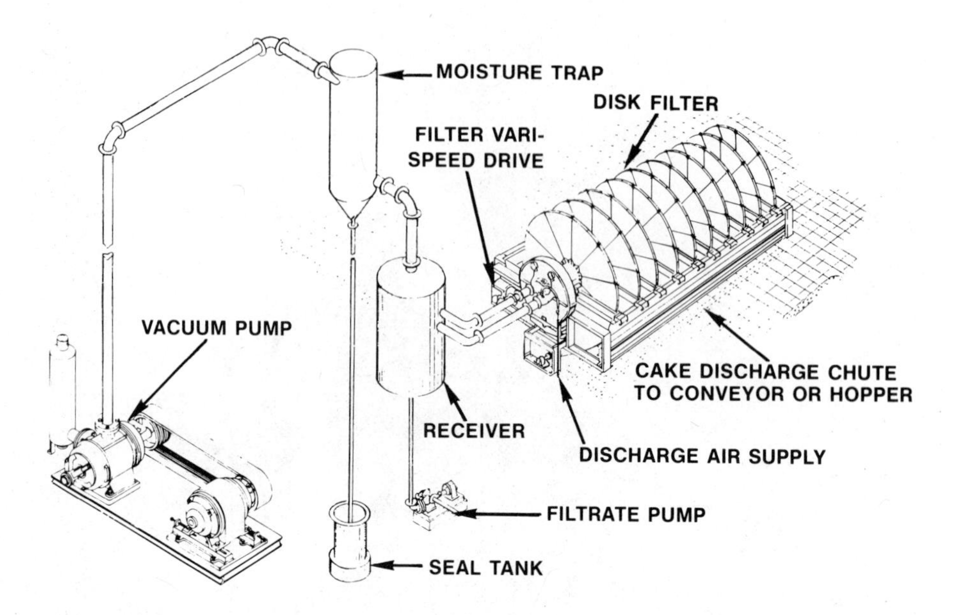

Figure 8.2 Schematic of coal Agidisc filter installation. (Courtesy of Envirotech Corporation.)

but also of a receiver which separates the filtrate from the air drawn through the cake. The receiver usually includes a moisture trap with a barometric leg to protect the vacuum pump from a slug of filtrate. In addition, cake is discharged by blowing air back through the filter valve, internal pipe, disk sector, and filter media. This cake discharge operation requires a compressor and normally a compressed air tank to minimize compressor horsepower.

B. Factors Affecting the Coal Surface Moisture

To fully understand how surface moisture can be removed from fine coal, it is necessary to identify how water exists in fine coal. Figure 8.3 shows a typical group of fine coal particles and identifies several forms of water associated with this mass of material [10]:

A. *Interior adsorption water*. Contained in micropores and microcapillaries within each coal particle, deposited during formation

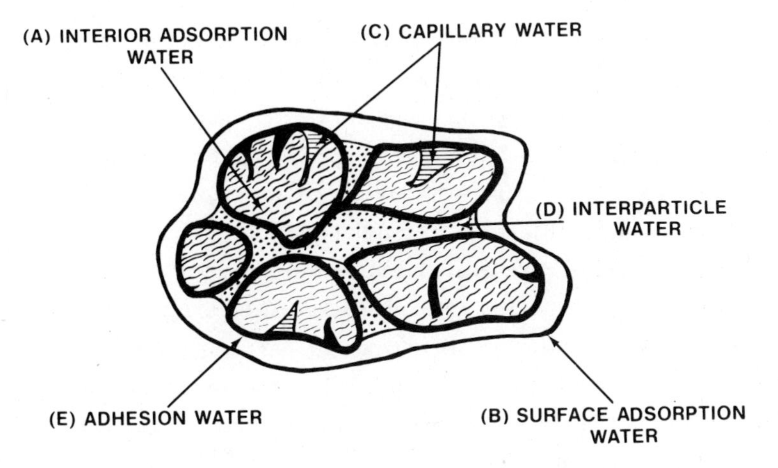

Figure 8.3 Types of water in coal particles.

B. *Surface adsorption water.* Forms a layer of water molecules adjacent to coal molecules, but on the particle surface only
C. *Capillary water.* Contained in capillaries and small crevices found in the surface of the particles
D. *Interparticle water.* Contained in capillaries and small crevices found between two or more particles
E. *Adhesion water.* Forms a layer or film around the surface of individual or agglomerated particles

The water which can be readily removed using mechanical fine coal dewatering devices, such as vacuum filters, is categorized by types D and E; and is generally termed surface moisture. Water type C can be removed partially, depending upon the size of the openings in the coal surface and the drying time available in the filter cycle. Inherent moisture is the general term used in a typical proximate analysis of coal to describe the water as defined by types A and B. ASTM test procedure D271 for gaseous fuels, coal and coke, used to measure surface moisture relys on ambient evaporation for water removal; heat is then applied to a temperature a few degrees above the boiling point of water to vaporize all forms of water on or in the coal yielding inherent moisture.

In coal processing, the surface moisture is eliminated by first removing the interparticle water formed in a filter cake; this is done by applying a pressure differential across the cake so the water flows from the compacted particles. Next, air is used to remove the adhesion water (type E). In this step, momentum is imparted to the water as the air passes over the water surface. Limited evaporation also occurs; the amount of water removed is dependent on the available energy or temperature of the coal and on the relative humidity of the air passing through the cake.

The amount of the adhesion and interparticle water found in a filter cake depends on the size and number of particles. Particle shape and size distribution of particles determine the void space between particles which contains water. This can be better understood by looking at the voidage expression for a bed of granular solid particles of mixed sizes given below [15, pp. 5-52 to 5-53]:

$$\varepsilon = 1 - \frac{S\phi_s D_p}{6} \tag{8.1}$$

where

$$\frac{1}{D_p} = \Sigma \frac{x}{D_{p,x}} \tag{8.2}$$

In Eqs. (8.1) and (8.2), the different symbols are defined as follows:

D_p = average particle diameter, defined as diameter of sphere of same volume as particle, ft
ε = voidage (fractional free volume), dimensionless
s = specific surface, or area of particle surface per unit volume of bed, ft^2/ft^3
ϕ_s = shape factor of solid defined as quotient of area of sphere equivalent to volume of particle divided by actual surface of particle, dimensionless (for natural coal dust, up to 3/8 in., ϕ_s = 0.65; for pulverized coal dust, ϕ_s = 0.73)
x = weight fraction, dimensionless, of particle diameter $D_{p,x}$ ft

Since the voidage is directly proportional to the average particle diameter, a greater amount of relatively large particles (+200 mesh) will create more interparticle water when the particles are formed as in a filter cake. The quantity of the adhesion water found in a filter cake or packed bed of solids is directly proportional to the total surface area of the solids. Equation (8.1) shows that surface area is inversely proportional to particle diameter.

To better understand the effect of particle size of coal on the ability to remove water from filter cakes, it is helpful to review the fluid flow characteristics through a packed bed [15,p. 5-52] as illustrated, for example, in Eq. (8.3):

$$G = \left[\frac{\Delta P\ D_p g_c \phi_s^{3-n} \varepsilon^3}{2f_m L(1-\varepsilon)^{3-n}}\right]^{1/2} \tag{8.3}$$

where

G = fluid superficial mass velocity based on empty cross section, lb/(sec)(ft^2)
ΔP = pressure drop, lb_f/ft^2
D_p = average particle diameter, ft
g_c = dimensional constant, 32.17 lb ft/(lb_f)(sec^2)
ρ = fluid density, lb/ft^3
ϕ_s = shape factor of solid (dimensionless)
ε = voidage (dimensionless)
f_m = friction factor (dimensionless)
L = bed depth or thickness, ft
n = exponent, dimensionless (a function of the Reynolds number)

This correlation, along with Eq. (8.1), shows that when the average particle diameter is decreased, which causes the voidage to decrease, the superficial mass velocity or flow of fluid from the packed bed or filter cake is decreased substantially. Although Eq. (8.3) does not exactly describe fluid characteristics of a compressible fluid, such as air, through the filter, the same general relationship exists between the mass flow and average particle size.

Removing the adhesion water from the surface of the coal is thought to be affected predominantly by the mass flow of air sweeping across the surface of the particles. The effect of a smaller average particle diameter increases the total surface area of the solids per unit volume, because the voidage is decreased. This increases adhesion water and at the same time limits the mass flow rate of both water and air through the filter cake. The net effect of particle size distribution on coal surface moisture is that the average particle size for a given volume of solids will control the rate of water removal and govern the amount and type of water present.

Ash content found in the clean fine coal dewatered by a vacuum disk filter is an indicator of the amount of micrometer-sized particles found in the sample. Generally, the ash is the remainder of clay, shale, or rock dust that was not removed in the cleaning process. Fine refuse found along with the fine coal usually is smaller in size than coal dust. Some clays can have particle diameters between essentially zero to 10 μm. If the cleaning operation of fine coal is not designed properly to remove the ultrafine refuse material, the moisture content in the filter cake can increase due to the decrease of the average particle size. Note that the variables which govern the moisture content in fine coal filtration are also basically applicable to fine refuse vacuum filtration. However, the coal content is normally small in the fine refuse material.

There are three types of moisture which are measured by the analytic procedure for surface moisture: the remaining interparticle water, adhesion water, and the capillary water. Capillary water is the type of water found in capillaries and small crevices in the surface of the coal particles. The quantity and size of these openings is dependent upon the physical structure of the coal.

Coal has certain physical properties which are used for classification according to a system that ranks the coal from a high to low rank, i.e., anthracite to bituminous to lignite. Basically, this system describes the rank using measurements such as volatility, hardness, moisture, carbon content, and heating value of the coal. The number of capillaries within and crevices on the surface of a coal particle are a function of the coal age, density, and hardness. Older or higher-rank coals are harder and more dense than those in lower ranks. Because of this relationship, the amount of the inherent moisture increases with decreasing rank since more internal capillaries exist.

There is a direct relation between the amount of surface capillary water and the performance of the vacuum filter measured by "surface moisture." If the capillary water is relatively large in proportion to adhesion water and interparticle water, then it will be measured as a significant portion of the surface moisture as defined in ASTM standard D271. Surface capillary water will not normally be removed during the filtration cycle due to the limited driving force and time in the vacuum filter cycle. Filtration performance, measured as surface moisture, will show a significant difference when fine coal product from two plants are compared, i.e., plant A processes coal with more surface capillary water than found in the coal processed in plant B. Plant A will produce a fine coal product from the filter with higher surface moisture than plant B due to the greater amount of surface capillary water. Surface capillary water increases with decreasing rank of coal.

Because different types of coal yield different amounts of surface moisture, it is necessary to use different flowsheets to dry the fine coal product, i.e., flowsheets with thermal drying, and others with drying on filters only. Thermal drying can remove both surface adsorption water and capillary water; it also has the potential to remove part of the interior adsorption water. Keeping the relationship of the different types of water in mind when designing vacuum filtration equipment for fine coal and refuse dewatering will allow more accurate prediction of the moisture content of the product.

C. Applied Filtration Theory for Fine Coal Dewatering

The proper sizing of vacuum disk-type filters for fine coal dewatering requires the determination of the rates of cake formation and cake dewatering. The latter are complex functions of such variables as filter solids concentration, filter cycle time, cake air requirements, filter media, and filter feed size consists, etc. In what follows, an applied filtration theory for fine coal dewatering is described in terms of cake formation and cake dewatering rates. Practical design procedures for sizing filters from bench-scale test results and plant-scale data correlations are also discussed.

Cake Formation Rate. The well-known Hagen-Poiseuille equation is generally adapted to relate the differential or instantaneous rate of filtration per unit area and the ratio of a driving force, pressure drop, to the product of viscosity by the sum of cake resistance and filter medium resistance. This yields the following expression for cake formation rate [12, p. 12-56]:

$$\frac{1}{A}\frac{dV}{d\theta} = \frac{\Delta P}{\mu(\gamma_{av} wV/A + \gamma_m)} \tag{8.4}$$

where

- V = volume of filtrate, ft^3 or m^3
- A = filter area, ft^2 or m^2
- θ = filtration time, sec, min, or hr
- ΔP = total pressure drop across the filter cake and medium, lb_f/ft^2, mmHg, or atm
- μ = viscosity of liquid filtrate, lb/(ft)(sec) or P
- γ_{av} = the average cake resistance, ft/lb or m/kg
- w = weight of dry solids in the filter cake per unit of filtrate, lb/ft^3 or kg/m^3
- γ_m = resistance of the filter medium and internal drainage network of the filter, ft^{-1} or m^{-1}

When the cake is composed of hard granular particles which make it rigid and incompressible, the filter medium resistance is often considered negligible compared to the filter cake resistance if the medium binding is not a characteristic of the application. Thus, if γ_m is neglected, Eq. (8.4) can be integrated from $V = 0$ when $\theta = 0$ to $V = V_f$ (volume of filtrate per filter cycle) when $\theta = \theta_f$ (form time, or cake formation time per filter cycle). The integrated result can be written as

$$Z_f = \frac{V_f w}{A\theta_f} = \left(\frac{2w\Delta P}{\mu\gamma_{av}\theta_f}\right)^{1/2} \tag{8.5}$$

In Eq. (8.5), Z_f is defined as the form filtration rate with a unit of weight of dry solids per unit area per unit time of cake formation, lb/(hr)(ft^2). Equation (8.5) indicates that Z_f is ideally a function of the square root of ΔP and w, and inversely proportional to the square root of the terms μ, γ_{av}, and θ_f.

It should be emphasized that Eq. (8.5) defines the form filtration rate Z_f as the rate per unit time of cake formation. In order to obtain the equivalent filtration rate for a total filter cycle, it is necessary to multiply Z_f by the fraction of the time during the filter cycle that is devoted to cake formation. Note that the apparent submergence of the filter disk is usually 35 to 40% to avoid stuffing boxes on the filter trunion. This leads to maximum effective submergence of about 30 to 35%. Also, to obtain the proper filtration rate in full-scale operations, Z_f is generally multiplied by a scale-up factor of about 80% or 0.8. This factor accounts for the fraction of the filter cycle needed for cake discharge, as well as the uncertainty involved in scale-up from bench-scale filter leaf tests [15, pp. 19-60 to 19-61; 21].

It is worthwhile to examine further the quantitative effect of the cake formation time (form time) θ_f, or the equivalent total cycle time, θ, on the form filtration rate, Z_f, according to Eq. (8.5). The latter expression indicates that when all other operational parameters are held constant, a log-log plot of Z_f as a function of θ_f should ideally yield a straight line of slope −0.5. In fine coal filtration, the temperature is generally constant, and the viscosity, μ, of the liquid filtrate is also constant. In addition, the total pressure drop across the filter cake and medium, P, is essentially maintained constant over the cake formation time θ_f; and economics will generally justify a vacuum level of 20 to 22 in. of mercury (508 to 559 mmHg). Actual full-scale filtration data, as illustrated in Fig. 8.4, indicate the coal slurries generally exhibit relatively ideal log-log plots of Z_f versus θ_f with slopes of −0.50 to −0.55, unless feed solids concentrations are too low. Figure 8.4 also shows that the filtration rate can be increased by reducing the cake formation time. However, a sufficient cake formation time must be employed to form at least a 3/8-in. (9.5 mm) cake as required by the disk-type filter. Normally, full-scale sizing will be based on a ½-in. (12.7 mm) cake for maintaining plant operational flexibility. Note that correlations of full-scale fine coal filtration data also show that

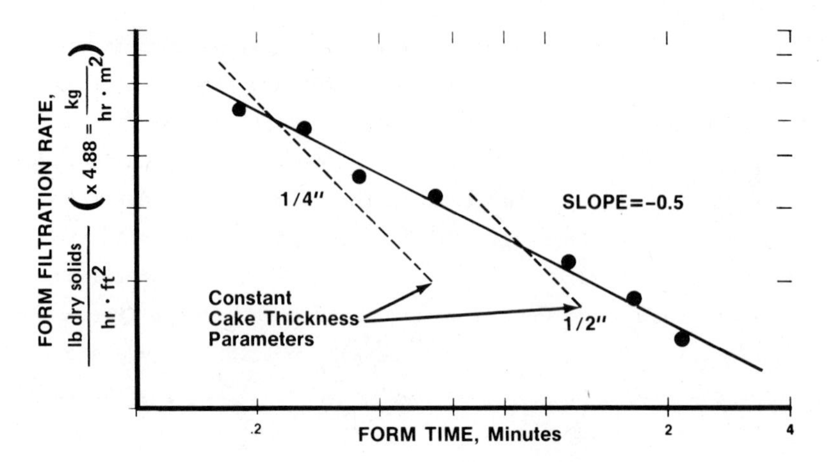

Figure 8.4 Typical plot log of form filtration rave vs. log of form time for fine coal.

a log-log plot of the dry cake filtration rate, or called the bone dry cake rate (pounds of dry solids per hour per square foot of filter area), versus the total filter cycle time generally yields a straight line of slope −0.5, when all other operational parameters are held constant. Thus, when increasing or decreasing the filter cake output as a function of changing the total filter cycle time, the resulting change in dry cake filtration rate is equal to the square root of the original total filter cycle time divided by the new total filter cycle time. Stated in mathematical terms, this relationship is as follows [12, p. 12-60]:

$$\begin{pmatrix}\text{New bone dry}\\ \text{cake rate}\end{pmatrix} = \begin{pmatrix}\text{old bone dry}\\ \text{cake rate}\end{pmatrix} \times \left(\frac{\text{old total filter cycle time}}{\text{new total filter cycle time}}\right)^{1/2} \tag{8.6}$$

Because of the importance of the cake formation time or the total filter cycle time in determining the cake filtration rate, nearly every filter is equipped with a variable-speed filter drive for controlling the cycle time. Whenever possible, it is recommended that the disk-type vacuum filters for fine coal dewatering be sized at a total filter cycle time of between 2 and 9 min/rev.

Cake Dewatering Rate. The cake dewatering rate is a complex function, which can best be portrayed by the following expression [13]:

$$\%M = f(F_a, \%M_r, X) \tag{8.7}$$

where

$\%M$ = weight % surface moisture in the filter cake
F_a = approach factor [see Eq. (8.8)]
$\%M_r$ = residual cake surface moisture, which is defined as the cake moisture at infinite time with saturated air being drawn through the cake at a pressure drop ΔP
X = particle shape, surface characteristics, and size distribution

Then term X is a complex one and can be best approximated by the parameter of size distribution. Generally, the particle shape and surface characteristics will remain constant as the size distribution changes. The residual cake surface moisture $\%M_r$ is the minimum moisture which can be obtained and is approached asymptotically as the dewatering time is increased.

The approach factor F_a has also been correlated in the literature according to the following equation:

$$F_a = \frac{\theta_d}{W} \Delta P \left(\frac{CFM}{ft^2}\right)_{\Delta P} \tag{8.8}$$

where

θ_d = drying or dewatering time during the total filter cycle, min
W = weight of dry solids in the filter cake per unit filter area, lb/(ft^2)(rev) or kg/(m^2)(rev)
$\left(\frac{CFM}{ft^2}\right)_{\Delta P}$ = cake air requirement, ft^3 of air per min per ft^2 of filter area drawn through the filter cake measured at a given pressure drop ΔP or vacuum condition

The subscript ΔP indicates that the parameter of pressure drop should be employed in determining the approach factor F_a. Figure 8.5 is a typical plot of the filter moisture %M as a function of the simplified approach factor θ_d/W at two different vacuum levels observed in fine coal dewatering. It can be seen that the filter cake moisture exhibits a sharply descending curve which passes through a "knee" and then becomes asymptotic to a minimum value corresponding to a residual

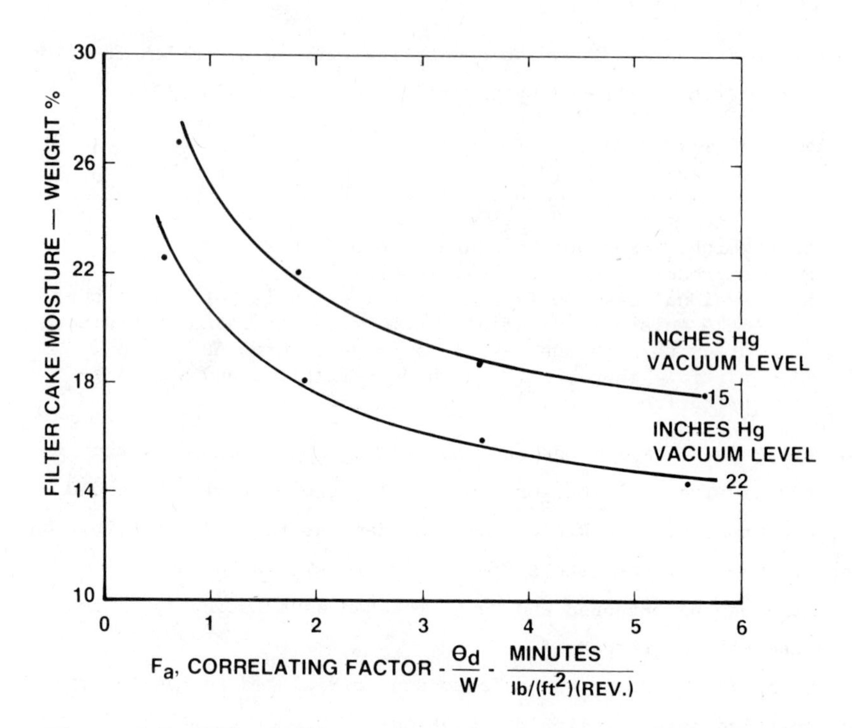

Figure 8.5 Typical filter cake moisture content as a function of the correlation factor with parameters of pressure drop for fine coal.

cake surface moisture $\%M_r$. This asymptotic dependence suggests that when using the same type of correlation for designing a full-scale filter for fine coal dewatering, the design value of θ_d/W shall be selected to the right of the "knee" of Fig. 8.5. In this way, normal plant fluctuations will only result in a small variation in moisture content of the discharged filter cake.

Plant-Scale Data Correlations. It is evident from the preceding discussion that the rates of cake formation and cake dewatering are complex functions of many variables such as the following:

1. Size distribution of filter feed solids
2. Feed solids concentration
3. Pressure drop across cake and medium, which is the driving force

4. Cake formation time per filter cycle
5. Cake dewatering time per filter cycle
6. Specific resistance of the filter cake
7. Filter medium
8. Filtration temperature
9. Filter cake thickness as it affects cake dewatering
10. Particle shape and surface characteristics
11. Cake air requirement
12. Inherent moisture of the solids
13. Viscosity and surface tension of the liquid filtrate
14. Types and amounts of polyelectrolytes used to coagulate fines

These are the major factors, although such things as plant operation and maintenance can be highly important if not practiced correctly.

Any attempt to quantitatively identify the effects of all variables influencing fine coal dewatering would be exceedingly difficult, if not impractical. However, the quantitative efforts of a number of major variables on both fine coal filtration rate and cake surface moisture content have been correlated based on plant-scale data obtained from numerous fine coal dewatering installations. These major variables include the following [12, pp. 12-57 to 12-58]:

1. Filter cycle time
2. Cake vacuum requirement
3. Filter medium
4. Filter feed solids concentration
5. Size distribution of filter feed solids

The effects of the first two major variables have already been discussed in the preceding section.

It has been suggested that under conditions of fixed filter cycle time and cake air requirement (and hence, filter vacuum level), the plant-scale fine coal filtration rates from numerous installations using open-mesh filter media can be correlated as smooth functions of an emperical factor, $(\% -200\text{ mesh})^{1/2}$ (% ash in the –200-mesh fraction), with parameters of filter feed solids concentrations. Figure 8.6 shows a typical plot for such a correlation [12, pp. 12-65 to 12-67]. The plant-scale data included in the figure have been normalized to a total filter cycle time of 3 min/rev for the disk-type filter. To

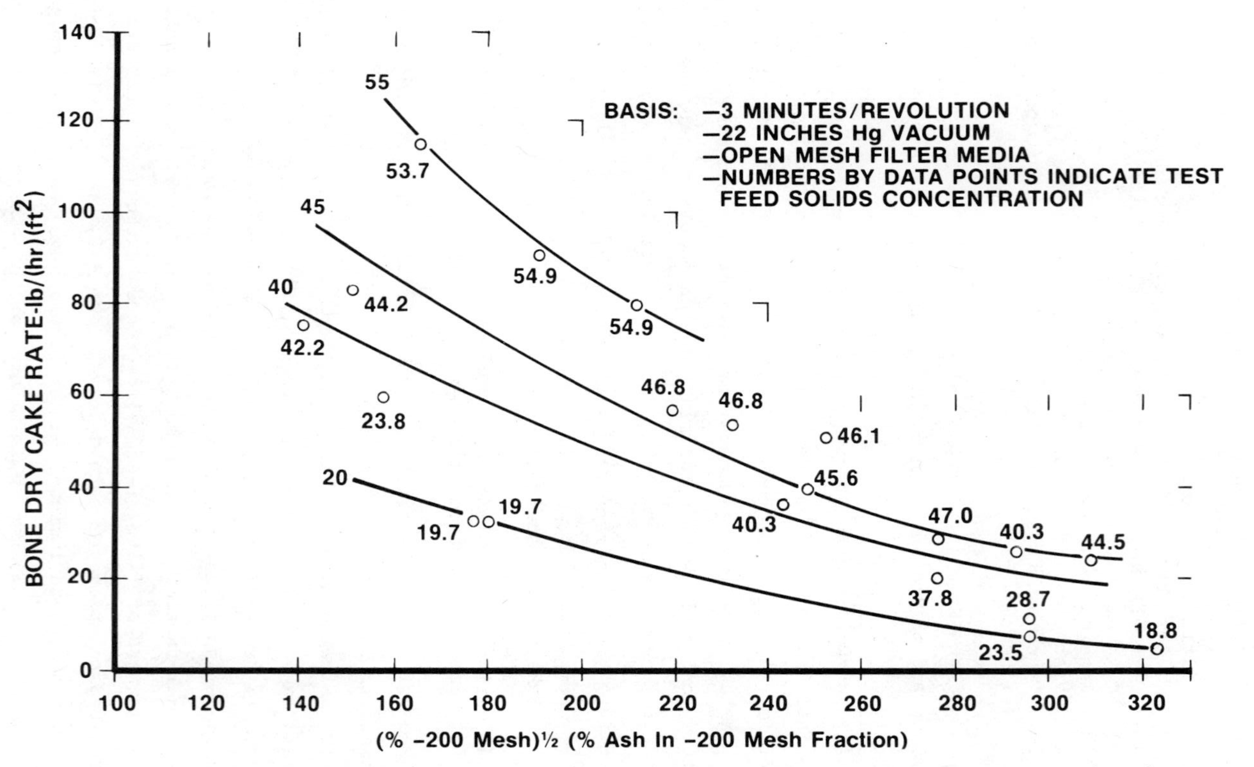

Figure 8.6 Continuous filtration rate for fine nonfloat coal vs. empirical correlating factor for specific permeability with parameters of feed solids concentration. (From Ref. 12.)

predict the results at different cycle times, the rate in the figure is multiplied by a factor of (3 per new cycle time)$^{\frac{1}{2}}$, as suggested by Eq. (8.6). This prediction is valid as long as at least a 3/8-in. cake is formed.

The correlating factor (% —200 mesh)$^{\frac{1}{2}}$(% ash in the —200-mesh fraction) is actually a function of the specific surface area of the filter feed solids. Correlations of fine coal filtration rates based on this factor, using actual plant results, are very dependable for design purposes. These correlations also indicate that the ash in the —200-mesh fraction is the predominant factor with respect to size distribution of filter feed solids, as it is an indirect measure of the amount of the colloidal-type clays and slimes in the feed.

Figure 8.6 also shows the importance of increasing filter feed solids concentration and the effect on increasing the filtration rate. Data for this plot were obtained in the late 1950s when flotation separation was in its infancy and cyclones were predominantly used to separate refuse from —28-mesh raw coal. Cyclones yield solids concentrations in streams going to disk filters that contain solids at concentrations between 20 and 55% by weight. Note that the filtration rates for the different coal slurries increase sharply above about 35% solids. In a cyclone flowsheet, the filtration rate can be increased or decreased by changing the solids concentration in the feed [12, p. 12-68].

Currently, flotation separation has replaced cyclones as the predominant means of cleaning —28-mesh coal. Flotation separation is a more efficient process to separate refuse from clean coal and typically produces a clean coal stream with a solids concentration of between 18 and 27% solids. The ash content of clean coal from flotation is usually lower than from cyclones. As a result of more efficient separation of clay-bearing refuse from coal, filtration rates for a typical —28-mesh coal can range between 60 and 100 lb. dry solids per hour per square foot [293 to 488 kg/(hr)(ft^2)] of filter area based on a large amount of published and operating data as shown in Fig. 8.6 [12, pp. 12-65 to 12-67]. Figure 8.7 illustrates

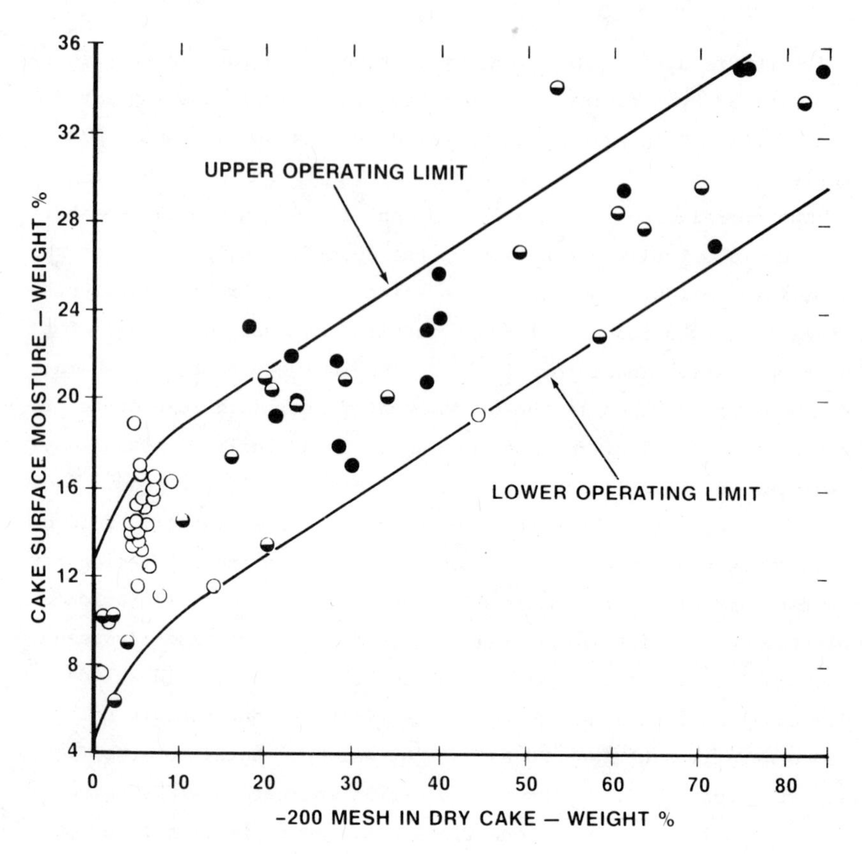

Figure 8.7 Filter cake surface moisture vs. % —200-mesh fuel-scale results for fine coal. (From Ref. 12.)

that a straight-line relationship results for moisture content data obtained with filter feed size distributions above 10% —200 mesh [12, p. 12-64]. An approximate spread of 8 percentage points between maximum and minimum value at any % —200 mesh can also be noted. This range is a function of the particle shape, the surface characteristics, the amount of surface capillary water of the coal, and the design efficiency of the filter station. Generally, the minimum values for all types of coal are approached if vacuum is maintained at 20 to 22 in. of mercury.

Sizing Filters from Bench-Scale Test Results. Bench-scale filtration tests can be conducted to establish filtration rates and surface moisture values for coal preparation plants that are processing coal from a coal seam that has no other operational experience available for comparative sizing. Such tests require relatively simple, bench-scale equipment which can be purchased, rented, or borrowed from most filter manufacturers. Also, specific procedures for carrying out such bench-scale tests are described in many manuals issued by filter manufacturers. For example, Ref. 22 contains the procedure for the determination of filtration characteristics of any given slurry by laboratory test leaf, which can be followed to obtain the necessary data for sizing a vacuum disk-type filter for fine coal dewatering. In particular, bench-scale experiments allow one to determine a cake filtration rate W in terms of the weight of dry solids per hour per square foot of filter area, needed in a given fine coal dewatering problem. To obtain the corresponding filtration rate for the full-scale plant, the required filter-cake surface moisture content specified by the design problem is used together with the correlation of Fig. 8.4 to determine the desired value of the simplified correlating factor θ_d/W at the selected vacuum level. Using the cake filtration rate W obtained from bench-scale tests, the desired dewatering time θ_d can then be determined. With the disk-type filter, dewatering time is normally about 45% of the total filter cycle so that the total filter cycle time can be found by dividing θ_d by 0.45 to give the desired filter cycle time in terms of minutes per revolution (filter cycle). Thus, the full-scale filtration rate becomes

$$\text{Filtration rate} = W \frac{60}{\theta_d/0.45} \times 0.8 \times \frac{\text{weight of dry solids}}{\text{hour} \times \text{unit area}} \tag{8.9}$$

In Eq. (8.9), the term 0.8 is again the scale-up factor. Naturally, the cake formation rate, as previously determined by bench-scale tests according to Eqs. (8.4) and (8.5), must be checked to see if it is larger on a full-scale basis than that determined by Eq. (8.9).

The surface moisture of a fine coal product can be determined directly by preparing a graph such as seen in Fig. 8.5. This graph can be used to show the effects of varying cycle time and tonnage to the filter represented by θ_d, the dry time, and W, the weight of dry solids in the filter cake per square foot of filter area.

Sizing Filters from Existing Installations and Experiences. Most vacuum disk filters used for both clean coal and refuse are sized based on comparing the anticipated fines characteristics to existing installations with similar characteristics to predict the performance of the filters. This method works very well because normally there are several plants that mine coal from the same seam and are located in about the same geological region. Filter performance in the new plant should be similar to existing installations. To effectively size a new filter, it is necessary to understand variables such as coal seam variability, mining technique used to extract the coal, flowsheet design, and equipment used for classifying and cleaning the coal, and how well the existing plants are maintained and operated.

The most important information obtained from existing installations is the range of variability of tonnage and quality of fines that will be dewatered using vacuum disk filters. The most common error committed when sizing a filter is to ignore the range of performance required to dewater all or most of the solids being fed to the filter. An undersized filter results in lost tonnage of recovered fine coal or reduced tonnage through the plant to maintain recovery efficiency. Reduced tonnage through the plant may also be due to the inability to dewater fine refuse at a rate equal to the rate of production of fine refuse. Excess fine refuse then accumulates in the thickner-clarifier until solids overflow in the recycle water back into the cleaning circuit, which causes control problems due to an increase in specific gravity of the water.

D. New Developments in Fine Coal Dewatering

Innovations to Improve Filter Performance. In recent years, there have been many improvements in the disk filter construction which can be cited. For example, the EIMCO Agidisc filter's internal drainage network has been improved through a better understanding of the flow of fluids, particularly two-phase flow, which occurs during portions of the filter cycle. Accordingly, the ports on filtrate and air passages within the filter have been increased so that pressure drop through the internal network is a maximum of 1 in. of mercury. Figure 8.8 is a picture of the fabricated center barrel that is designed for improved fluid flow.

A snap blow technique (see Fig. 8.1) has been developed to improve cake discharge. Compressed air of 20 to 50 lb/in.^{2} (1½ to 3½ atm) is timed to discharge when the blow port is opposite the trunion port. The compressed air is discharged back through the ports, the disk sectors, and filter medium. The resulting shock increases greatly

Figure 8.8 Fabricated center barrel with plastic grid sectors. (Courtesy of Envirotech Corporation.)

the cake discharge as the blow takes place in ½ to a maximum of 1 sec. This is particularly important with textile cloths, because the bag inflates almost instantaneously to jar the cake loose.

The design of the disk sector has also been improved. Formerly, either punched plates or wooden sectors were available. The former exhibit a relatively small open area (about 20%) and the latter have a tendency to plug with fine solids, especially when flocculants are used in the water circuit or in the filter. Molded plastic sectors are now utilized to give better drainage; and with the new plastics available they can be made highly abrasion resistant.

The recent development of synthetic filter media has also been important in obtaining greater resistance to blinding and longer cloth life. Polyethylene, polypropylene, and nylon find the widest use. Furthermore, a number of different weaves have been developed to improve resistance to blinding, while increasing strength and abrasion resistance.

Stainless steel screen has been predominantely employed as a filter medium (normally 40 × 60 mesh), which yields a very long life (up to 2 yr) without using the snap blow technique. For easily filtered coal, this will normally be employed due to the greater life and resulting lower operating cost per ton of coal filtered. Note that the operational lifetime of the stainless steel screen may be reduced if the snap blow discharge technique is used.

Automatic Control of the Filter Station. Because of the nature of the coal processing plant and its economics, automation of the filter station (see Fig. 8.2) has not been as great as in other industries. The filter station is normally the last processing step for fine coal and must treat whatever is fed to it. This increases the importance of designing proper capacity and flexibility into the station, so that it will be able to handle the various surges fed to it. In addition, as much care should be given to the filter feed, so that it is maintained at as high a solids concentration as possible, consistent with proper performance of preceding processing steps. Thus, thickener underflows which are fed to this filter can be automatically controlled by density measurements for controlling the underflow pump speeds. Likewise, when dewatering the flotation coal, the flotation cells are maintained at high froth solids concentrations.

The filter is equipped with a variable-speed drive (see Fig. 8.8) so that its cycle time can be easily changed. As seen earlier, filtration rate is essentially inversely proportional to the square root of the filter cycle time. While the latter is normally economically adjusted, it could be controlled automatically by a level controller in the filter tank changing the speed of the filter drive. If relatively consistent feed solids quality and concentration are maintained, this will work satisfactorily. However, if wider fluctuations are encountered, the level controller is instead employed to control the filtration rate by diluting the feed with process water. This, of course, entails design of the filter section for achieving the maximum required capacity. By this method, the filter tank level is measured continuously; and it increases or decreases dilution of the feed dependent on the feed rate. The operator can occasionally "fine tune" this by changing the speed on the variable-speed drive. Obviously, this same level control intelligence could be carried into the variable-speed drive, but the coal industry generally prefers some operator attention.

The vacuum pump (see Fig. 8.2) must be designed such that its capacity will generally maintain 20 to 22 inHg vacuum. This will normally be an air rate of 3 to 5 cfm/ft^2 at 20 to 22 inHg vacuum for nonflotation coal, and 3 to 7 cfm/ft^2 at 20 to 22 inHg for flotation coal. The vacuum pump is normally protected by an automatic relief valve such that if a certain high vacuum is realized, the relief valve opens so as not to consume excess horsepower. Specially developed centrifugal pumps are employed. These pumps are receiver mounted and have a low net positive suction head (NPSH) such that they will pump both air and filtrate if necessary.

Innovations to Improve Moisture Removal. As the cost of energy increases, the importance of moisture content in the coal also increases. A 1% increase in moisture content, for example, would mean approximately a 0.1% loss in kilowatt-hours generated by 1 ton of coal in a utility boiler. While this may seem small, in a 500-MW installation, this would mean enough additional coal would be required to generate about 3 million extra kilowatt-hours of energy per year at 70% average capacity. Based on a plant life of 30 yr, 90 million kWh, in effect, would be lost. With a 12,000 Btu/lb coal at 35% thermal efficiency, this would mean that about 37,000 tons of extra coal would have to be consumed over these 30 years. Additionally, of course, the extra water would increase freight rates another negative factor.

Much of the moisture content in the final coal product is found in the fine coal. Coarse coal of about ¼-in. will usually dewater on screens to 3 to 4% surface moisture. Centrifugal dewatering of the intermediate-sized coal (¼ in. by 28 mesh) will exhibit around 8 to 9% surface moisture. However, the fine coal (—28 mesh, typically 25 to 50% —200 mesh) will usually exhibit a 20 to 22% surface moisture. At the same time, thermal drying of fines is generally resisted, because of its fire and explosive hazards, air pollution problems and operating costs. Thus, improved dewatering methods should be considered as they can very well be less expensive than "drying" in the boiler.

The first thing which should be done is to design the filter station so that maximum dewatering can be achieved by the methods described earlier. This would mean (1) operating at the proper θ_d/W [min/(lb)(ft^2)(rev)] value (see Fig. 8.4), (2) maximizing the vacuum level to the economic operating point, (3) using correct filter media, (4) designing proper filter hydraulics to maximize the pressure drop across the cake, and (5) practicing good maintenance on the filter installation. The resulting dewatering operation would exhibit a minimum surface moisture level as indicated by Fig. 8.4.

The use of surfactants has been studied on several occasions in laboratory-scale as well as pilot-plant-scale experiments designed to demonstrate the effectiveness of different types of surfactants in reducing the surface moisture of fine coal. There are two general types of surfactants that improve the removal of water from coal:

(1) additives that change the surface tension of water, and (2) hydrophobic or oily substances that are surface active agents.

Surfactants that change the surface tension of water are further divided into three groups:

1. Anionic, for example, sulfonated aliphatic esters, aliphatic sulfates, aliphatic sulfonates, alkyaryl sulfonates, esters, and ethers
2. Cationic, for example, dodeclammonium bromide or octadecil-ammonium chloride
3. Nonionic, for example, the product of condensation of alcohols or phenols with a polymer from ethylene oxide

Baker [1] used an anionic surfactant at dosages up to 1.2 lb/ton (600 g/ton) and demonstrated a reduction in surface moisture from 20 wt % without surfactant to 10 wt % with surfactant. Work done by Nicol [14] showed that the residual moisture of an Australian coal could be reduced seven percentage points by the addition of about 75 ppm of an anionic surfactant, where the addition of 500 ppm of a cationic surfactant only reduced the residual moisture less than five percentage points. Similar results have been demonstrated by Silverblatt and Dahlstrom [14], as well as Mielecki and Kurzeja [10].

The effectiveness of any surfactant to remove more surface water from coal depends on the chemistry of the water as well as the type of water present on and in the coal, which is related to the type of coal. Adhesion water and interparticle water, as discussed earlier, are the predominant forms of water affected by surfactants. The amount of water that surfactants are capable of removing changes from coal to coal and must be evaluated for each possible application.

Surfactants are effective in reducing surface moisture in fine coal dewatering. However, the full-scale use of surfactants has not been popular in coal preparation plants because of cost and the secondary effects on other unit operations in the plant such as flotation and refuse thickening and water clarification. More developmental work needs to be done to demonstrate the benefits of surfactants on a full scale so that more accurate economic analyses can be completed, including secondary effects of surfactants on plant operation.

Another method which can be employed economically, particularly with plant effluent discharge requirements, is to operate the plant with a closed water circuit. By this means, a clarifier-thickener is employed so that the overflow is recirculated back to the plant at the same time, the tailings or fine refuse are concentrated to economize the tailings disposal. Temperatures of the various processing steps within the plant are maintained at around 70°F (21°C), thus keeping the viscosity of water at relatively low levels. This is particularly important where ambient outdoor temperatures are cold in fall, winter, and spring. In one plant, this accounted for up to 2 percentage points in fine coal moisture reduction during these months. Furthermore, this was experienced on the total plant product and not just the fine coal.

A relatively recent development which is practiced right on the filter station is steam drying. Here, dry steam is passed through the filter cake to heat the coal to about 190 to 200°F (88 to 93°C). Because of the capillary bed, heat transfer is very efficient, thus condensing the steam and causing a condensate front to pass through the filter cake. The condensate will normally issue from the cake at 70 to 90°F so that a very high thermal efficiency is experienced. Normally, steam requirements have been 0.8 to 1.2 lb of steam per pound of water eliminated. In addition, there is no danger of fire or explosion as the cake is in a steam atmosphere of 212°F (100°C), and dust problems found in fluid-bed dryers are eliminated. Figure 8.9 is a typical plot of filter cake moisture content as a function of the correlating factor, with the total drying time expressed as the sum of θ_{st} (steam time) plus θ_{ha} (hot air time) plus θ_{aa} (ambient air time). Parameters are ambient dewatering with no steam or hot air, steam drying followed by ambient drying, and steam drying followed by hot-air drying followed by ambient drying. A very sizeable reduction in moisture content is apparent from steam drying. In addition, another one or two percentage points can be removed by adiabatic or evaporative cooling of the filter cake after discharge. Two different methods of adiabatic cooling on the filter are indicated in Fig. 8.9. Adiabatic cooling can also be achieved by simple transfer

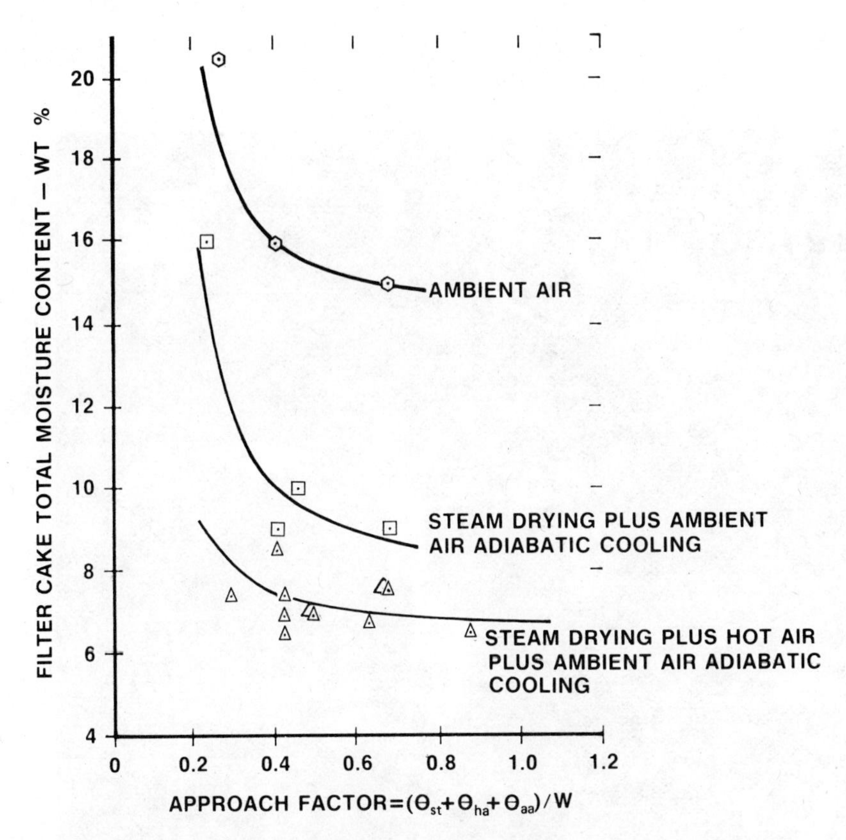

Figure 8.9 Filter cake moisture with and without steam drying as a function of approach factor for fine coal from pilot-plant filter.

points on belt conveyors, where the coal is allowed to fall in contact with air in a countercurrent fashion. This is the only time where evaporation takes place as the phenomenon of steam drying occurs primarily by better drainage from the filter cake due to the reduction in viscosity of the water. At 190 to 200°F, the water viscosity is only around 0.3 cP, or in other words, a 70% reduction from ambient temperatures of 68°F (20°C).

Figure 8.10 is a picture of a steam-drying filter of the disk type. A steam hood is placed partially around the disks to contain the steam. A seal is made on the rising side on the slurry in the filter tank while a special elastomer seal is applied on the discharge

Figure 8.10 Steam-drying Agidisc filter with steam hood located in the third quadrant. (Courtesy of Envirotech Corporation.)

side which rides on the filter cake. This seal is shown in Fig. 8.10 approximately 12 o'clock position. In this way, a 100% steam atmosphere is maintained in the steam hood.

It should also be noted that steam drying is automatically controlled so that only the required steam quantity is employed. However, it is stressed that the filter now must be operated similar to any dryer in that the feed rate and quality should be stabilized, as otherwise erratic results will occur. Filtration rates, as would be expected, will be reduced as compared to ambient filtration, to economically maximize the moisture reduction. For example, if the ambient filtration rate is around 75 lb dry solids/(hr)(ft^2) [366 kg/(hr)(m^2)] for a coal, the steam drying rate would probably be about 50 lb dry solids/(hr)(ft^2) [244 kg/(hr)(m^2)]. However, extra moisture removal would normally be in the range of 50 to 60%.

8.3 DEWATERING OF FINE COAL REFUSE

Separating the fine refuse from the raw coal to increase the recovery efficiency of coal in a preparation plant has made it necessary to size and purchase equipment and unit operations for the disposal of the fine refuse. The increasing importance of the fine refuse disposal is due to several factors:

1. Mechanization of underground mining equipment produced more fines.
2. Higher prices of coal now economically justify fine coal recovery.
3. Federal regulations control refuse disposal and water runoff quality.

Generally, there are two unit operations needed to dispose of fine refuse:

1. Thickening of fine material and, at the same time, clarifying recycle water to the plant
2. Dewatering the concentrated fine refuse material

A. Practical Design Considerations

The most important thing to remember when sizing any equipment for a fine coal refuse circuit is that the quality and quantity of the solids to be processed will change significantly during normal operation of the plant. In addition, it should be known that sizing will differ from plant to plant because of different flowsheets, mining techniques and coal seams. Sizing equipment and designing a flowsheet are more accurately done using both accumulated experience and scientific technology. Experience in relating existing flowsheets to proposed flowsheets is required so that a plant can be operated as originally designed.

Two alternative design concepts which could be used when sizing the fine refuse circuit are

1. To build a fine refuse circuit which normally will have an excess capacity, but when required, will handle the maximum amount and poorest quality of fine refuse and allow the plant to operate at the design tonnage.
2. To build a fine refuse circuit with no excess capacity for additional fines, which requires a reduced tonnage through the plant to maintain an adequate quality control and coal recovery efficiency.

The major problem created with an undersized fine refuse circuit is that the recycle water used in washing the coal becomes contaminated with slimes. The slimes carried in the thickener-clarifier overflow result in a change in specific gravity. This would cause control problems in separation equipment which functions by a difference in specific gravity. Also, the contaminated plant water would lead to both quality control problems and poor recovery of coal.

General practice in the coal industry is to first identify the variability of the fines coming into a preparation plant; this is usually done by obtaining data from existing plants which are mining coal from the same seam located in the same general region. Some of the variables to be identified which control the type and amount of fines are

1. Clay content
2. Seam characteristics, such as a split seam with refuse in the middle
3. Mining techniques (i.e., removing rock in the floor and/or ceiling or leaving coal in the floor and/or ceiling)
4. Type of mining, surface or underground

5. Amount of oxidized coal (not floatable in froth cells)
6. Number of seams to be mined for plant feed

These items represent some of the important factors connected with the coal alone. Information about these items can be obtained from core samples and are as reliable as the statistical number of samples drilled. The most reliable sources of information are at preparation plants that process coal from working mines.

Other site-specific limitations may also need to be satisfied when designing a fine refuse circuit. These are

1. Geographic location of the plant
2. Available capital

Coal preparation plants located in the Appalachian Mountains, for instance, can have land limitations at the plant site which limit equipment size and, most frequently, restrict refuse disposal; while plants in the midwest or western United States usually have more land area for refuse disposal.

Fine refuse can be disposed of in several different ways, such as

1. Dewatered to a solid consistency and blended in with coarse refuse
2. Thickened to a slurry consistency and pumped to a pond for further dewatering
3. Thickened and pumped into an abandoned mine for final disposal

Using vacuum disk filters to dewater refuse is becoming the most popular means of disposal if the fine material can be blended with coarse material to form a stable refuse pile. Constructing a fine refuse pond can be expensive because impoundment dams must meed federal specifications to ensure against failure.

B. Thickening of Fine Coal Refuse

High-Rate Thickeners for the Coal Refuse. The thickening of the fine coal refuse is done in either a conventional thickener or in one of the new high-rate thickeners. Conventional thickeners have been used for many years. Originally they were sized based on the settling characteristics of the fine refuse flowing into them, with occasional use of coagulants, such as starch or alum to help clarify the overflow water. However, over the last 10 to 15 years, much effort has been put into the development of synthetic polyelectrolytic polymers of increasingly larger molecular weights with better flocculation characteristics. Consequently, many polymers are now available to be used to flocculate practically any kind of fine material found in the fine refuse circuit of a coal preparation plant.

Note that the feed stream to a thickener-clarifier will have a solids content ranging from 2 to 10 wt %. The concentrated solids in the underflow will usually be at a concentration of 25 to 40 wt %. Overflow, or plant recycle water, will typically have a solids content of 150 to 5000 ppm. The total polymer dosage varies typically from 0.01 to 0.5 lb polymer per ton of solids.

A new type of high-rate thickener-clarifier has been developed which operates on the principle of using the polyelectrolyte polymers more efficiently than in a conventional thickener. As a result, these new thickener-clarifiers are 1/2 to 1/10 of the area used in a conventional thickener handling the same flow and tonnage of solids.

Figure 8.11 illustrates the schematic of one type of high-rate thickeners, namely, the Hi-Capacity Thickener currently being manufactured by the EIMCO Process Machinery Division of the Envirotech Corporation. The important features which distinguish the Hi-Capacity Thickener from conventional thickeners are the flocculant mix chamber, inclined settling plates, and automatic sludge level control system [7]. These features are briefly discussed below.

As mentioned earlier, the reason that the high-rate thickeners exist is because of the development of new and better polymers which are used to flocculate the fine refuse particles into relatively fast settling flocculi. One of the ways the polymer is used more efficiently is by adding a carefully controlled dosage of polymer into the mixing chamber, where the polymer is thoroughly dispersed into the feed stream.

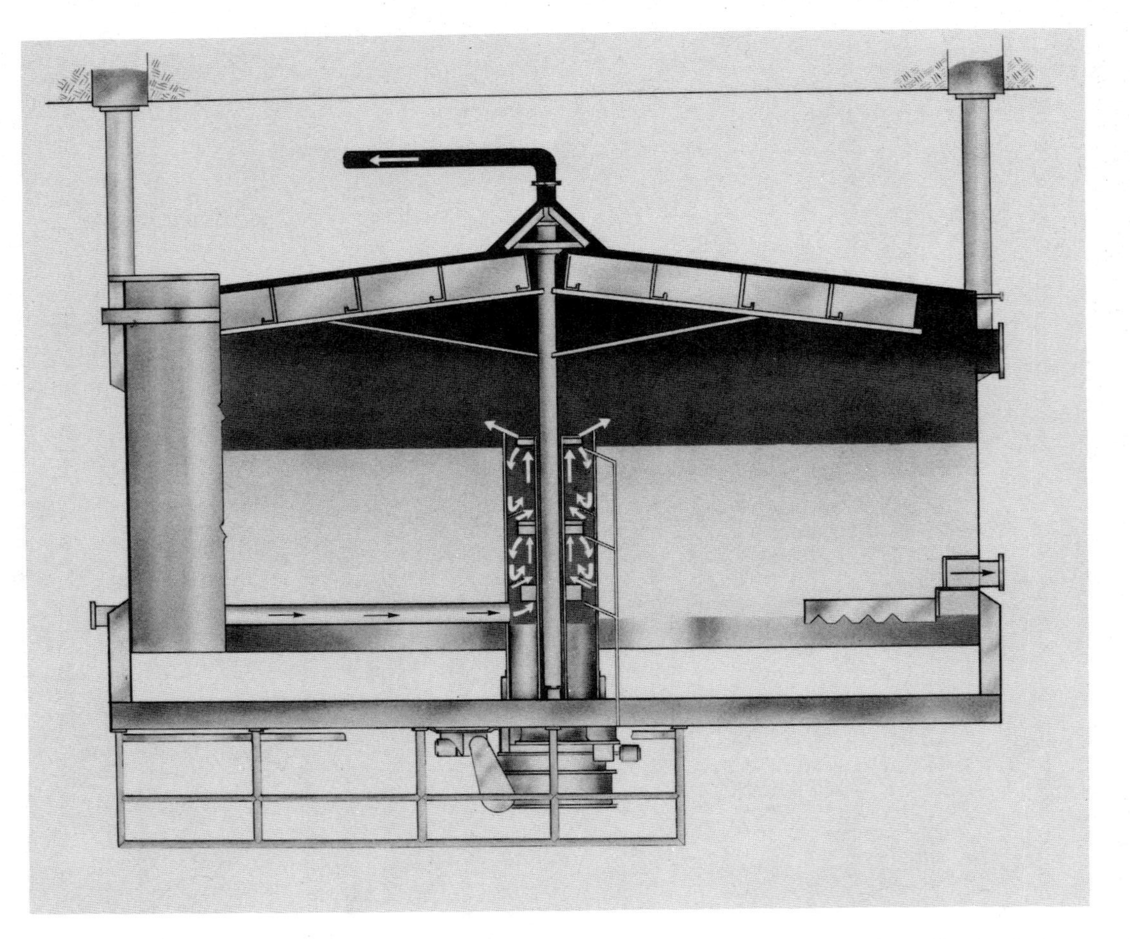

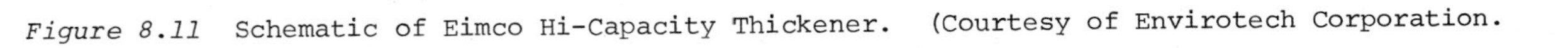

Figure 8.11 Schematic of Eimco Hi-Capacity Thickener. (Courtesy of Envirotech Corporation.

Flocculation is a function of time in that fine refuse particles must come in contact with polymer in a relatively dilute mixture. To ensure that most of the fine refuse particles contact polymer molecules, the feed enters the thickener at the bottom of the mixing chamber where it flows into a sludge bed composed of floccules and residual polymer. The solids concentration in this zone is substantially higher than that in the feed stream. Consequently, the unattached refuse particles rapidly form floccules.

Controlling the sludge bed level is very important, since minimum depth of sludge must be present above the feed inlet, so that an efficient capture of fine refuse particles occurs. To accomplish the sludge level control, two devices are used. The first device is a sludge level detector which is used to control the sludge withdrawal rate. The detector senses the interface location and transmits a signal to a controller which changes the underflow pumping rate to compensate for either a raising or falling sludge bed level.

The second device is a set of inclined settling plates positioned inside the thickener to help stabilize the inevitable variation in solids loading and water flow rate into the thickener. Basically, the inclined plates perform two basic functions that enhance the performance of the unit. First, water rising to the surface vertically is intercepted along the bottom side of each plate and escapes from the sludge bed by flowing up the underside of the plate in a channel. Secondly, the solids above the plates settle to the top of the plate and slide down the plate into the thickening section of the unit below the feed inlet. Rubin and Zahavi [18] describe the theory in detail.

Thickener Sizing for the Fine Refuse Circuit. Sizing both conventional and high-rate thickeners can be done confidently if a sample of fine refuse is obtained from an existing plant with a nearly identical flowsheet where coal is extracted using similar mining techniques from the same coal seam. The idea is to increase the confidence level of the results by comparing experimental data with actual operations. Once a sample has been obtained, it is extremely important that the experimental tests be conducted as soon as possible.

This is because the refuse material, specifically clay, will change physically and/or chemically due to the hydration of individual particles. The resulting effect in the experimental tests will be different flocculating characteristics, leading to false settling rates and polymer consumption data. Generally, it is felt that tests should be conducted within a day after the sample has been generated. The other source of samples are from washability tests or bench-scale flotation tests from core samples. These samples must also be used promptly when employed for sizing tests.

Basically, there are two conditions which must be considered when selecting the size of a thickener:

1. The thickener overflow clarity as measured by the suspended solids concentration
2. The thickener underflow solids concentration

In the coal industry, the limiting factor which determines the size of the thickener is the rise rate which is normally defined in units of gallons per minute per square foot since it is the other major factor in determining the refuse thickener size.

The basic strategy used in conducting 2-liter cylinder settling tests is structured to develop performance information by changing polymer dosage and initial solids concentration. It is important that the solids used in the test be characterized by sieving at 28, 100, 200, and −200 mesh, with ash analyses done for each fraction. By characterizing the sample, it is possible to compare the settling rates of different solids fractions (i.e., with and without clay and with different amounts of coal).

Traditionally, the data collected from 2-liter settling tests have been analyzed by the Coe-Clevenger [3] or modified Kynch (Talmage and Fitch) methods [11]. Each of these methods requires a subjective decision during the procedure, specifically the selection of a point that is critical to determining the size, or has specific restrictions connected with the procedure. Recently, Naide and Wilhelm [21] have developed a method that uses direct calculation based on the settling curve generated by plotting the interfacial height of the sludge in a 2-liter cylinder versus the settling time (see Fig. 8.12). Several

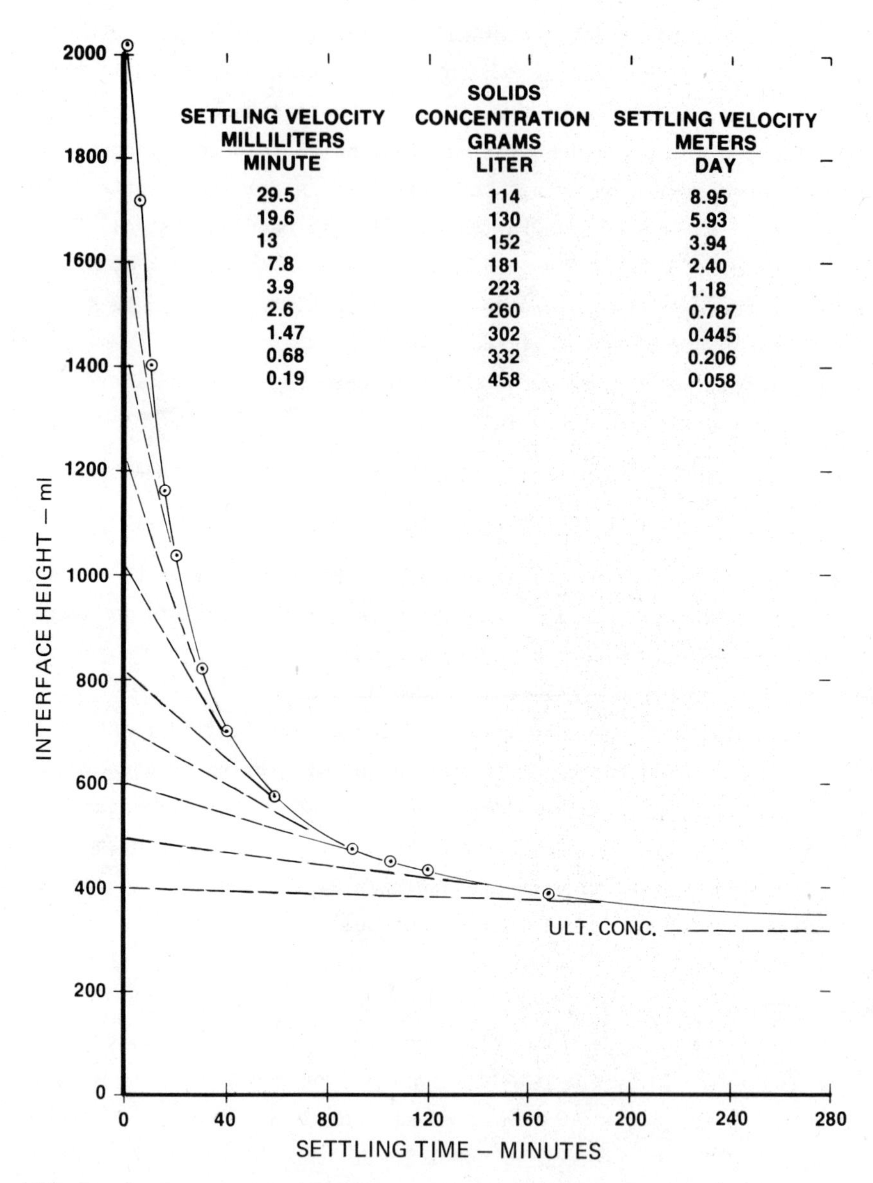

Figure 8.12 Determination of settling velocity vs concentration relationship for thickening of fine coal refuse: C_0 = 90 g/l. (From Ref. 21.)

settling tests are conducted at various conditions such as different initial solids concentrations and different flocculant dosages. In this method, the interfacial height as a function of settling time for each of the settling tests is plotted on a linear graph paper. Tangents are drawn from the curve to the vertical axis. The settling velocities (m/day) equal to the slope of the tangent lines are plotted as a function of solids concentrations (g/l) on a log-log graph paper (see Fig. 8.13). On the log-log plot, two to three straight lines can be drawn to closely approximate the curve of settling velocity versus solids concentration. Each of these straight lines leads to a general correlation represented by the following equation:

$$V = aC^{-b} \tag{8.10}$$

where

V = settling velocity, m/day
C = concentration, t/m^3
a = constant
$-b$ = slope of each straight line (dimensionless)

The constant a can be calculated by selecting any point on one of the lines.

For each of the straight lines found on Fig. 8.13, an equation for unit area [m^2/(t)(day)] can be written as:

$$\text{Unit area} = \frac{(b-1)^{b-1}/b}{ab} \, C_u^{\,b-1} \tag{8.11}$$

where

C_u = solids concentration of underflow at the corresponding unit area, t/m^3

Equation (8.11) is then plotted on a log-log graph paper for each set of values of a and b. The resulting curve as shown in Fig. 8.14 can be used to determine the unit area required to obtain a desired underflow solids concentration.

Conventional and high-rate thickeners can be sized using the above procedure. When sizing a high-rate thickener, special attention must be given to the flocculant addition during the 2-liter

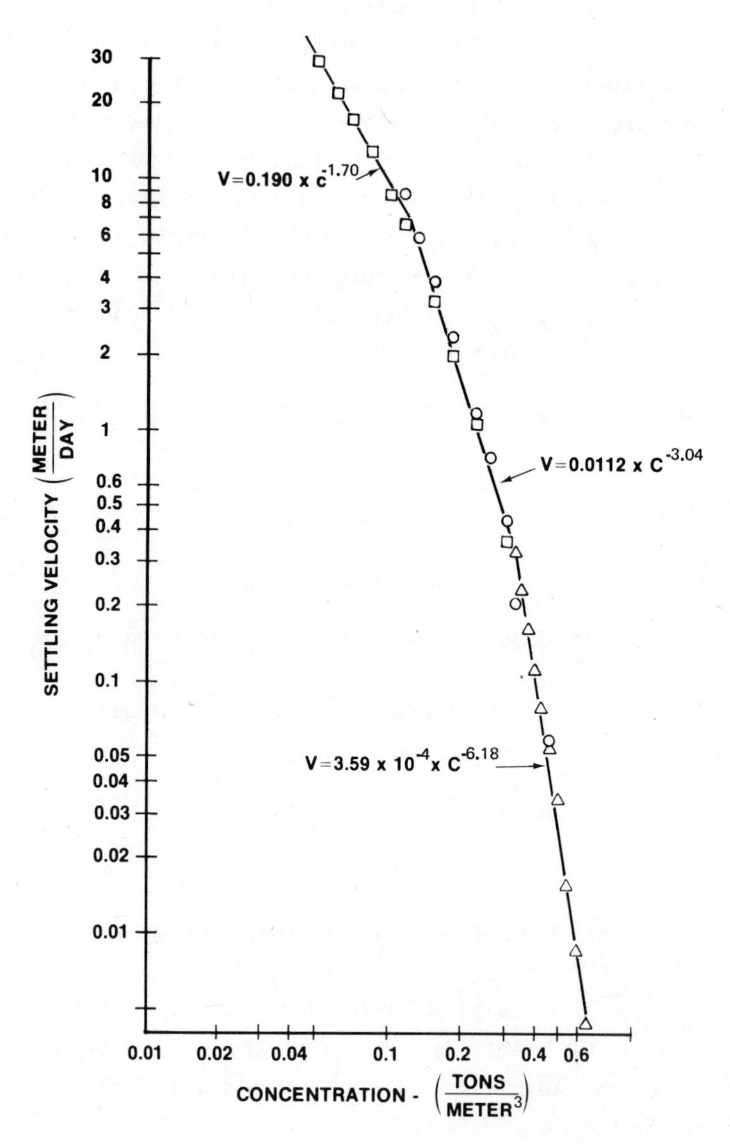

Figure 8.13 Settling velocity vs concentration for thickening of fine coal refuse. (From Ref. 21.)

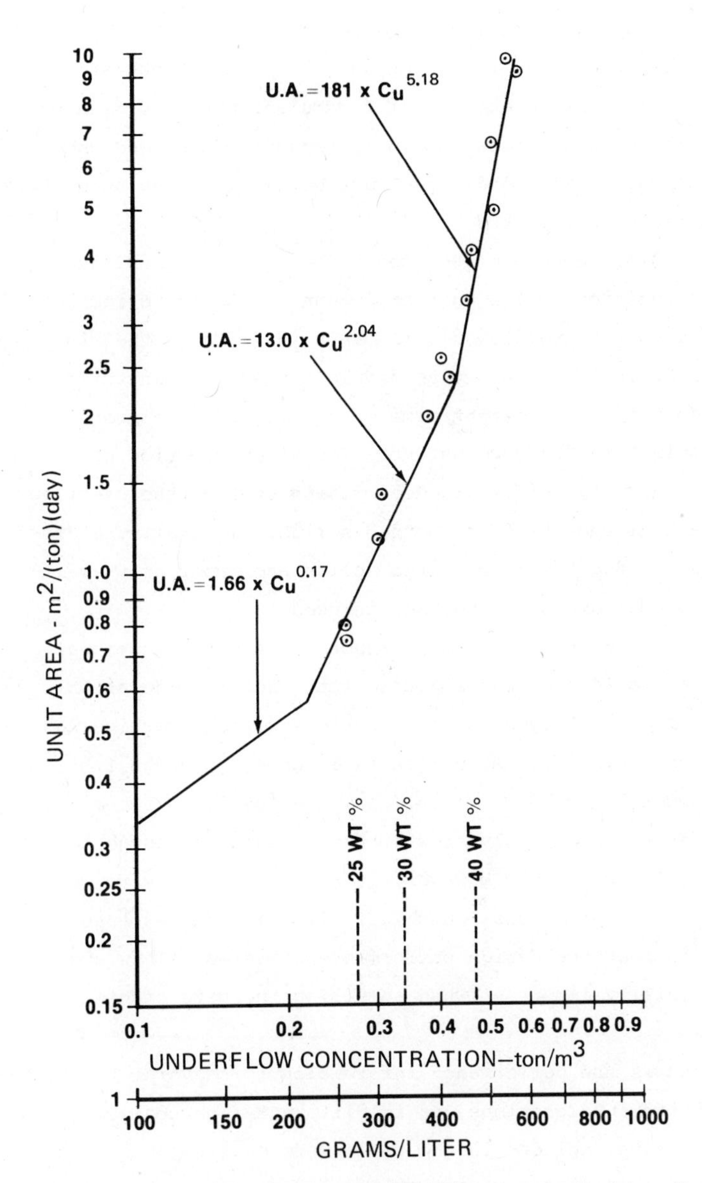

Figure 8.14 Unit area vs. underflow concentration for thickening of fine coal refuse. (From Ref. 21.)

cylinder tests to ensure that the flocculant is distributed evenly throughout the cylinder and flocculi are not destroyed by an excessive mixing. One of the most important aspects of the high-rate thickeners is its more efficient use of flocculant by dispersion of the flocculant in the feed stream and by introducing the feed under the sludge bed. These same conditions must be simulated as accurately as possible in bench-scale tests.

The above sizing technique is based on the underflow solids concentration as a function of the unit area, but the design parameters can be converted to the overflow clarity as a function of the rise rate. To do this, it is necessary to sample the supernatant of the 2-liter settling test at different time intervals and then measure the suspended solids in the supernatant. Visual observation of the supernatant will usually suffice in determining whether the overflow contains substantial amounts of suspended solids. Typically, a distinct interface between the flocculated solids and supernatant will exist when the proper amount of polymer is used.

The size of thickeners today is a function of the traditional variables such as solids loading and rise rate, but consideration should also be given to polymer cost. Previous sizing practice was to size a thickener-clarifier at a rise rate between 0.5 and 1.0 gpm/ft^2, or between 5 and 10 ft^2/(ton)(day), and use the criteria that gave the largest area; polymer was then added at a dosage to control the overflow clarity. However, it is now necessary to weigh the capital cost of a particular thickener size with the expense of polymer needed to meet the design performance criteria. This cost analysis is especially important when selecting the size of high-rate thickeners.

Table 8.1 shows the performance information for several thickener-clarifiers at different locations and in different fine coal circuits. The first four plants included in the table were participants in a pilot plant program involving the EIMCO Process Machinery Division's Hi-Capacity Thickeners. The last six plants are conventional thickeners with the performance data presented as typical of normal plant operation. Plants disposing of clay are noted by large amounts of

Table 8.1 Thickener-Clarifier Performance on Fine Coal Refuse

Plant location	Fines circuit	−200-mesh solids, wt %	Wt % ash in −200 mesh	Wt % solids	
				Feed	Underflow
Central Utah	−28 mesh flotation	~83	N.A.	2-4	33-37
Western Pennsylvania	−65 mesh flotation	~65	~74	9-12	34-39
Central West Virginia	−100 mesh flotation	59-69	~70	8-15	29-35
Eastern Kentucky	−100 mesh flotation	60-70	~59	6-8	25-35
Southeastern Ohio	−150 mesh cyclone and jig	~85	~75	~3.4	~36
Central Pennsylvania	−28 mesh flotation	~46	~42	~3.7	~33
Central West Virginia	−28 mesh flotation	~50	~73	~3.6	~39
Southern West Virginia	−28 mesh flotation	~54	~41	~1.6	~29
Southern Illinois	−16 mesh flotation and jig	~45	~50	~2.9	19[b]
Northern Virginia	−28 mesh flotation	~43	~40	~4.1	18[c]

[a]Results obtained from EIMCO Hi-Capacity Thickener pilot-plant tests.
[b]Underflow pump oversized at time of data acquisition, underflow typically ~35 wt %.
[c]Underflow diluted.

−200-mesh material along with high ash content. Note that obtaining a concentrated underflow from a thickener depends first of all on the size of the unit and the polymer dosage, but if these two criteria are met, then the underflow pumping rate will determine the underflow solids concentration. The underflow slurry is pumped using a constant speed centrifugal slurry pump which removes a relatively constant volume from the thickener. The volume removed must be nearly the same as solids input. Otherwise, the solids will accumulate in the thickener, or the thickener will empty of solids with the underflow becoming

Solids overflow, ppm	Unit area, ft^2/tpd	Rise rate, gpm/ft^2	Polymer type	Polymer in total feed, ppm	Lb polymer/ ton solids
50-150	1.4-1.0[a]	5.0-6.5	Anionic	10-16	1.0-1.6
150-300	1.3-1.0[a]	1.5-2.0[a]	Nonionic	20-30	0.3-0.5
175-400	0.8-0.4[a]	1.6-8.1	Anionic Cationic	13-20 N.A.	0.2-0.5 N.A.
150-400	0.7-0.4[a]	2.7-3.5[a]	Anionic Cationic	1-10 N.A.	0.03-0.3 N.A.
100-10,000	18.3	0.49	Anionic Cationic	35-200+ ~5	1.0-5.8 ~0.3
4000	21.9	0.56	Anionic	3-5	0.1-0.3
100	9.7	0.47	Anionic	N.A.	N.A.
N.A.	23.6	0.50	Cationic Anionic	8-9 0.6-1.0	0.9-1.2 0.08-0.13
1500	8.3	0.67	Cationic Anionic	~3 N.A.	~0.2 N.A.
6000	17.5	0.41	Anionic	~4	~0.19

dilute as seen at the last two plants listed in the table. Normally, the underflow pump will be slightly oversized and will be operated intermittently to maintain a concentrated underflow. This point also has been noted in Table 8.1.

C. Dewatering of Fine Coal Refuse

Several alternatives can be used for mechanical dewatering of fine coal refuse: vacuum disk filters, filter presses (pressure filters), or solid bowl centrifuges. Each of the alternatives has both strong and weak points, as described briefly below.

Vacuum Agidisc Filters. Vacuum disk filters are effectively used as the primary mechanical dewatering device of fine coal refuse in the United States. Table 8.2 (pp. 440-441) shows several different installations and characteristics of the material being dewatered [9].

Many of the problems and variables connected with the selection of a thickener, as discussed previously, are applicable to the selection or sizing of a vacuum disk filter. For example, variations in the quality and quantity of solids going to the filter are reflected in the filtration rates and cake moistures in Table 8.2. Also, control of the thickener-clarifier and the refuse filter are interrelated, since the solids going to the filter for dewatering came from the thickener underflow.

Usually, the refuse filter receives minimal operator attention, because the only requirement of the filter is to dewater all the fine refuse into a solid mass that can either be trucked or moved to a disposal area on a conveyor. It is also important that the cake characteristics are such that the fine refuse, when blended with coarser material, is compactible in a stable refuse pile.

All of the technology described for the fine coal filtration also applies for the refuse filtration. Typical refuse filtration rates without clay are between 20 to 40 lb/(hr)(ft^2). But, as discussed earlier, it is extremely important to provide enough filtration area to remove refuse at a rate which will allow the preparation plant to produce a certain minimum tonnage. When substantial amounts of clay exist in the refuse, filtration rates of 10 to 20 lb/(hr)(ft^2) are typically used to size the filter. The limitation of a disk filter is whether the cake can be discharged at a minimum thickness of 3/8 in.

Filter Presses. Because of the nature of the material, fine coal refuse which contains a large amount of colloidal clay usually will not dewater well in a centrifuge or on a vacuum disk filter. The mechanical dewatering device which will dewater this "difficult-to-dispose-of refuse" is the filter press (see Fig. 8.15).

Filter presses have been used for years in Europe, because many times there is no land available for large sludge ponds. Within the

Table 8.2 Vacuum Disk Filter Performance on Fine Coal Refuse

Plant location	Fines circuit	−325 mesh solids, wt %	Wt % ash in −325 mesh	Wt % solids feed	Wt % moisture cake	Filtration rate, lb/(hr)(ft^2)
Central West Virginia	−28 mesh flotation	~71	~56	25-30	34-38	30-50[a]
Central Pennsylvania	−28 mesh flotation	~52	~31	18-31	18-31	10-96[b]
Eastern Kentucky	−28 mesh cyclone	~64	~59	~33	35-40	19-27[c]
Eastern Kentucky	−65 mesh cyclone	~77	~61	~30	32-36	8-15
Central Pennsylvania	−65 mesh flotation	~38	~45	~32	~21	8-15[d]
Southern West Virginia	−28 mesh flotation	~50	~40	~29	21-26	34-38
Northern Virginia	−28 mesh flotation	~32	N.A.	~18[e]	19-22	27-36

[a] Rates are maximum and vary due to solids characteristics; this is a drum filter.

[b] Rates vary due to seam changes and mining technique.

[c] All fines go to the refuse filter because oxidized coal will not float.

[d] Rates are low because there is an excessive filter area.

[e] Feed was diluted to lower filtration rate because fine refuse tonnage was lower than normal.

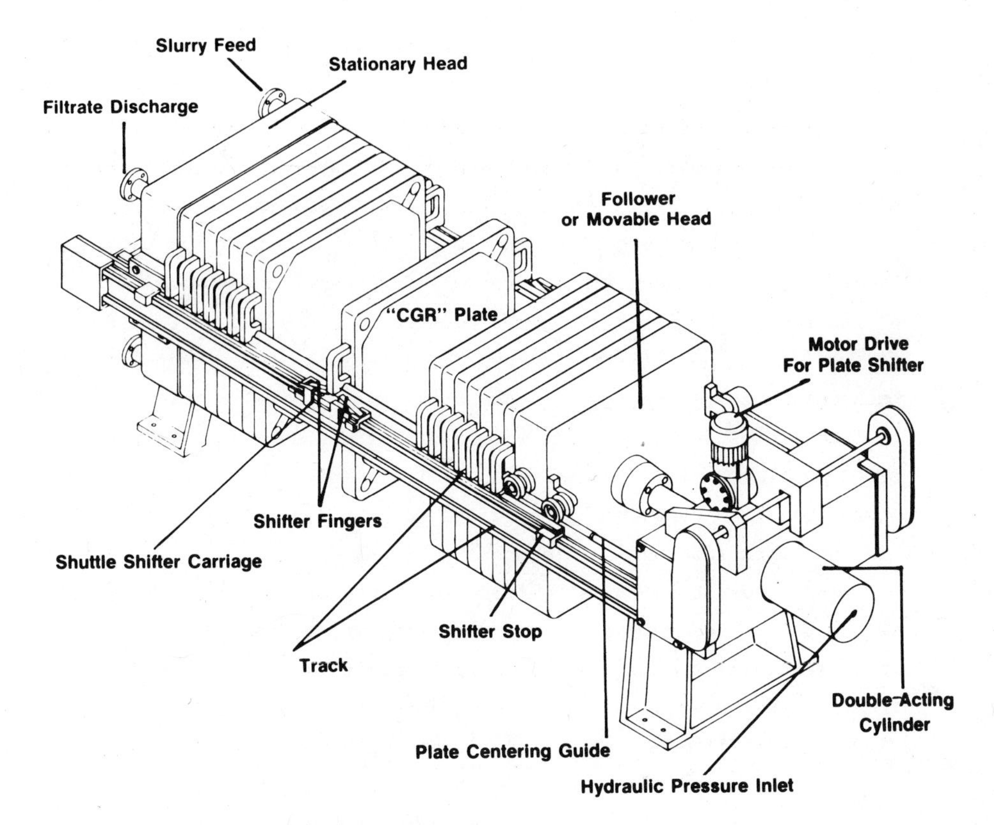

Figure 8.15 Filter press used for high-clay content refuse. (Courtesy of Envirotech Corporation.)

last 5 years, several plants in the United States rather than build sludge ponds, have elected to use filter presses to dewater high-clay refuse [17]. The need for filter presses in the United States is a result of governmental regulations promulgated to make refuse disposal areas safer; these regulations follow the collapse of refuse disposal areas in Great Britain and West Virginia.

The filter press is considered to be a sturdy piece of equipment and is capable of dewatering a wide variation of feed solids. The moisture content of the filter cake is often very consistent, producing an easy to handle material. Feed flow rate usually has very little effect on the filtrate quality. The maximum feed rate is determined by the pressure drop across the filter or by the feed pump capacity.

One of the reasons why filter presses have not been widely used in the coal industry is the relatively high labor requirement for their operations [11]. However, the filter press manufacturers have developed mechanical opening and closing operators that are either hydraulically or electrically controlled. These innovations have helped reduce the labor and time required to empty and prepare the filter for the next filtration cycle. Materials of construction have been expanded from the traditional cast iron to new light weight plastic plates. Reducing the weight of the plates minimizes the problem of mechanical damage to the filter media during the closing operation. New synthetic fiber filter cloths have also helped to prolong cloth life and to aid in cake discharge.

The alternative of using filter presses for dewatering fine refuse can be economically attractive when sludge pond construction in certain site-specific analyses is prohibitive. Because the coal industry is making a conscientious effort to safely dispose of high-clay refuse, filter presses will be used much more in the future.

Typical overall filtration rates can range from 5 to 15 lb/(hr)(ft^2), which includes the total cycle time between discharge of each batch of refuse [8]. Control of the discharge sequence and closing operation has helped reduce the labor requirements. The ability of the pressure filter to remove water from fine refuse and produce a handleable and stable material usually overshadows some of the inconvenience associated with the filter operation.

In all fine refuse dewatering applications, polyelectrolyte polymers are used to capture the ultrafine refuse and to minimize the separation of coarse and fine particles, which causes blinding on the filter. The ability to filter fine refuse is due to the flocculation of the slime material, such that the latter is dewatered along with filter particles which allows the flow to continue through the filter cake.

Solid-Bowl Centrifuges. Solid-bowl centrifuges are used infrequently to dewater fine refuse in the United States, because much of the fine refuse contains substantial amounts of −325-mesh or colloidal solids. The capture of colloidal material in a solid-bowl centrifuge is difficult, but can be successful if flocculant is used to agglomerate the ultrafine solids, and other variables such as feed rate can be controlled.

The physical size of solid-bowl centrifuges is usually not very large due to the balance problems created at high rpm and high gravitation forces. Therefore, it is usually necessary to install several centrifuges to handle the average tonnage of fine refuse as well as peak loads. The centrifuges to handle the average tonnage of fine refuse so that fines do not accumulate in the plant water system.

A possible alternative to starting and stopping machines is to provide surge capacity either in the refuse thickener or in storage tanks. Fine refuse can then be dewatered at a constant rate.

8.4 DESIGN CONSIDERATIONS FOR FILTRATION AND DEWATERING SYSTEMS IN PHYSICAL COAL CLEANING PROCESSES

The design of the fine coal circuit, generally accepted as −28-mesh material, must be considered as a complete process. Included in the fine coal circuit are size classification equipment, separation units, and dewatering devices. The performance of each successive unit operation depends on the performance of the preceding piece of equipment.

For example, a typical fine coal circuit will first have raw coal screens and prewet screens to remove all −1/4- or 3/8-in. raw coal. The coal slurry usually is pumped to desliming screens where the solids are classified at 28 mesh with the −28-mesh slurry flowing to either froth flotation units or cyclones for separation of refuse and coal. Screen maintenance will control the tonnage of solids as well as the size of solids being fed to the separator for refuse and coal. Screen maintenance will control solids being fed to the separation equipment. Worn screens will allow more solids and larger solids into equipment downstream.

The separation equipment divides the refuse and coal into separate streams. The efficiency of the separation depends on different characteristics of the coal and refuse for froth flotation units and cyclones. Cyclone performance is determined by the specific gravity difference of the solids, particle diameter, solids concentration, operating pressure, and geometry of the cyclone. All of latter parameters will vary considerably, with the cyclone generally producing more refuse with the coal product when adverse conditions arise. The quality of clean coal is usually more variable when cyclones are used because the separation efficiency of cyclones changes.

Froth flotation equipment has been used more frequently in recent coal preparation plants because a high-quality product coal is recovered from this unit operation. The flotation operation is more selective in separating coal. Operation is controlled by matching specific conditioning chemical, frothing chemicals, as well as flotation cell variables such as conditioning time, air rate, froth depth, and through put rate to different types of coal. If operating conditions are not favorable, coal simply will not float and separate from the refuse. Therefore, it is imparitive that different types of coal be identified by mining records and processed under conditions that allow the coal to be recovered.

Clean coal filters should be selected by thoroughly understanding the potential variability of the raw coal being processed in the plant and how the equipment upstream of the filter will perform with the variability in quality and quantity of fines. Normally it is more profitable to purchase an excess filter capacity to insure that all the clean coal recovered can be dewatered, rather than throwing fine clean coal away with refuse. This follows because the filter is short of capacity during periods when the fine coal tonnage is above the average throughput. If the clean coal filter is undersized, then plant production is made less efficient, either because fine coal product is lost with refuse or the plant operates at a reduced tonnage limited by the clean coal filter. It is also adviseable to purchase spare filter sectors to replace damaged or worn sectors when necessary. Multiple vacuum pumps might also be purchased to have an adequate vacuum capacity during maintenance periods.

Fine refuse disposal from either cyclones or flotation tailings requires the identification of the variability of the quality and quantity of material to be disposed, so that the refuse thickener and refuse dewatering device can be sized to maintain an adequate plant capacity by providing clean recycle water. In a flotation flowsheet, operational problems with the flotation cells will cause all or most of the fine coal to go into the refuse circuit. The major problem is the inability to float oxidized coal. The latter is found in many surface mining operations, and when coal has had prolonged storage prior to cleaning. It is important to provide an adequate capacity to remove refuse from the plant water circuit. Failure to accomplish the refuse removal may require that the feed to the plant be reduced to prevent the thickener from filling with solids. Selecting the thickener should be based on economic considerations as discussed earlier in this chapter. The costs to be considered are site preparation, polymer cost and equipment cost.

Integral to the thickener operation is the capacity of the underflow pump to remove thickened refuse. The thickener underflow pump links the thickener with the refuse dewatering device. The underflow pumping rate is used to control the solids inventory within the thickener as well as controlling the feed rate to the dewatering device which is usually a vacuum disk filter. The ability to control both the thickener and the dewatering device requires that the pump be properly sized to conform with operational characteristics of both pieces of equipment.

The dewatering device should be designed to dewater refuse at a rate that will maintain a relatively constant solids inventory within the thickener. The refuse solids should include the maximum expected fine refuse tonnage plus an estimated tonnage of unrecovered fine coal which might get into the refuse circuit due to operating variations in the equipment upstream of the filter.

A design basis and operating strategy for the coal preparation plant should be established before equipment is selected and costs are established, because these criteria affect capital investments, operating costs and production output. Flexibility in plant operation can be achieved by spending a small amount of capital for fines dewatering equipment. Finally, it should be mentioned that the technical information and economic considerations discussed in this chapter can be readily applied for the design of filtration and dewatering systems for various new physical coal cleaning processes covered in this book. A specific example of this application can be found in a recent article by Slaughter and Karlson [23], which describes the test results and equipment selection for the design of filtration and dewatering systems for the cleaning of wet pulverized coal by high-gradient magnetic separation (HGMS).

8.5 CONCLUSIONS

Future energy needs of the country will be met by using coal as fuel for electric power generation and as the source for synthetic fuels to supplement the shrinking supplies of natural gas and crude oil. Coal-cleaning technology improvements will be used to remove more sulfur, ash, and other impurities to minimize the contamination of products and the environment. Filtration and sedimentation technologies are available to be used in the future as integral segments of coal-cleaning flowsheets.

Innovations in filter and thickener design and operation used in other mineral processing industries can and will be used in coal-cleaning operations. Steam drying, surfactants, and still undiscovered innovations are possible improvements that can be used on fine coal products to economically reduce the surface water.

ACKNOWLEDGMENTS

The authors wish to thank Mr. S. Erickson, Ms. N. Selu, Ms. G. Newell, and Ms. S. Winger for their patience in preparing the final copy and Mr. R. Player for his skill in preparing the illustrations.

REFERENCES

1. A. F. Baker, Hot Surfactant Solution as a Dewatering Aid During Filtration, paper presented at Second Symposium on Coal Preparation, NCA/BCR Coal Conference and Expo III, Louisville, Ky, Oct. 19-21, 1977.

2. G. Burton, A New Process for Reducing the Moisture Content of Filter Cake, *Fourth International Coal Preparation Congress*, Harrogate, England, 1972.

3. H. S. Coe and G. H. Clevenger, Methods for Determining the Capacities of Slime Settling Tanks, *Trans. Amer. Inst. Min. Eng., 55,* 356-384 (1916).

4. D. A. Dahlstrom and C. E. Silverblatt, Production of Low Moisture Fine Coal Without Thermal Drying, *Min. Congr. J.,* 32-40, Dec. (1973).

5. R. C. Emmett, Jr. and D. A. Dahlstrom, Preparation Plant Water Control and Refuse Dewatering Systems, paper presented at American Mining Congress, Coal Convention, Pittsburgh, Pa., May 1969.

6. R. R. English and B. C. Radford, Filter Presses in Coal Preparation, *Filtration and Separation, 14*(5), 492 (1977).

7. R. C. Emmett, Jr. and R. P. Klepper, The Technology and Performance of the Hi-Capacity Thickener, paper presented at the Annual Meeting of Society of Mining Engineers of AIME, Denver, Colo., Feb. 26 to Mar. 2, 1978.

8. D. W. Hutchinson and E. M. Duralia, Pilot Plant Pressure Filtration of Coal Preparation Plant Refuse Slurry, paper presented at the Annual Meeting American Institute of Mining Metallurgical and Petroleum Engineers, Atlanta, Ga., Mar. 6-10, 1977.

9. R. P. Klepper, New Trends in Fine Refuse Vacuum Filtration, *Proceedings Fourth Kentucky Coal Refuse Disposal and Utilization Seminar*, University of Kentucky, Lexington, Ky., 1978, pp. 51-54.

10. G. Kurzeja and T. Mielecki, Attempts to Remove Excessive External Water Contents from Washed Coal with Use of Chemicals, *Selected Translations on Coal Processing,* translated from Polish, The Scientific Publications Foreign Cooperation Center of Central Institute for Scientific, Technical and Economic Information, Warsaw, Poland, 1965, pp. 199-223.

11. G. J. Kynch, A Theory of Sedimentation, *Trans. Faraday Soc.,* 45 and 166 (1952).

12. J. W. Leonard (ed.), *Coal Preparation,* 4th ed., AIME, New York, 1979.

13. P. A. Nelun and D. A. Dahlstrom, Moisture Content Correlation of Rotary Vacuum Filter Cakes, *Chem. Eng. Prog.*, 320-327, July (1957).

14. S. K. Nicol, The Effects of Surfactants on the Dewatering of Fine Coal, *Proc. Australas, Inst. Min. Metal.*, *260*, 37-44 (1976).

15. J. H. Perry and C. H. Chilton (eds.), *Chemical Engineers' Handbook*, 5th ed., McGraw-Hill, New York, 1973.

16. Derek B. Purchus (ed.), Continuous Filters, *Solid/Liquid Separation Equipment Scale-Up*, Upland Press, Croyden, England, 1977, Chap. 11, pp. 445-492.

17. J. M. Regan, Filter Presses--Still Alive and Kicking! *Filtration and Separation*, *14*(5), 485 (1977).

18. E. Rubin and E. Zahavi, Enhanced Settling Rates of Solids Suspensions in the Presence of Inclined Plates, paper presented at the AIChE National Meeting, Houston, Tex., Mar. 16-20, 1975.

19. C. E. Silverblatt and D. A. Dahlstrom, Moisture Content of Fine-Coal Filter Cake: Effect of Viscosity and Surface Tension, *Ind. Eng. Chem.*, *46*(6), 1201-1207 (1954).

20. C. E. Silverblatt and D. A. Dahlstrom, Theory and Practice of Continuous Filtration of Fine Coal and Slimes, paper presented at Second International Coal Preparation Congress, Essen, Germany, 1954.

21. J. H. Wilhem and Y. Naide, Sizing Thickeners from Bench-Scale Data, paper presented at AIME Annual Meeting, New Orleans, La., Feb. 18-22, 1979.

22. Dorr-Oliver Inc., *Filtration Leaf Test Procedures*, Stanford, Conn., Aug. 1972.

23. W. W. Slaughter and F. V. Karlsan, Process Design Considerations for the Cleaning of Wet Pulverized Coal by High Gradient Magnetic Separation, in *Industrial Applications of Magnetic Separation* (Y. A. Liu, ed.), IEEE Publication No. 78CH1447-2MAG, IEEE, New York, 1979, pp. 89-90.

9

Economic Assessment of Selected Coal Beneficiation Methods

SUMAN P. N. SINGH

Oak Ridge National Laboratory
Oak Ridge, Tennessee

9.1 INTRODUCTION

The energy situation in the industrialized Western world is deteriorating. In the United States, for example, domestic supplies of oil and natural gas (which have up to now been the primary source of energy) are being rapidly depleted, while the demand for energy remains strong. Increasing quantities of oil are being imported from the oil-producing nations to satisfy the apparently insatiable national energy demand. To decrease the growing dependence on imported oil, it is imperative that the nation turn to using coal, its most abundant fossil fuel resource.

Coal at one time in the not too distant past was the world's dominant fuel. It even helped nurture the Industrial Revolution. However, during the past two decades, coal was supplanted by oil and natural gas as the primary source of energy because

1. It is a solid fuel that is generally dirty and relatively more difficult to handle than oil and natural gas.
2. Oil and natural gas were cheap, convenient and readily available.

However, the oil embargo and the subsequent dramatic increase in crude oil prices in 1973-1974 served to rivet worldwide attention on two critical issues:

1. The world supply of crude oil and natural gas is limited and, at projected rates of exploitation, is expected to be depleted within the next 25 to 30 years.
2. If worldwide economic development is to continue, alternate fuel resources (such as coal) need to be more fully exploited to meet the anticipated world energy requirements.

Coal is the world's most abundant fossil fuel resource. World coal reserves have been estimated to be about 2.5×10^{12} tons which, on a common energy basis, is approximately six times the estimated recoverable world oil resources [1]. Coal is also relatively widely distributed worldwide even though estimates made in 1972-1973 indicated that the United States, USSR, and China collectively possessed over 70% of the measured world coal reserves. The United States, for example, has better than one-third of the known economically recoverable world coal reserves.

Although sufficient coal reserves exist to meet the projected energy requirements for the next several hundred years, yet, much of the readily available coal cannot be used directly because it is too dirty (that is, high in sulfur and/or mineral impurities). Burning this coal directly will result in environmentally unacceptable sulfur oxides (SO_x) emission levels. One partial solution to the problem of using dirty run-of-mine (ROM) coal to meet the energy needs in an environmentally acceptable manner is to beneficiate the coal prior to its use.

Coal beneficiation is a generic term that is used to designate the various operations performed on the ROM coal to prepare it for specific end uses such as feed to a coke oven or a coal-fired boiler or to a coal conversion process. It is also referred to as coal cleaning or coal preparation.

Coal beneficiation has come to encompass the entire spectrum of operations ranging from the relatively simple crushing and size classification operations (that are almost routinely performed on all coals used today) to rather elaborate chemical and microbiological processes that are used or are being developed to render the ROM coal more suitable for the end-use process. Coal beneficiation processes prepare the ROM coal for its end use by removing the undesirable constituents associated with the coal without destroying the physical identity of the coal. However, liquefaction-type processes (such as the solvent-refined coal process) that upgrade the coal to yield a clean fuel product are generally not regarded as coal beneficiation processes primarily because they alter the physical identity of the coal; that is, the coal is liquefied and then upgraded.

A recent study by the Hoffman-Munter Corporation for the U.S. Bureau of Mines and the Environmental Protection Agency [2] (as described in Chap. 10) indicated that physical coal beneficiation combined with flue gas desulfurization (FGD) appeared to offer the most economical means of achieving sulfur oxides emission control for coal burning facilities. The study further stated that for some coals, beneficiation could even obviate the need for FGD systems in order to achieve acceptable sulfur oxides emission control.

Up to the present, commercial practice has largely relied on physical coal cleaning processes to beneficiate coals. Chemical, microbiological, and other novel coal beneficiation processes are of recent origin and still at various levels of process development. The chemical and other novel beneficiation processes, though generally capable of producing a higher yield of a cleaner coal product from the ROM coal, have not been used on a commercial scale primarily because they have not yet proven to be economical.

A number of specific advantages can be obtained from beneficiating the coal prior to its end use, and are summarized as follows:

1. The heating value of the cleaned coal is higher since the impurities are either partially or completely removed.
2. The product coal from coal beneficiation is more uniform in size, composition, and heating value, thereby resulting in, for example, more uniform and steady combustion than would otherwise be possible.
3. By removing the sulfur impurities present in the coal, beneficiation contributes to reduced slagging and fouling in the boiler combustion chamber. This increases the boiler's on-stream availability and reduces its maintenance costs.
4. Removal of the associated mineral matter from the ROM coal can result in lower transportation costs, higher combustion efficiency, and reduced ash disposal and flue gas cleanup requirements.
5. Reduction of the coal moisture content by beneficiation results in improved coal handling and burning characteristics. It also leads to more efficient use of the fuel, since less energy is expended in the furnace to dry the coal.

A. Scope of the Chapter

The objective of this chapter is to provide a brief introduction and rationale for coal beneficiation and to develop on a common and consistent basis the costs associated with conceptual plants using some of the new and conventional coal beneficiation processes. The assessment should result in indicating the range of costs and benefits for the selected conceptual plants and thereby provide a means of assessing the relative merits of the processes considered.

To achieve the above objective, the succeeding sections of this chapter will consist of the following:

1. Overview of conventional coal beneficiation
2. General (technical) assessment of new physical coal beneficiation processes
3. Brief descriptions of the processes chosen for the economic evaluation, namely:
   a. Wet high-gradient magnetic beneficiation
   b. Chemical comminution
   c. Level 2 wet mechanical beneficiation
   d. Level 4 wet mechanical beneficiation
   e. Meyers fine coal process
4. Economic assessment of the selected processes
5. Conclusions

9.2 BACKGROUND

In order to provide the background for the economic assessment of several new and conventional coal beneficiation methods, a brief discussion of several currently used coal beneficiation processes and new coal cleaning approaches is presented below.

A. Current Coal Beneficiation Practices

Currently in the United States and other coal-producing countries, the ROM coal is prepared by a physical beneficiation process. The degree of preparation can vary from no beneficiation to a very thorough treatment of the raw coal. The process(es) used and the degree of cleaning employed are very dependent on the type of coal and the product coal specifications desired. However, in general, physical beneficiation processes rely on the use of gravitational and/or centrifugal forces to effect the separation of the clean coal from the accompanying impurities. Physical coal cleaning generally consists of dry or wet beneficiation methods. According to Ref. 3, in 1975, only 2.5% of the coals cleaned in the United States were beneficiated using dry separation methods; the other 97.5% were cleaned by using wet beneficiation methods.

In general, physical beneficiation processes consist of various combinations of some or all of the following unit operations.

1. *Size reduction*. This consists of reducing the size of the coal received from the mine (often 24 in. × 0) to more manageable sizes. Size reduction is usually accomplished by using equipment

such as rotary breakers, impact mills, and single and double roll crushers.

2. *Size classification*. This consists of segregating the coal into various size fractions to facilitate downstream processing. Both the ROM coal and the crushed product may be classified into different size fractions. Equipment for size classification includes stationary, vibrating, and cross-flow screens and classifying cyclones.

3. *Cleaning*. This is the heart of many coal beneficiation (preparation) plants. It involves mainly the separation of the physically attached sulfur and/or mineral impurities of higher specific gravities from the coal of lower specific gravity. This step is often accomplished by using jigs, cyclones, and concentration tables, which utilize a combination of frictional and/or gravity or centrifugal forces to effect an apparent density differential between the coal and its sulfur and mineral impurities. Another commonly used, cleaning method is the heavy-medium separation, which employs fine heavy minerals, such as magnetite or sand of an intermediate specific gravity, dispersed in water to effect the desired separation. In general, heavy-medium separation results in a fairly high recovery of the clean coal, although the latter fraction has to be separated from the heavy medium before it can be either used or processed further. Because of this additional processing step required, heavy-medium separation incurs higher operating costs than similar beneficiation processes using only clear water. Finally, froth flotation processes are generally used to beneficiate very fine-size (28 mesh × 0) fractions. In froth flotation, the coal is beneficiated in a liquid medium (usually water) by air bubbles (injected into the coal bath) that float the very fine clean coal particles to the liquid surface where the coal particles are mechanically skimmed. A surfactant is generally added to the coal bath to render the coal more hydrophobic and thereby facilitate the flotation of the coal. The impurities associated with the coal sink to the bottom of the vessel from where they are removed for eventual disposal.

4. *Drying*. This unit operation involves the reduction of the moisture content in the coal to the desired value. Various types of

equipment such as screens, filters, centrifuges and thermal dryers are used to dry the coal, depending upon the desired moisture content in the product coal.

In general, coal beneficiation plants use various combinations of all or some of the above unit operations to beneficiate different size fractions of the raw coal, depending upon the level of beneficiation desired. The latter is greatly dependent on the desired specifications of the coal to be produced. The various levels of coal beneficiation are discussed below.

B. Levels of Coal Beneficiation

ROM coal may be beneficiated at various levels ranging from level 1, which involves essentially no beneficiation, up to level 4, which implies a very thorough beneficiation of the coal. Of course, the cost of beneficiation also increases correspondingly from level 1 to level 4. Level 4 cleaning is generally intended for coals to be used in metallurgical operations (for coke production, for example), although some Eastern and Interior Basin coals (intended for steam production) may also require this thorough level of beneficiation in order to meet environmental restrictions. The four levels of coal beneficiation are briefly described below.

Level 1. This is a very basic level of beneficiation consisting of size reduction and classification with some attendant removal of refuse and mine dilutions such as pieces of timber, stray machine parts, etc., which can cause problems with downstream processing equipment. Level 1 beneficiation is practiced on essentially all coal burned. Calorific recovery or recovery of the ROM coal heating value is about 100%. However, there is essentially no reduction in the mineral impurities present in the coal.

Level 2. This involves level 1 preparation and wet beneficiation of the coarse coal (generally 3/8 in. size) fraction only. The fines fraction generated in the process is generally collected and shipped as is with the product coal. Calorific recovery at this level of treatment is generally high (>90%), but there is little to no reduction in the mineral impurities in the coal. Figure 9.1 is a sketch of a conceptual level 2 coal beneficiation plant.

Level 3. This involves level 2 preparation and beneficiation of all coal down to +28-mesh size fraction. The −28-mesh coal is either dewatered and shipped with the plant product or disposed of as refuse, provided environmental regulations permit such disposal. Calorific recovery is generally good (>80%), and there is a significant reduction in the sulfur and mineral impurities in the product coal.

Level 4. This involves a full-scale or thorough beneficiation of the coal. Figure 9.2 is a sketch of one version of a conceptual level 4 coal beneficiation plant. In the version shown, only one product stream is shown for simplicity. However, level 4 cleaning can usually yield several coal product streams containing varying levels of sulfur and mineral matter. The ultraclean fraction with the lowest level of sulfur and ash is generally routed to metallurgical operations. This stream may amount to as little as 25% of the raw coal feed to the plant [4]. Several intermediate fractions are also produced which contain coal with significantly reduced sulfur and ash levels but not low enough to meet metallurgical coal specifications. These intermediate streams are often referred to as "middlings" and are generally suitable for steam generation purposes. Material yields at this level of beneficiation generally range between 60 and 80%, while calorific recovery is generally between 85 and 98% of the incoming coal [4].

Regardless of the levels of cleaning achieved, physical coal beneficiation processes have limitations in that they can only remove the inorganic sulfur (mainly the pyritic sulfur) and the extraneous mineral impurities from the ROM coal. These processes are unable to reduce the organic sulfur content of the coal. Inorganic sulfur reductions by conventional physical beneficiation processes can range between 0 and 60 wt %. This generally corresponds to a 0 to 50 wt % reduction of the total sulfur content of the coal. Material recoveries or clean coal yields for the currently used beneficiation processes generally vary between 60 to 80 wt %.

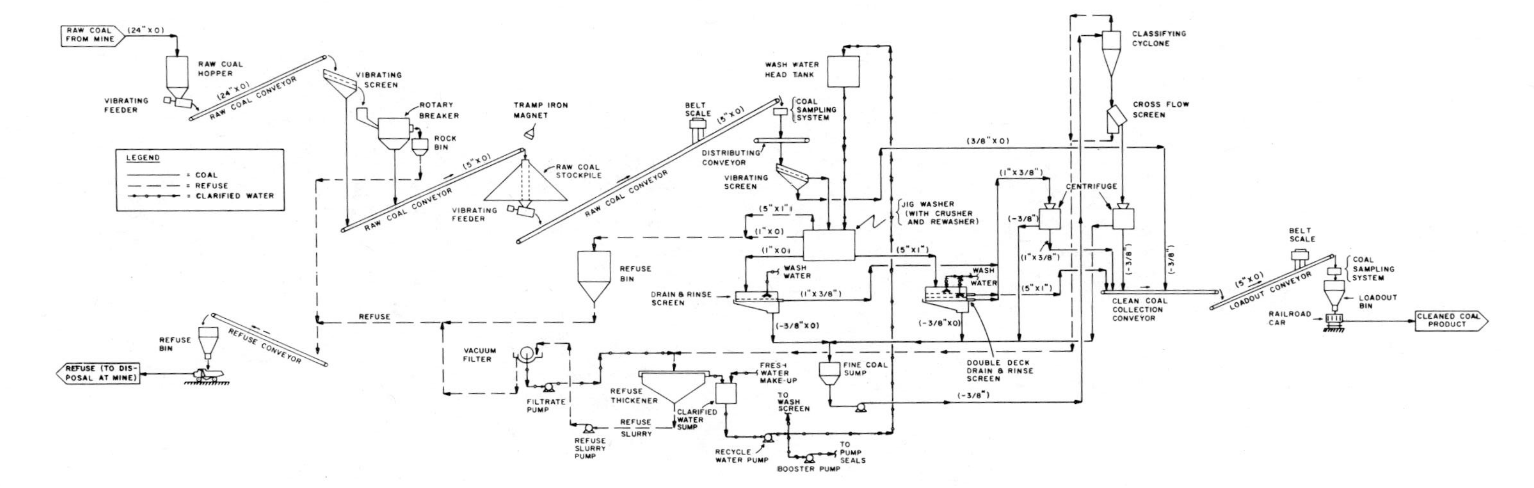

Figure 9.1 Process flow diagram of one version of a conceptual level 2 wet mechanical coal beneficiation plant.

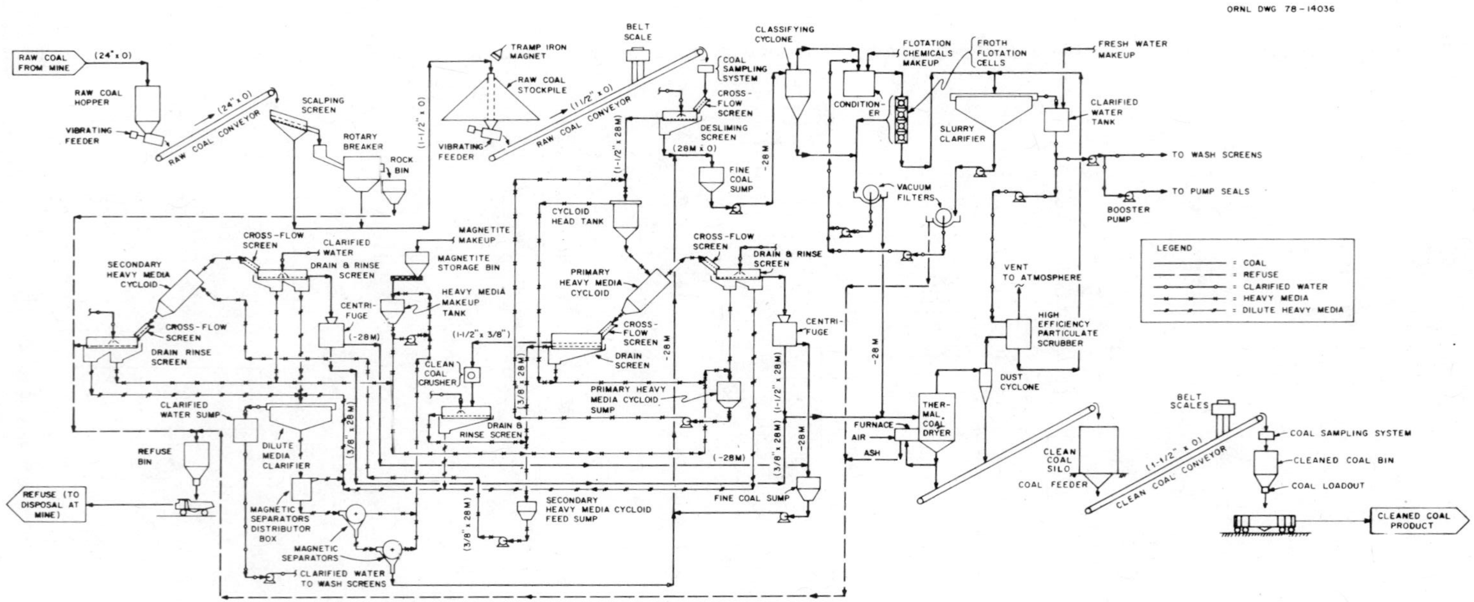

ORNL DWG 78-14036

Figure 9.2 Process flow diagram of one version of a conceptual level 4 wet mechanical coal beneficiation plant.

C. New Physical Beneficiation Processes

Historically, coals (other than those intended for metallurgical operations) were given perfunctory beneficiation with the objective of recovering as much lump coal as possible. The fines generated during the upgrading operations were often discarded as plant refuse. However, during the past few years, the needs of the coal markets have greatly changed because of the following factors:

1. The increased emphasis to use coal instead of oil or natural gas to generate electric power
2. The increasingly stringent environmental controls being promulgated for the burning of coal and for the disposal of refuse from coal processing plants
3. Increased mechanization in the coal mines, which results in higher volumes of fines in the ROM coal
4. Current coal economics, which almost mandates that as much clean and marketable coal be recovered from every ton of raw product mined

As a result, more and more emphasis is being placed on processing the fine-size (generally −1/4 in.) coal to recover usable coal and minimize the amount discarded as refuse. In particular, considerable research effort is currently being expended on developing new processes that either minimize the production of fines during the cleaning operations (such as the chemical comminution process) or clean the fine coal to yield a marketable product and thereby concurrently minimize the refuse to be processed from the beneficiation plant.

Most of the new physical coal beneficiation processes currently being developed or demonstrated have already been described in detail in the preceding chapters. Hence, it is not necessary to discuss further each of the processes included in this book. Instead, a number of general characteristics, which relate to the prospective of the commercial applicability of these processes to coal beneficiation operations, are summarized below.

1. All the processes described in the book can be classified as developmental-type processes. The processes are at various levels of process development ranging from bench-scale to pilot-scale evaluations. None of the processes have been evaluated or used on a commercial scale. However, efforts are being made to commercialize several of the new physical methods.

2. All the processes exhibit high recoveries ranging up to 90 to 95% of the incoming coal. These recoveries, if achieved in commercial practice, indicate significant improvements over current coal beneficiation processes.

3. The new processes offer significantly higher sulfur and/or ash reductions than can be achieved even by the highest level of the conventional wet beneficiation processes practiced today. Inorganic sulfur reductions as high as 90 wt % have been reported for some coals [5]. However, with the exception of the combined physiochemical approach of using microwave treatment, none of the new processes can affect the organic sulfur level of the coal. Although this limits the potential applicability of the new processes to cleaning coals having high inorganic sulfur content and low organic sulfur level, this is not a severe handicap, as conventional physical beneficiation processes are also restricted to the removal of the inorganic sulfur and/or mineral impurities of the coal.

4. Except the chemical comminution process, the new physical coal cleaning methods can achieve the high sulfur reductions mentioned above by beneficiating the fine and ultrafine coal (—28 mesh and smaller). This appears to be acceptable commercially since most of the coal used in utility boilers today is fired in pulverized coal-fired boilers. However, this may necessitate that these beneficiation processes be located near the end-use facility so as to avoid either the excessive losses in fine coal transportation or the increased costs associated with briquetting operations.

5. The application of the new physical coal cleaning methods described in this book in commercial operations, though potentially yielding higher recoveries of cleaner coal, will undoubtedly raise the price of the cleaned coal. However, the extent of the increase may well be less than the increased cost that may be necessitated due to the installation of alternate flue gas desulfurization systems.

6. Most of the processes described can be considered to be add-on type processes which, on a commercial scale, could be added to a conventional coal beneficiation plant to clean, for example, the

fines generated in the process. However, several of the processes (such as the high-gradient magnetic beneficiation process) can also be considered as stand-alone-type processes which may be used to clean the entire raw coal feed to the beneficiation plant.

D. New Chemical Beneficiation Methods

Chemical cleaning methods presently being developed beneficiate the coal under much more severe operating conditions that the physical methods. Examples of the chemical methods are the Meyers [6], Battelle hydrothermal [7,8], Ledgemont [9-11], KVB [12,13], and the PERC [14,15] processes. The chemical methods generally process pulverized coal (coal sizes ranging from —14 mesh to —200 mesh) at elevated temperatures and pressures, and in the presence of chemical reagents such as ferric sulfate, sulfuric acid, sodium hydroxide, and nitrogen dioxide. The majority of the methods can effect a significant reduction (up to 90 to 95%) in the pyritic sulfur and also remove varying amounts of the organic sulfur present in the coal.

Most of the chemical methods are at the laboratory-scale level of development. However, an 8 ton/day process development unit has been built to demonstrate key steps of the Meyers process and to determine its commercial viability. For illustrative purposes and to provide a background for the chemical beneficiation methods, a brief description of the Meyers process is presented below. The Meyers process was chosen because it is the most developed of the new chemical beneficiation processes. More detailed information regarding the various chemical methods can be obtained from the references cited above with each process and also from Refs. 6 and 16.

The Meyers process beneficiates the coal by treating it with an aqueous ferric sulfate solution at temperatures ranging between 194 and 266°F (90 and 130°C) and at pressures up to 120 psig (0.9 MPa). The coal matter goes through the process virtually unchanged, but the pyritic sulfur present in the coal is leached out by the ferric

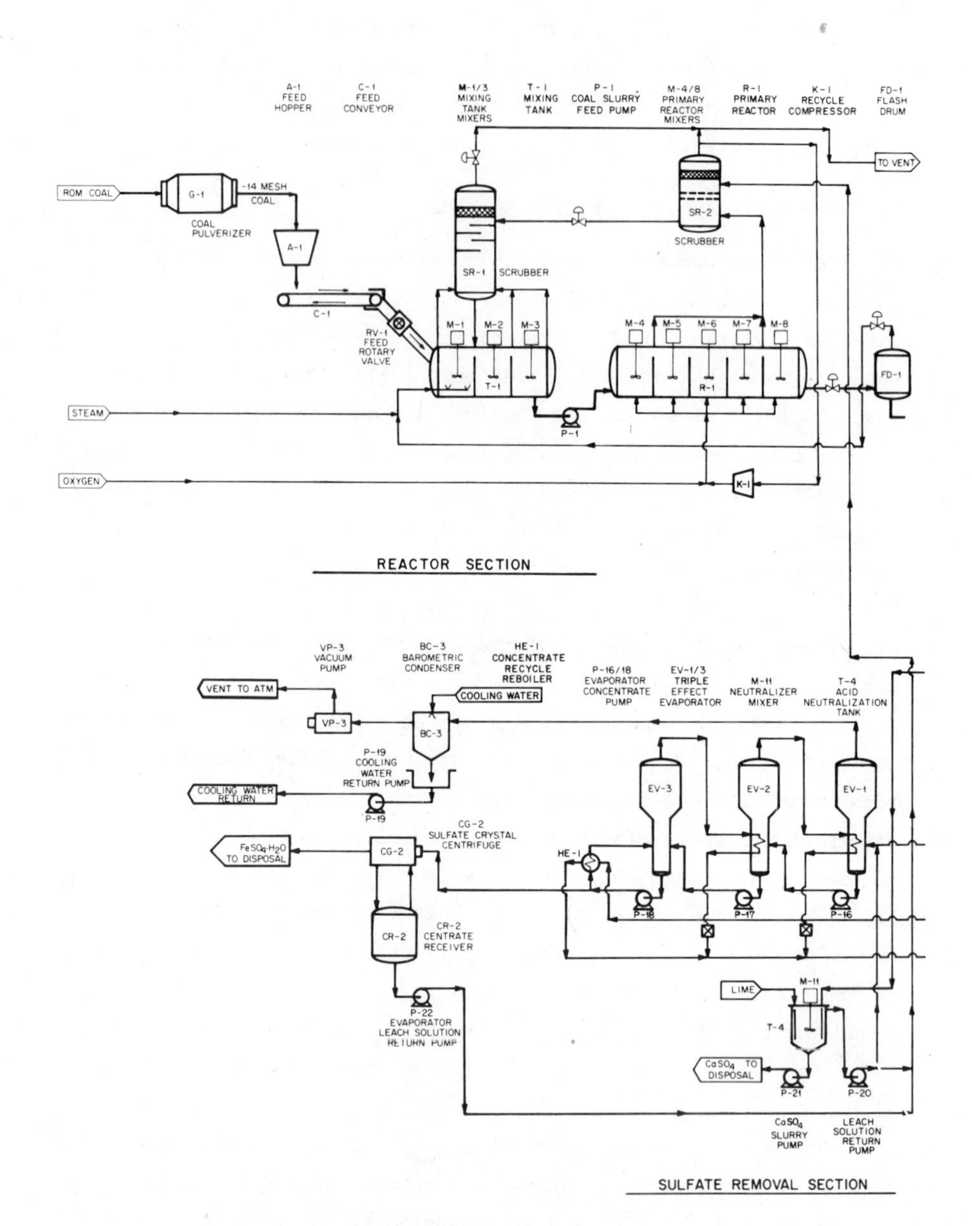

Figure 9.3 Process flow diagram of the conceptual commercial version of the Meyers fine coal process.

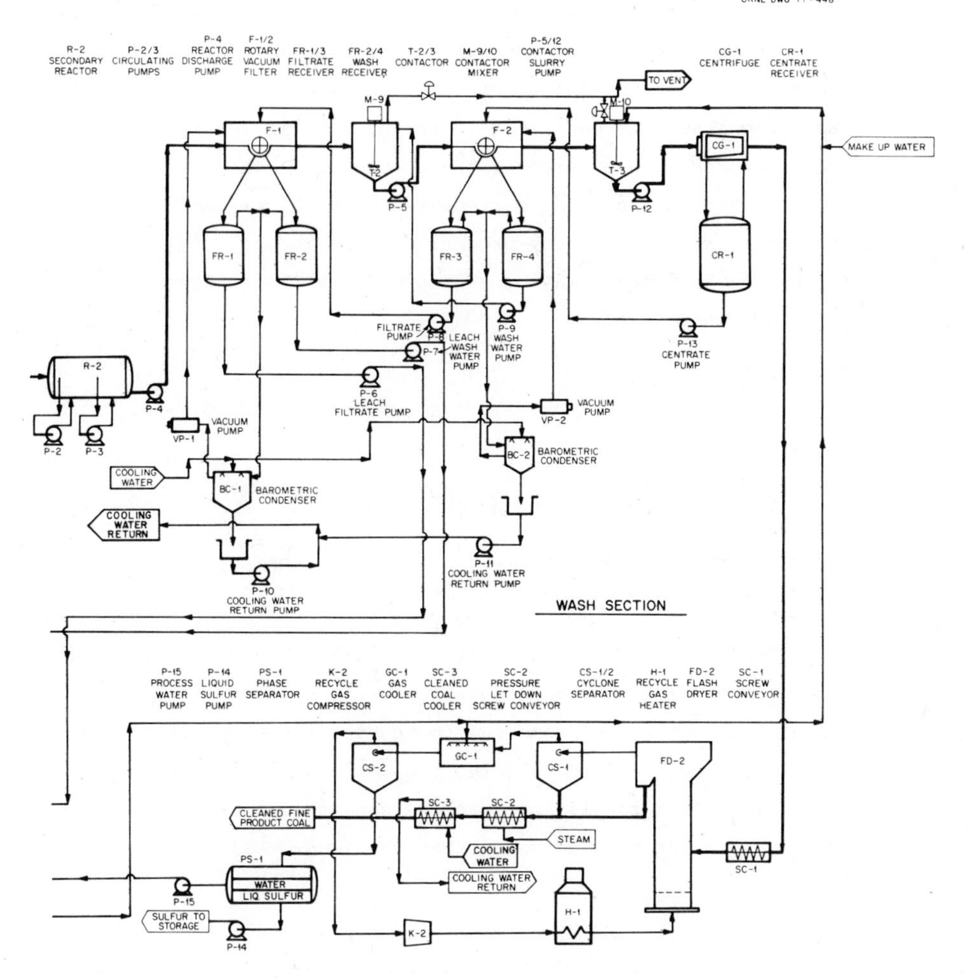

ORNL DWG 77-448
R-2 SECONDARY REACTOR
P-2/3 CIRCULATING PUMPS
P-4 REACTOR DISCHARGE PUMP
F-1/2 ROTARY VACUUM FILTER
FR-1/3 FILTRATE RECEIVER
FR-2/4 WASH RECEIVER
T-2/3 CONTACTOR
M-9/10 CONTACTOR MIXER
P-5/12 CONTACTOR SLURRY PUMP
CG-1 CENTRIFUGE
CR-1 CENTRATE RECEIVER
TO VENT
MAKE UP WATER
FILTRATE PUMP
P-8 LEACH WASH WATER PUMP
P-9 WASH WATER PUMP
P-13 CENTRATE PUMP
P-6 LEACH FILTRATE PUMP
VP-1 VACUUM PUMP
VP-2 VACUUM PUMP
COOLING WATER
BC-1 BAROMETRIC CONDENSER
BC-2 BAROMETRIC CONDENSER
COOLING WATER RETURN
P-10 COOLING WATER RETURN PUMP
P-11 COOLING WATER RETURN PUMP
WASH SECTION
P-15 PROCESS WATER PUMP
P-14 LIQUID SULFUR PUMP
PS-1 PHASE SEPARATOR
K-2 RECYCLE GAS COMPRESSOR
GC-1 GAS COOLER
SC-3 CLEANED COAL COOLER
SC-2 PRESSURE LET DOWN SCREW CONVEYOR
CS-1/2 CYCLONE SEPARATOR
H-1 RECYCLE GAS HEATER
FD-2 FLASH DRYER
SC-1 SCREW CONVEYOR
CLEANED FINE PRODUCT COAL
STEAM
COOLING WATER
COOLING WATER RETURN
WATER
LIQ. SULFUR
SULFUR TO STORAGE
SULFUR REMOVAL SECTION
TRW-MEYERS FINE COAL DESULFURIZATION PROCESS
PROCESS SCHEMATIC BY SUMAN P.N. SINGH
DATE 1/21/77
ISSUE 1

sulfate solution to form ferrous sulfate and elemental sulfur. The ferric sulfate which is consumed in the process is continuously regenerated by reacting the spent leach solution with oxygen or air. Excess ferrous sulfate generated in the process is separated out and may be recovered as a by-product of the process. The elemental sulfur formed in the reaction is also recovered by either steam distillation or solvent extraction as a by-product of the process. The underlying chemistry of the process may be represented by the following stoichiometric equations:

Treating reaction:

$$FeS_2 + 4.6Fe_2(SO_4)_3 + 4.8H_2O \rightarrow 10.2FeSO_4 + 4.8H_2SO_4 + 0.8S \quad (9.1)$$

Regeneration reaction:

$$2.4O_2 + 9.6FeSO_4 + 4.8H_2SO_4 \rightarrow 4.8Fe_2(SO_4)_3 + 4.8H_2O \quad (9.2)$$

Net overall reaction:

$$2.4O_2 + FeS_2 \rightarrow 0.2Fe_2(SO_4)_3 + 0.6FeSO_4 + 0.8S \quad (9.3)$$

Figure 9.3 is a schematic flow diagram of the Meyers process for cleaning fine coal. The process is especially suited to processing coals containing a high percentage of pyritic sulfur. The process (by itself) is capable of removing up to 95% of the pyritic sulfur present in the coal and in reducing the ash content of the feed coal by 10 to 30%. As a result, the process can yield a product coal that has up to 5% higher heating value than the feed coal.

9.3 PROCESS DESCRIPTIONS

Brief descriptions are provided below for the five conceptual processes selected for the economic assessment. Details regarding the processes may be obtained from the references cited.

A. Wet High-Gradient Magnetic Beneficiation

In the wet magnetic beneficiation process, coal is beneficiated by passing a water slurry of finely pulverized coal (generally 70% –200 mesh) through a container where it is subjected to a high-intensity, high-gradient magnetic field. Coal slurries up to 30 wt % coal and magnetic field intensities up to 20 kOe have been used. The container is generally packed with a "capture" matrix made of stainless steel wool. In the presence of the magnetic field, the paramagnetic material present in the coal slurry, such as pyrite and mineral matter, becomes magnetized and is trapped in the matrix; while the diamagnetic coal particles, unaffected by the magnetic field, pass through the container for further downstream processing. When the matrix is loaded to its magnetic capacity, the slurry feed is stopped and the electric power is cut off. The matrix is then backwashed to remove the trapped pyrite and mineral matter. Following the backwashing step, the feed and the power are resumed, and the entire process is repeated.

In a more recent version of the process given in Ref. 17, a continuous high gradient magnetic separator developed by Sala Magnetics, Inc. is used. In the new design, the "capture" matrix is a continuous, segmented metal belt. The belt moves continuously through the magnetic separator and the washing section. Figure 9.4 is a schematic flow diagram of the conceptual wet magnetic beneficiation process using the continuous magnetic separator described above. The magnetic beneficiation section of the flow scheme was developed based on information given in Ref. 17. In the continuous separator, the belt first passes through the magnetic section of the separator where the coal slurry is introduced and beneficiated. The belt then travels into the washing section (devoid of the magnetic field) where the pyrite and the mineral matter trapped on the belt are washed off with water. The cleaned belt section then returns to the head end of the magnetic separator where the entire process is repeated. The pyrite and the mineral matter removed from the belt are thickened to recover the water and disposed off as plant refuse, while the collected water is treated and reused in the process.

The cleaned coal slurry from the magnetic separator is vacuum filtered to recover moist, cleaned coal. The coal is then thermally dried and shipped as plant product. Water recovered in the vacuum filtration step is treated and reused in the process. Small amounts

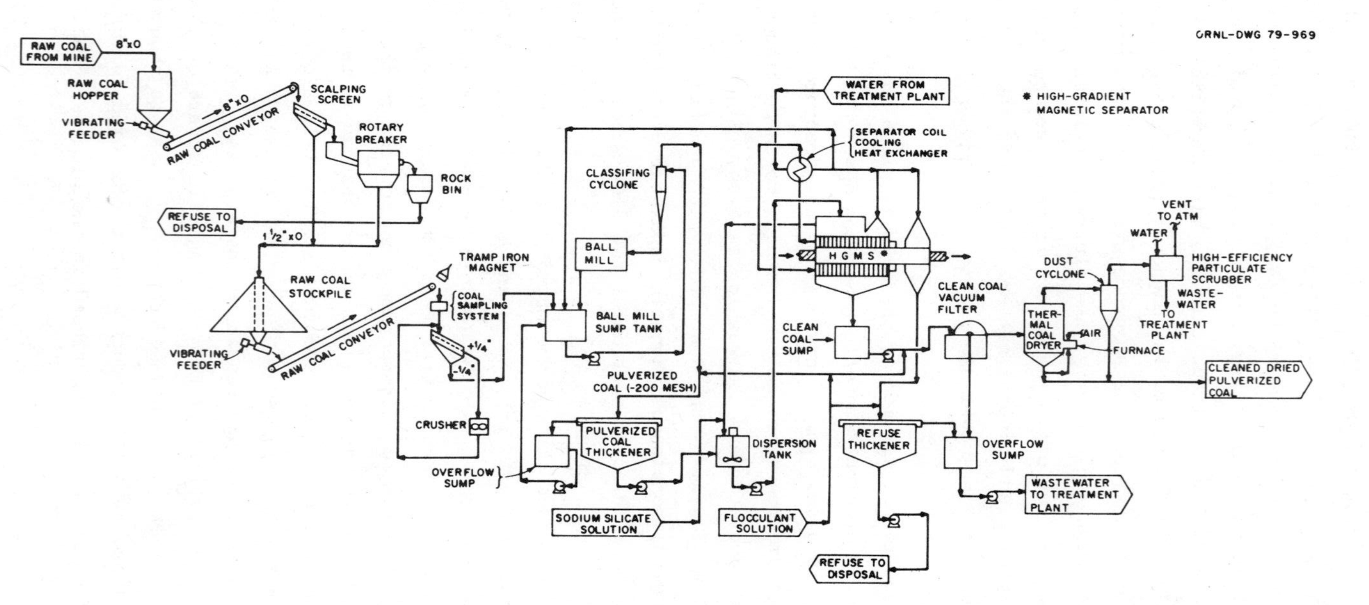

Figure 9.4 Process flow diagram of the conceptual wet magnetic beneficiation process using the continuous high-gradient magnetic separator.

of flocculant and dispersant (sodium silicate) solutions are added to the coal slurry to facilitate the beneficiation process. Further details regarding the high-gradient magnetic beneficiation process may be obtained from Chap. 4 and Refs. 5 and 16 to 18.

B. Chemical Comminution

In the chemical comminution process, the raw coal is reduced in size by using chemical agents rather than mechanical crushing methods. This provides a more selective breaking of the coal along the internal fault system in the coal. The chemicals (generally concentrated aqueous ammonia solutions or gaseous ammonia) induce a fracture of the coal along already existing boundaries which contain the pyrite and mineral matter, thereby liberating the impurities present in the coal. The chemically comminuted coal is then subjected to conventional wet beneficiation methods wherein the coal matter is separated from the liberated impurities.

Figure 9.5 is a schematic flow diagram of the conceptual process scheme utilizing chemical comminution and heavy-medium wet beneficiation to separate the coal from the attendant impurities. In the figure, the chemical comminution step has been shown as a block because either the batch or the continuous version of the process may be used. In the conceptual flow scheme, the raw coal is initially reduced to 1½ in. × 0 size and fed to the chemical comminution step where it is reacted with ammonia. Debris and large rocks present with the raw coal are removed in the rotary breakers upstream of the chemical comminution vessels and disposed off as plant refuse. The chemically comminuted coal is screened at 100 mesh. The 3/8 in. × 100 mesh size fraction is cleaned using heavy-medium separation while the −100-mesh fraction is cleaned using froth flotation to recover additional coal. Fine coal recovered from the magnetic separators in the heavy-medium cleaning operation is combined with the above −100-mesh size fraction from the chemical comminution step and is also subjected to froth flotation. Refuse from the froth flotation cells is concentrated using clarifiers and vacuum filters and is disposed off as

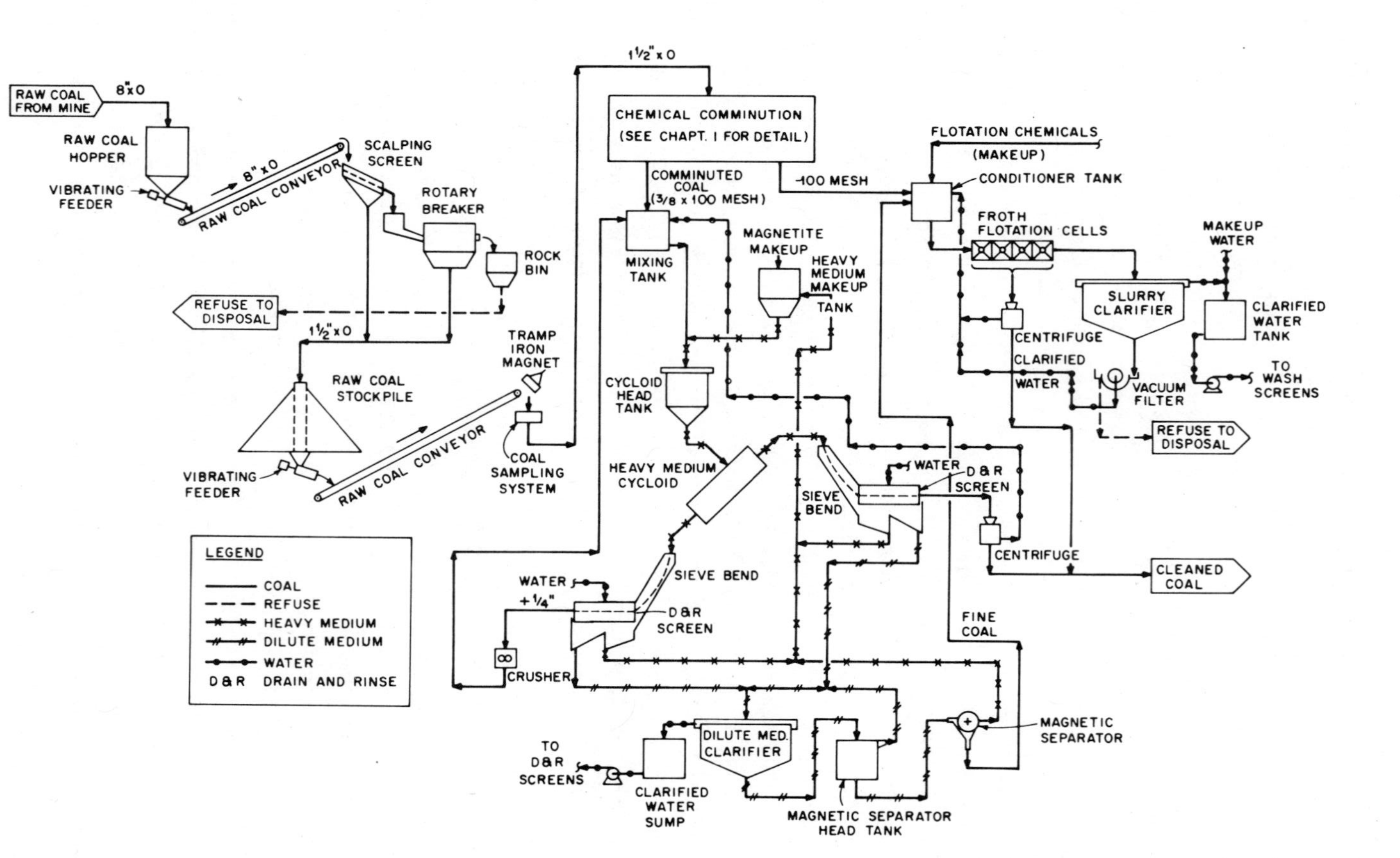

Figure 9.5 Schematic flow diagram of the conceptual beneficiation process using chemical comminution and heavy-medium separation.

plant waste. Water recovered from the clarifiers and the vacuum filters is recycled in the process.

Further details regarding the chemical comminution process may be obtained from Chap. 1.

C. Level 2 Wet Mechanical Beneficiation

This cleaning method was briefly described in Sec. 9.2. Basically, level 2 wet beneficiation of the coal involves size reduction, screening, and wet cleaning (using water only) of the coarse coal (+3/8 in.) fraction. Several versions of the process are possible, depending upon the raw coal properties and the cleaned coal specifications desired. A sketch of one version of the process is shown in Fig. 9.1.

Referring to Fig. 9.1, the raw coal is mechanically crushed and screened to yield a 5 in. × 0 size fraction. Rock and debris released in the crushing process are removed in the rotary breakers and sent to the refuse bin for disposal. The 5 in. × 0 coal is classified using vibrating screens to yield a coarse coal (+3/8 in.) and a fine coal (—3/8 in.) fraction. The fine coal fraction by-passes the cleaning ciruit and reports as product coal on the clean coal conveyor.

The coarse coal is washed and cleaned (using jigs and drain and rinse screens) to remove the associated and liberated mineral matter and pyrites. Fines generated during the washing operations are collected and cleaned using classifying cyclones and crossflow screens. Centrifuges are used to remove the excess moisture from the cleaned coal. The cleaned coal then reports at the cleaned coal collection conveyor from where it is conveyed to the clean coal silo for eventual loadout as plant product.

The refuse slurry generated in the washing steps is collected and thickened using clarifiers (thickeners) and vacuum filters. The clarified water that is recovered is reused in the process while the moist refuse is conveyed to the refuse bin for disposal.

D. Level 4 Wet Mechanical Beneficiation

This cleaning method has been briefly described in Sec. 9.2. A schematic flow diagram of one version of the process is given in

Fig. 9.2. Other versions of the process are possible depending upon the feed coal properties and the product coal specifications desired. The process is similar to the level 2 beneficiation method described previously except that

1. The feed coal is mechanically crushed to 1½ in. × 0 size to liberate more of the imbedded pyrites and mineral matter.
2. The coal is cleaned more intensely using heavy medium (magnetite-water slurry) rather than water only.
3. The entire coal fraction is cleaned and dried in a thermal dryer.
4. Froth flotation is used to clean the —28-mesh size fraction.
5. Additional processing steps are required. However, the product coal has a higher value (that is, it contains less mineral matter and pyrite).

A detailed discussion of the process is given in Ref. 16.

E. Meyers Fine Coal Process

The Meyers fine coal process consists of chemically beneficiating the coal to remove up to 95% of the pyritic sulfur present in the coal. A schematic flow diagram of the conceptual commercial-scale version of the process is given in Fig. 9.3, and the chemistry of the process is indicated in Sec. 9.2D.

Basically, raw coal ground to —14 mesh is reacted with an aqueous ferric sulfate solution for about 8 hr at temperatures ranging up to 266°F (130°C) and at pressures up to 120 psig (0.9 MPa). The heat of the reaction is used to heat the recycle ferric sulfate solution. The majority of the pyrites present in the coal are converted to ferrous sulfate, sulfuric acid, and elemental sulfur. High-purity oxygen is also added to the reaction mixture to simultaneously regenerate the ferric sulfate from the spent ferrous sulfate solution. After about 8 hr, the reaction mixture is transferred to the secondary reactor where the residual pyrites in the coal are converted.

Following the depyritization reaction, the coal is filtered from the spent solution. The coal is then washed with recycle water and dried using centrifuges. The moist coal from the centrifuges is then flash-dried by high-temperature steam. Flash-drying the coal also vaporizes the elemental sulfur accompanying the coal from the depyritization reactors. The dried coal is then separated from the steam-sulfur vapor stream using cyclone separators and is cooled to yield the cleaned product coal.

Most of the water and the ferrous sulfate generated is reused in the process. The excess ferrous sulfate produced is either sold as a by-product (if a market exists) or is disposed of as plant refuse. A portion of the ferrous sulfate solution is neutralized by the addition of lime. The neutralized wash water is separated from the gypsum and is recycled to the depyritization reactors. The gypsum produced in the process is appropriately disposed of.

More detailed descriptions of the Meyers process are given in Refs. 6 and 16.

9.4 ECONOMIC ASSESSMENT

Since most of the new physical coal cleaning methods described in this book are currently under bench-scale or pilot-plant development, much data needed to perform detailed economic assessment on the processes are not available. Nonetheless, preliminary economic assessments have been made for two of the new methods described in the book for conceptual plants capable of processing 200 tons of raw feed coal per hour. These two methods are

1. Wet high-gradient magnetic beneficiation process
2. Chemical comminution process

In addition, for comparison purposes, economic assessments have also been made for conceptual level 2 and level 4 conventional, wet mechanical beneficiation plants and for a chemical pyrite removal plant based on the Meyers fine coal process. The plants were also designed to process 200 tons of raw feed coal per hour.

The major objective of the economic assessment is to indicate the range of costs associated with conceptual plants using the selected new and conventional coal beneficiation methods. Since the detailed design data (such as the type of coal processed, its physical and washability characteristics, and the cleaned product coal specifications) are not fully available and included in the following assessment, the specific results to be reported should be considered as being of a preliminary nature.

A. Evaluation Procedure

Preliminary economic analyses were performed for the five conceptual, new and conventional coal beneficiation plants using an economic analysis program PRP developed by the Oak Ridge National Laboratory [19]. Briefly stated, the PRP computer program calculates the price of a product from a processing plant for a given set of input economic parameters using the discounted cash flow (DCF) method.

1. The conceptual plants were evaluated as battery-limits facilities wherein all utilities were assumed to be available and were purchased.

2. Two financing schemes were used to finance the conceptual plants: (a) 100% equity capital and (b) 70% debt and 30% equity capital. The latter corresponds to a utility-financed venture. Values of 12 and 15% annual after-tax rates of return on equity capital were used in the analyses, and interest rate on the debt fraction of the plant capital was computed at 9% per annum.

3. All costs were developed based on second-quarter 1978 dollars. No forward escalation beyond second-quarter 1978 was included in the cost estimates. Time escalation of plant costs (when obtained from a study conducted in a different time frame, for example) was made according to the escalation factors given in Table 9.1. No cost was included for a captive coal mine. Instead, the feed coal costs were varied from $10 to $50/ton in $10/ton increments.

4. The project life for each of the conceptual plants was assumed to consist of a 2-year construction phase followed by a 20-year operating life. During the construction phase, 40 and 60% of the depreciable capital investment were spent within year 1 and year 2, respectively. The conceptual plants were premised to operate 14 hr/day for 329 days/year. The remaining time was allocated to providing the necessary maintenance on the plants. The annual plant service factor during the operating life of the conceptual plants was assumed to be 52.6% from the second year of the plant operations.

Table 9.1 Escalation Factors Used in the Economic Analyses of the Conceptual Beneficiation Plants

Year	Escalation index (1958 = 100)
1961	103
1962	104
1963	105
1964	107
1965	109
1966	114
1967	118
1968	124
1969	135
1970	144
1971	155
1972	163
1973	172
1974	226
1975	276
1976	330
1977	364
1978	374
1979	399
1980	431

The service factor for the first year of operation was assumed to be 46.7% to account for startup problems. A depreciable life of 16 years for the process facilities, commencing at the end of the construction period was assumed in the analyses. Also, the salvage value of the plants was taken as zero at the end of the project life.

5. The working capital was assumed to remain invested throughout the project and to be recovered intact at the end of the project life. The capital invested during the construction phase of the conceptual plants was considered to earn interest at the same rate as

during the years following the startup of the facilities. However, since there is no income generated by the plants during the period prior to startup, the interest that accrued on the capital was accumulated and added to the outstanding capital that was to be recovered during the plant operating years. For tax purposes, the interest on the debt was treated as a deductible expense in the year in which it was paid. The procedure outlined above is accounted for internally in the computer program PRP.

6. Taxes were accounted for in the computer program PRP as part of the DCF analysis. The following tax rates were assumed: (a) federal income tax, 48%; (b) state income tax, 3%; (c) state revenue tax, 0%; and (d) local property tax, 2%. An investment tax credit of 10% of the depreciable capital was also included in the economic analyses.

B. Estimates of Investment Cost and Working Capital

Tables 9.2 and 9.3 show the estimates of investment cost and working capital, respectively, for the five conceptual, battery-limits 200 ton/hr coal beneficiation plants. Some specific assumptions leading to these cost estimates have been included in the table. In addition, a number of remarks related to the tabulated results are presented as follows:

1. The direct on-site plant costs (DOPC) for the conceptual beneficiation plants (item 3 in Table 9.2) were determined as follows:
   a. The investment cost for the wet magnetic beneficiation plant, particularly the high-gradient magnetic separation units, was scaled from the cost reported in Ref. 17 using 0.7 as the exponent. In addition, estimated investment costs developed based on in-house information were added for the size reduction and thermal drying facilities.
   b. The investment cost for the chemical comminution plant was scaled from the costs obtained from Ref. 20, using 0.7 as the exponent.
   c. The investment costs for the level 2 and level 4 wet mechanical beneficiation plants were estimated from in-house information on coal beneficiation. Some of the in-house information is available in the literature [16].

Table 9.2 Capital Investment Estimates for the Conceptual Battery-Limits 200 ton/hr Coal Beneficiation Plants (Second-Quarter 1978 Costs in $\$10^6$)

	1	2	3	4	5
			Wet mechanical beneficiation		
Component	Wet magnetic beneficiation	Chemical comminution	Level 2	Level 4	Meyers process
1. Installed plant cost (IPC)	21.98	13.69	6.40	8.02	29.23
2. Installed cost for environmental control equipment, 10% IPC	2.20	1.37	0.64	0.80	2.92
3. Direct on-site plant cost (DOPC)	24.18	15.06	7.04	8.82	32.15
4. Engineering and contractors' fees, 15% DOPC	3.63	2.26	1.06	1.32	4.82
5. Project and/or process contingency, 15% (cases 3,4) or 20% (cases 1,2,5) of DOPC	4.84	3.01	1.06	1.32	6.43
6. Startup plant modifications, 2% DOPC	0.48	0.30	0.14	0.18	0.64
7. Working capital (see Table 9.3)	4.98	4.87	4.12	4.29	5.45
8. Land costs, $3,000/acre	0.18	0.21	0.15	0.18	0.30
Total capital investment	38.29	25.71	13.57	16.11	49.79

Table 9.3 Working Capital Estimates for the Conceptual Battery-Limits 200 ton/hr Coal Beneficiation Plants (Second-Quarter 1978 Costs in $\$10^6$)

Component	1	2	3	4	5
			Wet mechanical beneficiation		
	Wet magnetic beneficiation	Chemical comminution	Level 2	Level 4	Meyers process
1. 30 days of feedstock coal cost, at $25/ton	2.10	2.10	2.10	2.10	2.10
2. 30 days of product coal inventory at $30/ton	2.14	2.24	1.81	1.86	2.17
3. 30 days of gross operating costs (see Table 9.4)	0.74	0.53	0.21	0.33	1.18
Total working capital	4.98	4.87	4.12	4.29	5.45

   d. The cost for the Meyers fine coal process was scaled from the costs obtained from Ref. 21 using 0.7 as the exponent. In addition, estimated investment costs were added (based on in-house information) for size reduction equipment to reduce the coal feed to —14 mesh.

2. Contingency costs (item 5 in Table 9.2) were estimated to be 15% of the DOPC for the two wet mechanical beneficiation processes since these are commercially available processes, and 20% of the DOPC for the wet magnetic beneficiation, chemical comminution, and Meyers processes because these are more developmental-type processes. Project and/or process contingencies (where applicable) are included in this cost. However, since the estimates were made in terms of constant dollars, the contingency does not include any allowance for the time escalation of costs.
3. The working capital (item 7 of Table 9.2 and Table 9.3) was assumed to remain invested throughout the project and to be recovered at the end of the project life.
4. Land requirements (item 8 in Table 9.2) were estimated to be 50 acres for case 3, 60 acres for cases 1 and 4, 70 acres for case 2, and 100 acres for case 5.

C. Estimates of Annual Operating Costs

Table 9.4 presents the estimates of annual operating costs for the five conceptual coal beneficiation plants. The tabulated operating costs do not include feedstock cost, depreciation, and taxes. The latter costs are accounted for separately in the PRP computer program. A number of assumptions were made in obtaining the tabulated estimates, and these are as follows:

1. Process operating labor costs (item 1a in Table 9.4) were estimated at $8.50/man-hr. The operating labor requirements for the various processes were assumed to be as follows:

Process	Operating labor operators/shift
Wet magnetic beneficiation	12
Chemical comminution	15
Level 2 wet mechanical	8
Level 4 wet mechanical	13
Meyers	16

Table 9.4 Annual Operating Cost Estimates for the Conceptual Battery-Limits 200 ton/hr Coal Beneficiation Plants (Second-Quarter 1978 Costs in $\$10^6$)

	1	2	3	4	5
			Wet mechanical beneficiation		
Component	Wet magnetic beneficiation	Chemical comminution	Level 2	Level 4	Meyers process
1. Labor costs					
a. Process operating labor	0.83	1.04	0.56	0.90	1.11
b. Supervisory personnel cost (20% of 1a)	0.17	0.21	0.11	0.18	0.22
2. Operating labor burden (35% of 1a and 1b)	0.35	0.44	0.24	0.38	0.47
3. Plant maintenance cost (5% of DOPC in Table 9.2)	1.21	0.75	0.35	0.44	1.61
4. Operating supplies (30% of 1a)	0.25	0.31	0.17	0.27	0.33
5. General administrative overhead costs (40% of the sum of 1 to 4)	0.92	0.84	0.43	0.64	1.22
6. Utilities and chemicals cost (see Table 9.5)	4.71	2.50	0.33	0.82	8.80
7. Waste disposal cost	0.26	0.19	0.29	0.27	0.25
8. Property insurance cost	0.25	0.15	0.07	0.09	0.33
Gross annual operating cost	8.95	6.43	2.55	3.99	14.34

Table 9.5 Utility Costs Used in the Economic Evaluation of the Conceptual Coal Beneficiation Plants (Second-Quarter 1978 Costs)

Utilities	Unit[a]	Cost, $/unit
Electric power	kWh	0.030
Steam		
50 psig	M lb	2.50
150 psig	M lb	3.00
Coal (as fuel)	MMBtu	1.00
Raw water	M gal	0.05
Treated water (cooling tower makeup water)	M gal	0.30

[a]k and M = thousand; MM = million.

2. The operating labor burden (item 2 in Table 9.4) includes the associated costs of social security and unemployment insurance contributions, sick and vacation pay, and other fringe benefits. The plant maintenance cost (item 3 in Table 9.4) includes the cost of maintenance materials, maintenance labor and supervision, and the labor burden for the maintenance crew. The general administrative overhead costs (item 5 in Table 9.4) include the costs for services that are generally necessary for the efficient operation of the plant (other than those directly related to the plant operation) such as medical, secretarial, janitorial services, and general engineering, etc.

3. Utilities requirements for the wet magnetic beneficiation, chemical comminution, and the Meyers processes were scaled for the plant sizes evaluated from information reported in the literature [17,20,21], and the required operating costs were determined by using the unit costs given in Table 9.5. Costs for process chemicals for the above processes were scaled from the costs reported in the same references [17,20,21]. The costs for utilities and chemicals for the level 2 and level 4 wet mechanical processes were determined from in-house information, which has been summarized in Ref. 16. Waste disposal costs (item 7 in Table 9.4) were estimated at $1/ton of waste product.

D. Evaluation Results and Discussion

Based on the above estimates of investment cost, working capital and operating cost, the unit costs for the five conceptual, battery-limits 200 ton/hr coal beneficiation plants were obtained by using the PRP

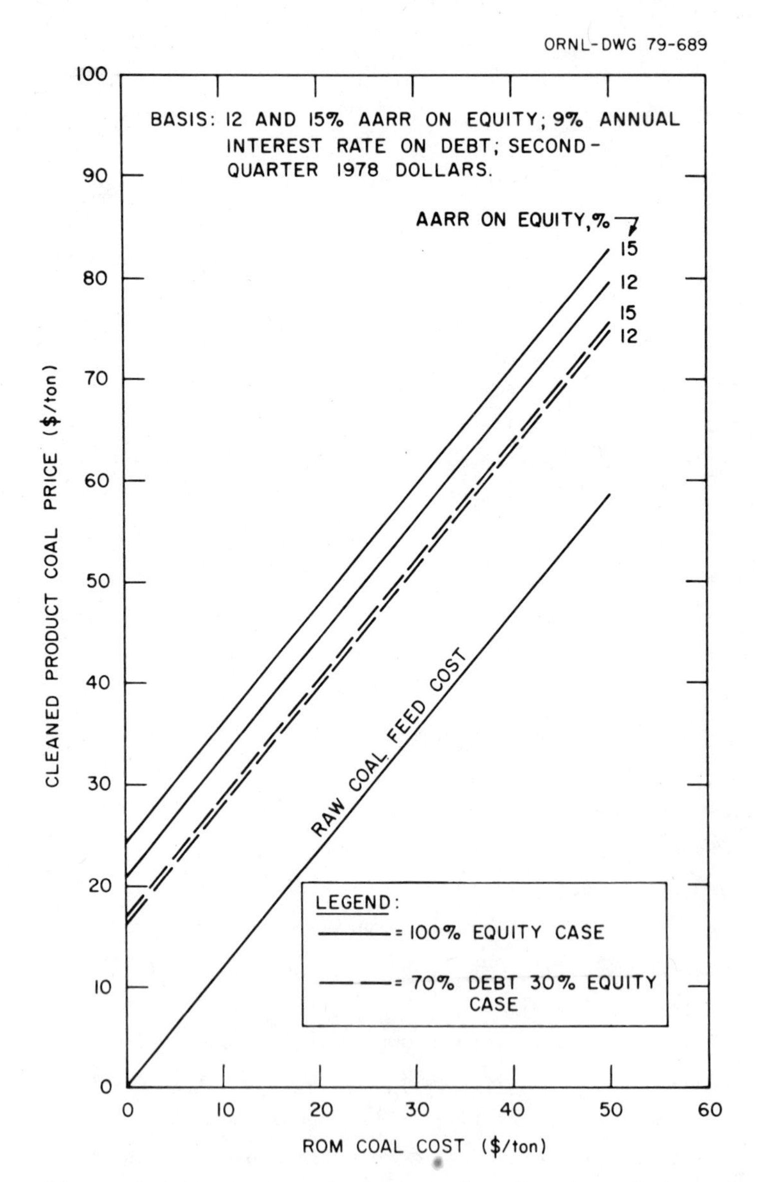

Figure 9.6 Variation of the calculated price of cleaned coal product with ROM coal cost for the 200 ton/hr conceptual battery-limits wet magnetic beneficiation plant.

computer program. Typical results obtained are illustrated in Figs. 9.6 to 9.10 for the four financing scenarios considered. These figures show the variation in the calculated price of the cleaned product coal for different feed coal costs. The coal price shown includes

ORNL-DWG 79-693

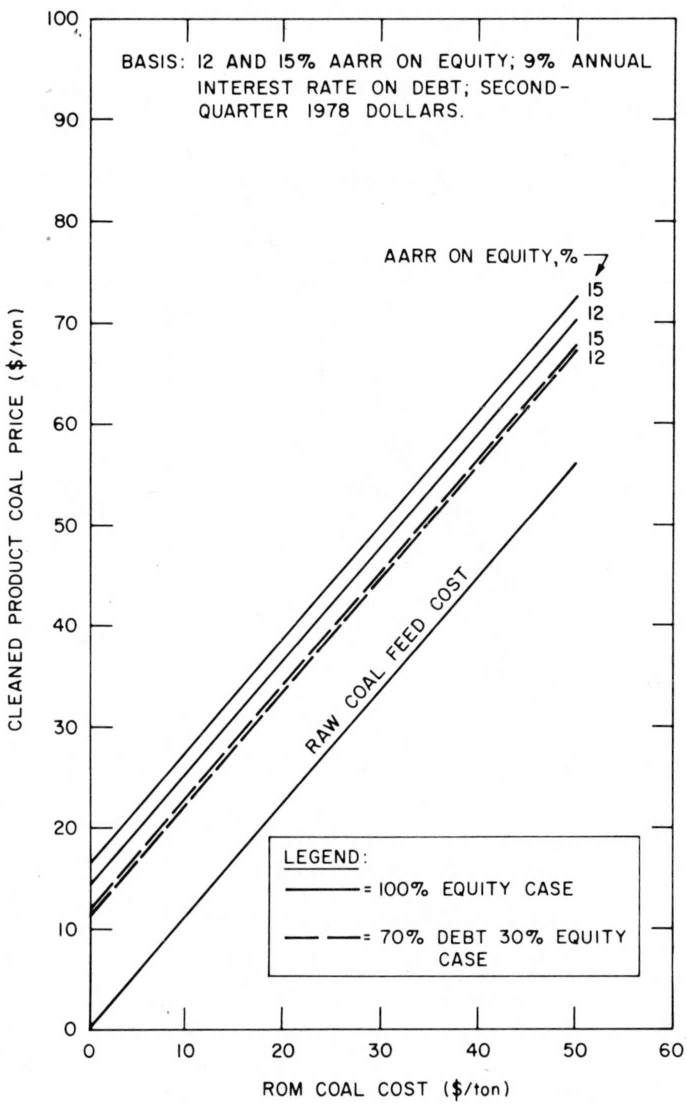

Figure 9.7 Variation of the calculated price of cleaned coal product with ROM coal cost for the 200 ton/hr conceptual battery-limits chemical comminution beneficiation plant.

the cost of the raw feed coal used to produce the cleaned product. For example, for 100% equity financing, 15% AARR on equity, and for \$20/ton feed coal, the cleaned coal price varies from a low of about \$38/ton of product coal from the level 2 mechanical beneficiation

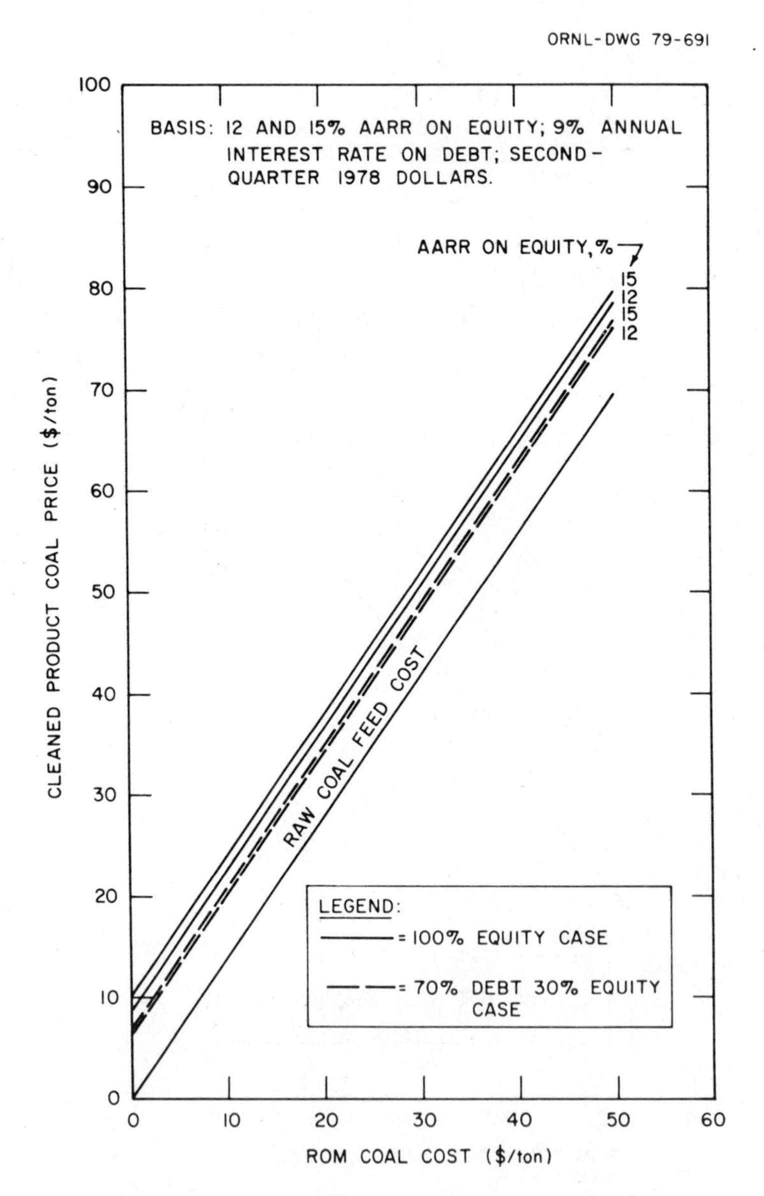

Figure 9.8 Variation of the calculated price of cleaned coal product with ROM coal cost for the 200 ton/hr conceptual battery-limits level 2 wet mechanical beneficiation plant.

process to about $56/ton of product coal from the Meyers process. Similar variations in the product coal price can be discerned for the other financing scenarios.

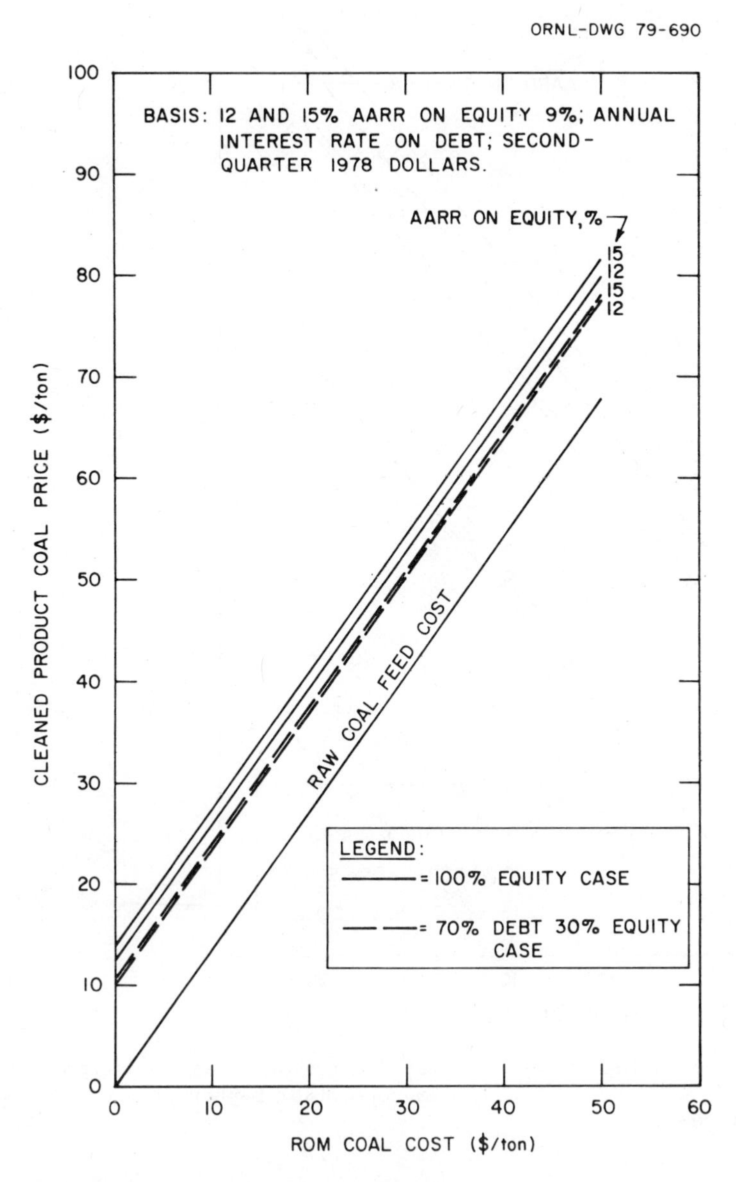

Figure 9.9 Variation of the calculated price of cleaned coal product with ROM coal cost for the 200 ton/hr conceptual battery-limits level 4 wet mechanical beneficiation plant.

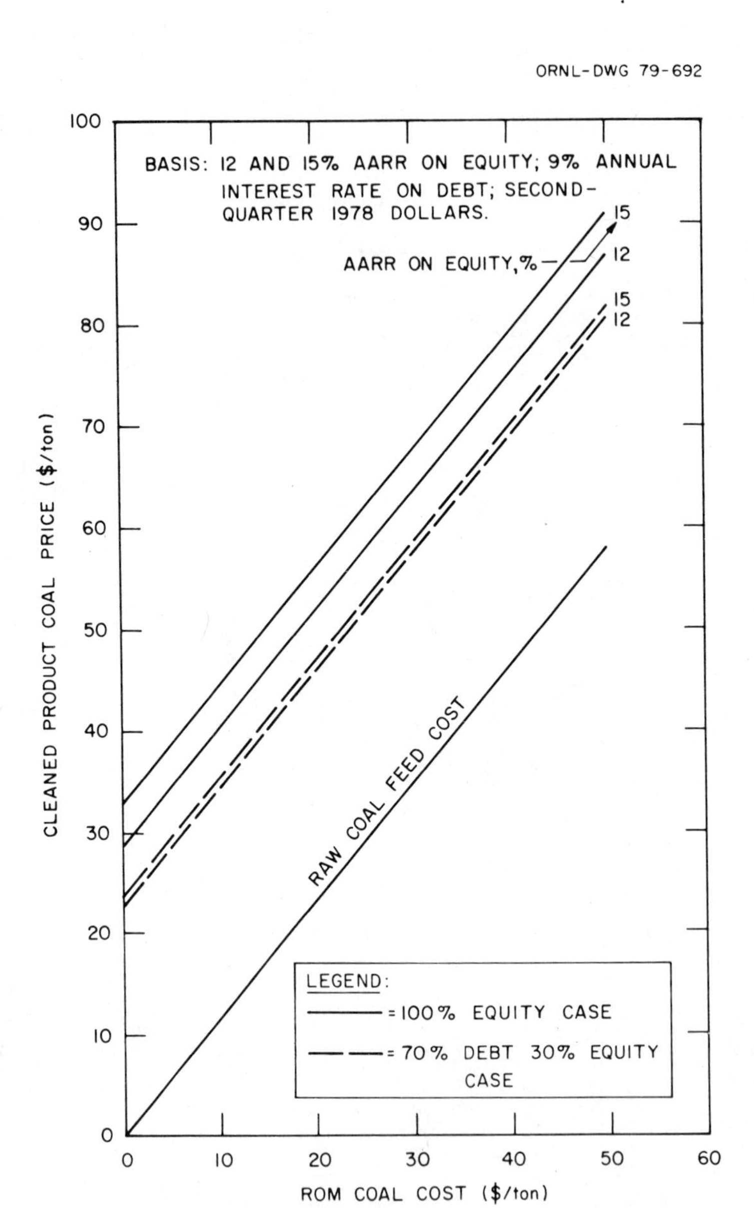

Figure 9.10 Variation of the calculated price of cleaned coal product with ROM coal cost for the 200 ton/hr conceptual battery-limits Meyers fine coal process.

The estimated material recoveries and the cost of beneficiating the coal by the five conceptual processes evaluated are illustrated in Table 9.6 for the 100% equity capital, 15% AARR on equity, and $20/ton ROM coal cost case. Similar information can be developed for the other financing schemes and other ROM coal costs. As can be seen from Table 9.6, the beneficiation costs for the newer processes appear to be about two to four times the cost for the conventional wet mechanical beneficiation processes. Of course, the costs compared are those for pioneer plants (for the developing technologies) compared with the costs of currently available technology. It is quite likely that as the newer technologies develop and mature, the gap between the two can be considerably decreased. Also, although the beneficiation costs for the three new coal cleaning processes are higher, these processes are capable of higher yields of better quality (cleaner) coal than the level 2 and even the level 4 wet mechanical beneficiation processes. This is a significant factor that must be borne in mind especially considering the continually increasing demand for cleaner coal and the increasingly stringent emission control requirements being placed on coal processing facilities.

The sensitivity of the calculated beneficiation costs to step changes in several economic parameters of interest was investigated. The results of the analyses are summarized in Table 9.7 for the five beneficiation processes.

It can be seen from Table 9.7 that the beneficiation costs for all the five processes are most sensitive to the change in the hours per day of plant operation, the newer processes appearing to benefit more than the conventional wet cleaning processes. This is not altogether surprising since these processes are essentially round-the-clock-type processes. However, in order to perform the economic assessments in a consistent manner, all the processes were premised to operate 14 hr/day, which is the current industry practice for wet mechanical beneficiation plants (due to the increased maintenance requirements of these plants).

The effect of step changes in other parameters evaluated can be discerned from Table 9.7.

Table 9.6 Comparison of the Material Recoveries and Beneficiation Costs for the Five Conceptual Coal Beneficiation Processes (Second-Quarter 1978 Costs)

Process	Material recovery,[a] %	Beneficiation cost per ton coal processed,[b,c] $/ton	Incremental cost over wet mechanical beneficiation: Level 2, $/ton	Incremental cost over wet mechanical beneficiation: Level 4, $/ton
Wet magnetic beneficiation	85	20.42	13.17	10.12
Chemical comminution	89	14.46	7.21	4.16
Meyers process	86	28.23	20.98	17.93
Level 2 wet mechanical	72	7.25		−3.05
Level 4 wet mechanical	74	10.30	3.05	

[a]Derived as tons of cleaned product coal per tons of ROM coal feed.

[b]Derived as (product price per ton − feed coal cost per ton of product coal) × material recovery, %/100.

[c]The values reported are for $20/ton ROM coal, 100% equity financing, and 15% AARR on equity.

Table 9.7 Sensitivity Analyses of Beneficiation Costs ($ per Ton Coal Processed) for the Conceptual Coal Cleaning Processes Evaluated

	1		2		3		4		5	
	Wet magnetic beneficiation		Chemical comminution		Wet mechanical beneficiation				Meyers fine coal process	
					Level 2		Level 4			
	$/ton	% change	$/ton	% change	$/ton	% change	$/ton	% change	$/ton	% change
1. Base case: beneficiation cost for $20/ton ROM coal, 100% equity capital, and 15% AARR on equity; 14 hr/day plant operation	20.42	0	14.46	0	7.25	0	10.30	0	28.23	0
2. Capital investment decreased by 20%	18.15	−11.1	13.05	−9.8	6.61	−8.8	9.51	−7.7	25.21	−10.7
3. For 24 hr/day plant operation	14.63	−28.4	10.88	−24.8	5.70	−21.4	8.30	−19.4	20.45	−27.6
4. Operating costs increased by 20%	21.85	+7.0	15.58	+7.8	7.78	+7.3	11.26	+9.3	30.41	+7.7
5. Working capital decreased by 20%	20.09	−1.6	14.14	−2.2	6.98	−3.7	10.02	−2.7	27.87	−1.3

9.5 CONCLUSIONS

Several conclusions may be drawn from the information presented in this chapter. Among them are the following:

1. Coal is likely to play an increasingly significant role in meeting the near-term future energy needs of the industrialized world. To meet the increased demands in an environmentally acceptable manner, more coal will have to be more thoroughly cleaned than currently practiced.
2. Current coal cleaning methods have limitations regarding their ability to reject mineral matter and pyrites from the raw coal. Newer methods that can clean the coal more efficiently are needed.
3. Newer techniques to clean the raw coal more efficiently are being developed. These processes have the potential for recovering more coal containing less mineral matter and pyrites than currently possible. However, most of the promising techniques are at the bench-scale and pilot-scale level of development. Much developmental effort needs to be expended before they can be considered to be commercially viable.
4. The new coal beneficiation processes will cost more than current methods. The costs appear to increase with the degree of processing involved. However, the additional costs should result in significantly higher yields of better quality (coal containing less impurities) product, which in turn should result in higher returns to the plant.
5. The economic evaluations presented in this chapter should be regarded as being preliminary in nature since no specific coal or plant situation was evaluated. The intent of the assessment was to provide a comparison between the potential costs of the selected newer coal cleaning techniques and the current methods, on a common and consistent basis. The detailed evaluation of a specific plant situation or a coal type may alter the results from those presented.
6. Operating the beneficiation plants on a round-the-clock basis (with sufficient spares) may prove to be more economic than current industrial practice of operating for two shifts per day.

REFERENCES

1. W. F. Martin (ed.), *Workshop on Alternate Energy Strategies, Energy Supply to the Year 2000, Global and National Studies*, MIT Press, Cambridge, Mass., 1977, p. 65.

2. L. Hoffman, S. J. Aresco, and E. C. Holt, Jr., *Engineering/Economic Analyses of Coal Preparation with SO_2 Cleanup Process for Keeping High Sulfur Coals in the Energy Market*, Report No. EPA-600/7-78-002, U.S. Environmental Protection Agency, Washington, D.C., Jan. 1978.

3. J. A. Cavallaro and A. W. Deurbrouck, An Overview of Coal Preparation, in *Coal Desulfurization: Chemical and Physical Methods* (T. D. Wheelock, ed.), *ACS Symp. Ser.*, No. 64, pp. 35-57, American Chemical Society, Washington, D.C., 1977.

4. P. J. Phillips, *Coal Preparation for Combustion and Conversion Final Report*, EPRI Report No. EPRI-AF-791, Electric Power Research Institute, Palo Alto, Calif., May 1978.

5. Y. A. Liu and C. J. Lin, High Gradient Magnetic Separation in Coal Beneficiation, *CEP Technical Manual*, Vol. 4, *Coal Processing Technology*, American Institute of Chemical Engineers, New York, 1978, p. 205.

6. R. A. Meyers, *Coal Desulfurization*, Marcel Dekker, New York, 1977.

7. E. P. Stambaugh, Environmentally Acceptable Solid Fuels by the Battelle Hydrothermal Coal Process, Second Symposium on Coal Utilization, Louisville, Ky., October 21-23, 1975.

8. W. Worthy, Hydrothermal Process Cleans Up Coal, *Chem. Eng. News*, 24-25, July 7 (1975).

9. S. S. Sareen, The Use of Oxygen/Water for Removal of Sulfur from Coals, 80th National A.I.Ch.E. Meeting, Boston, Mass., Sept. 1975.

10. J. C. Agarwal, R. A. Gilberti, P. F. Irminger, L. F. Petrovic, and S. S. Sareen, Chemical Desulfurization of Coal, *Min. Congr. J.*, *61*(3), 40-43, 1975.

11. S. S. Sareen, Sulfur Removal from Coals: Ammonia/Oxygen System, 173rd National ACS Meeting, Division of Fuel Chemistry, New Orleans, La., Mar. 1977.

12. A. F. Diaz and E. D. Guth, Coal Desulfurization Process, U.S. Patent 3,909,211, Sept. 30, 1975.

13. E. D. Guth and J. M. Robinson, *Coal Desulfurization Process*, KVB Handout No. B-3, KVB Engineering, Feb. 15, 1977.

14. S. Friedman, R. B. LaCount, and R. P. Warzinski, Oxidative Desulfurization of Coal, *Amer. Chem. Soc., Div. Fuel Chem. Prepr.*, *22*(2), 132 (1977).

15. S. Friedman and R. P. Warzinski, Chemical Cleaning of Coal, ASME Winter Annual Meeting, New York, Dec. 1976.

16. S. P. N. Singh and G. R. Peterson, *Survey and Evaluation of Current and Potential Coal Beneficiation Processes*, Report No. ORNL/TM-5953, Oak Ridge National Laboratory, U.S. Department of Energy, Oak Ridge, Tenn., Mar. 1979.

17. F. V. Karlson, H. Huettenhain, W. W. Slaughter, and K. L. Clifford, The Potential of Magnetic Separation in Coal Cleaning, EPA Symposium on Coal Cleaning to Achieve Energy and Environmental Goals, Hollywood, Fla., Sept. 1978.

18. C. J. Lin and Y. A. Liu, Desulfurization of Coals by High-Intensity, High-Gradient Magnetic Separation: Conceptual Process Design and Cost Estimation, in *Coal Desulfurization: Chemical and Physical Methods* (T. D. Wheelock, ed.), *ACS Symp. Ser.*, No. 64, American Chemical Society, Washington, D.C., 1977, pp. 121-139.

19. R. Salmon, *PRP--A Discounted Cash Flow Program for Calculating the Production Cost (Product Price) of the Product from a Process Plant*, Report No. ORNL-5251, Oak Ridge National Laboratory, U.S. Department of Energy, Oak Ridge, Tenn., Mar. 1977.

20. M. N. Caraway, Catalytic, Inc., Philadelphia, Pennsylvania, personal communication, Oct. 24, 1978.

21. E. P. Koutsoukos, M. L. Kraft, R. A. Orsini, R. A. Meyers, M. J. Santy, and L. J. Van Nice, *Meyers Process Development for Chemical Desulfurization of Coal, Vol. I*, Report No. EPA-600/2-76-143a, U.S. Environmental Protection Agency, Washington, D.C., May 1976.

10

Physical Coal Cleaning with Scrubbing for Sulfur Control

LAWRENCE HOFFMAN
ELMER C. HOLT, JR.

Hoffman-Holt, Inc.
Silver Spring, Maryland

10.1 INTRODUCTION AND BACKGROUND

A concern for preserving the quality of the environment resulted in the Air Quality Act of 1963 which initiated a concerted effort by federal, state, and local governments for the preservation of the nation's air quality. This act called for an expanded federal research and development program and placed special emphasis on the problem of sulfur oxides (SO_x) emissions from the combustion of coal and oil in stationary plants.

Currently and potentially available methods for controlling sulfur oxides emissions from coal-fired stationary combustion sources fall into the following categories:

1. The use of low-sulfur coal either naturally occurring or physically cleaned
2. Removal of sulfur oxides from the combustion flue gas via stack gas scrubbing (flue gas desulfurization)
3. Chemical treatment to extract sulfur from coal
4. Conversion of coal to a clean fuel by such processes as gasification and liquefaction

Of these methods, for certain coals, physical removal of pyritic sulfur (principally FeS_2) is the lowest cost and has the most developed technology. Even so, the potential total sulfur reduction is limited since the organic sulfur cannot be physically removed by this process. Another approach to controlling SO_x emissions could be the combined use of physical coal cleaning (also referred to as coal beneficiation, preparation or washing) followed by stack gas scrubbing (i.e., flue gas desulfurization). The potential benefits derived from combined physical cleaning and flue gas desulfurization include (1) the ability to utilize high-sulfur coal with minimal to moderate physical cleaning, (2) the availability of a more uniform coal, (3) lower effective transportation costs, (4) reduced operating and maintenance costs, (5) lower coal grinding and ash disposal costs, and (6) lower emission control costs.

Physical cleaning of coal has been used for many years. In the past, its principal purpose was to reduce the amount of ash-forming impurities. However, today cleaning is of significant value in reducing the sulfur content of certain coals. Its applicability in this regard is not universal due to the various forms that sulfur occurs in coal. Sulfur exists in coal in two principal forms: organic and inorganic. Organic sulfur is one of those impurities referred to earlier which is chemically bound to the coal and cannot be removed by physical means. On the other hand, inorganic sulfur (mainly pyritic sulfur) is not bound chemically and may be physically removed to varying degrees from the coal. The extent to which pyritic sulfur can be removed economically is a function of pyrite size and distribution. Once this information is obtained through careful laboratory analysis, then economic considerations will influence the level to which the coal is crushed and subjected to the cleaning operation.

Physical cleaning processes, while removing impurities from the coal, also reduce the total Btu recovery (i.e., a portion of the heat content of the raw coal is lost with the impurities). However, the Btu content per unit weight of the cleaned coal increases due to the removal of the lower heat-value impurities. In practice, an economic balance must be achieved between the Btu loss and the improvement in coal quality for various coals and applications. Certainly, this balance is further influenced by market and environmental considerations.

In 1965, EPA sponsored a study to quantify the impact that coal cleaning, optimized for pyrite removal, could have on the control of SO_x emissions. This study suggested that it was impossible to quantify the impact of coal cleaning on SO_x emissions because of large gaps in available information. The study identified the following areas where required information was either not available or was inadequate for appraising this impact:

1. Knowledge of the distribution of sulfur forms in major utility-coal producing coalbeds in the United States;
2. Effectiveness of available commercial coal preparation methods for pyrite separation, together with the development and/or modification of these techniques to maximize sulfur reduction; and
3. Assessment and identification of processes that could economically utilize coal cleaning reject material for by-product recovery, thereby aiding the overall cleaning economics and reducing potential air, water, and solid pollution.

The findings of the 1965 study led EPA to proceed with implementation of a comprehensive program designed to define the role of coal cleaning in controlling SO_x emissions from coal-fired sources. An important part of the program was to determine the extent to which the sulfur content of U.S. coals could be reduced by coal cleaning processes based on differences in physical properties. A good indicator of the "cleanability" of pyritic sulfur from a coal is the specific gravity analysis, or float-and-sink test, in which individual size fractions of the crushed coal are tested at various specific gravities to effect a separation between coal and impurities. Such float-and-sink tests form an important element of the total program by providing the basic data for determining the sharpness of the separation potentially achievable [1,2].

Increased interest in the utilization of coal together with the realization that substantial quantities of Interior and Appalachian coals exhibit reasonable sulfur reductions on physical cleaning has led to considerations of physical cleaning as a total or partial step in meeting environmental standards. One approach would be to clean selected coals that could provide for reasonable sulfur reductions at an acceptable cost followed by a minimal, economically attractive, flue gas desulfurization system.

This concept of physical coal cleaning combined with flue gas desulfurization is not new [3]. For some time there have been discussions, speculations, and some very preliminary assessments addressing the possible benefits of physical coal desulfurization followed by flue gas desulfurization. Past opinions based on a general appreciation of some of the cost and benefit factors associated with such a combined approach have led to the expressed belief that economic advantage in many instances could be attained. Even so, contradictory expressions are also evident.

The U.S. Bureau of Mines, therefore, decided to proceed with an analytical assessment that would more fully define the potential economics of physical coal cleaning followed by flue gas desulfurization as a means for increasing the potential of utilizing some of our higher sulfur-content coals.

This chapter provides an overview of that study performed for the U.S. Bureau of Mines in 1976 by the Hoffman-Muntner Corporation. The study, based on reasonable and realistic parameters, examined the economics of physical coal cleaning followed by flue gas desulfurization as a means of satisfying environmental sulfur-related emission standards. Economics-associated parameters were based on current and past conditions. It should be noted that industry economic factors exhibit considerable spread. In this regard, study values chosen tended to provide the conservative (i.e., least attractive) economics of physical coal cleaning followed by flue gas desulfurization.

10.2 GENERAL APPROACH

Some coals can be physically cleaned to a total sulfur-content level consistent with governing environmental standards. In some cases, this can be accomplished at reasonable cost and acceptable loss in total heat-content value (i.e., on a per-ton basis). Even though many coals can be cleaned to provide a substantial reduction in total sulfur, the beneficiated coal product is often too high in sulfur to meet environmental requirements. In many cases, the beneficiated coal is close enough to meeting standards that a smaller scale, flue gas desulfurization system would enable the coal to be used in an environmentally acceptable fashion at a lower total cost. The required degree of SO_x emission control can often be achieved by employing a flue gas desulfurization system of a low SO_x removal efficiency or a highly efficient flue gas desulfurization system treating only a portion of the total stack gas. This may be particularly applicable in the case of existing power plants which in many states are still permitted to operate at sulfur dioxide emission levels substantially above that required of new facilities.

The relationship between the cleaning efficiency of flue gas desulfurization (FGD) n, the portion of the total gas cleaned x, and the normalized emission level y is defined by the following:

$$y = 1 - nx$$

where

y = (total SO_x emitted with FGD)/(total SO_x emitted without FGD)
n = flue gas desulfurization efficiency of treated portion of flue gas (expressed as a decimal)
x = decimal fraction of flue gas cleaned

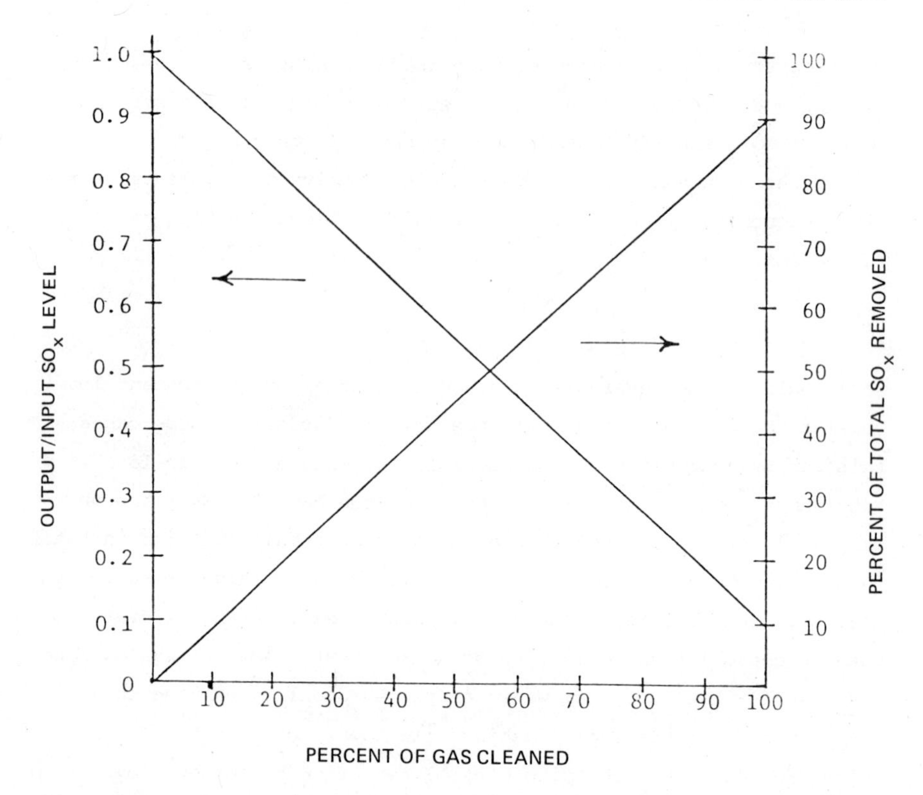

Figure 10.1 SO_x characteristics after cleaning a portion of total stack gas: SO_x removal efficiency, 90% (for treated gas). (From Ref. 3.)

Therefore, for a given cleaning efficiency, the relationship between the portion of gas cleaned x and the normalized emitted SO_x emission level y is defined by a straight line. In a similar manner, the relationship between the portion of gas cleaned and the normalized amount removed is also described by a straight line (see Fig. 10.1).

Results from an extensive Bureau of Mines investigation of coal washability (e.g., float-and-sink tests) show that the organic sulfur and the effects of crushing vary widely when the washability of coals throughout the United States are examined. The regions containing some higher-sulfur-content coals, which indicate a potential to wash to a total sulfur content of approximately 1%, are principally the Appalachian regions, particularily the northern region. Figures 10.2 through 10.7 [4] are a sampling of the washing characteristics

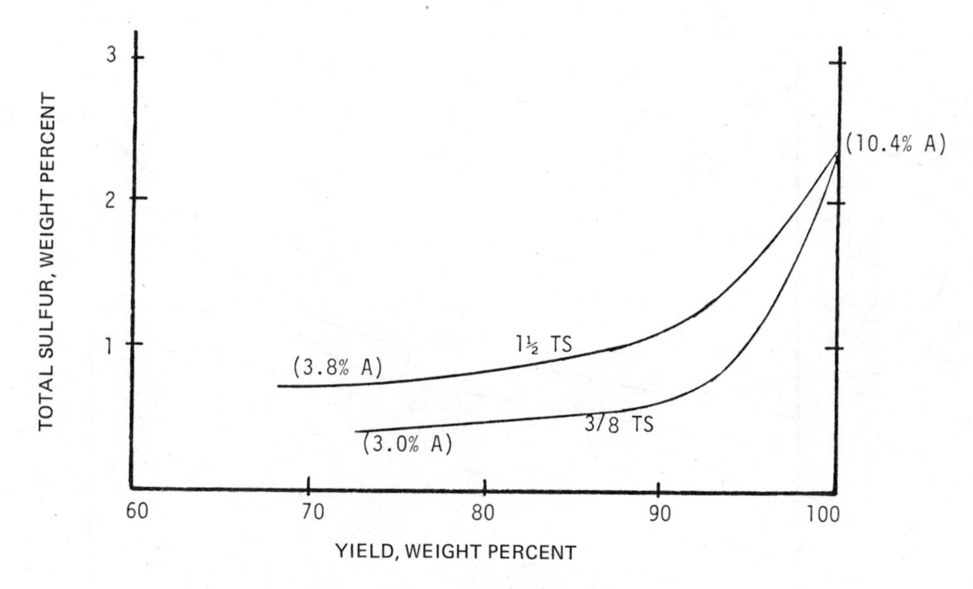

Figure 10.2 Washability test: Middle Kittanning coalbed, Columbiana County, Ohio. (From Ref. 4.)

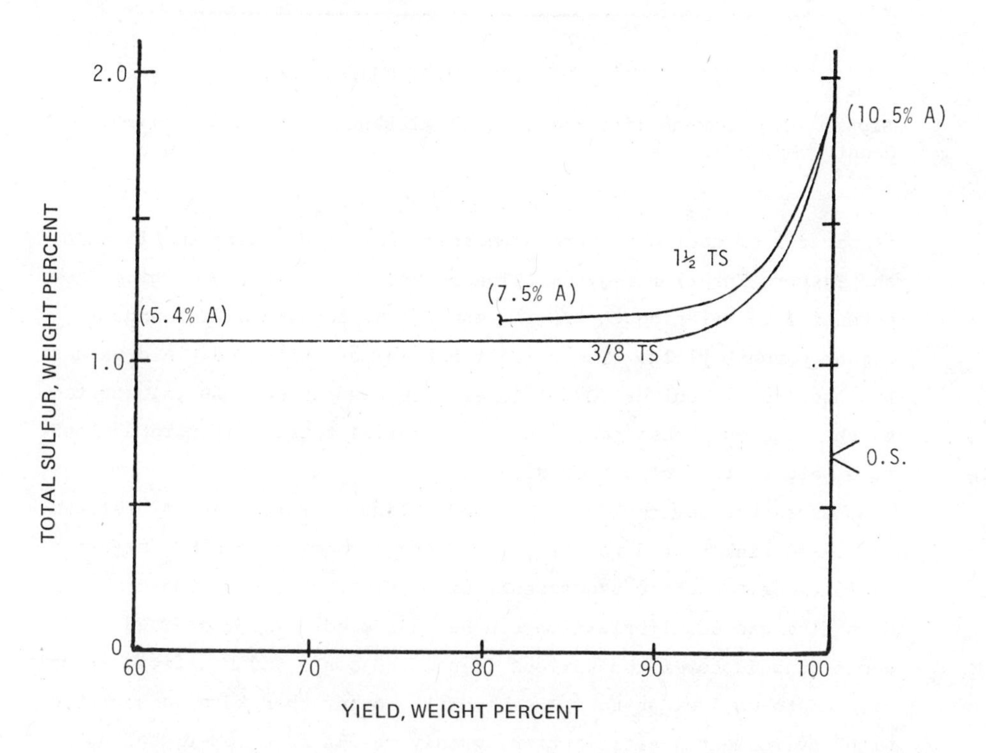

Figure 10.3 Washability test: coalbed No. VII, Sullivan County, Ind. (From Ref. 4.)

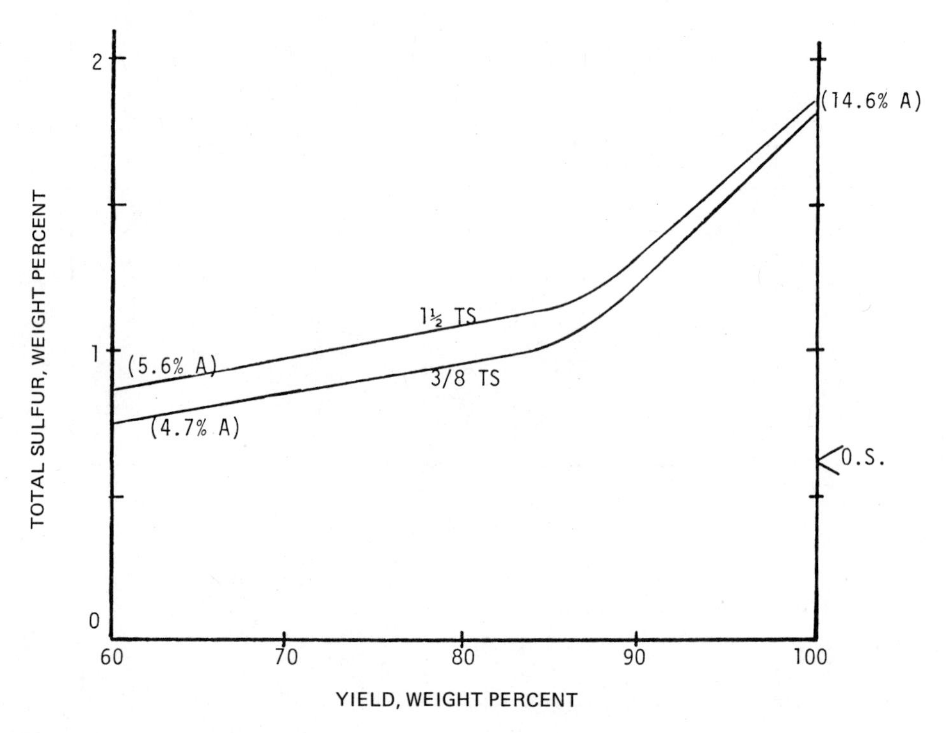

Figure 10.4 Washability test: Lower Kittanning coalbed, Garrett County, Md. (From Ref. 4.)

of several of the more attractive coals in the Northern Appalachian and Eastern Interior regions. The curves show sulfur reduction potential in relation to 3/8 in. and 1½-in. top sizes (TS) and weight percent yield. The percent ash (A) at indicated yield values and the organic sulfur (O.S.) levels are also shown. As illustrated by the figures, these coals show substantial total sulfur reductions at a weight yield of 90% or greater.

Given the raw coal ash, Btu and sulfur characteristics, related float-and-sink test data, transportation economics, and SO_x emission limitations, economic assessments of physical coal cleaning combined with flue gas desulfurization can be addressed. The addressed economics should cover the various identified costs and benefits directly related to coal cleaning. The benefits, other than those associated with environmental satisfaction, mainly result from lower coal ash

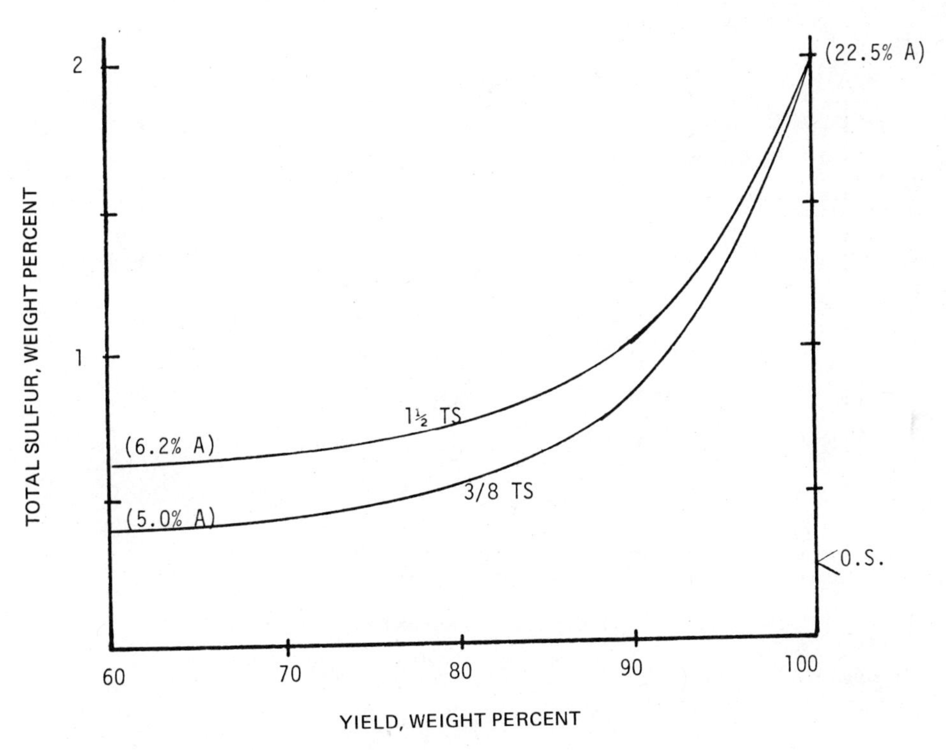

Figure 10.5 Washability test: Upper Freeport coalbed, Allegheny County, Pa. (From Ref. 4.)

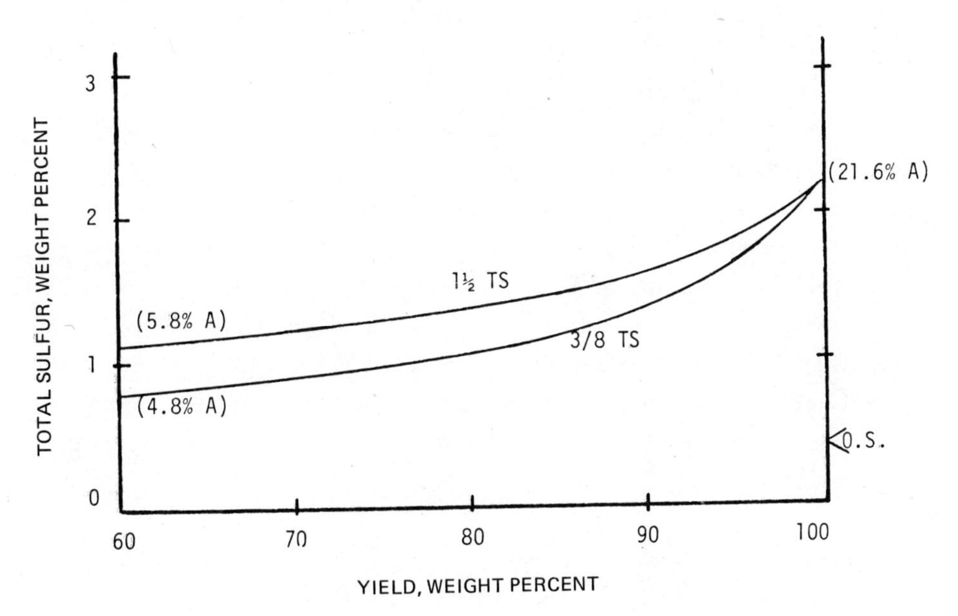

Figure 10.6 Washability test: Upper Freeport coalbed, Armstrong County, Pa. (From Ref. 4.)

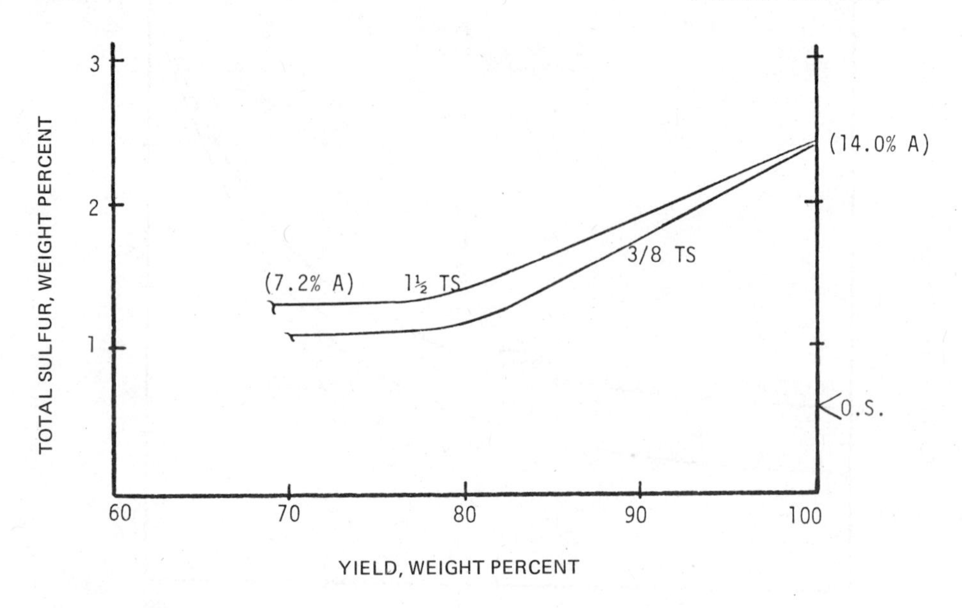

Figure 10.7 Washability test: Upper Freeport coalbed, Garrett County, Md. (From Ref. 4.)

and sulfur contents via cleaning. The cleaning cost-to-benefit assessments can then be modified to encompass the SO_x cleanup process.

The overall economics will be sensitive to the achieved ash and sulfur reductions, the resulting coal yield and cost of physical cleaning. The amount of physical desulfurization will relate to the level of sulfur prior to combustion and will therefore help define the required degree of flue gas desulfurization. In this regard, it should be noted that in general for a given coal top size, coal essentially exhibits a breakpoint in attainable ash and sulfur reduction as a function of weight yield. If coal beneficiation has any attractiveness when applied to a given situation, the optimum level of coal beneficiation would most likely be that level where a substantial change in ash and sulfur reduction potential versus coal yield occurs as shown in Fig. 10.8 [3].

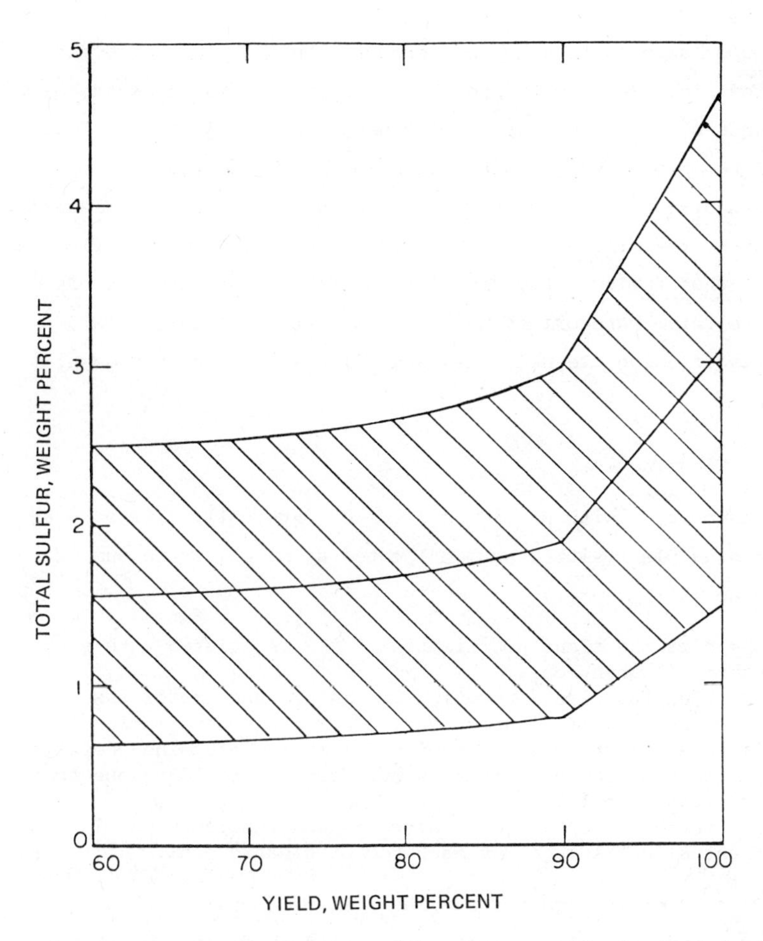

Figure 10.8 Average total sulfur content, ±1 standard deviation at 3/8-in. top size, Northern Appalachian region coals. (From Ref. 3.)

10.3 ECONOMIC ANALYSIS

In order to determine the relative economics of meeting environmental standards via physical coal cleaning combined with partial flue gas desulfurization (FGD) versus full-scale FGD, consideration must be given to the factors influencing the costs and benefits of each approach. For physical coal cleaning, the costs are principally dependent upon such factors as coal composition and plant size and cost. Offsetting these costs are the benefits of using cleaned coal. The magnitude of these benefits is influenced by the increased heat content of the processed product brought on by the removal of some of the low heat-content impurities. In the case of flue gas desulfurization, costs are influenced by the sulfur composition of the flue gas stream. In what follows, the key cost-benefit factors are discussed according to three categories, namely, (1) cost of cleaning coal, (2) benefits of using cleaned coal, and (3) cost of flue gas desulfurization.

A. Cost of Cleaning Coal

The costs associated with providing a ton of physically cleaned coal at the cleaning plant site (assumed located at mine) is the sum of the following [5]:

1. The cleaning plant amortization costs associated with each ton of cleaned coal.
2. The cleaning plant operating and maintenance costs associated with each ton of cleaned coal. Operating costs include the disposal of the waste associated with the cleaning operations.
3. The cost of raw coal that is utilized in providing one ton of cleaned coal.
4. The share of state and local taxes and insurance on cleaning plant allocated against each ton of cleaned coal.

Cleaning Plant Amortization Costs. The cost of a coal cleaning plant can vary over a wide range. Plant cost basically depends on plant capacity, equipment composition, and top size of prepared coal assuming no undue site preparation charges. In general, for a given cleaning plant capacity, the smaller the coal particles, the greater the cleaning plant costs. The increase in plant costs required to clean fine coal is due to the greater capital equipment cost for handling and cleaning the finer coal as well as the significant capital and operating costs associated with fines dewatering, especially if thermal drying has to be used.

There are a variety of approaches to spreading the preparation plant capital cost over the material processed. Whether you are allocating this cost to each ton of raw coal fed to the plant or each ton of clean product, the figure arrived at under all approaches is sensitive to the following factors:

1. Total capital required
2. Plant capacity
3. Operating hours per period (utilization)
4. Cost of money
5. Plant write-off period

Certainly, the first two factors will be directly related to the magnitude and complexity of the particular preparation plant. For most operations today, the third factor, that of plant utilization, will usually vary between 30 and 40%.

With regard to the cost of money, a range of values will occur in practice whether the plant is purchased with borrowed or internal funds. If the funds are acquired outside the firm, their cost will be a function of the current prime interest rate, term of loan, and the credit of the borrower. Even larger firms experience loan rates of 2 to 3 percentage points over prime for purchases of this type. If the plant is to be funded directly by the firm, considerations must be given to the rate of return which might be realized by alternative uses.

The final variable in determining capital amortization is the plant write-off period. Although this is influenced by the anticipated life of the plant and Internal Revenue Service depreciation guidelines, its final determination is made on the basis of individual company fiscal policy.

Operating and Maintenance Costs. Operating and maintenance (O&M) costs are a function of:

1. Coal yield (i.e., weight percent of plant input coal that becomes clean coal)
2. Cleaning plant labor rates and requirements
3. Maintenance costs
4. Cost of operating supplies
5. Power costs
6. Refuse handling and disposal costs

For a given plant, the throughput time required to produce a ton of "clean" coal depends on weight yield (i.e., the weight percentage of run-of-mine coal that ends up as cleaned coal). As an example, if a cleaning plant operating at input capacity takes X hours with a given weight yield to process or provide a ton of clean coal, then the same plant when operating at half the previous weight yield will take essentially 2X hours to provide a ton of clean coal. Therefore, the cleaning plant operating costs attributable to a ton of cleaned coal are inversely proportional to the cleaning plant yield. Likewise, maintenance, supply, and power costs will also be inversely proportional to the weight yield. Refuse handling and disposal costs, however, do not follow this same relationship. This is evident from the fact that for 100% weight yield, refuse will not exist, and therefore refuse costs will be zero.

Operational and maintenance costs can therefore be considered to be composed of the sum of two cost categories. One cost category (i.e., sum of plant labor, maintenance, supplies, and power costs) being inversely proportional to yield and the other cost category (refuse handling and disposal costs) being directly proportional to the percent of refuse.

Cost of Raw Coal Required per Ton of Cleaned Coal. Operator's cost is defined as the operator's break-even cost for providing 1 ton of raw coal input to the cleaning plant. Such cost would include all appropriate expenses (e.g., royalties, labor and equipment, fair share of insurance, taxes, and mine development costs). For the purposes of this analysis, profit was not included in the break-even cost of providing 1 ton of raw coal input to the cleaning plant. This was done since the analysis treats the mine and the coal preparation plant as an integrated operation under common ownership and therefore the coal is treated as work in process until it has been cleaned. However, one might argue that under different business management and/or accounting arrangements, profit should be included in the raw coal input cost to the cleaning plant. To allow for such variations, a range of mine operator's costs to provide one ton of raw coal input was used in the analysis [3].

Since the clean coal yield is less than 100 wt % of the raw coal input, it takes more than one ton of raw coal to provide one ton of cleaned coal. The cost of raw coal that is used to produce one ton of cleaned coal is simply equal to the per-ton operator's cost divided by the weight yield of the cleaning plant expressed as a decimal. The "additional" raw coal cost to provide one ton of cleaned coal is the cost of the raw coal lost during the cleaning process and is equal to the operator's coal cost required to produce one ton of cleaned coal less the operator's per-ton coal cost.

The philosophy behind applying this cost is based on the fact that the material which is discarded does in fact contain some of the heat content (Btu) of the original raw coal feed. However, there is much discussion on whether or not it is appropriate to apply the "cost" of these "lost" Btu to coal preparation. This question arises since the material discarded in cleaning, which contained these lost Btu, is essentially large quantities of the undesirable raw coal constituents such as ash and sulfur, whose removal was the very purpose of the preparation process. It is for this reason that the individual process should be evaluated to determine whether or not a maximum economic Btu recovery point has been reached. If the process is performing reasonably close to well conceived design limits for Btu recovery, it might be more appropriate to neglect these lost Btu as a cost. However, if the process is either poorly conceived from the standpoint of economic Btu recovery or the plant is not performing close to its design capabilities, then an inefficiency "penalty" might be assessed to reflect this discarded heat content which was lost unnecessarily. This being the case, the cost of the lost heat content has been treated as a separate element which the reader may apply or not as seen fit [6].

State and Local Taxes and Insurance Allocated Against Each Ton of Cleaned Coal. Each ton of cleaned coal must support its share of cleaning plant insurance and state and local taxes. The insurance and tax (I&T) load attributable against each ton of cleaned coal (assuming cleaning plant operates at its design capacity) is

$$\frac{\text{I\&T}}{\text{ton clean coal}} = \frac{\text{cleaning plant cost per ton hour input capacity} \times \text{I\&T rate (decimal value)}}{\text{operating hours per year} \times \text{yield (decimal value)}}$$

B. Benefits of Using Cleaned Coal

The readily identifiable monetary benefits attributable to a power plant burning cleaned coal are as follows [5]:

1. Cleaned coal has a higher heat content per ton than the raw coal from which it is produced. Therefore, less cleaned coal will be required to provide a given heat input to the boiler.
2. Less cleaned coal needs to be shipped to provide a given heat input for a boiler because the heat content per unit weight of coal has been increased through cleaning. Therefore, the cost of shipping an equivalent amount of energy will be less.
3. Cleaning reduces the ash content of coal. The amount of ash requiring disposal by the utility will be less due to the lower ash content per ton burned and to the fact that fewer tons of coal will be burned to provide an equivalent heat input.
4. Less cleaned coal needs to be pulverized for a given heat content value, thus reducing the operating and maintenance costs of the pulverizer. Further, cleaning often reduces the amount of harder particles in the coal. However, this reduction in hardness is difficult to quantify and, therefore, was not considered.
5. There is evidence that savings in power plant maintenance costs will be obtained by using coal with reduced sulfur and ash contents. Since physical cleaning reduces the ash and sulfur contents of coal, lower power plant maintenance cost may result. In addition, for a given heat content value, fewer tons of cleaned coal need to be handled by the power plant. Therefore, power plant equipment, which is tonnage (utilization) limited, will have an increased operational lifetime.

Increased Heat Content. An economic benefit readily identifiable to the purchaser of the cleaned coal is equivalent to the free-on-board (FOB) raw coal price (i.e., price of unprepared coal) times the fractional increase in heat value of the cleaned coal over that of the raw coal. This is the equivalent FOB mine raw coal price associated with the amount of coal that provides the equivalent heat increase.

If raw coal has an initial heat content of A per ton and if, upon physical cleaning the heat content decimal increase is x, then the physically cleaned coal will have a heat content of $A(1 + x)$ per ton. The increase over the initial value is $A(1 + x) - A$ or Ax per ton. The fractional amount of cleaned coal that need not be shipped (i.e., the amount that offsets the heat content increase) is $Ax/A(1 + x)$ or $x/(1 + x)$. For the purposes of this analysis, the factor $x/(1 + x)$ shall be termed the multiplier factor K. Therefore, the per cleaned ton benefit gained, a function of the raw coal FOB price and the increase in heat content, is equal to the FOB raw coal price (per ton) times $x/(1 + x)$.

Savings in Transportation Costs. Since cleaning increases the heat content of coal, less clean coal need be shipped to supply a given heat value. The amount of transport savings attainable by cleaning is a function of the increase in the coal heat content and the coal shipping costs per unit weight.

The effective transport savings per ton of cleaned coal is equal to

$$TC \times K$$

where

TC = transportation cost per ton of coal
K = the multiplier factor, $x/(1 + x)$

Savings in Ash Disposal Costs. Savings in ash disposal costs will be realized when cleaned coal is used (i.e., as opposed to raw coal). This saving is due to the fact that cleaning lowers the ash content of coal, thereby raising the heat content of the cleaned coal. Thus, less coal needs to be burned for a given heat content value.

The effective savings in ash disposal cost per ton of cleaned coal burned is equal to the ash disposal saving per ton of coal burned, resulting from lower ash content, times a multiplier factor to account for the higher heat content of the cleaned coal. This multiplier factor is $1 + x$, where x is the fractional increase in heat content. Assume that a x fractional increase in heat content results from a x fractional reduction in ash content, then the total effective ash disposal savings per ton of coal burned is equal to the disposal cost per ton of ash times $x(1 + x)$.

Savings in Coal Pulverizing Costs. Evidence exists that in some cases reducing the amount of the harder impurities (e.g., pyrite) makes coal easier to pulverize. However, this effect does not appear to be universal, and the impact on pulverizing cost is not always discernible.

The identifiable savings associated with this benefit factor results from the increased heat content of a ton of cleaned coal. Therefore, less tonnage need be pulverized to supply a given heat value. The cost savings relationship (on a per ton basis) is

$$\text{Cost savings} = \text{pulverizing cost per ton} \times K$$

where K is the multiplier factor $x/(1 + x)$.

As an example, consider that cleaning raises the heat content of coal by 10%, e.g., from 10,000 to 11,000 Btu/lb. In this case, the fractional decrease in the amount of coal that would result from this increase in heat content is $0.1/(1 + 0.1) = 0.1/1.1$. The saving would be equal to grinding cost multiplied by 0.1/1.1 for each ton of cleaned coal burned.

Maintenance Savings. Only two data sources were identified and considered by this analysis in which operating conditions and collected data are such as to permit a judgment relating sulfur and ash contents in coal to power plant maintenance costs. The two sets of data are for identical steam power plants which burn coals of different ash and sulfur contents. The available data, however, do not permit firm assessments of variations in maintenance savings with variations in coal ash and sulfur contents.

These data are derived from two Tennessee Valley Authority (TVA) generating boiler units, each having a 200-MW generating capacity. These two units, both pulverized-coal-fired boilers, were placed in operation in the middle 1950s. Information on the coal-associated maintenance costs for the two plants is given in Tables 10.1 and 10.2. The data cover approximately 74 million tons of consumed coal. The economic data are contained in a February 1969 report [7] and are assumed to represent economic relationships during 1968. The available maintenance data suggest that

1. Maintenance costs for all the items listed in Table 10.2 are greater for the plant burning coal of higher ash and sulfur contents.

Table 10.1 Average Analyses of As-Burned Coals Utilized in Two Identical Steam Electric Plants

	Plant A	Plant B
Moisture, wt %	4.9	5.1
Volatile matter, wt %	32.4	33.8
Fixed carbon, wt %	52.1	47.7
Ash, wt %	10.8	13.4
Sulfur, wt %	1.0	2.7
Btu content (per pound)	12,680	12,053

Source: Ref. 7.

2. The difference in maintenance cost in 1968 amounted to 11.63¢/ton of coal. This difference in maintenance cost for the two units is almost 2 to 1. This difference is not accounted for by the 5% difference in heat content value between the two coals.

As indicated in Table 10.1, the average difference in ash content of the coals consumed by the two plants was 2.6 wt % and the

Table 10.2 1968 Maintenance Costs in Identical Power Plants Burning Coals of Different Ash and Sulfur Contents

Items with coal associated maintenance costs	Maintenance costs, ¢/ton	
	Plant A	Plant B
Primary coal crushing	0.34	0.76
Coal conveyors	1.24	1.65
Boilers	3.72	6.42
Soot blowers	0.80	2.55
Pulverizers	3.45	5.38
Burners	0.80	1.59
Air preheaters	0.41	0.62
Bottom-ash hoppers	0.55	1.73
Fly-ash collectors	0.80	1.52
Ash disposal system	2.70	3.60
Coal piping	0.80	1.42
Totals	15.61	27.24

Source: Ref. 7.

average difference in the sulfur content was 1.7 wt %. No information is available on maintenance costs with variations in ash and sulfur contents for a given plant. However, these results suggest that substantial maintenance savings per ton of coal burned may be expected in steam plants capable of substituting coal of lower sulfur and ash contents for their current steam coal.

Maintenance cost savings used in the analysis were based on 1968 values conservatively updated to 1975. It was assumed that maintenance savings would vary from 13 to 30¢/ton of coal burned for additive reductions in ash and sulfur contents of 2 to 15%.

C. Flue Gas Desulfurization Cost Factors

One of the most common FGD systems accomplishes its mission by forcing the SO_2 in the flue gas to react with a limestone slurry. This type of FGD system was assumed for the present analysis. The cost of these systems goes up substantially with the amount of flue gas treated, which in turn, is a function of the sulfur concentration of the flue gas stream. Therefore, significant savings can be realized if the sulfur concentration of the stream can be limited to the point where less than half of the total flue gas needs to be treated by the FGD system to meet standards. This becomes the basic economic concept of combining the two sulfur reduction technologies, since such a combined approach could mean the ability to use higher sulfur-content coals with minimal to moderate stack gas scrubbing.

FGD associated costs include

1. Amortization of capital cost of FGD system;
2. Fuel and electricity cost associated with FGD system; and
3. Operating and maintenance costs of FGD system, including chemical reagent, fixation materials, operating labor, maintenance labor and materials, supplies, and overhead.

Capital and operating costs are dependent on many factors including FGD system size, SO_x concentration of untreated flue gas, SO_x removal efficiency, new or retrofit unit, etc. Capital and operating cost values are obtainable from numerous sources including industrial and governmental reports [8,9].

Based on capital and operating costs and the selected amortization parameters, total FGD costs allocated per ton of coal burned can be defined.

D. Major Analysis Factors

Using the key cost-benefit factors associated with the two approaches as identified in the previous section, case analyses were performed based, in part, upon the following major factors identified from actual industry performance and practice.

1. The coals considered were those for which the Bureau of Mines had performed float-sink analyses [2,4]. Bureau of Mines estimates of ash and sulfur levels versus top size and weight yield values were used
2. The top size coal considered was 3/8 in.
3. Economic assessments were based on a range of selling prices of raw coal and a range of mining costs for a ton of coal
4. A coal cleaning plant cost of $18,000/(ton)(hr) input capacity for plants of 500 tons/hr or greater
5. A coal cleaning plant financed by a 15-year equal payment self-liquidating loan
6. A coal cleaning plant use factor of 38.5% (i.e., plant operates 260 days/year at 13 hr/day)
7. A coal cleaning plant property tax and insurance level equal to 2% of the initial investment
8. A power plant ash disposal cost of $4/ton of ash
9. A power plant coal pulverizing cost of $0.50/ton
10. Annual stack gas scrubber capital charges that are based on the original investment and the remaining boiler life
11. Stack gas scrubber size-cost values based on recent EPA funded studies [8,9] and a size-cost exponential relationship of 0.8
12. Stack gas scrubber operating labor cost that does not vary with plant size (i.e., there is a minimum practical labor level)
13. General power plant locations that are consistent with actual conditions
14. Coal producing areas serving assumed user locations consistent with past patterns
15. Coal transportation parameters consistent with historical patterns and economical coal delivery (to insure conservative economic relationships)
16. An assumed power plant boiler size of 500 MW with an annual utilization of 7000 hr for a new plant and 5000 hr for a 10-year-old plant
17. Stack gas scrubber systems of the minimum "practical" sizes necessary to meet standards

In addition to the many cost and benefit factors that were included as part of this analysis, there were several others which were considered but omitted owing to either their inability to be reasonably defined and/or quantified at this time, or their inconsequential impact upon the overall findings of the effort. The following is a brief summary of those factors.

1. *Crushing and Screening Cost Savings*. When raw coal is not physically cleaned, it is normal practice in the coal industry to crush and screen the coal to be sold to utility plants to a size of approximately 2 × 0 in. When coal is physically cleaned, it is not necessary to perform the crushing and screening separately, since these functions are now integrated with the overall cleaning process. For purposes of this analysis, coal cleaning costs were not reduced by the cost of crushing and screening, even though there are definite savings to the mine operator by not having to perform these separately. If these savings had been considered a benefit by the analysis, the net cost of physical cleaning would have been reduced, and thus the relative economics of using physically cleaned coal followed by FGD as compared to FGD alone would have been improved.

2. *Reduction in FGD Installation Time*. By using the physically cleaned coal, less amount of the flue gas needs to be processed by the FGD system and, therefore, fewer or smaller FGD units are required to meet environmental standards. This has not only the benefit (as covered by the analysis) of reducing the FGD cost, but it also reduces the FGD construction time. Additionally, the total sludge disposal at the plant site is reduced substantially from what it would be in the absence of physical coal cleaning. This will, in some cases, reduce the time required for and the difficulty with obtaining legal permits and site acquisition for ponding and/or landfill.

3. *Stack Gas Reheating Cost Savings*. When coals being burned have not been physically cleaned, it may be necessary to process all or nearly all of the flue gas through the FGD system in order to meet environmental standards. In these situations, the stack gas must be reheated following FGD. When only a small portion of the total flue gas is cleaned, reheat would not be required. The absence of reheat

would save energy and consequently amount to a substantial benefit for the combined approach of physical coal cleaning followed by FGD.

4. *Less Derating of Power Plant Output*. FGD systems require power to operate. The smaller FGD systems that could result from using physically cleaned coal would require less operating power. This would result in more marketable power for the same fuel input level to a given power plant.

5. *Cost of Land and Working Capital*. When considering the cost of a coal cleaning plant as well as a FGD system, neither the cost of the working capital needed to handle the larger operation nor the cost of the land area to be occupied was included in the analysis. The reasoning behind this exclusion relates to their very limited impact on a per-million-Btu basis. Additionally, it is not uncommon to have sufficient land area available at the mine site to handle a coal preparation plant and very little land available at the power plant for a retrofit FGD installation. If quantified, this factor would tend to provide additional benefit to the approach of physical coal cleaning followed by less-than-full-scale FGD.

6. *Possible Creation of a Market or the Establishment of a Long-Term Sales Contract*. In many cases, physical cleaning of coal could create markets for coals which are relatively unattractive in their raw form. Providing cleaned coals compatible with particular use situations could lead to the creation of long-term sales contracts matching specific cleaned coals with specific users. This arrangement works to the advantage of all concerned by giving the mine operator a long-term source for his product and the utility a predictably uniform product over an extended period which has been efficiently matched to operating requirements.

E. Case Studies

A total of 48 case analyses were performed. Each case examines one of twelve selected coals and possible use areas from the standpoint of both a new and an existing utility plant using either a combination of physically coal cleaning followed by flue gas desulfurization or sulfur cleanup exclusively by flue gas desulfurization. The cases are grouped according to coal and use area. This permits ease of comparison between similar plants using the same coal in the same area but utilizing two different approaches to meeting existing or projected environmental emission standards. In most cases, the emission standards used were applicable as of January 1976. However, where knowledge of projected changes was present, those standards were used and identified accordingly.

The overall analysis covering new utility plants using physically cleaned coal followed by FGD indicated a savings of 2 to 112% as compared to meeting standards by FGD alone. The results were even more impressive for existing plants, where study assessments indicated a 13 to 140% savings for physical coal cleaning followed by FGD as compared to FGD alone.

A sample case study is illustrated in Table 10.3 for cases 2A and 2B. The key element, which shows the economic advantage in favor of the combined approach, is in the significant difference in FGD costs. This significant difference results from only 21% of the flue gas being processed by the FGD system as opposed to 85% when physically cleaned coal is not used. A graphic representation of these two approaches is presented as Fig. 10.9 [10].

F. Summary of Case Results (Table 10.4)

Included as part of this section are summaries of each of the forty-eight cases analyzed. These summaries are grouped in sets for comparing the costs associated with meeting emission standards by the two alternate approaches, and are listed below. For example, cases 1A, 1B, 1C, and 1D are based on an actual coal coming from Sullivan County, Indiana, which is assumed being used in a utility plant in the Knoxville, Tennessee area. Cases 1A and 1B approach the analysis based on an assumed new facility, whereas cases 1C and 1D assume an existing facility. However, cases 1A and 1C address the analysis using a combination of physically cleaned coal followed by FGD, whereas cases 1B and 1D approach the situation from the standpoint of using FGD alone. As can be readily seen from the summary of

Table 10.3 Case Study

Problem:

Determine most cost-effective approach for new coal burning utility plant to meet emission standard of 1.2 lbs SO_2 per million Btu (MBTU)

Case Conditions:

Coal Use Area: Tonawanda (Buffalo), New York

Coal Source Area: Cambria County, Pennsylvania Coalbed: Lower Freeport

Raw Coal Characteristics: 11.4% Ash, 2.4% Sulfur

Cleaned Coal Characteristics: 6.7% Ash, 1.01% Sulfur

Costs of Alternate Approaches to Meeting Standard:

	Physical Cleaning Followed by FGD	*FGD Alone*
Coal Cleaning Cost		
Amortization of Cleaning Plant Capital Cost	$ 0.68/Ton	$ -0-
O & M Cost of Cleaning Plant	0.75/Ton	-0-
Cost of Coal Lost During Cleaning	1.56 To 2.00/Ton	-0-
Taxes & Insurance of Cleaning Plant	0.12/Ton	-0-
Total Cleaning Cost	$ 3.11 To 3.55/Ton	$ -0-
Cost of Flue Gas Desulfurization (FGD)		
Amortization of FGD Capital Cost	$ 0.92/Ton	$2.92/Ton
Fuel & Electricity of FGD System	0.17/Ton	0.66/Ton
O & M Cost of FGD System	0.72/Ton	2.24/Ton
Total FGD Cost	$ 1.81/Ton*	$5.82/Ton*
Benefits of Using Cleaned Coal		
Increased Heat Content	$ 1.21 To 1.40/Ton	$ -0-
Transportation Savings	0.30/Ton	-0-
Ash Disposal Savings	0.21/Ton	-0-
Pulverizing Savings	0.02/Ton	-0-
Maintenance & Other Savings	0.23/Ton	-0-
Total Benefit of Cleaning	$ 1.97 To 2.16/Ton	$ -0-
Net Cost (Costs Less Benefits)	$ 2.76 To 3.39/Ton	$5.82/Ton
Converted to per MBTU	$ 0.10 To 0.12	$0.22

SOLUTION:

The combined approach is the most cost-effective, compared to a 100% additional cost for meeting the standard by FGD alone.

*Significant difference in cost of FGD results from 21% of the flue gas being processed versus 85% in the more expensive approach to meet the required emission standard.

Source: Ref. 10.

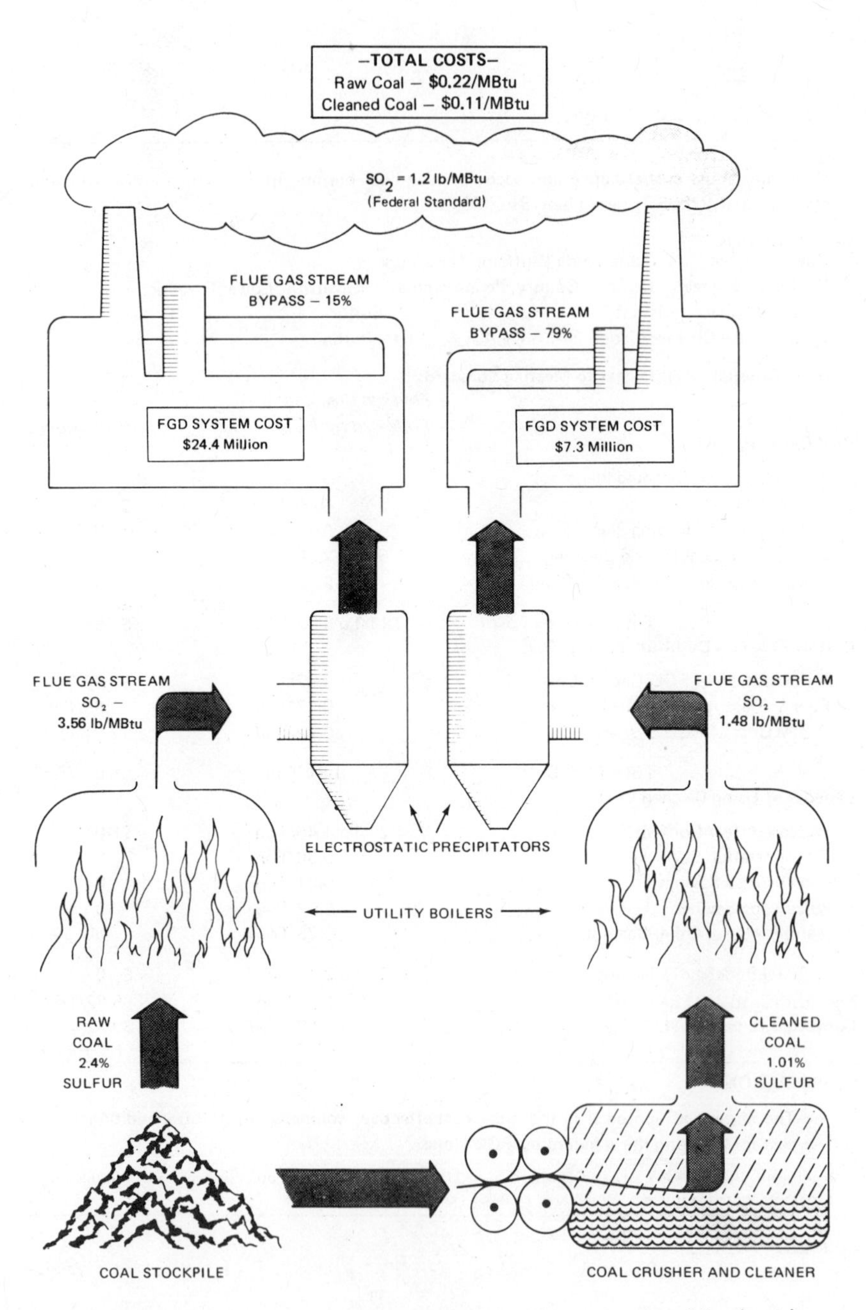

Figure 10.9 Physical coal cleaning with flue gas desulfurization vs. flue gas desulfurization alone. (From Ref. 10.)

Table 10.4 Case Results

Case Numbers: 1A, 1B, 1C, & 1D

Case Conditions

Coal Use Area: Knoxville (Clinton), Tennessee

Coal Source Area: Sullivan County, Indiana Coalbed: Number VII

Raw Coal Characteristics: 10.5 % Ash 1.87 % Sulfur

Clean Coal Characteristics: 7.3 % Ash 1.11 % Sulfur

Comparison of Costs

CASE NO.	*TYPE OF PLANT & APPROACH*	*EMISSION STANDARD*	*COST TO MEET EMISSION STD.*	*ECONOMIC ADVANTAGE*
1A –	New Plant PC Followed by FGD	1.2 lbs SO_2 per MBTU	\$0.15-0.17 per MBTU	16% less than by FGD alone
1B –	New Plant FGD Alone	1.2 lbs SO_2 per MBTU	\$0.19 per MBTU	19% more than by PC & FGD
1C –	Existing Plant PC Followed by FGD	1.2 lbs SO_2 per MBTU	\$0.23-0.25 per MBTU	25% less than FGD alone
1D –	Existing Plant FGD Alone	1.2 lbs SO_2 per MBTU	\$0.32 per MBTU	33% more than by PC & FGD

Case Numbers: 2A, 2B, 2C, & 2D

Case Conditions

Coal Use Area: Tonawanda (Buffalo), New York

Coal Source Area: Cambria County, Pennsylvania Coalbed: Lower Freeport

Raw Coal Characteristics: 11.4 % Ash 2.4 % Sulfur

Clean Coal Characteristics: 6.7 % Ash 1.01 % Sulfur

Comparison of Costs

CASE NO.	*TYPE OF PLANT & APPROACH*	*EMISSION STANDARD*	*COST TO MEET EMISSION STD.*	*ECONOMIC ADVANTAGE*
2A –	New Plant PC Followed by FGD	1.2 lbs SO_2 per MBTU	\$0.10-0.12 per MBTU	50% less than by FGD alone
2B –	New Plant FGD Alone	1.2 lbs SO_2 per MBTU	\$0.22 per MBTU	100% more than by PC & FGD
2C –	Existing Plant PC Followed by FGD	1.4 lbs SO_2 per MBTU	\$0.06-0.09 per MBTU	50% less than by FGD alone
2D –	Existing Plant FGD Alone	1.4 lbs SO_2 per MBTU	\$0.15 per MBTU	100% more than by PC & FGD

Table 10.4 Continued

Case Numbers: 3A, 3B, 3C & 3D

Case Conditions

Coal Use Area: Essexville (Saginaw), Michigan

Coal Source Area: Harrison County, Ohio Coalbed: Lower Freeport

Raw Coal Characteristics: 10.4 % Ash 2.30 % Sulfur

Clean Coal Characteristics: 4.8 % Ash 1.26 % Sulfur

Comparison of Costs

CASE NO.	*TYPE OF PLANT & APPROACH*	*EMISSION STANDARD*	*COST TO MEET EMISSION STD.*	*ECONOMIC ADVANTAGE*
3A –	New Plant PC Followed by FGD	1.2 lbs SO_2 per MBTU	$0.12-0.14 per MBTU	38% less than by FGD alone
3B –	New Plant FGD Alone	1.2 lbs SO_2 per MBTU	$0.21 per MBTU	62% more than by PC & FGD
3C –	Existing Plant PC Followed by FGD	1.6 lbs SO_2 per MBTU	$0.11-0.14 per MBTU	58% less than FGD alone
3D –	Existing Plant FGD Alone	1.6 lbs SO_2 per MBTU	$0.30 per MBTU	140% more than by PC & FGD

Case Numbers: 4A, 4B, 4C, & 4D

Case Conditions

Coal Use Area: Boston, Massachusetts

Coal Source Area: Clearfield County, Pennsylvania Coalbed: Upper Kittanning

Raw Coal Characteristics: 9.3 % Ash 0.85 % Sulfur

Clean Coal Characteristics: 7.0 % Ash 0.45 % Sulfur

Comparison of Costs

CASE NO.	*TYPE OF PLANT & APPROACH*	*EMISSION STANDARD*	*COST TO MEET EMISSION STD.*	*ECONOMIC ADVANTAGE*
4A –	New Plant PC Followed by FGD	0.28 lbs sul. per MBTU	$0.12-0.14 per MBTU	24% less than by FGD alone
4B –	New Plant FGD Alone	0.28 lbs sul. per MBTU	$0.17 per MBTU	31% more than by PC & FGD
4C –	Existing Plant PC Followed by FGD	0.28 lbs sul. per MBTU	$0.17-0.19 per MBTU	38% less than by FGD alone
4D –	Existing Plant FGD Alone	0.28 lbs sul. per MBTU	$0.29 per MBTU	61% more than by PC & FGD

Table 10.4 Continued

Case Numbers: 5A, 5B, 5C, & 5D

Case Conditions

Coal Use Area: Grand Rapids, Michigan

Coal Source Area: Preston County, W. Virginia Coalbed: Upper Freeport

Raw Coal Characteristics: 18.5 % Ash 2.24 % Sulfur

Clean Coal Characteristics: 11.9 % Ash 1.25 % Sulfur

Comparison of Costs

CASE NO.	TYPE OF PLANT & APPROACH	EMISSION STANDARD	COST TO MEET EMISSION STD.	ECONOMIC ADVANTAGE
5A –	New Plant PC Followed by FGD	1.6 lbs SO_2 per MBTU	\$0.07-0.10 per MBTU	53% less than by FGD alone
5B –	New Plant FGD Alone	1.6 lbs SO_2 per MBTU	\$0.18 per MBTU	112% more than by PC & FGD
5C –	Existing Plant PC Followed by FGD	1.6 lbs SO_2 per MBTU	\$0.12-0.14 per MBTU	57% less than FGD alone
5D –	Existing Plant FGD Alone	1.6 lbs SO_2 per MBTU	\$0.30 per MBTU	131% more than by PC & FGD

Case Numbers: 6A, 6B, 6C, & 6D

Case Conditions

Coal Use Area: Springfield, Massachusetts

Coal Source Area: Armstrong County, Penn. Coalbed: Upper Freeport

Raw Coal Characteristics: 13.0 % Ash 2.53 % Sulfur

Clean Coal Characteristics: 7.2 % Ash 1.09 % Sulfur

Comparison of Costs

CASE NO.	TYPE OF PLANT & APPROACH	EMISSION STANDARD	COST TO MEET EMISSION STD.	ECONOMIC ADVANTAGE
6A –	New Plant PC Followed by FGD	0.55 lbs sulfur per MBTU	\$0.11-0.13 per MBTU	48% less than by FGD alone
6B –	New Plant FGD Alone	0.55 lbs sulfur per MBTU	\$0.23 per MBTU	92% more than by PC & FGD
6C –	Existing Plant PC Followed by FGD	0.55 lbs sulfur per MBTU	\$0.18-0.21 per MBTU	49% less than by FGD alone
6D –	Existing Plant FGD Alone	0.55 lbs sulfur per MBTU	\$0.38 per MBTU	95% more than by PC & FGD

Table 10.4 Continued

Case Numbers: 7A, 7B, 7C, & 7D
Case Conditions
Coal Use Area: Lansing, Michigan
Coal Source Area: Jefferson County, Ohio Coalbed: Pittsburgh
Raw Coal Characteristics: 9.8 % Ash 2.82 % Sulfur
Clean Coal Characteristics: 6.0 % Ash 2.03 % Sulfur

Comparison of Costs

CASE NO.	TYPE OF PLANT & APPROACH	EMISSION STANDARD	COST TO MEET EMISSION STD.	ECONOMIC ADVANTAGE
7A –	New Plant PC Followed by FGD	1.6 lbs SO_2 per MBTU	\$0.19-0.22 per MBTU	6.8% less than by FGD alone
7B –	New Plant FGD Alone	1.6 lbs SO_2 per MBTU	\$0.22 per MBTU	7.3% more than by PC & FGD
7C –	Existing Plant PC Followed by FGD	1.6 lbs SO_2 per MBTU	\$0.30-0.32 per MBTU	11% less than FGD alone
7D –	Existing Plant FGD Alone	1.6 lbs SO_2 per MBTU	\$0.35 per MBTU	13% more than by PC & FGD

Case Numbers: 8A, 8B, 8C, & 8D
Case Conditions
Coal Use Area: Nashville (Gallatin), Tennessee
Coal Source Area: Vigo County, Indiana Coalbed: Number VII
Raw Coal Characteristics: 12.0 % Ash 1.54 % Sulfur
Clean Coal Characteristics: 7.7 % Ash 0.90 % Sulfur

Comparison of Costs

CASE NO.	TYPE OF PLANT & APPROACH	EMISSION STANDARD	COST TO MEET EMISSION STD.	ECONOMIC ADVANTAGE
8A –	New Plant PC Followed by FGD	1.2 lbs SO_2 per MBTU	\$0.11-0.13 per MBTU	25% less than by FGD alone
8B –	New Plant FGD Alone	1.2 lbs SO_2 per MBTU	\$0.16 per MBTU	33% more than by PC & FGD
8C –	Existing Plant PC Followed by FGD	1.2 lbs SO_2 per MBTU	\$0.16-0.18 per MBTU	39% less than by FGD alone
8D –	Existing Plant FGD Alone	1.2 lbs SO_2 per MBTU	\$0.28 per MBTU	65% more than by PC & FGD

Table 10.4 Continued

Case Numbers: 9A, 9B, 9C, & 9D

Case Conditions

Coal Use Area: Burlington, New Jersey

Coal Source Area: Garrett County, Maryland Coalbed: Upper Freeport

Raw Coal Characteristics: 13.8 % Ash 2.37 % Sulfur

Clean Coal Characteristics: 8.8 % Ash 1.6 % Sulfur

Comparison of Costs

CASE NO.	*TYPE OF PLANT & APPROACH*	*EMISSION STANDARD*	*COST TO MEET EMISSION STD.*	*ECONOMIC ADVANTAGE*
9A –	New Plant PC Followed by FGD	0.30 lbs SO_2 per MBTU	This coal will not meet emission standards for new plants in the State of New Jersey with either combined physical cleaning and FGD or FGD alone.	
9B –	New Plant FGD Alone	0.30 lbs SO_2 per MBTU		
9C –	Existing Plant PC Followed by FGD	1% sulfur by weight per MBTU	$0.23-0.25 per MBTU	25% less than FGD alone
9D –	Existing Plant FGD Alone	1% sulfur by weight per MBTU	$0.32 per MBTU	33% more than by PC & FGD

Case Numbers: 10A, 10B, 10C, & 10D

Case Conditions

Coal Use Area: Milwaukee, Wisconsin

Coal Source Area: Franklin County, Illinois Coalbed: Number 6

Raw Coal Characteristics: 14.8 % Ash 1.12 % Sulfur

Clean Coal Characteristics: 7.1 % Ash 0.95 % Sulfur

Comparison of Costs

CASE NO.	*TYPE OF PLANT & APPROACH*	*EMISSION STANDARD*	*COST TO MEET EMISSION STD.*	*ECONOMIC ADVANTAGE*
10A –	New Plant PC Followed by FGD	1.2 lbs SO_2 per MBTU	$0.07-0.10 per MBTU	29% less than by FGD alone
10B –	New Plant FGD Alone	1.2 lbs SO_2 per MBTU	$0.12 per MBTU	41% more than by PC & FGD
10C –	Existing Plant PC Followed by FGD	1.2 lbs SO_2 per MBTU	$0.12-0.15 per MBTU	33% less than by FGD alone
10D –	Existing Plant FGD Alone	1.2 lbs SO_2 per MBTU	$0.20 per MBTU	48% more than by PC & FGD

Table 10.4 Continued

Case Numbers: 11A, 11B, 11C, & 11D
Case Conditions
Coal Use Area: Concord, New Hampshire
Coal Source Area: Greene County, Pennsylvania Coalbed: Sewickley
Raw Coal Characteristics: 11.4 % Ash 3.45 % Sulfur
Clean Coal Characteristics: 8.1 % Ash 2.20 % Sulfur

Comparison of Costs

CASE NO.	*TYPE OF PLANT & APPROACH*	*EMISSION STANDARD*	*COST TO MEET EMISSION STD.*	*ECONOMIC ADVANTAGE*
11A –	New Plant PC Followed by FGD	1.2 lbs SO_2 per MBTU (projected standard)	$0.25-0.28 per MBTU	2% less than by FGD alone
11B –	New Plant FGD Alone	1.2 lbs SO_2 per MBTU (projected standard)	$0.27 per MBTU	2% more than by PC & FGD
11C –	Existing Plant PC Followed by FGD	1.5 lbs sulfur per MBTU	$0.12-0.14 per MBTU	54% less than FGD alone
11D –	Existing Plant FGD Alone	1.5 lbs sulfur per MBTU	$0.28 per MBTU	115% more than by PC & FGD

Case Numbers: 12A, 12B, 12C, & 12D
Case Conditions
Coal Use Area: Dickerson (Montgomery County), Maryland
Coal Source Area: Marion County, W. Virginia Coalbed: Pittsburgh
Raw Coal Characteristics: 11.0 % Ash 3.80 % Sulfur
Clean Coal Characteristics: 5.9 % Ash 2.16 % Sulfur

Comparison of Costs

CASE NO.	*TYPE OF PLANT & APPROACH*	*EMISSION STANDARD*	*COST TO MEET EMISSION STD.*	*ECONOMIC ADVANTAGE*
12A –	New Plant PC Followed by FGD	1% sulfur by weight	$0.20-0.23 per MBTU	20% less than by FGD alone
12B –	New Plant FGD Alone	1% sulfur by weight	$0.27 per MBTU	26% more than by PC & FGD
12C –	Existing Plant PC Followed by FGD	1% sulfur by weight	$0.32-0.34 per MBTU	23% less than by FGD alone
12D –	Existing Plant FGD Alone	1% sulfur by weight	$0.43 per MBTU	30% more than by PC & FGD

Source: Ref. 10.

these cases, the cost to meet the applicable sulfur emission standard is less in cases 1A and 1C, which are the plants using physically cleaned coal followed by FGD in new and existing facilities, respectively. In each set of cases, the relative economic advantage (or disadvantage) is expressed as a percent for comparative purposes. The manner of stating economic advantage (less costly) or disadvantage (more costly) is dependent upon what action the coal-using plant has taken to meet environmental standards. For example, if, as in case 1C, the utility is using physically cleaned coal followed by FGD, then the economic advantage could be stated as their cost is 25% less than by FGD alone. If, as in case 1D, the utility is using FGD alone, then the economic disadvantage could be stated as their cost is 33% more than by the combined approach of physical coal cleaning and FGD.

10.4 STUDY FINDINGS AND CONTINUING PROSPECTS

A. Study Findings

The economic analyses covering physical coal cleaning followed by flue gas desulfurization, and flue gas desulfurization used alone, for selected coal-user combinations indicate that economic generalizations must be approached with caution. The ranges of variables are such that each case must be individually assessed. This is evident from the 48 case studies summarized.

The potential attractiveness of combined physical coal cleaning followed by flue gas desulfurization is highly dependent on both the emission standards and the achievable reduction in coal ash and sulfur contents upon physical cleaning. Specifically, the potential attractiveness depends on the ability to reduce the flue gas desulfurization requirement by economically removing sulfur from coal through physical cleaning and from other economic benefits accruing from shipping and burning coal of reduced ash and sulfur contents. The reduction in ash content provides for coal of a higher Btu content, thereby providing benefits such as reduced effective transportation costs and lower ash disposal costs to the user. An examination of the overall economic factors indicate that it is economically preferable to clean coal at the mine rather than after shipping at the user location.

Every attempt was made to insure a conservative analysis. The intent was to understate rather than overstate economic benefits. In this regard, attempts were made to employ cost-related factors such as transportation, ash disposal, grinding, cleaning plant utilization, cleaning plant operation and maintenance, etc. that would tend toward conservative economic findings. As indicated, the case study results display a range of values for the cost of meeting emission standards for use of physical coal cleaning followed by flue gas desulfurization. This spread in costs resulted from considerations of a range of prices for both the mine operator's cost to provide a ton of raw coal to the coal cleaning plant and a range of FOB mine raw coal prices.

To further insure reasonable comparisons of the cost to meet SO_x emission standards with flue gas desulfurization as compared to use of physical coal cleaning followed by flue gas desulfurization, the following guidelines were used.

1. The same coal was considered for both cases.
2. The amount of flue gas desulfurization considered was only that needed (with small margin) to achieve SO_x emission standards.
3. Where a range of economic values exist (e.g., coal transportation costs), only those values that would provide conservative savings were used.

It should be noted that for this analysis, coals were essentially selected at random. The principal criteria were that (1) upon physical desulfurization, they did not display overly attractive sulfur reduction potentials, and (2) they indicated reasonable sulfur reduction levels with a topsize of 3/8 in. or larger with attractive weight yields.

B. Continuing Prospects

The federal government as well as the coal industry has established a production goal of more than one billion tons of coal by 1985.

Although this means an increase of roughly 400 million tons over current production, the majority of this additional coal will be consumed by the approximately 250 new coal-fired facilities scheduled to be on line by that time. This substantial projected increase in coal utilization has precipitated much discussion concerning the possibility of revising EPA's new source performance standard (NSPS) for coal-fired plants. The extent and form of any such change are unknown at this time. One possible way in which the standard could be tightened would be to reduce the current 1.2 lb of SO_2 per million Btu to 0.8 lb. Whereas now a coal having a heat content of approximately 25 million Btu/ton and a sulfur content of 0.8% can meet standard, only those coals of 0.5% or less sulfur content could meet the 0.8 lb of SO_2 per million Btu standard. If the standard was revised in this manner, there would be a significant reduction in the amount of coal which can be consumed without the use of SO_2 emission control devices. This would include both the coals naturally occurring with a low enough sulfur content and those capable of meeting the current standard through physical cleaning alone.

Another possible way in which the emission standard could be changed would be to require removal of a designated percentage (e.g., 85 to 90%) of the sulfur contained in the raw coal at some point before the stack gases were released by the utility plant. It is assumed that if the revised NSPS took this form, there would be a maximum SO_2 emission limit set (e.g., 1.2 lb per million Btu) and some minimum point (0.2 lb per million Btu) at which no sulfur control would be required. Obviously, such a revised NSPS would essentially eliminate the burning of raw or cleaned coal without the use of some level of sulfur emission control equipment.

However, should the standard be amended in either of the foregoing ways, it will not negate the benefits associated with coal cleaning as set forth in section 10.3. What it will mean is a reevaluation of emission control strategy and a comprehensive assessment of the overall merits of burning cleaned coal on the part of potential coal users. In this latter case, coal preparation might still play a role in establishing the most cost effective approach to sulfur emission control as long as credit was given for the sulfur reduction achieved through cleaning. However, even if no credit was given for the washing step, the value of coal preparation remains significant when applied in many situations.

In summary, depending on the exact nature of any new standard(s), it is conceivable that the combined use of physical coal cleaning and flue gas desulfurization could be attractively employed to satisfy a future SO_x emission limitation. Whether the standard takes the form of a reduction in pounds of SO_2 emitted or an overall percentage reduction as outlined above, the major change in the attractiveness of combined physical coal cleaning and flue gas desulfurization approach would most likely be that fewer coal use situations would economically benefit from the combined process. Even so, for a number of selected coal-user situations, the combined use of coal cleaning and flue gas cleaning should be economically attractive. Furthermore, if economic credits are given to such benefits as savings in stack gas reheat cost and additional marketable electric power (i.e., less derating of power plant), the attractiveness of the combined approach, even with more restrictive emission standards, would significantly increase.

10.5 CONCLUSIONS

Available data indicate that many coals can be cleaned to remove ash and sulfur at an attractive net cost. These cleaned coals with reduced ash and sulfur contents are often not too far removed from the sulfur contents required to meet environmental standards in some areas traditionally served by burning such coals. In these situations, flue gas desulfurization only treating a portion of the flue gas would provide for environmental satisfaction. In many applications, the net cost of physical coal cleaning followed by flue gas desulfurization is substantially less than that through using flue gas desulfurization alone. This is due to the net economics associated with physical coal cleaning combined with the substantially lower flue gas desulfurization costs. In essence, the net cost (i.e., costs minus benefits) associated with physical coal cleaning would be less than the additional cost if flue gas desulfurization was used alone.

For existing power plants, the real cost for flue gas desulfurization systems are especially expensive due to the higher capital costs and the shorter economic written-off periods of the systems. In many such cases, the use of physical coal cleaning followed by flue gas desulfurization can be particularly attractive.

The case studies carried out indicate that for many potential situations, the economic advantage of a combined approach is quite significant. Key elements in the economic advantage are related to (1) the availability of coals capable of significant reductions in ash and sulfur at reasonable weight yields and (2) a sulfur level of the cleaned coal that is compatible with significantly less than full-scale scrubbing requirements. Even so, the wide range of net costs which resulted from different case studies is such that each coal source-user combination must be individually assessed. In this regard, it should be noted the results can unrealistically be weighted to indicate excessively attractive economics by employing unrealistic factors (e.g., shipping coal further than is normally warranted).

The assessments imply that the attractiveness of many of our medium to high sulfur-content coals can be enhanced by cleaning to provide an assured supply of coal that can be used with more economic flue gas desulfurization. This is true for many as-mined medium-sulfur-content coals and some higher-sulfur-content coals that could serve areas with less restrictive environmental standards.

Finally, it should be noted that even though these analyses were based on emission limitation levels, physical coal cleaning followed by flue gas desulfurization could equally be applied to provide for a given percentage reduction in the potential SO_2 emission level. As indicated, generalities are not warranted and each case would require independent analysis.

REFERENCES

1. L. Hoffman, J. B. Truett, and S. J. Aresco, *An Interpretative Compilation of EPA Studies Related to Coal Quality and Cleanability*, U.S. Environmental Protection Agency, Washington, D.C., *EPA Report No. EPA-650/2-74-030*, May 1974.

2. J. A. Cavallaro, M. T. Johnson, and A. W. Deurbrouck, *Sulfur Reduction Potential of the Coals of the United States*, Report of Investigation, No. 8118, U.S. Bureau of Mines, Washington, D.C., 1976.

3. L. Hoffman, S. J. Aresco, and E. C. Holt, Jr., *Engineering/Economic Analyses of Coal Preparation with SO_2 Clean Up Process*, Hoffman-Muntner Corporation, EPA Report No. EPA-600/7-78-002, U.S. Environmental Protection Agency, Washington, D.C., Jan. 1978.

4. A. W. Deurbrouck, *Sulfur Reduction Potential of Coals of the United States*, Report of Investigation, No. 7633, U.S. Bureau of Mines, Washington, D.C., 1972.

5. L. Hoffman, and K. E. Yeager, *The Physical Desulfurization of Coal: Major Considerations for SO_2 Emission Control*, The Mitre Corporation, Report No. MTR-4151, prepared for U.S. Environmental Protection Agency, Washington, D.C., Nov. 1970.

6. E. C. Holt, Jr., *An Engineering/Economic Analysis of Coal Preparation Plant Operation and Cost*, Hoffman-Muntner Corporation, EPA Report No. EPA-600/7-78-124, U.S. Environmental Protection Agency, Washington, D.C., July 1978.

7. J. G. Holmes, Jr., The Effect of Coal Quality on the Operation and Maintenance of Large Central Station Boilers, paper presented at Annual Meeting of the American Institute of Mining, Metallurgical and Petroleum Engineers, Washington, D.C., Feb. 16-20, 1969.

8. G. G. McGlamery, R. L. Torstick, W. J. Broadfoot, J. P. Simpson, L. J. Henson, S. V. Tomlinson, and J. F. Young, *Detailed Cost Estimates for Advanced Effluent Desulfurization Process*, U.S. Environmental Agency, Washington, D.C., Report No. EPA-600/2-75-006, Jan. 1975.

9. PEDCo--Environmental Specialists, Inc., *Flue Gas Desulfurization Process Cost Estimates*, Special report prepared for U.S. Environmental Protection Agency, Washington, D.C., May 1975.

10. E. C. Holt, Jr., and A. W. Deurbrouck, *Coal Cleaning with Scrubbing for Sulfur Control: An Engineering/Economic Summary*, U.S. Environmental Protection Agency, Washington, D.C., EPA Report No. EPA-600/9-77-017, Aug. 1977.

Index